SELECTED SOLUTIONS MANUAL

Joseph Topich

Virginia Commonwealth University

CHEMISTRY

Eighth Edition

Jill K. Robinson

John E. McMurry

Robert C. Fay

 **Pearson**

Courseware Portfolio Manager: Jeanne Zalesky
Director of Portfolio Management: Terry Haugen
Content Producer: Shercian Kinosian
Managing Producer: Kristen Flathman
Courseware Director, Content Development: Barbara Yien
Courseware Analysts: Cathy Murphy, Coleen Morrison, Jay McElroy
Courseware Editorial Assistant: Harry Misthos
Full-Service Vendor: SPi Global
Main Text Cover Designer: Gary Hespeneide
Manufacturing Buyer: Stacey Weinberger
Director of Product Marketing: Allison Rona
Senior Product Marketing Manager: Elizabeth Bell

Cover Photo Credit: Beauty of Science/Science Source

ScoutAutomatedPrintCode

ISBN 10: **0-135-43192-1**
ISBN 13: **978-0-135-43192-4**

Contents

Preface

Chapter 1:   Chemical Tools: Experimentation and Measurement   1

Chapter 2:   Atoms, Molecules, and Ions   13

Chapter 3:   Mass Relationships in Chemical Reactions   25

Chapter 4:   Reactions in Aqueous Solution   43

Chapter 5:   Periodicity and the Electronic Structure of Atoms   63

Chapter 6:   Ionic Compounds: Periodic Trends and Bonding Theory   77

Chapter 7:   Covalent Bonding and Electron-Dot Structures   87

Chapter 8:   Covalent Compounds: Bonding Theories and Molecular Structure 111

Chapter 9:   Thermochemistry: Chemical Energy   125

Chapter 10: Gases: Their Properties and Behavior   141

Chapter 11: Liquids and Phase Changes   161

Chapter 12: Solids and Solid-State Materials   171

Chapter 13: Solutions and Their Properties   187

Chapter 14: Chemical Kinetics   209

Chapter 15: Chemical Equilibrium   237

Chapter 16: Aqueous Equilibria: Acids and Bases   263

Chapter 17: Applications of Aqueous Equilibria   291

Chapter 18: Thermodynamics: Entropy, Free Energy, and Spontaneity   325

Chapter 19: Electrochemistry   345

Chapter 20: Nuclear Chemistry   377

Chapter 21: Transition Elements and Coordination Chemistry   395

Chapter 22: The Main-Group Elements   415

Chapter 23: Organic and Biological Chemistry   431

Preface

The *Selected Solutions Manual* to accompany CHEMISTRY, 8/e by Robinson, McMurry, and Fay contains the solutions to all in-chapter, and the even-numbered conceptual, and end-of-chapter questions and problems.

Critical thinking and problem solving are your keys to success in chemistry! CHEMISTRY, 8/e by Robinson, McMurry, and Fay contains thousands of questions and problems to challenge you. Develop a problem-solving strategy that works for you. Carefully read each problem. Identify what the problem is asking for. List all relevant information contained in the problem. Use chemistry concepts and principles to connect the problem's information to the solution you are looking for. Set up and solve the problem. Look at your solution. Did you answer the question? Are the units correct? Is your answer reasonable? Only then, check your answer with the *Selected Solutions Manual*.

I have worked to ensure that the solutions in this manual are as error free as possible. Solutions have been double-checked (in some cases triple-checked). Small differences in numerical answers between student results and those in the *Selected Solutions Manual* may result from rounding or significant figure differences. Also remember that there is, in many instances, more than one acceptable setup for a problem.

I would like to thank Jill Robinson, John McMurry, and Robert Fay for the opportunity to contribute to their CHEMISTRY package. I also want to thank them for their helpful comments as I worked on this solutions manual. I also want to acknowledge and thank Alton Hassell (accuracy checker) and the entire Pearson staff. Finally, I want to thank in a very special way my wife, Ruth, and our daughter, Judy, for their constant encouragement and support as I worked on this project.

Joseph Topich
Department of Chemistry
Virginia Commonwealth University

1.1 5.0×10^{-8} m; 5.0×10^{-8} m $= 50 \times 10^{-9}$ m $= 50$ nm

1.2   (a) 7×10^{-5} m   (b) 2×10^{13} kg

1.3 $°C = \dfrac{5}{9} \times (°F - 32) = \dfrac{5}{9} \times (1474 - 32) = 801\ °C$

         $K = °C + 273.15 = 801 + 273.15 = 1074.15$ K or 1074 K

1.4   The melting point of gallium is converted from 302.91 K to °F for comparison.
        $°C = K - 273.15 = 302.91 - 273.15 = 29.76\ °C$

         $°F = (\dfrac{9}{5} \times °C) + 32 = (\dfrac{9}{5} \times 29.76) + 32 = 85.57\ °F$

        The temperature in the compartment (88 °F) is above the melting point, so the liquid state exists.

1.5   Volume $= 9.37$ g $\times \dfrac{1\ \text{mL}}{1.483\ \text{g}} = 6.32$ mL

1.6   Bracelet mass $= 80.0$ g

        Bracelet volume $= 17.61$ mL $- 10.0$ mL $= 7.61$ mL

        Bracelet density $= \dfrac{80.0\ \text{g}}{7.61\ \text{mL}} = 10.5$ g/mL

        The density of the bracelet matches the density of silver. Since density is one way to identify an unknown substance, it is likely that the bracelet is made of pure silver.

1.7 6.6×10^{-24} g $= 6.6 \times 10^{-27}$ kg

        $E_K = \tfrac{1}{2}mv^2 = \tfrac{1}{2}(6.6 \times 10^{-27}\ \text{kg})\left(\dfrac{1.5 \times 10^7\ \text{m}}{\text{s}}\right)^2 = 7.4 \times 10^{-13}\ \dfrac{\text{kg·m}^2}{\text{s}^2} = 7.4 \times 10^{-13}$ J

1.8 450 g $= 0.450$ kg; $E_K = 406$ J $= 406\ \dfrac{\text{kg·m}^2}{\text{s}^2}$

        $E_K = \tfrac{1}{2}mv^2$

        $v = \sqrt{\dfrac{2 \times E_K}{m}} = \sqrt{\dfrac{2 \times 406\ \text{kg·m}^2/\text{s}^2}{0.450\ \text{kg}}} = 42.5$ m/s

1.9 (a) 0.003 00 mL has 3 significant figures because zeros at the beginning of a number are not significant and zeros at the end of a number and after the decimal point are always significant.

(b) 2070 mi has 3 or 4 significant figures because a zero in the middle of a number is significant and a zero at the end of a number and before the decimal point may or may not be significant.

(c) 47.60 mL has 4 significant figures because a zero at the end of a number and after the decimal point is always significant.

1.10 To indicate the uncertainty in a measurement, the value you record should use all the digits you are sure of plus one additional digit that you estimate. The volume can be read to the tenths place and therefore the hundredths place should be estimated. The volume reported to the correct number of significant figures is 4.55 mL.

1.11 In figure (c) darts are scattered (low precision) and are away from the bull's-eye (low accuracy).

1.12 The three measurements are 0.7783 g, 0.7780 g, and 0.7786 g. There is little variation between the three measurements so they have fairly high precision. However, the measurements are all lower than the true value and therefore, the accuracy is low.

1.13 (a)

$$\begin{array}{r} 24.567 \text{ g} \\ + \quad 0.044\ 78 \text{ g} \\ \hline 24.611\ 78 \text{ g} \end{array}$$

This result should be expressed with 3 decimal places. Because the digit to be dropped (7) is greater than 5, round up. The result is 24.612 g (5 significant figures).

(b) 4.6742 g / 0.003 71 L = 1259.89 g/L

0.003 71 has only 3 significant figures so the result of the division should have only 3 significant figures. Because the digit to be dropped (first 9) is greater than 5, round up. The result is 1260 g/L (3 significant figures), or 1.26×10^3 g/L.

1.14 NaCl mass = 36.2365 g − 35.6783 g = 0.5582 g
NaCl concentration = 0.5582 g/25.0 mL = 0.0223 g/mL = 2.23×10^{-2} g/mL

1.15 1 carat = 200 mg = 200×10^{-3} g = 0.200 g

Mass of Hope Diamond in grams = 44.4 carats $\times \dfrac{0.200 \text{ g}}{1 \text{ carat}}$ = 8.88 g

1 ounce = 28.35 g

Mass of Hope Diamond in ounces = 8.88 g $\times \dfrac{1 \text{ ounce}}{28.35 \text{ g}}$ = 0.313 ounces

1.16 Volume of Hope Diamond = 8.88 g $\times \dfrac{1 \text{ cm}^3}{3.52 \text{ g}}$ = 2.52 cm^3

1.17 area = 1.08×10^4 m^2 $\times \left(\dfrac{3.28 \text{ ft}}{1 \text{ m}}\right)^2$ = 1.16×10^5 ft^2

1.18   Volume of a cylinder = $\pi r^2 h$
Cell radius = 6×10^{-6} m/2 = 3×10^{-6} m
Cell volume = $\pi(3 \times 10^{-6}$ m$)^2(2 \times 10^{-6}$ m$)$ = 6×10^{-17} m^3

$$\text{Cell volume} = 6 \times 10^{-17}\text{ m}^3 \times \left(\frac{1\text{ cm}}{1 \times 10^{-2}\text{ m}}\right)^3 = 6 \times 10^{-11}\text{ cm}^3 = 6 \times 10^{-11}\text{ mL}$$

$$\text{Cell volume} = 6 \times 10^{-11}\text{ mL} \times \frac{1 \times 10^{-3}\text{ L}}{1\text{ mL}} = 6 \times 10^{-14}\text{ L} = 0.06 \times 10^{-12}\text{ L} = 0.06\text{ pL}$$

1.19   Only (b), a cell.

1.20   Both (c), a virus, and (d), a molecule.

1.21   The diameter of a human hair ($\sim 1 \times 10^{-5}$ m) is approximately 1,000 times larger than the diameter of a 10 nm nanoparticle. (b) A red blood cell ($\sim 1 \times 10^{-6}$ m) is approximately 10,000 times larger than a glucose molecule (1×10^{-10} m).

1.22   (a), (b) and (c)

1.23   (a)

1.24   radius = 5.0 nm/2 = 2.5 nm = 2.5×10^{-9} m = 0.0025×10^{-6} m = 0.0025 μm
(a) SA = $4\pi r^2$ = $4\pi(0.0025$ μm$)^2$ = 7.9×10^{-5} μm^2
(b) Volume = $\frac{4}{3}\pi r^3$ = $\frac{4}{3}\pi(0.0025$ μm$)^3$ = 6.5×10^{-8} μm^3

(c) $\dfrac{\text{SA}}{\text{Volume}} = \dfrac{7.9 \times 10^{-5}\text{ μm}^2}{6.5 \times 10^{-8}\text{ μm}^3} = 1{,}200\text{ μm}^{-1}$

(d) 1,200 μm^{-1}/0.6 μm^{-1} = 2,000 times

1.25   (a) An atom on the surface of a nanoparticle is more reactive.

1.26   Assume that individual atoms pack as cubes in the nanoparticle.
5.0 nm = 5.0×10^{-9} m; 10.0 nm = 10.0×10^{-9} m; 250 pm = 250×10^{-12} m
Atom volume = $(250 \times 10^{-12}$ m$)^3$ = 1.6×10^{-29} m^3
(a) Particle volume = $(5.0 \times 10^{-9}$ m$)^3$ = 1.3×10^{-25} m^3

$$\text{Atoms/particle} = \frac{\text{particle volume}}{\text{atom volume}} = \frac{1.3 \times 10^{-25}\text{ m}^3\text{/particle}}{1.6 \times 10^{-29}\text{ m}^3\text{/atom}} = 8125\text{ atoms/particle}$$

Atom face area = $(250 \times 10^{-12}$ m$)^2$ = 6.25×10^{-20} m^2
Particle face area = $(5.0 \times 10^{-9}$ m$)^2$ = 2.5×10^{-17} m^2

$$\text{Atoms/particle face} = \frac{\text{particle face area}}{\text{atom face area}} = \frac{2.5 \times 10^{-17}\text{ m}^2\text{/particle}}{6.25 \times 10^{-20}\text{ m}^2\text{/atom}} = 400\text{ atoms/particle face}$$

$$\% \text{ atoms on surface} = \frac{(6\text{ faces})(400\text{ atoms/face})}{8125\text{ atoms}} \times 100 = 30\%$$

(b) Particle volume = $(10.0 \times 10^{-9} \text{ m})^3 = 1.0 \times 10^{-24} \text{ m}^3$

$$\text{Atoms/particle} = \frac{\text{particle volume}}{\text{atom volume}} = \frac{1.0 \times 10^{-24} \text{ m}^3/\text{particle}}{1.6 \times 10^{-29} \text{ m}^3/\text{atom}} = 62,500 \text{ atoms/particle}$$

Atom face area = $(250 \times 10^{-12} \text{ m})^2 = 6.25 \times 10^{-20} \text{ m}^2$

Particle face area = $(10.0 \times 10^{-9} \text{ m})^2 = 1.0 \times 10^{-16} \text{ m}^2$

$$\text{Atoms/particle face} = \frac{\text{particle face area}}{\text{atom face area}} = \frac{1.0 \times 10^{-16} \text{ m}^2/\text{particle}}{6.25 \times 10^{-20} \text{ m}^2/\text{atom}} = 1,600 \text{ atoms/particle face}$$

$$\% \text{ atoms on surface} = \frac{(6 \text{ faces})(1,600 \text{ atoms/face})}{62,500 \text{ atoms}} \times 100\% = 15\%$$

Conceptual Problems

1.28   The level of the liquid in the thermometer is just past the 32 °C mark on the thermometer. The temperature is 32.2°C (3 significant figures).

1.30

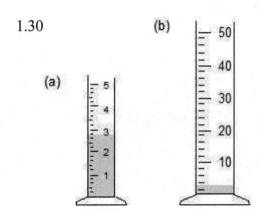

The 5 mL graduated cylinder is marked every 0.2 mL and can be read to ± 0.02 mL. The 50 mL graduated cylinder is marked every 2 mL and can only be read to ± 0.2 mL. The 5 mL graduated cylinder will give more accurate measurements.

Section Problems
Scientific Method (Section 1.1)

1.32   (a) experiment   (b) hypothesis   (c) observation

1.34   (a), (b) and (d) are quantitative.   (c) and (e) are qualitative.

1.36   A theory.

SI Units and Scientific Notation (Section 1.2)

1.38   (a) kilogram, kg   (b) meter, m   (c) kelvin, K

(d) cubic meter, m^3 (e) joule, $(\text{kg} \cdot \text{m}^2)/\text{s}^2$ (f) kg/m^3 or g/cm^3

1.40   (a) 1 km = 10^3 m = 1000 m   (b) 1m = 10^{-3} km = 0.001 km

(c) 1mmol = 10^{-3} mol = 0.001 mol   (d) 1 mol = 10^3 mmol = 1000 mmol

1.42   cL is centiliter (10^{-2} L)

1.44 (a) Convert pm to cm and compare the two quantities.

$$154 \text{ pm} \times \frac{1 \times 10^{-12} \text{ m}}{1 \text{ pm}} \times \frac{1 \text{ cm}}{1 \times 10^{-2} \text{ m}} = 15.4 \times 10^{-9} \text{ cm}$$

7.7×10^{-9} cm is smaller.

(b) Convert μm to km and compare the two quantities.

$$1.86 \times 10^{11} \text{ μm} \times \frac{1 \times 10^{-6} \text{ m}}{1 \text{ μm}} \times \frac{1 \text{ km}}{1000 \text{ m}} = 1.86 \times 10^{2} \text{ km}$$

1.86×10^{11} μm is smaller.

(c) Convert GA to μA and compare the two quantities.

$$2.9 \text{ GA} \times \frac{1 \times 10^{9} \text{ A}}{1 \text{ GA}} \times \frac{1 \text{ μA}}{1 \times 10^{-6} \text{ A}} = 2.9 \times 10^{15} \text{ μA}$$

2.9 GA is smaller.

1.46 $1 \text{ μL} = 10^{-6} \text{ L}$ $\qquad \dfrac{1 \text{ μL}}{10^{-6} \text{ L}} = 10^{6} \text{ μL/L}$

$20 \text{ mL} = 20 \times 10^{-3} \text{ L}$ $\qquad \dfrac{20 \times 10^{-3} \text{ L}}{20 \text{ mL}} \times \dfrac{1 \text{ μL}}{10^{-6} \text{ L}} = 2 \times 10^{4} \text{ μL/20 mL}$

1.48 (a) To convert 453.32 mg to scientific notation, move the decimal point 2 places to the left and include an exponent of 10^{2}. The result is 4.5332×10^{2} mg.
(b) To convert 0.000 042 1 mL to scientific notation, move the decimal point 5 places to the right and include an exponent of 10^{-5}. The result is 4.21×10^{-5} mL.
(c) To convert 667,000 g to scientific notation, move the decimal point 5 places to the left and include an exponent of 10^{5}. The result is 6.67×10^{5} g.

Measurement of Mass, Length, and Temperature (Sections 1.3–1.5)

1.50 (b) 0.2500

1.52 A Celsius degree is larger than a Fahrenheit degree by a factor of $\dfrac{9}{5}$.

1.54 $^{\circ}\text{F} = (\dfrac{9}{5} \times {}^{\circ}\text{C}) + 32$

$^{\circ}\text{F} = (\dfrac{9}{5} \times 39.9{}^{\circ}\text{C}) + 32 = 103.8\,{}^{\circ}\text{F}$ (goat)

$^{\circ}\text{F} = (\dfrac{9}{5} \times 22.2{}^{\circ}\text{C}) + 32 = 72.0\,{}^{\circ}\text{F}$ (Australian spiny anteater)

1.56 $^{\circ}\text{F} = (\dfrac{9}{5} \times {}^{\circ}\text{C}) + 32 = (\dfrac{9}{5} \times 175) + 32 = 347\,{}^{\circ}\text{F}$

1.58

Ethanol boiling point	78.5 °C	173.3 °F	200 °E
Ethanol melting point	−117.3 °C	−179.1 °F	0 °E

(a) $\dfrac{200\,°E}{[78.5\,°C - (-117.3\,°C)]} = \dfrac{200\,°E}{195.8\,°C} = 1.021\,°E/°C$

(b) $\dfrac{200\,°E}{[173.3\,°F - (-179.1\,°F)]} = \dfrac{200\,°E}{352.4\,°F} = 0.5675\,°E/°F$

(c) $°E = \dfrac{200}{195.8} \times (°C + 117.3)$

H_2O melting point $= 0°C$; $°E = \dfrac{200}{195.8} \times (0 + 117.3) = 119.8\,°E$

H_2O boiling point $= 100°C$; $°E = \dfrac{200}{195.8} \times (100 + 117.3) = 222.0\,°E$

(d) $°E = \dfrac{200}{352.4} \times (°F + 179.1) = \dfrac{200}{352.4} \times (98.6 + 179.1) = 157.6\,°E$

(e) $°F = \left(°E \times \dfrac{352.4}{200}\right) - 179.1 = \left(130 \times \dfrac{352.4}{200}\right) - 179.1 = 50.0\,°F$

Because the outside temperature is 50.0°F, I would wear a sweater or light jacket.

1.60   NaCl melting point $= 1074\,K$
$°C = K - 273.15 = 1074 - 273.15 = 800.85\,°C = 801\,°C$

$°F = (\dfrac{9}{5} \times °C) + 32 = (\dfrac{9}{5} \times 800.85) + 32 = 1473.53\,°F = 1474\,°F$

NaCl boiling point $= 1686\,K$
$°C = K - 273.15 = 1686 - 273.15 = 1412.85\,°C = 1413\,°C$

$°F = (\dfrac{9}{5} \times °C) + 32 = (\dfrac{9}{5} \times 1412.85) + 32 = 2575.13\,°F = 2575\,°F$

Derived Units: Volume and Density (Sections 1.6–1.7)

1.62   There are only seven fundamental (base) SI units for scientific measurement. A derived SI unit is some combination of two or more base SI units.
Base SI unit: Mass, kg;   Derived SI unit: Density, kg/m^3

1.64 $7.0\,dm = 7.0 \times 10^{-1}\,m = 0.70\,m$ and $1\,L = 10^{-3}\,m^3$
$V = (0.70\,m)^3 \times \dfrac{1\,L}{10^{-3}\,m^3} = 343\,L = 340\,L$

1.66 $d = \dfrac{m}{V} = \dfrac{27.43\,g}{12.40\,cm^3} = 2.212\,g/cm^3$

1.68 $3.10\,g/cm^3 = 3.10\,g/mL$
mass $= 3.10\,g/mL \times \dfrac{1\,kg}{1000\,g} \times \dfrac{1\,mL}{1 \times 10^{-3}\,L} \times 4.67\,L = 14.5\,kg$

1.70   For H_2: $V = 1.0078 \text{ g} \times \dfrac{1 \text{ L}}{0.0899 \text{ g}} = 11.2 \text{ L}$

For Cl_2: $V = 35.45 \text{ g} \times \dfrac{1 \text{ L}}{3.214 \text{ g}} = 11.03 \text{ L}$

1.72 $d = \dfrac{m}{V} = \dfrac{220.9 \text{ g}}{(0.50 \times 1.55 \times 25.00) \text{ cm}^3} = 11.4 \dfrac{\text{g}}{\text{cm}^3} = 11 \dfrac{\text{g}}{\text{cm}^3}$

1.74   Silverware mass = 80.56 g

Silverware volume = 15.90 mL – 10.00 mL = 5.90 mL

Silverware density = $\dfrac{80.56 \text{ g}}{5.90 \text{ mL}} = 13.7 \text{ g/mL}$

The density of the silverware and pure silver are different. The silverware is not pure silver.

1.76 $V = 112.5 \text{ g} \times \dfrac{1 \text{ mL}}{1.4832 \text{ g}} = 75.85 \text{ mL}$

Energy (Section 1.8)

1.78   Car: $E_K = \frac{1}{2}(1400 \text{ kg})\left(\dfrac{115 \times 10^3 \text{ m}}{3600 \text{ s}}\right)^2 = 7.1 \times 10^5 \text{ J}$

Truck: $E_K = \frac{1}{2}(12{,}000 \text{ kg})\left(\dfrac{38 \times 10^3 \text{ m}}{3600 \text{ s}}\right)^2 = 6.7 \times 10^5 \text{ J}$

The car has more kinetic energy.

1.80   1 oz = 28.35 g

energy = $0.450 \text{ oz} \times \dfrac{28.35 \text{ g}}{1 \text{ oz}} \times \dfrac{2498 \text{ kJ}}{45.0 \text{ g}} \times \dfrac{1 \text{ kcal}}{4.184 \text{ kJ}} = 169 \text{ kcal}$

1.82   (a) $540 \text{ Cal} \times \dfrac{1000 \text{ cal}}{1 \text{ Cal}} \times \dfrac{4.184 \text{ J}}{1 \text{ cal}} \times \dfrac{1 \text{ kJ}}{1000 \text{ J}} = 2259 \text{ kJ} = 2300 \text{ kJ}$

(b) 100 watts = 100 J/s

time = $2259 \text{ kJ} \times \dfrac{1000 \text{ J}}{1 \text{ kJ}} \times \dfrac{1 \text{ s}}{100 \text{ J}} \times \dfrac{1 \text{ min}}{60 \text{ s}} \times \dfrac{1 \text{ h}}{60 \text{ min}} = 6.275 \text{ h} = 6.3 \text{ h}$

Accuracy, Precision, and Significant Figures (Sections 1.9–1.10)

1.84   (a) and (b) are exact numbers because they are both definitions.
(c) and (d) are not exact numbers because they result from measurements.

1.86   (a) 35.0445 g has 6 significant figures because zeros in the middle of a number are significant.

(b) 59.0001 cm has 6 significant figures because zeros in the middle of a number are significant.

(c) 0.030 03 kg has 4 significant figures because zeros at the beginning of a number are not significant and zeros in the middle of a number are significant.

(d) 0.004 50 m has 3 significant figures because zeros at the beginning of a number are not significant and zeros at the end of a number and after the decimal point are always significant.

(e) 67,000 m^2 has 2, 3, 4, or 5 significant figures because zeros at the end of a number and before the decimal point may or may not be significant.

(f) 3.8200 x 10^3 L has 5 significant figures because zeros at the end of a number and after the decimal point are always significant.

1.88 To convert 3,666,500 m^3 to scientific notation, move the decimal point 6 places to the left and include an exponent of 10^6. The result is 3.6665 x 10^6 m^3. Because the digit to be dropped is 5 with nothing following, round down. The result is 3.666 x 10^6 m^3 (4 significant figures). Because the digit to be dropped (the second 6) is greater than 5, round up. The result is 3.7 x 10^6 m^3 (2 significant figures).

1.90 (a) Because the digit to be dropped (0) is less than 5, round down. The result is 3.567 x 10^4 or 35,670 m (4 significant figures).
Because the digit to be dropped (the second 6) is greater than 5, round up. The result is 35,670.1 m (6 significant figures).
(b) Because the digit to be dropped is 5 with nonzero digits following, round up. The result is 69 g (2 significant figures).
Because the digit to be dropped (0) is less than 5, round down. The result is 68.5 g (3 significant figures).
(c) Because the digit to be dropped is 5 with nothing following, round down. The result is 4.99 x 10^3 cm (3 significant figures).
(d) Because the digit to be dropped is 5 with nothing following, round down. The result is 2.3098 x 10^{-4} kg (5 significant figures).

1.92 (a) 4.884 x 2.05 = 10.012
The result should contain only 3 significant figures because 2.05 contains 3 significant figures (the smaller number of significant figures of the two). Because the digit to be dropped (1) is less than 5, round down. The result is 10.0.
(b) 94.61 / 3.7 = 25.57
The result should contain only 2 significant figures because 3.7 contains 2 significant figures (the smaller number of significant figures of the two). Because the digit to be dropped (second 5) is 5 with nonzero digits following, round up. The result is 26.
(c) 3.7 / 94.61 = 0.0391
The result should contain only 2 significant figures because 3.7 contains 2 significant figures (the smaller number of significant figures of the two). Because the digit to be dropped (1) is less than 5, round down. The result is 0.039.

(d)   5502.3   This result should be expressed with no decimal places. Because the
        24   digit to be dropped (3) is less than 5, round down. The result is
   +   0.01   5526.
    5526.31

(e)   86.3   This result should be expressed with only 1 decimal place. Because
  + 1.42   the digit to be dropped (3) is less than 5, round down. The result is
  − 0.09   87.6.
   87.63

(f) $5.7 \times 2.31 = 13.167$
The result should contain only 2 significant figures because 5.7 contains 2 significant figures (the smaller number of significant figures of the two). Because the digit to be dropped (second 1) is less than 5, round down. The result is 13.

Unit Conversions (Section 1.11)

1.94   (a) $0.25 \text{ lb} \times \dfrac{453.59 \text{ g}}{1 \text{ lb}} = 113.4 \text{ g} = 110 \text{ g}$

     (b) $1454 \text{ ft} \times \dfrac{12 \text{ in.}}{1 \text{ ft}} \times \dfrac{2.54 \text{ cm}}{1 \text{ in.}} \times \dfrac{1 \times 10^{-2} \text{ m}}{1 \text{ cm}} = 443.2 \text{ m}$

     (c) $2{,}941{,}526 \text{ mi}^2 \times \left(\dfrac{1.6093 \text{ km}}{1 \text{ mi}} \right)^2 \times \left(\dfrac{1000 \text{ m}}{1 \text{ km}} \right)^2 = 7.6181 \times 10^{12} \text{ m}^2$

1.96   1 mile = 1.6093 km; The time is 1 h, 5 min, and 26.6 s.
     Convert the time to seconds and then hours.

$$\text{time} = \left(1 \text{ hr} \times \dfrac{60 \text{ min}}{1 \text{ h}} \times \dfrac{60 \text{ s}}{1 \text{ min}} \right) + \left(5 \text{ min} \times \dfrac{60 \text{ s}}{1 \text{ min}} \right) + 26.6 \text{ s} = 3926.6 \text{ s}$$

$$\text{time} = 3926.6 \text{ s} \times \dfrac{1 \text{ min}}{60 \text{ s}} \times \dfrac{1 \text{ h}}{60 \text{ min}} = 1.0907 \text{ h}$$

     Convert meters to miles.

$$20{,}000 \text{ m} \times \dfrac{1 \text{ km}}{1000 \text{ m}} \times \dfrac{1 \text{ mi}}{1.6093 \text{ km}} = 12.4278 \text{ mi}$$

$$\text{average speed} = \dfrac{12.4278 \text{ mi}}{1.0907 \text{ h}} = 11.394 \text{ mi/h}$$

1.98   (a) $1 \text{ acre-ft} \times \dfrac{1 \text{ mi}^2}{640 \text{ acres}} \times \left(\dfrac{5280 \text{ ft}}{1 \text{ mi}} \right)^2 = 43{,}560 \text{ ft}^3$

     (b) $116 \text{ mi}^3 \times \left(\dfrac{5280 \text{ ft}}{1 \text{ mi}} \right)^3 \times \dfrac{1 \text{ acre-ft}}{43{,}560 \text{ ft}^3} = 3.92 \times 10^8 \text{ acre-ft}$

1.100 $8.65 \text{ stones} \times \dfrac{14 \text{ lb}}{1 \text{ stone}} = 121 \text{ lb}$

1.102 $160 \text{ lb} \times \dfrac{1 \text{ kg}}{2.2046 \text{ lb}} = 72.6 \text{ kg}$

$72.6 \text{ kg} \times \dfrac{20 \text{ μg}}{1 \text{ kg}} \times \dfrac{1 \text{ mg}}{1 \times 10^3 \text{ μg}} = 1.452 \text{ mg} = 1.5 \text{ mg}$

1.104 (a) A liter is just slightly larger than a quart.
   (b) A mile is about twice as long as a kilometer.
   (c) An ounce is about 30 times larger than a gram.
   (d) An inch is about 2.5 times larger than a centimeter.

1.106 $d = 0.55 \dfrac{\text{oz}}{\text{in}^3} \times \dfrac{1 \text{ lb}}{16 \text{ oz}} \times \dfrac{453.59 \text{ g}}{1 \text{ lb}} \times \left(\dfrac{1 \text{ in}}{2.54 \text{ cm}}\right)^3 = 0.95 \text{ g/cm}^3$

1.108 amount of chocolate =

$2.0 \text{ cups coffee} \times \dfrac{105 \text{ mg caffeine}}{1 \text{ cup coffee}} \times \dfrac{1.0 \text{ ounce chocolate}}{15 \text{ mg caffeine}} = 14 \text{ ounces of chocolate}$

14 ounces of chocolate is just under 1 pound.

Multiconcept Problems

1.110 volume of sphere $= \dfrac{4}{3} \pi r^3$; sphere radius = 7.60 cm/2 = 3.80 cm

   (a) $d = \dfrac{m}{V} = \dfrac{313 \text{ g}}{\dfrac{4}{3} \pi (3.80 \text{ cm})^3} = 1.36 \text{ g/cm}^3 = 1.36 \text{ g/mL}$

   (b) Because the density is greater than 1.0 g/mL, the sphere will sink in water.
   (c) Because the density is less than 1.48 g/mL, the sphere will float in chloroform.

1.112 (a) number of Hershey's Kisses =
   $2.0 \text{ lb} \times \dfrac{453.59 \text{ g}}{1 \text{ lb}} \times \dfrac{1 \text{ serving}}{41 \text{ g}} \times \dfrac{9 \text{ kisses}}{1 \text{ serving}} = 199 \text{ kisses} = 200 \text{ kisses}$

   (b) Hershey's Kiss volume $= \dfrac{41 \text{ g}}{1 \text{ serving}} \times \dfrac{1 \text{ serving}}{9 \text{ kisses}} \times \dfrac{1 \text{ mL}}{1.4 \text{ g}} = 3.254 \text{ mL} = 3.3 \text{ mL}$

   (c) Calories/Hershey's Kiss $= \dfrac{230 \text{ Cal}}{1 \text{ serving}} \times \dfrac{1 \text{ serving}}{9 \text{ kisses}} = 25.55 \text{ Cal/kiss} = 26 \text{ Cal/kiss}$

   (d) % fat Calories =
   $\dfrac{13 \text{ g fat}}{1 \text{ serving}} \times \dfrac{9 \text{ Cal from fat}}{1 \text{ g fat}} \times \dfrac{1 \text{ serving}}{230 \text{ Cal total}} \times 100\% = 51\% \text{ Calories from fat}$

1.114 $°C = \dfrac{5}{9} \times (°F - 32)$

Set $°C = °F$: $°C = \dfrac{5}{9} \times (°C - 32)$

Solve for $°C$: $°C \times \dfrac{9}{5} = °C - 32$

$\left(°C \times \dfrac{9}{5}\right) - °C = -32$

$°C \times \dfrac{4}{5} = -32$

$°C = \dfrac{5}{4}(-32) = -40\,°C$

The Celsius and Fahrenheit scales "cross" at $-40\,°C$ ($-40\,°F$).

1.116 Ethyl alcohol density $= \dfrac{19.7325\ \text{g}}{25.00\ \text{mL}} = 0.7893$ g/mL

total mass = metal mass + ethyl alcohol mass $= 38.4704$ g

ethyl alcohol mass = total mass – metal mass = 38.4704 g – 25.0920 g = 13.3784 g

ethyl alcohol volume $= 13.3784\ \text{g} \times \dfrac{1\ \text{mL}}{0.7893\ \text{g}} = 16.95$ mL

metal volume = total volume – ethyl alcohol volume = 25.00 mL – 16.95 mL = 8.05 mL

metal density $= \dfrac{25.0920\ \text{g}}{8.05\ \text{mL}} = 3.12$ g/mL

1.118 $35\ \text{sv} = 35 \times 10^9\ \dfrac{\text{m}^3}{\text{s}}$

(a) gulf stream flow $= \left(35 \times 10^9\ \dfrac{\text{m}^3}{\text{s}}\right)\left(\dfrac{1\ \text{cm}}{1 \times 10^{-2}\ \text{m}}\right)^3\left(\dfrac{1\ \text{mL}}{1\ \text{cm}^3}\right)\left(\dfrac{60\ \text{s}}{1\ \text{min}}\right) = 2.1 \times 10^{18}$ mL/min

(b) mass of $H_2O = \left(2.1 \times 10^{18}\ \dfrac{\text{mL}}{\text{min}}\right)\left(\dfrac{60\ \text{min}}{1\ \text{h}}\right)(24\ \text{h})\left(\dfrac{1.025\ \text{g}}{1\ \text{mL}}\right) = 3.1 \times 10^{21}$ g $= 3.1 \times 10^{18}$ kg

(c) time $= \left(1.0 \times 10^{15}\ \text{L}\right)\left(\dfrac{1\ \text{mL}}{1 \times 10^{-3}\ \text{L}}\right)\left(\dfrac{1\ \text{min}}{2.1 \times 10^{18}\ \text{mL}}\right) = 0.48$ min

2 Atoms, Molecules, and Ions

2.1 It is a metal, and most likely near the end of the transition metals because it can be found in nature in its pure form. A likely candidate is the element marked "b", which is silver.

2.2 (a) K, potassium, metal (b) shiny, metallic solid
(c) It would deform and not crack since metals are malleable.
(d) It would conduct electricity, but it is not a good choice for wiring because it is reacts when exposed to oxygen and humidity (water) in the atmosphere. Potassium is also a soft metal which, makes it unsuitable for use as a wire.

2.3 First, find the S:O ratio in each compound.
Substance A: S:O mass ratio = (6.00 g S) / (5.99 g O) = 1.00
Substance B: S:O mass ratio = (8.60 g S) / (12.88 g O) = 0.668
$$\frac{\text{S:O mass ratio in substance A}}{\text{S:O mass ratio in substance B}} = \frac{1.00}{0.668} = 1.50 = \frac{3}{2}$$
Yes, the mass ratios in the two compounds are simple multiples of each other.

2.4 In compound A the O/S mass ratio is 1. In compound B, the O/S mass ratio is 3/2. This means there is 3/2 times more O in compound B. To find the formula of Compound B multiply the subscript on O in Compound A by 3/2. Compound B is SO_3.

2.5 $0.005 \text{ mm} \times \dfrac{1 \times 10^{-3} \text{ m}}{1 \text{ mm}} \times \dfrac{1 \text{ Au atom}}{2.9 \times 10^{-10} \text{ m}} = 2 \times 10^4 \text{ Au atoms}$

2.6 $1 \times 10^{19} \text{ C atoms} \times \dfrac{1.5 \times 10^{-10} \text{ m}}{\text{C atom}} \times \dfrac{1 \text{ km}}{1000 \text{ m}} \times \dfrac{1 \text{ time}}{40,075 \text{ km}} = 37.4 \text{ times} \sim 40 \text{ times}$

2.7 $^{75}_{34}$ Se has 34 protons, 34 electrons, and (75 − 34) = 41 neutrons.

2.8 The element with 24 protons is Cr. The mass number is the sum of the protons and the neutrons, 24 + 28 = 52. The isotope symbol is $^{52}_{24}$Cr.

2.9 atomic mass = (62.93)(0.6915) + (64.93)(0.3085) = 63.55
(63.546 from periodic table)

2.10 (a) The atomic weight of Ga is 69.7231. This mass is closer to the mass of gallium-69 than to gallium-71; therefore gallium-69 must be more abundant.
(b) The total abundance of both isotopes must be 100.00%. Let Y be the natural abundance of ^{69}Ga and [1 − Y] the natural abundance of ^{71}Ga.
(68.9256)(Y) + (70.9247)([1 − Y]) = 69.7231

Solve for Y. $Y = \dfrac{-1.2016}{-1.9991} = 0.6011$

^{69}Ga natural abundance = 60.11% and ^{71}Ga natural abundance = 100.00 − 60.11 = 39.89%

2.11 Pt, 195.078

mol Pt = 9.50 g Pt x $\dfrac{1 \text{ mol Pt}}{195.078 \text{ g Pt}}$ = 0.0487 mol Pt

atoms Pt = 0.0487 mol Pt x $\dfrac{6.022 \times 10^{23} \text{ atoms Pt}}{1 \text{ mol Pt}}$ = 2.93 x 10^{22} atoms Pt

2.12 atomic mass in g = $\dfrac{1.50 \text{ g}}{2.26 \times 10^{22} \text{ atoms}}$ x 6.022 x 10^{23} atoms = 40.0 g; Y = Ca

2.13 Total intensity = 100.00 + 92.90 = 192.90
mass of isotope 1 = 106.905; mass of isotope 2 = 108.905

fractional abundance of isotope 1 = $\dfrac{100.00}{192.90}$ = 0.5184

fractional abundance of isotope 2 = $\dfrac{92.90}{192.90}$ = 0.4816

atomic weight = (106.905)(0.5184) + (108.905)(0.4816) = 107.87
The element is Ag.

2.14 Total intensity = 2.672 + 45.992 + 42.176 + 100.000 = 190.840
mass of isotope 1 = 203.9730;   mass of isotope 2 = 205.9744
mass of isotope 3 = 206.9758;   mass of isotope 4 = 207.9766

fractional abundance of isotope 1 = $\dfrac{2.672}{190.840}$ = 0.01400

fractional abundance of isotope 2 = $\dfrac{45.992}{190.840}$ = 0.24100

fractional abundance of isotope 3 = $\dfrac{42.176}{190.840}$ = 0.22100

fractional abundance of isotope 4 = $\dfrac{100.000}{190.840}$ = 0.524000

atomic weight = (203.9730)(0.01400) + (205.9744)(0.24100) + (206.9758)(0.22100) +
(207.9766)(0.524000) = 207.217

The element is Pb.

2.15 Figure (b) represents a collection of pure hydrogen peroxide (H_2O_2) molecules.

2.16 (a) Figures (b) and (d) illustrate pure substances.
(b) Figures (a) and (c) illustrate mixtures.
(c) Figures (b) and (d) illustrate the law of multiple proportions.

2.17   thymine, $C_5H_6N_2O_2$

2.18   adrenaline, $C_9H_{13}NO_3$

2.19   (a) LiBr is an ionic compound.

2.20   Figure (a) most likely represents an ionic compound because there are no discrete molecules, only a regular array of two different chemical species (ions). Figure (b) most likely represents a molecular compound because discrete molecules are present.

2.21   (a) magnesium fluoride, MgF_2 (b) tin(IV) oxide, SnO_2

2.22   red – potassium sulfide, K_2S;   green – strontium iodide, SrI_2;   blue – gallium oxide, Ga_2O_3

2.23   iron(III) carbonate, $Fe_2(CO_3)_3$

2.24   Drawing 1 represents ionic compounds with one cation and two anions. Only (c) $CaCl_2$ is consistent with drawing 1.
        Drawing 2 represents ionic compounds with one cation and one anion. Both (a) LiBr and (b) $NaNO_2$ are consistent with drawing 2.

2.25 N_2O_5, dinitrogen pentoxide

2.26   (a) PCl_5, phosphorus pentachloride   (b) N_2O, dinitrogen monoxide

2.27   (e)

2.28   (a) false   (b) true except for 2 or 3 years prior to 1940   (c) true   (d) true

2.29 ^{18}O, 8 protons, 10 neutrons, 8 electrons; ^2H, 1 proton, 1 neutron, 1 electron

2.30   Warm seawater near the equator has a higher ratio of $^{18}O/^{16}O$ than snow falling in Antarctica.

2.31   (a) ~14 degrees   (b) ~70%

2.32   (a) % ^{18}O = 0.1995% = 0.001 995
        % ^{16}O = 100.0000% − 0.1995% = 99.8005% = 0.998 005
        seawater O atomic weight = (15.994 914 6)(0.998 005) + (17.999 161 0)(0.001 995)
                                         = 15.998 91 u
        (b) % ^{18}O = 0.1971% = 0.001 971
        % ^{16}O = 100.0000% − 0.1971% = 99.8029% = 0.998 029
        polar ice O atomic weight = (15.994 914 6)(0.998 029) + (17.999 161 0)(0.001 971)
                                         = 15.998 86 u

2.33 H_2O, 18.015 g/mol

(a) mol H_2O = 1.00 x 10^{-6}g H_2O x $\dfrac{1\ mol\ H_2O}{18.015\ mol\ H_2O}$ = 5.55 x 10^{-8} mol H_2O

(b) % deuterium = 0.0156% = 0.000 156

deuterium atoms = (5.55 x 10^{-8} mol H_2O)$\left(\dfrac{6.022\ x10^{23}\ H_2O\ molecules}{1\ mol\ H_2O}\right)$ x

$\left(\dfrac{3\ atoms}{1\ H_2O\ molecule}\right)\left(\dfrac{0.000\ 156\ D\ atoms}{H_2O\ atoms}\right)$ = 1.56 x 10^{13} D atoms

Conceptual Problems

2.34

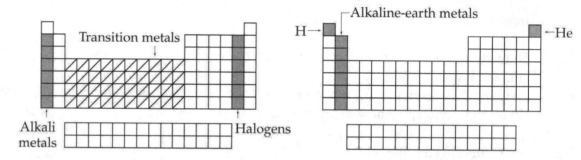

2.36   The element is americium (Am) with atomic number = 95. It is in the actinide series.

2.38   Drawing (a) represents a collection of SO_2 units. Drawing (d) represents a mixture of S atoms and O_2 units.

2.40   Figures (b) and (c) both contain two protons but different numbers of neutrons. They are isotopes of the same element. Figure (a) contains only one proton. It is a different element than (b) and (c).

2.42   Figure (b) represents a pure substance consisting of a compound.
Figure (c) represents a pure substance consisting of an element.
Figure (a) represents a mixture of elements.

2.44   (a) alanine, $C_3H_7NO_2$ (b) ethylene glycol, $C_2H_6O_2$ (c) acetic acid, $C_2H_4O_2$

Section Problems
Elements and the Periodic Table (Sections 2.1–2.3)

2.46   118 elements are presently known. About 90 elements occur naturally.

2.48   (a) gadolinium, Gd   (b) germanium, Ge   (c) technetium, Tc   (d) arsenic, As

2.50   (a) Te, tellurium   (b) Re, rhenium   (c) Be, beryllium   (d) Ar, argon
       (e) Pu, plutonium

2.52   (a) Tin is Sn. Ti is titanium.   (b) Manganese is Mn. Mg is magnesium.
       (c) Potassium is K. Po is polonium.
       (d) The symbol for helium is He. The second letter is lowercase.

2.54   (a) Mendeleev used the experimentally observed chemistry of the elements to organize
       them in his table.
       (b) Mendeleev realized that there were "holes" in the table below aluminum and silicon.
       The chemical behavior of aluminum (relative mass ≈ 27.3) is similar to that of boron
       (relative mass ≈ 11). In the same way, silicon (relative mass ≈ 28) is similar to carbon
       (relative mass ≈ 12). Using these observations, Mendeleev predicted with remarkable
       accuracy what the properties of these unknown elements would be. The element
       immediately below aluminum should have a relative mass near 68 and should have a low
       melting point. The element below silicon should have a relative mass near 72 and should
       be dark gray in color.

2.56   The rows are called periods, and the columns are called groups.

2.58   Elements within a group have similar chemical properties.

2.60
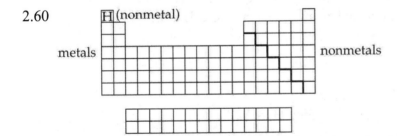

2.62   (a) Ti, metal   (b) Te, semimetal   (c) Se, nonmetal
       (d) Sc, metal   (e) Si, semimetal

2.64   (a) The alkali metals are shiny, soft, low-melting metals that react rapidly with water to
       form products that are alkaline.
       (b) The noble gases are gases of very low reactivity.
       (c) The halogens are nonmetallic and corrosive. They are found in nature only in
       combination with other elements.

2.66   F, Cl, Br, and I

2.68   An element that is a soft, silver-colored solid that reacts violently with water and is a
       good conductor of electricity has the characteristics of a metal.

2.70   An element that is a yellow crystalline solid, does not conduct electricity, and when hit
       with a hammer it shatters has the characteristics of a nonmetal.

2.72   All match in groups 2A and 7A.

2.74   Intensive properties do not depend on the amount of substance present.

Atomic Theory (Sections 2.4–2.5)

2.76   The law of mass conservation in terms of Dalton's atomic theory states that chemical reactions only rearrange the way that atoms are combined; the atoms themselves are not changed.
The law of definite proportions in terms of Dalton's atomic theory states that the chemical combination of elements to make different substances occurs when atoms join together in small, whole-number ratios.

2.78   In any chemical reaction, the combined mass of the final products equals the combined mass of the starting reactants.
mass of reactants = mass of products
mass of reactants = mass of Hg + mass of O_2 = 114.0 g + 12.8 g = 126.8 g
mass of products = mass of HgO + mass of left over O_2 = 123.1 g + mass of left over O_2
126.8 g = 123.1 g + mass of left over O_2
mass of left over O_2 = 126.8 g − 123.1 g = 3.7 g

2.80   For the "other" compound: C:H mass ratio = (32.0 g C) / (8.0 g H) = 4
The "other" compound is not methane because the methane C:H mass ratio is 3.
$$\frac{\text{C:H mass ratio in "other"}}{\text{C:H mass ratio in methane}} = \frac{4}{3}$$

2.82   First, find the C:H ratio in each compound.
Benzene: C:H mass ratio = (4.61 g C) / (0.39 g H) = 12
Ethane: C:H mass ratio = (4.00 g C) / (1.00 g H) = 4.00
Ethylene: C:H mass ratio = (4.29 g C) / (0.71 g H) = 6.0
$$\frac{\text{C:H mass ratio in benzene}}{\text{C:H mass ratio in ethane}} = \frac{12}{4.00} = \frac{3}{1}$$
$$\frac{\text{C:H mass ratio in benzene}}{\text{C:H mass ratio in ethylene}} = \frac{12}{6.0} = \frac{2}{1}$$
$$\frac{\text{C:H mass ratio in ethylene}}{\text{C:H mass ratio in ethane}} = \frac{6.0}{4.00} = \frac{3}{2}$$

2.84   Assume a 100.0 g sample for each compound and then find the O:C ratio in each compound.
Compound 1: O:C mass ratio = (57.1 g O)/(42.9 g C) = 1.33
Compound 2: O:C mass ratio = (72.7 g O)/(27.3 g C) = 2.66
$$\frac{\text{O:C mass ratio in compound 2}}{\text{O:C mass ratio in compound 1}} = \frac{2.66}{1.33} = \frac{2}{1}$$

If compound 1 is CO and the O:C mass ratio is 2 times that of compound 1, then the formula for compound 2 is CO_2.

Elements and Atoms (Sections 2.6–2.8)

2.86   electron

2.88   (a) true   (b) true   (c) true   (d) true   (e) false   (f) true

2.90   (a) -1.010×10^{-18} C because it is not an integer multiple of the electron charge.

2.92   (a) The alpha particles would pass right through the gold foil with little to no deflection.

2.94   350 pm = 350×10^{-12} m

$$\text{Pb atoms} = 0.25 \text{ in} \times \frac{2.54 \text{ cm}}{1 \text{ in}} \times \frac{1 \times 10^{-2} \text{ m}}{1 \text{ cm}} \times \frac{1 \text{ Pb atom}}{350 \times 10^{-12} \text{ m}} = 1.8 \times 10^7 \text{ Pb atoms thick}$$

2.96   The atomic number is equal to the number of protons.
        The mass number is equal to the sum of the number of protons and the number of neutrons.

2.98   The subscript giving the atomic number of an atom is often left off an isotope symbol because one can readily look up the atomic number in the periodic table.

2.100   (a) carbon, C   (b) argon, Ar   (c) vanadium, V

2.102   (a) $^{220}_{86}$Rn   (b) $^{210}_{84}$Po   (c) $^{197}_{79}$Au

2.104   (a) $^{15}_{7}$N, 7 protons, 7 electrons, $(15 - 7) = 8$ neutrons

        (b) $^{60}_{27}$Co, 27 protons, 27 electrons, $(60 - 27) = 33$ neutrons

        (c) $^{131}_{53}$I, 53 protons, 53 electrons, $(131 - 53) = 78$ neutrons

        (d) $^{142}_{58}$Ce, 58 protons, 58 electrons, $(142 - 58) = 84$ neutrons

2.106   (a) $^{24}_{12}$Mg, magnesium   (b) $^{58}_{28}$Ni, nickel

        (c) $^{104}_{46}$Pd, palladium   (d) $^{183}_{74}$W, tungsten

2.108 $^{12}_{5}$C, the atomic number for carbon is 6, not 5.

        $^{33}_{35}$Br, the mass number must be greater than the atomic number.

        $^{11}_{5}$Bo, the element symbol for boron is B.

2.110 Deuterium is ^2H and deuterium fluoride is ^2HF.
^2H has 1 proton, 1 neutron, and 1 electron.
F has 9 protons, 10 neutrons, and 9 electrons.
^2HF has 10 protons, 11 neutrons, and 10 electrons.
Chemically, ^2HF is like HF and is a weak acid.

Atomic Weight, Moles, and Mass Spectrometry (Sections 2.9–2.10)

2.112 (b) ^{12}C

2.114 (a) atomic mass   (b) atomic number   (c) molar mass
(d) mass number   (e) atomic weight

2.116 An element's atomic mass is the weighted average of the isotopic masses of the element's naturally occurring isotopes. The atomic mass for Cu (63.546) must fall between the masses of its two isotopes. If one isotope is ^{65}Cu, the other isotope must be ^{63}Cu, and not ^{66}Cu. If the other isotope was ^{66}Cu, the atomic mass for Cu would be greater than 65.

2.118 $(10.0129)(0.199) + (11.009\ 31)(0.801) = 10.8$ for B

2.120 $24.305 = (23.985)(0.7899) + (24.986)(0.1000) + (Z)(0.1101)$
Solve for Z. $Z = 25.982$ for ^{26}Mg.

2.122 atomic weight $= (62.93)(0.6915) + (64.93)(0.3085) = 63.55$

2.124 (a) $g\ Ti = 1.505\ mol\ Ti \times \dfrac{47.867\ g\ Ti}{1\ mol\ Ti} = 72.04\ g\ Ti$

(b) $g\ Na = 0.337\ mol\ Na \times \dfrac{22.989\ 770\ g\ Na}{1\ mol\ Na} = 7.75\ g\ Na$

(c) $g\ U = 2.583\ mol\ U \times \dfrac{238.028\ 91\ g\ U}{1\ mol\ U} = 614.8\ g\ U$

2.126 The mass of 6.02×10^{23} atoms is its atomic mass expressed in grams. If the atomic mass of an element is X, then 6.02×10^{23} atoms of this element weighs X grams.

2.128 The mass of 6.02×10^{23} atoms is its atomic mass expressed in grams. If the mass of 6.02×10^{23} atoms of element Y is 83.80 g, then the atomic mass of Y is 83.80. Y is Kr.

2.130 (a) The purpose of bombarding gaseous atoms with an electron beam is to knock electrons off of the atoms, leaving them with a positive charge.
(b) A lighter ion will be deflected to a greater degree by the magnetic field.
(c) The magnetic field strength is adjusted so that a different mass-to-charge ratio ion strikes the detector.

2.132　Total intensity = 100.00 + 74.80 = 174.80

mass of isotope 1 = 120.9038; mass of isotope 2 = 122.9042

$$\text{fractional abundance of isotope 1} = \frac{100.00}{174.80} = 0.5721$$

$$\text{fractional abundance of isotope 2} = \frac{74.80}{174.80} = 0.4279$$

atomic weight = (120.9038)(0.5721) + (122.9042)(0.4279) = 121.76

The element is Sb.

Chemical Compounds (Sections 2.11–2.12)

2.134　A covalent bond results when two atoms share several (usually two) of their electrons. An ionic bond results from a complete transfer of one or more electrons from one atom to another. The C–H bonds in methane (CH_4) are covalent bonds. The bond in NaCl (Na^+Cl^-) is an ionic bond.

2.136　Element symbols are composed of one or two letters. If the element symbol is two letters, the first letter is uppercase and the second is lowercase. CO stands for carbon and oxygen in carbon monoxide.

2.138　(a) Be^{2+}, 4 protons and 2 electrons　　　　(b) Rb^+, 37 protons and 36 electrons
　　　　(c) Se^{2-}, 34 protons and 36 electrons　　　(d) Au^{3+}, 79 protons and 76 electrons

2.140　C_3H_8O

2.142

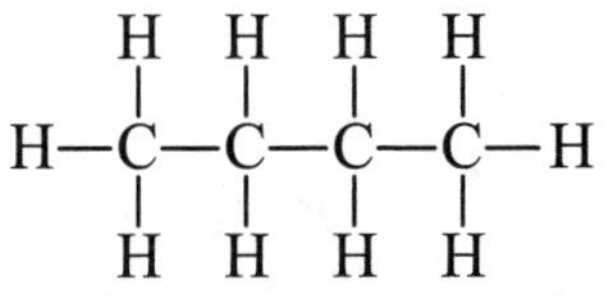

2.144

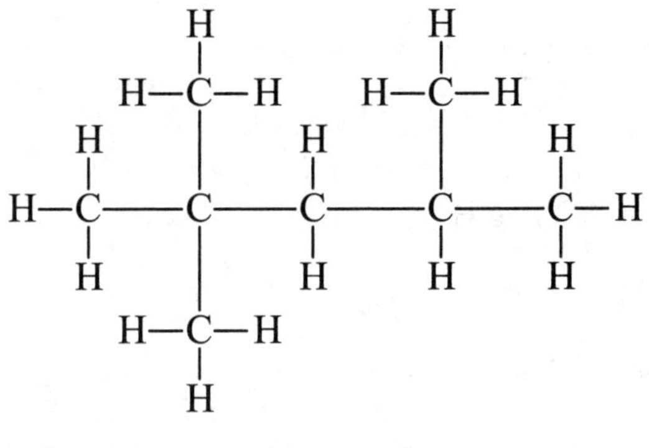

Naming Compounds (Section 2.13)

2.146　(a) CsF, cesium fluoride　　(b) K_2O, potassium oxide　　　(c) CuO, copper(II) oxide

2.148 (a) potassium chloride, KCl (b) tin(II) bromide, $SnBr_2$ (c) calcium oxide, CaO
(d) barium chloride, $BaCl_2$ (e) aluminum hydride, AlH_3

2.150 (a) calcium acetate, $Ca(CH_3CO_2)_2$ (b) iron(II) cyanide, $Fe(CN)_2$
(c) sodium dichromate, $Na_2Cr_2O_7$ (d) chromium(III) sulfate, $Cr_2(SO_4)_3$
(e) mercury(II) perchlorate, $Hg(ClO_4)_2$

2.152 (a) $Ca(ClO)_2$, calcium hypochlorite
(b) $Ag_2S_2O_3$, silver(I) thiosulfate or silver thiosulfate
(c) NaH_2PO_4, sodium dihydrogen phosphate (d) $Sn(NO_3)_2$, tin(II) nitrate
(e) $Pb(CH_3CO_2)_4$, lead(IV) acetate (f) $(NH_4)_2SO_4$, ammonium sulfate

2.154 (a) $CaBr_2$ (b) $CaSO_4$ (c) $Al_2(SO_4)_3$

2.156 (a) $CaCl_2$ (b) CaO (c) CaS

2.158 (a) sulfite ion, SO_3^{2-} (b) phosphate ion, PO_4^{3-} (c) zirconium(IV) ion, Zr^{4+}
(d) chromate ion, CrO_4^{2-} (e) acetate ion, $CH_3CO_2^-$ (f) thiosulfate ion, $S_2O_3^{2-}$

2.160 (a) CCl_4, carbon tetrachloride (b) ClO_2, chlorine dioxide
(c) N_2O, dinitrogen monoxide (d) N_2O_3, dinitrogen trioxide

2.162 (a) NO, nitrogen monoxide (b) N_2O, dinitrogen monoxide (c) NO_2, nitrogen dioxide
(d) N_2O_4, dinitrogen tetroxide (e) N_2O_5, dinitrogen pentoxide

2.164 (a) Na^+ and SO_4^{2-}; therefore the formula is Na_2SO_4
(b) Ba^{2+} and PO_4^{3-}; therefore the formula is $Ba_3(PO_4)_2$
(c) Ga^{3+} and SO_4^{2-}; therefore the formula is $Ga_2(SO_4)_3$

Multiconcept Problems

2.166 For NH_3, $(2.34 \text{ g N})\left(\dfrac{3 \times 1.0079 \text{ H}}{14.0067 \text{ N}}\right) = 0.505 \text{ g H}$

For N_2H_4, $(2.34 \text{ g N})\left(\dfrac{4 \times 1.0079 \text{ H}}{2 \times 14.0067 \text{ N}}\right) = 0.337 \text{ g H}$

2.168 $\dfrac{12.0000}{15.9994} = \dfrac{X}{16.0000}$; $X = 12.0005$ for ^{12}C prior to 1961.

2.170 molecular mass $= (8 \times 12.011) + (9 \times 1.0079) + (1 \times 14.0067)$
$+ (2 \times 15.9994) = 151.165 \text{ g/mol}$

2.172 (a) Arrange the droplet charges in order of increasing charge.
2.21×10^{-16} C
4.42×10^{-16} C
4.98×10^{-16} C
6.64×10^{-16} C
7.74×10^{-16} C

Find the smallest charge difference between droplet charges.
$(4.42 - 2.21) \times 10^{-16}$ C $= 2.21 \times 10^{-16}$ C
$(4.98 - 4.42) \times 10^{-16}$ C $= 0.56 \times 10^{-16}$ C
$(6.64 - 4.98) \times 10^{-16}$ C $= 1.66 \times 10^{-16}$ C
$(7.74 - 6.64) \times 10^{-16}$ C $= 1.10 \times 10^{-16}$ C
The smallest difference (0.56×10^{-16} C) is the approximate value for the charge on one blorvek.

To get a more accurate value, divide the droplet charges by the 0.56×10^{-16} C to determine the total number of blorveks.
2.21×10^{-16} C$/0.56 \times 10^{-16}$ C $= 4$
4.42×10^{-16} C$/0.56 \times 10^{-16}$ C $= 8$
4.98×10^{-16} C$/0.56 \times 10^{-16}$ C $= 9$
6.64×10^{-16} C$/0.56 \times 10^{-16}$ C $= 12$
7.74×10^{-16} C$/0.56 \times 10^{-16}$ C $= 14$
There are $(4 + 8 + 9 + 12 + 14) = 47$ blorveks on the 5 droplets with a total charge of $(2.21 + 4.42 + 4.98 + 6.64 + 7.74) \times 10^{-16}$ C $= 25.99 \times 10^{-16}$ C.

The charge on one blorvek is $\dfrac{25.99 \times 10^{-16} \text{ C}}{47 \text{ blorveks}} = 5.53 \times 10^{-17}$ C/blorvek.

(b) Again arrange the droplet charges in increasing order.
2.21×10^{-16} C
4.42×10^{-16} C
4.98×10^{-16} C
5.81×10^{-16} C
6.64×10^{-16} C
7.74×10^{-16} C

Find the smallest charge difference between droplet charges.
$(4.42 - 2.21) \times 10^{-16}$ C $= 2.21 \times 10^{-16}$ C
$(4.98 - 4.42) \times 10^{-16}$ C $= 0.56 \times 10^{-16}$ C
$(5.81 - 4.98) \times 10^{-16}$ C $= 0.83 \times 10^{-16}$ C
$(6.64 - 5.81) \times 10^{-16}$ C $= 0.83 \times 10^{-16}$ C
$(7.74 - 6.64) \times 10^{-16}$ C $= 1.10 \times 10^{-16}$ C
The new charge difference (0.83×10^{-16} C) is not an integer multiple of 5.53×10^{-17} C, which means the charge on the blorvek must be smaller than 5.53×10^{-17} C.
The difference between 0.83×10^{-16} C and 0.56×10^{-16} C $= 0.27 \times 10^{-16}$ C is the approximate value for the charge on one blorvek.

To get a more accurate value, divide the droplet charges by the 0.27×10^{-16} C to determine the total number of blorveks.

2.21×10^{-16} C$/0.27 \times 10^{-16}$ C = 8

4.42×10^{-16} C$/0.27 \times 10^{-16}$ C = 16

4.98×10^{-16} C$/0.27 \times 10^{-16}$ C = 18

5.81×10^{-16} C$/0.27 \times 10^{-16}$ C = 21

6.64×10^{-16} C$/0.27 \times 10^{-16}$ C = 24

7.74×10^{-16} C$/0.27 \times 10^{-16}$ C = 28

There are $(8 + 16 + 18 + 21 + 24 + 28) = 115$ blorveks on the 6 droplets with a total charge of $(2.21 + 4.42 + 4.98 + 5.81 + 6.64 + 7.74) \times 10^{-16}$ C = 31.80×10^{-16} C.

The charge on one blorvek is $\dfrac{31.80 \times 10^{-16}\ \text{C}}{115\ \text{blorveks}} = 2.77 \times 10^{-17}$ C/blorvek.

3

Mass Relationships in Chemical Reactions

3.1 $3 A_2 + 2 B \rightarrow 2 BA_3$

3.2   reactants, box (d), and products, box (c)

3.3 $4 NH_3 + Cl_2 \rightarrow N_2H_4 + 2 NH_4Cl$

3.4 $8 KClO_3 + C_{12}H_{22}O_{11} \rightarrow 8 KCl + 12 CO_2 + 11 H_2O$

3.5 H_2SO_4 $2(1.01) + 1(32.07) + 4(16.00) = 98.1$

3.6   sucrose, $C_{12}H_{22}O_{11}$ $12(12.01) + 22(1.01) + 11(16.00) = 342.3$
       molar mass $= 342.3$ g/mol

3.7 $NaHCO_3$, 84.0

$$5.26 \text{ g NaHCO}_3 \times \frac{1 \text{ mol NaHCO}_3}{84.0 \text{ g NaHCO}_3} = 0.0626 \text{ mol NaHCO}_3$$

3.8   glucose, $C_6H_{12}O_6$, 180.0

(a) 0.0833 mol glucose $\times \dfrac{180.0 \text{ g glucose}}{1 \text{ mol glucose}} = 15.0$ g glucose

(b) 15.0 g glucose $\times \dfrac{1 \text{ glucose tablet}}{3.75 \text{ g glucose}} = 4$ glucose tablets

(c) 0.0833 mol glucose $\times \dfrac{6.022 \times 10^{23} \text{ glucose molecules}}{1 \text{ mol glucose}} = 5.02 \times 10^{22}$ glucose molecules

3.9   salicylic acid, $C_7H_6O_3$, 138.1;   acetic anhydride, $C_4H_6O_3$, 102.1

$$4.50 \text{ g } C_7H_6O_3 \times \frac{1 \text{ mol } C_7H_6O_3}{138.1 \text{ g } C_7H_6O_3} \times \frac{1 \text{ mol } C_4H_6O_3}{1 \text{ mol } C_7H_6O_3} \times \frac{102.1 \text{ g } C_4H_6O_3}{1 \text{ mol } C_4H_6O_3} = 3.33 \text{ g } C_4H_6O_3$$

3.10   salicylic acid, $C_7H_6O_3$, 138.1;   acetic anhydride, $C_4H_6O_3$, 102.1
       aspirin, $C_9H_8O_4$, 180.2;   acetic acid, CH_3CO_2H, 60.1

(a) $10.0 \text{ g } C_9H_8O_4 \times \dfrac{1 \text{ mol } C_9H_8O_4}{180.2 \text{ g } C_9H_8O_4} \times \dfrac{1 \text{ mol } C_7H_6O_3}{1 \text{ mol } C_9H_8O_4} \times \dfrac{138.1 \text{ g } C_7H_6O_3}{1 \text{ mol } C_7H_6O_3} = 7.66 \text{ g } C_7H_6O_3$

(b) $10.0 \text{ g } C_9H_8O_4 \times \dfrac{1 \text{ mol } C_9H_8O_4}{180.2 \text{ g } C_9H_8O_4} \times \dfrac{1 \text{ mol } CH_3CO_2H}{1 \text{ mol } C_9H_8O_4} \times \dfrac{60.1 \text{ g } CH_3CO_2H}{1 \text{ mol } CH_3CO_2H} = 3.34 \text{ g } CH_3CO_2H$

3.11 C_2H_4, 28.1; C_2H_6O, 46.1

$$4.6 \text{ g } C_2H_4 \times \frac{1 \text{ mol } C_2H_4}{28.1 \text{ g } C_2H_4} \times \frac{1 \text{ mol } C_2H_6O}{1 \text{ mol } C_2H_4} \times \frac{46.1 \text{ g } C_2H_6O}{1 \text{ mol } C_2H_6O} = 7.5 \text{ g } C_2H_6O \text{ (theoretical yield)}$$

$$\text{Percent yield} = \frac{\text{Actual yield}}{\text{Theoretical yield}} \times 100\% = \frac{4.7 \text{ g}}{7.5 \text{ g}} \times 100\% = 63\%$$

3.12 C_2H_6O, 46.1; $C_4H_{10}O$, 74.1

(a) $$40.0 \text{ g } C_2H_6O \times \frac{1 \text{ mol } C_2H_6O}{46.1 \text{ g } C_2H_6O} \times \frac{1 \text{ mol } C_4H_{10}O}{2 \text{ mol } C_2H_6O} \times \frac{74.1 \text{ g } C_4H_{10}O}{1 \text{ mol } C_4H_{10}O} = 32.1 \text{ g } C_4H_{10}O \text{ (theoretical yield)}$$

actual yield = (32.1 g)(0.870) = 27.9 g $C_4H_{10}O$

(b) 100.0 g = (theoretical yield)(0.870)
theoretical yield = 100.0 g/0.870 = 115 g $C_4H_{10}O$

$$115 \text{ g } C_4H_{10}O \times \frac{1 \text{ mol } C_4H_{10}O}{74.1 \text{ g } C_4H_{10}O} \times \frac{2 \text{ mol } C_2H_6O}{1 \text{ mol } C_4H_{10}O} \times \frac{46.1 \text{ g } C_2H_6O}{1 \text{ mol } C_2H_6O} = 143 \text{ g } C_2H_6O$$

3.13 (a) $A + B_2 \rightarrow AB_2$
There is a 1:1 stoichiometry between the two reactants. A is the limiting reactant because there are fewer reactant A's than there are reactant B_2's.
(b) 1.0 mol of AB_2 can be made from 1.0 mol of A and 1.0 mol of B_2.

3.14

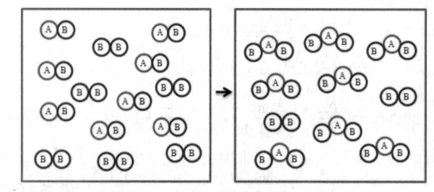

B_2 is in excess, AB is the limiting reactant.

3.15 Li_2O, 29.9: 65 kg = 65,000 g; H_2O, 18.0: 80.0 kg = 80,000 g; LiOH, 23.9

(a) $$65,000 \text{ g } Li_2O \times \frac{1 \text{ mol } Li_2O}{29.9 \text{ g } Li_2O} = 2.17 \times 10^3 \text{ mol } Li_2O$$

$$80,000 \text{ g } H_2O \times \frac{1 \text{ mol } H_2O}{18.0 \text{ g } H_2O} = 4.44 \times 10^3 \text{ mol } H_2O$$

The reaction stoichiometry between Li_2O and H_2O is one to one. There are twice as many moles of H_2O as there are moles of Li_2O. Therefore, Li_2O is the limiting reactant.

(b) $(4.44 \times 10^3 \text{ mol} - 2.17 \times 10^3 \text{ mol}) = 2.27 \times 10^3 \text{ mol H}_2\text{O remaining}$

$$2.27 \times 10^3 \text{ mol H}_2\text{O} \times \frac{18.0 \text{ g H}_2\text{O}}{1 \text{ mol H}_2\text{O}} = 40{,}860 \text{ g H}_2\text{O} = 40.9 \text{ kg} = 41 \text{ kg H}_2\text{O}$$

3.16 LiHCO_3, 68.0;   LiOH, 23.9

$$500.0 \text{ g LiHCO}_3 \times \frac{1 \text{ mol LiHCO}_3}{68.0 \text{ g LiHCO}_3} \times \frac{1 \text{ mol LiOH}}{1 \text{ mol LiHCO}_3} \times \frac{23.9 \text{ g LiOH}}{1 \text{ mol LiOH}} = 175.7 \text{ g LiOH}$$

You start with 400.0 g of LiOH. 500.0 g of LiHCO_3 are produced from 175.7 g of LiOH. Over 200 g of LiOH remain. Additional CO_2 can be removed.

3.17   Assume a 100.0 g sample. From the percent composition data, a 100.0 g sample contains 14.25 g C, 56.93 g O, and 28.83 g Mg.

$$14.25 \text{ g C} \times \frac{1 \text{ mol C}}{12.011 \text{ g C}} = 1.19 \text{ mol C}$$

$$56.93 \text{ g O} \times \frac{1 \text{ mol O}}{15.999 \text{ g O}} = 3.56 \text{ mol O}$$

$$28.83 \text{ g Mg} \times \frac{1 \text{ mol Mg}}{24.305 \text{ g Mg}} = 1.19 \text{ mol Mg}$$

$\text{Mg}_{1.19}\text{C}_{1.19}\text{O}_{3.56}$; divide each subscript by the smallest, 1.19.
$\text{Mg}_{1.19/1.19}\text{C}_{1.19/1.19}\text{O}_{3.56/1.19}$
The empirical formula is MgCO_3.

3.18   glucose, $\text{C}_6\text{H}_{12}\text{O}_6$, 180.0
Divide each subscript by the smallest, 6, to get the empirical formula.
$\text{C}_{6/6}\text{H}_{12/6}\text{O}_{6/6}$
The empirical formula is CH_2O.
Percent composition:

$$\% \text{ C} = \frac{6 \times 12.0 \text{ g}}{180.0 \text{ g}} \times 100\% = 40.0\%$$

$$\% \text{ H} = \frac{12 \times 1.01 \text{ g}}{180.0 \text{ g}} \times 100\% = 6.7\%$$

$$\% \text{ O} = \frac{6 \times 16.0 \text{ g}}{180.0 \text{ g}} \times 100\% = 53.3\%$$

3.19 $1.161 \text{ g H}_2\text{O} \times \dfrac{1 \text{ mol H}_2\text{O}}{18.0 \text{ g H}_2\text{O}} \times \dfrac{2 \text{ mol H}}{1 \text{ mol H}_2\text{O}} = 0.129 \text{ mol H}$

$2.818 \text{ g CO}_2 \times \dfrac{1 \text{ mol CO}_2}{44.0 \text{ g CO}_2} \times \dfrac{1 \text{ mol C}}{1 \text{ mol CO}_2} = 0.0640 \text{ mol C}$

$0.129 \text{ mol H} \times \dfrac{1.01 \text{ g H}}{1 \text{ mol H}} = 0.130 \text{ g H}$

$$0.0640 \text{ mol C} \times \frac{12.0 \text{ g C}}{1 \text{ mol C}} = 0.768 \text{ g C}$$

$$1.00 \text{ g total} - (0.130 \text{ g H} + 0.768 \text{ g C}) = 0.102 \text{ g O}$$

$$0.102 \text{ g O} \times \frac{1 \text{ mol O}}{16.0 \text{ g O}} = 0.006 \ 38 \text{ mol O}$$

$C_{0.0640}H_{0.129}O_{0.006\ 38}$; divide each subscript by the smallest, 0.006 38.
$C_{0.0640\ /\ 0.006\ 38}H_{0.129\ /\ 0.006\ 38}O_{0.006\ 38\ /\ 0.006\ 38}$
$C_{10.03}H_{20.22}O_1$
The empirical formula is $C_{10}H_{20}O$.

3.20 $0.697 \text{ g H}_2\text{O} \times \dfrac{1 \text{ mol H}_2\text{O}}{18.0 \text{ g H}_2\text{O}} \times \dfrac{2 \text{ mol H}}{1 \text{ mol H}_2\text{O}} = 0.0774 \text{ mol H}$

       $1.55 \text{ g CO}_2 \times \dfrac{1 \text{ mol CO}_2}{44.0 \text{ g CO}_2} \times \dfrac{1 \text{ mol C}}{1 \text{ mol CO}_2} = 0.0352 \text{ mol C}$

$C_{0.0352}H_{0.0774}$; divide each subscript by the smaller, 0.0352.
$C_{0.0352\ /\ 0.0352}H_{0.0774\ /\ 0.0352}$
CH_2
The empirical formula is CH_2, 14.0.
$142.0 \ / \ 14.0 = 10.1 = 10$; molecular formula $= C_{(10\ \times\ 1)}H_{(10\ \times 2)} = C_{10}H_{20}$

3.21   The empirical formula is C_6H_5, 77.0.
The peak with the largest mass in the mass spectrum occurs near a mass of 154, which would be the molecular weight of the compound.
$154/77.0 = 2$;   molecular formula $= C_{(2\ \times\ 6)}H_{(2\ \times\ 5)} = C_{12}H_{10}$

3.22 $0.57 \text{ g H}_2\text{O} \times \dfrac{1 \text{ mol H}_2\text{O}}{18.0 \text{ g H}_2\text{O}} \times \dfrac{2 \text{ mol H}}{1 \text{ mol H}_2\text{O}} = 0.063 \text{ mol H}$

       $2.79 \text{ g CO}_2 \times \dfrac{1 \text{ mol CO}_2}{44.0 \text{ g CO}_2} \times \dfrac{1 \text{ mol C}}{1 \text{ mol CO}_2} = 0.063 \text{ mol C}$

$0.063 \text{ mol H} \times \dfrac{1.01 \text{ g H}}{1 \text{ mol H}} = 0.063 \text{ g H}$; $0.063 \text{ mol C} \times \dfrac{12.0 \text{ g C}}{1 \text{ mol C}} = 0.76 \text{ g C}$

$$1.00 \text{ g total} - (0.063 \text{ g H} + 0.76 \text{ g C}) = 0.18 \text{ g N}$$

$$0.18 \text{ g N} \times \frac{1 \text{ mol N}}{14.0 \text{ g N}} = 0.013 \text{ mol N}$$

$C_{0.063}H_{0.063}N_{0.013}$; divide each subscript by the smallest, 0.013.
$C_{0.063\ /\ 0.013}H_{0.063\ /\ 0.013}N_{0.013\ /\ 0.013}$
$C_{4.85}H_{4.85}N$
The empirical formula is C_5H_5N, 79.0.
The peak with the largest mass in the mass spectrum occurs near a mass of 79, which would be the molecular weight of the compound, therefore the empirical formula and molecular formula are the same.

3.23   (b) Design safer chemical products and processes that reduce or eliminate the generation of hazardous substances.

3.24   (a) percent yield   (b) percent atom economy

3.25   (a) Reaction 1 has the higher % atom economy because all reactant atoms are used.
(b) rxn 1: $C_2H_4 + Cl_2 \rightarrow C_2H_4Cl_2$
There are no undesired products, therefore the % Atom Economy = 100%.

rxn 2: $CH_3Cl + Br^- \rightarrow CH_3Br + Cl^-$

Reactants:	Desired Product:
1 C = (1)(12.0) = 12.0	1 C = (1)(12.0) = 12.0
3 H = (3)(1.0) = 3.0	3 H = (3)(1.0) = 3.0
1 Cl = (1)(35.5) = 35.5	1 Br = (1)(80.0) = 80.0
1 Br = (1)(80.0) = 80.0	Σ = 93.0
Σ = 130.5	

$$\% \text{ Atom Economy} = \frac{\Sigma \text{ Atomic Weight}_{(\text{atoms in desired product})}}{\Sigma \text{ Atomic Weight}_{(\text{atoms in all reactants})}} = \frac{93.0}{130.5} \times 100 = 71\%$$

3.26   (a) C_3H_8O, 60.1; C_3H_6, 42.1

$$23.50 \text{ g } C_3H_8O \times \frac{1 \text{ mol } C_3H_8O}{60.1 \text{ g } C_3H_8O} = 0.39 \text{ mol } C_3H_8O$$

$$\text{theoretical yield} = (0.39 \text{ mol } C_3H_8O)\left(\frac{1 \text{ mol } C_3H_6}{1 \text{ mol } C_3H_8O}\right)\left(\frac{42.1 \text{ g } C_3H_6}{1 \text{ mol } C_3H_6}\right) = 16.4 \text{ g } C_3H_6$$

$$\text{percent yield} = \frac{\text{actual yield}}{\text{theoretical yield}} \times 100\% = \frac{10.15 \text{ g}}{16.4 \text{ g}} \times 100\% = 61.9\%$$

(b) $C_3H_8O \rightarrow C_3H_6 + H_2O$

Reactants:	Desired Product:
3 C = (3)(12.0) = 36.0	3 C = (3)(12.0) = 36.0
8 H = (8)(1.0) = 8.0	6 H = (6)(1.0) = 6.0
1 O = (1)(16.0) = 16.0	Σ = 42.0
Σ = 60.0	

$$\% \text{ Atom Economy} = \frac{\Sigma \text{ Atomic Weight}_{(\text{atoms in desired product})}}{\Sigma \text{ Atomic Weight}_{(\text{atoms in all reactants})}} = \frac{42.0}{60.0} \times 100 = 70\%$$

3.27   (a) Ibuprofen, $C_{13}H_{18}O_2$
(b) (13 x 12.0) + (18 x 1.0) + (2 x 16.0) = 206.0

$$(c) \ \% \text{ C} = \frac{13 \times 12.0 \text{ g}}{206.0 \text{ g}} \times 100\% = 75.7\%$$

$$\% \text{ H } = \frac{18 \times 1.01 \text{ g}}{206.0 \text{ g}} \times 100\% = 8.8\%$$

$$\% \text{ O } = \frac{2 \times 16.0 \text{ g}}{206.0 \text{ g}} \times 100\% = 15.5\%$$

3.28 (a) $4 \text{ mol H} \times \dfrac{1.0 \text{ g H}}{1 \text{ mol H}} = 4.0 \text{ g H}$

$2 \text{ mol C} \times \dfrac{12.0 \text{ g C}}{1 \text{ mol C}} = 24.0 \text{ g C}$

$2 \text{ mol O} \times \dfrac{16.0 \text{ g O}}{1 \text{ mol O}} = 32.0 \text{ g O}$

$4.0 \text{ g} + 24.0 \text{ g} + 32.0 \text{ g} = 60.0 \text{ g}$

(b) Ibuprofen, $C_{13}H_{18}O_2$, 206.0

$\text{mass Ibuprofen} = 30 \times 10^6 \text{ lbs} \times \dfrac{1 \text{ kg}}{2.20 \text{ lbs}} \times \dfrac{1000 \text{ g}}{1 \text{ kg}} = 1.4 \times 10^{10} \text{ g Ibuprofen}$

$\text{mol Ibuprofen} = 1.4 \times 10^{10} \text{ g Ibuprofen} \times \dfrac{1 \text{ mol Ibuprofen}}{206.0 \text{ g Ibuprofen}} = 6.6 \times 10^7 \text{ mol Ibuprofen}$

(c) $6.6 \times 10^7 \text{ mol Ibuprofen} \times \dfrac{60.0 \text{ g waste}}{1 \text{ mol Ibuprofen}} \times \dfrac{1 \text{ kg}}{1000 \text{ g}} = 4.0 \times 10^6 \text{ kg waste}$

Conceptual Problems

3.30 $2 A_5B_2 \rightarrow A_2 + 4 A_2B$

3.32 (a) $A_2 + 3 B_2 \rightarrow 2 AB_3$; B_2 is the limiting reactant because it is completely consumed.
(b) For 1.0 mol of A_2, 3.0 mol of B_2 are required. Because only 1.0 mol of B_2 is available, B_2 is the limiting reactant.

$1 \text{ mol B}_2 \times \dfrac{2 \text{ mol AB}_3}{3 \text{ mol B}_2} = 2/3 \text{ mol AB}_3$

3.34 The molecular formula for cytosine is $C_4H_5N_3O$.

$\text{mol CO}_2 = 0.001 \text{ mol cyt} \times \dfrac{4 \text{ C}}{\text{cyt}} \times \dfrac{1 \text{ CO}_2}{\text{C}} = 0.004 \text{ mol CO}_2$

$\text{mol H}_2\text{O} = 0.001 \text{ mol cyt} \times \dfrac{5 \text{ H}}{\text{cyt}} \times \dfrac{1 \text{ H}_2\text{O}}{2 \text{ H}} = 0.0025 \text{ mol H}_2\text{O}$

Section Problems
Balancing Equations (Section 3.2)

3.36 Equation (b) is balanced, (a) is not balanced.

3.38 (a) $Mg + 2 HNO_3 \rightarrow H_2 + Mg(NO_3)_2$
      (b) $CaC_2 + 2 H_2O \rightarrow Ca(OH)_2 + C_2H_2$
      (c) $2 S + 3 O_2 \rightarrow 2 SO_3$
      (d) $UO_2 + 4 HF \rightarrow UF_4 + 2 H_2O$

3.40 (a) $SiCl_4 + 2 H_2O \rightarrow SiO_2 + 4 HCl$
      (b) $P_4O_{10} + 6 H_2O \rightarrow 4 H_3PO_4$
      (c) $CaCN_2 + 3 H_2O \rightarrow CaCO_3 + 2 NH_3$
      (d) $3 NO_2 + H_2O \rightarrow 2 HNO_3 + NO$

3.42 (a) $4 C_6H_5NO_2 + 29 O_2 \rightarrow 24 CO_2 + 10 H_2O + 4 NO_2$
      (b) $2 Au + 6 H_2SeO_4 \rightarrow Au_2(SeO_4)_3 + 3 H_2SeO_3 + 3 H_2O$
      (c) $6 NH_4ClO_4 + 8 Al \rightarrow 4 Al_2O_3 + 3 N_2 + 3 Cl_2 + 12 H_2O$

Molecular Weights and Molar Mass (Section 3.3)

3.44 (a) Hg_2Cl_2: $2(200.59) + 2(35.45) = 472.1$
      (b) $C_4H_8O_2$: $4(12.01) + 8(1.01) + 2(16.00) = 88.1$
      (c) CF_2Cl_2: $1(12.01) + 2(19.00) + 2(35.45) = 120.9$

3.46 (a) $C_{33}H_{35}FN_2O_5$: $33(12.01) + 35(1.01) + 1(19.00) + 2(14.01) + 5(16.00) = 558.7$
      (b) $C_{22}H_{27}F_3O_4S$: $22(12.01) + 27(1.01) + 3(19.00) + 4(16.00) + 1(32.06) = 444.5$
      (c) $C_{16}H_{16}ClNO_2S$: $16(12.01) + 16(1.01) + 1(35.45) + 1(14.01) + 2(16.00) + 1(32.06) = 321.8$

3.48 One mole equals the atomic weight or molecular weight in grams.
      (a) Ti, 47.87 g   (b) Br_2, 159.81 g   (c) Hg, 200.59 g   (d) H_2O, 18.02 g

3.50 There are 3 ions (one Mg^{2+} and 2 Cl^-) per formula unit of $MgCl_2$.
      $MgCl_2$, 95.2

$$27.5 \text{ g } MgCl_2 \times \frac{1 \text{ mol } MgCl_2}{95.2 \text{ g } MgCl_2} \times \frac{3 \text{ mol ions}}{1 \text{ mol } MgCl_2} = 0.867 \text{ mol ions}$$

3.52 $\text{Molar mass} = \dfrac{3.28 \text{ g}}{0.0275 \text{ mol}} = 119 \text{ g/mol}$; molecular weight = 119.

3.54 $FeSO_4$, 151.9; 300 mg = 0.300 g

$$0.300 \text{ g } FeSO_4 \times \frac{1 \text{ mol } FeSO_4}{151.9 \text{ g } FeSO_4} = 1.97 \times 10^{-3} \text{ mol } FeSO_4$$

$$1.97 \times 10^{-3} \text{ mol } FeSO_4 \times \frac{6.022 \times 10^{23} \text{ Fe(II) atoms}}{1 \text{ mol } FeSO_4} = 1.19 \times 10^{21} \text{ Fe(II) atoms}$$

3.56 $C_8H_{10}N_4O_2$, 194.2; 125 mg = 0.125 g

$$0.125 \text{ g caffeine} \times \frac{1 \text{ mol caffeine}}{194.2 \text{ g caffeine}} = 6.44 \times 10^{-4} \text{ mol caffeine}$$

$$0.125 \text{ g caffeine} \times \frac{1 \text{ mol caffeine}}{194.2 \text{ g caffeine}} \times \frac{6.022 \times 10^{23} \text{ molecules}}{1 \text{ mol}} = 3.88 \times 10^{20} \text{ caffeine molecules}$$

3.58   By definition, 6.022×10^{23} particles are 1.000 mole of particles.

mol Ar = 0.2500 mol and mol "other" = 0.7500 mol

$$\text{mass Ar} = 0.2500 \text{ mol Ar} \times \frac{39.95 \text{ g Ar}}{1 \text{ mol Ar}} = 9.99 \text{ g Ar}$$

mass "other" = total mass – mass Ar = 25.12 g – 9.99 g = 15.13 g "other"

$$\text{molar mass "other"} = \frac{15.13 \text{ g "other"}}{0.7500 \text{ mol "other"}} = 20.17 \text{ g/mol}$$

The "other" element is neon (Ne).

3.60 TiO_2, 79.87; $100.0 \text{ kg Ti} \times \dfrac{79.87 \text{ kg } TiO_2}{47.87 \text{ kg Ti}} = 166.8 \text{ kg } TiO_2$

Stoichiometry (Section 3.4)

3.62   (a) $2 \, Fe_2O_3 + 3 \, C \rightarrow 4 \, Fe + 3 \, CO_2$

(b) Fe_2O_3, 159.7; $525 \text{ g } Fe_2O_3 \times \dfrac{1 \text{ mol } Fe_2O_3}{159.7 \text{ g } Fe_2O_3} \times \dfrac{3 \text{ mol C}}{2 \text{ mol } Fe_2O_3} = 4.93 \text{ mol C}$

(c) $4.93 \text{ mol C} \times \dfrac{12.01 \text{ g C}}{1 \text{ mol C}} = 59.2 \text{ g C}$

3.64   (a) $2 \, Mg + O_2 \rightarrow 2 \, MgO$

(b) Mg, 24.30; O_2, 32.00; MgO, 40.30

$$25.0 \text{ g Mg} \times \frac{1 \text{ mol Mg}}{24.30 \text{ g Mg}} \times \frac{1 \text{ mol } O_2}{2 \text{ mol Mg}} \times \frac{32.00 \text{ g } O_2}{1 \text{ mol } O_2} = 16.5 \text{ g } O_2$$

$$25.0 \text{ g Mg} \times \frac{1 \text{ mol Mg}}{24.30 \text{ g Mg}} \times \frac{2 \text{ mol MgO}}{2 \text{ mol Mg}} \times \frac{40.30 \text{ g MgO}}{1 \text{ mol MgO}} = 41.5 \text{ g MgO}$$

(c) $25.0 \text{ g } O_2 \times \dfrac{1 \text{ mol } O_2}{32.00 \text{ g } O_2} \times \dfrac{2 \text{ mol Mg}}{1 \text{ mol } O_2} \times \dfrac{24.30 \text{ g Mg}}{1 \text{ mol Mg}} = 38.0 \text{ g Mg}$

$$25.0 \text{ g } O_2 \times \frac{1 \text{ mol } O_2}{32.00 \text{ g } O_2} \times \frac{2 \text{ mol MgO}}{1 \text{ mol } O_2} \times \frac{40.30 \text{ g MgO}}{1 \text{ mol MgO}} = 63.0 \text{ g MgO}$$

3.66   (a) $2 \, HgO \rightarrow 2 \, Hg + O_2$

(b) HgO, 216.6; Hg, 200.6; O_2, 32.0

$$45.5 \text{ g HgO} \times \frac{1 \text{ mol HgO}}{216.6 \text{ g HgO}} \times \frac{2 \text{ mol Hg}}{2 \text{ mol HgO}} \times \frac{200.6 \text{ g Hg}}{1 \text{ mol Hg}} = 42.1 \text{ g Hg}$$

$$45.5 \text{ g HgO} \times \frac{1 \text{ mol HgO}}{216.6 \text{ g HgO}} \times \frac{1 \text{ mol } O_2}{2 \text{ mol HgO}} \times \frac{32.00 \text{ g } O_2}{1 \text{ mol } O_2} = 3.36 \text{ g } O_2$$

(c) $33.3 \text{ g } O_2 \times \dfrac{1 \text{ mol } O_2}{32.00 \text{ g } O_2} \times \dfrac{2 \text{ mol HgO}}{1 \text{ mol } O_2} \times \dfrac{216.6 \text{ g HgO}}{1 \text{ mol HgO}} = 451 \text{ g HgO}$

3.68 $2.00 \text{ g Ag} \times \dfrac{1 \text{ mol Ag}}{107.9 \text{ g Ag}} = 0.0185 \text{ mol Ag}; \quad 0.657 \text{ g Cl} \times \dfrac{1 \text{ mol Cl}}{35.45 \text{ g Cl}} = 0.0185 \text{ mol Cl}$

$Ag_{0.0185}Cl_{0.0185}$; divide both subscripts by 0.0185. The empirical formula is AgCl.

3.70 N_2H_4, 32.05; I_2, 253.8; HI, 127.9

(a) $36.7 \text{ g } N_2H_4 \times \dfrac{1 \text{ mol } N_2H_4}{32.05 \text{ g } N_2H_4} \times \dfrac{2 \text{ mol } I_2}{1 \text{ mol } N_2H_4} \times \dfrac{253.8 \text{ g } I_2}{1 \text{ mol } I_2} = 581 \text{ g } I_2$

(b) $115.7 \text{ g } N_2H_4 \times \dfrac{1 \text{ mol } N_2H_4}{32.05 \text{ g } N_2H_4} \times \dfrac{4 \text{ mol HI}}{1 \text{ mol } N_2H_4} \times \dfrac{127.9 \text{ g HI}}{1 \text{ mol HI}} = 1847 \text{ g HI}$

Reaction Yield and Limiting Reactants (Sections 3.5–3.6)

3.72 H_2SO_4, 98.08; $NiCO_3$, 118.7; $NiSO_4$, 154.8

(a) $14.5 \text{ g } NiCO_3 \times \dfrac{1 \text{ mol } NiCO_3}{118.7 \text{ g } NiCO_3} \times \dfrac{1 \text{ mol } H_2SO_4}{1 \text{ mol } NiCO_3} \times \dfrac{98.08 \text{ g } H_2SO_4}{1 \text{ mol } H_2SO_4} = 12.0 \text{ g } H_2SO_4$

(b) $14.5 \text{ g } NiCO_3 \times \dfrac{1 \text{ mol } NiCO_3}{118.7 \text{ g } NiCO_3} \times \dfrac{1 \text{ mol } NiSO_4}{1 \text{ mol } NiCO_3} \times \dfrac{154.8 \text{ g } NiSO_4}{1 \text{ mol } NiSO_4} \times 0.789 = 14.9 \text{ g } NiSO_4$

3.74 $3.44 \text{ mol } N_2 \times \dfrac{3 \text{ mol } H_2}{1 \text{ mol } N_2} = 10.3 \text{ mol } H_2$ required.

Because there is only 1.39 mol H_2, H_2 is the limiting reactant.

$1.39 \text{ mol } H_2 \times \dfrac{2 \text{ mol } NH_3}{3 \text{ mol } H_2} \times \dfrac{17.03 \text{ g } NH_3}{1 \text{ mol } NH_3} = 15.8 \text{ g } NH_3$

$1.39 \text{ mol } H_2 \times \dfrac{1 \text{ mol } N_2}{3 \text{ mol } H_2} \times \dfrac{28.01 \text{ g } N_2}{1 \text{ mol } N_2} = 13.0 \text{ g } N_2$ reacted

$3.44 \text{ mol } N_2 \times \dfrac{28.01 \text{ g } N_2}{1 \text{ mol } N_2} = 96.3 \text{ g } N_2$ initially

$(96.3 \text{ g} - 13.0 \text{ g}) = 83.3 \text{ g } N_2$ left over

3.76 C_2H_4, 28.05; Cl_2, 70.91; $C_2H_4Cl_2$, 98.96

$15.4 \text{ g } C_2H_4 \times \dfrac{1 \text{ mol } C_2H_4}{28.05 \text{ g } C_2H_4} = 0.549 \text{ mol } C_2H_4$

$3.74 \text{ g } Cl_2 \times \dfrac{1 \text{ mol } Cl_2}{70.91 \text{ g } Cl_2} = 0.0527 \text{ mol } Cl_2$

Because the reaction stoichiometry between C_2H_4 and Cl_2 is one to one, Cl_2 is the limiting reactant.

$$0.0527 \text{ mol } Cl_2 \text{ x } \frac{1 \text{ mol } C_2H_4Cl_2}{1 \text{ mol } Cl_2} \text{ x } \frac{98.96 \text{ g } C_2H_4Cl_2}{1 \text{ mol } C_2H_4Cl_2} = 5.22 \text{ g } C_2H_4Cl_2$$

3.78 $CaCO_3$, 100.1; HCl, 36.46

$$CaCO_3 \text{ + } 2 \text{ HCl } \rightarrow \text{ } CaCl_2 \text{ + } H_2O \text{ + } CO_2$$

$$2.35 \text{ g } CaCO_3 \text{ x } \frac{1 \text{ mol } CaCO_3}{100.1 \text{ g } CaCO_3} = 0.0235 \text{ mol } CaCO_3$$

$$2.35 \text{ g } HCl \text{ x } \frac{1 \text{ mol } HCl}{36.46 \text{ g } HCl} = 0.0645 \text{ mol } HCl$$

The reaction stoichiometry is 1 mole of $CaCO_3$ for every 2 moles of HCl. For 0.0235 mol $CaCO_3$, we only need 2(0.0235 mol) = 0.0470 mol HCl. We have 0.0645 mol HCl, therefore $CaCO_3$ is the limiting reactant.

$$0.0235 \text{ mol } CaCO_3 \text{ x } \frac{1 \text{ mol } CO_2}{1 \text{ mol } CaCO_3} \text{ x } \frac{22.4 \text{ L}}{1 \text{ mol } CO_2} = 0.526 \text{ L } CO_2$$

3.80 $CH_3CO_2H \text{ + } C_5H_{12}O \rightarrow C_7H_{14}O_2 \text{ + } H_2O$
       CH_3CO_2H, 60.05; $C_5H_{12}O$, 88.15; $C_7H_{14}O_2$, 130.19

$$3.58 \text{ g } CH_3CO_2H \text{ x } \frac{1 \text{ mol } CH_3CO_2H}{60.05 \text{ g } CH_3CO_2H} = 0.0596 \text{ mol } CH_3CO_2H$$

$$4.75 \text{ g } C_5H_{12}O \text{ x } \frac{1 \text{ mol } C_5H_{12}O}{88.15 \text{ g } C_5H_{12}O} = 0.0539 \text{ mol } C_5H_{12}O$$

Because the reaction stoichiometry between CH_3CO_2H and $C_5H_{12}O$ is one to one, isopentyl alcohol ($C_5H_{12}O$) is the limiting reactant.

$$0.0539 \text{ mol } C_5H_{12}O \text{ x } \frac{1 \text{ mol } C_7H_{14}O_2}{1 \text{ mol } C_5H_{12}O} \text{ x } \frac{130.19 \text{ g } C_7H_{14}O_2}{1 \text{ mol } C_7H_{14}O_2} = 7.02 \text{ g } C_7H_{14}O_2$$

7.02 g $C_7H_{14}O_2$ is the theoretical yield. Actual yield = (7.02 g)(0.45) = 3.2 g.

3.82 $CH_3CO_2H \text{ + } C_5H_{12}O \rightarrow C_7H_{14}O_2 \text{ + } H_2O$
       CH_3CO_2H, 60.05; $C_5H_{12}O$, 88.15; $C_7H_{14}O_2$, 130.19

$$1.87 \text{ g } CH_3CO_2H \text{ x } \frac{1 \text{ mol } CH_3CO_2H}{60.05 \text{ g } CH_3CO_2H} = 0.0311 \text{ mol } CH_3CO_2H$$

$$2.31 \text{ g } C_5H_{12}O \text{ x } \frac{1 \text{ mol } C_5H_{12}O}{88.15 \text{ g } C_5H_{12}O} = 0.0262 \text{ mol } C_5H_{12}O$$

Because the reaction stoichiometry between CH_3CO_2H and $C_5H_{12}O$ is one to one, isopentyl alcohol ($C_5H_{12}O$) is the limiting reactant.

$$0.0262 \text{ mol } C_5H_{12}O \text{ x } \frac{1 \text{ mol } C_7H_{14}O_2}{1 \text{ mol } C_5H_{12}O} \text{ x } \frac{130.19 \text{ g } C_7H_{14}O_2}{1 \text{ mol } C_7H_{14}O_2} = 3.41 \text{ g } C_7H_{14}O_2$$

3.41 g $C_7H_{14}O_2$ is the theoretical yield.

$$\% \text{ Yield} = \frac{\text{Actual yield}}{\text{Theoretical yield}} \text{ x } 100\% = \frac{2.96 \text{ g}}{3.41 \text{ g}} \text{ x } 100\% = 86.8\%$$

3.84 WCl_6, 396.6; W_6Cl_{12}, 1528

(a) $6 \text{ WCl}_6 + 8 \text{ Bi} \rightarrow W_6Cl_{12} + 8 \text{ BiCl}_3$

(b) $150.0 \text{ g WCl}_6 \text{ x } \dfrac{1 \text{ mol WCl}_6}{396.6 \text{ g WCl}_6} \text{ x } \dfrac{8 \text{ mol Bi}}{6 \text{ mol WCl}_6} \text{ x } \dfrac{209.0 \text{ g Bi}}{1 \text{ mol Bi}} = 105.4 \text{ g Bi}$

(c)   Compute the theoretical yield for W_6Cl_{12} from each reactant to determine the limiting reactant.

$$228 \text{ g WCl}_6 \text{ x } \frac{1 \text{ mol WCl}_6}{396.6 \text{ g WCl}_6} \text{ x } \frac{1 \text{ mol W}_6Cl_{12}}{6 \text{ mol WCl}_6} \text{ x } \frac{1528 \text{ g W}_6Cl_{12}}{1 \text{ mol W}_6Cl_{12}} = 146 \text{ g W}_6Cl_{12}$$

$$175 \text{ g Bi } \text{ x } \frac{1 \text{ mol Bi}}{209.0 \text{ g Bi}} \text{ x } \frac{1 \text{ mol W}_6Cl_{12}}{8 \text{ mol Bi}} \text{ x } \frac{1528 \text{ g W}_6Cl_{12}}{1 \text{ mol W}_6Cl_{12}} = 160 \text{ g W}_6Cl_{12}$$

The smaller theoretical yield (146 g) means that WCl_6 is the limiting reactant and 146 g W_6Cl_{12} are produced.

Percent Composition and Empirical Formulas (Section 3.7)

3.86 CH_4N_2O, 60.1

$$\% \text{ C} = \frac{12.0 \text{ g C}}{60.1 \text{ g}} \text{ x } 100\% = 20.0\%$$

$$\% \text{ H} = \frac{4 \text{ x } 1.01 \text{ g H}}{60.1 \text{ g}} \text{ x } 100\% = 6.72\%$$

$$\% \text{ N} = \frac{2 \text{ x } 14.0 \text{ g N}}{60.1 \text{ g}} \text{ x } 100\% = 46.6\%$$

$$\% \text{ O} = \frac{16.0 \text{ g O}}{60.1 \text{ g}} \text{ x } 100\% = 26.6\%$$

3.88   (a) Assume a 100.0 g sample of aspirin. From the percent composition data, a 100.0 g sample contains 60.00 g C, 35.52 g O, and 4.48 g H.

$$60.00 \text{ g C } \text{ x } \frac{1 \text{ mol C}}{12.01 \text{ g C}} = 4.996 \text{ mol C}$$

$$35.52 \text{ g O } \text{ x } \frac{1 \text{ mol O}}{16.00 \text{ g O}} = 2.220 \text{ mol O}$$

$$4.48 \text{ g H } \text{ x } \frac{1 \text{ mol H}}{1.01 \text{ g H}} = 4.44 \text{ mol H}$$

$C_{4.996}H_{4.44}O_{2.220}$; divide each subscript by the smallest, 2.220.
$C_{4.996 / 2.220}H_{4.44 / 2.220}O_{2.220 / 2.220}$
$C_{2.25}H_2O_1$; multiply each subscript by 4 to obtain integers.
The empirical formula is $C_9H_8O_4$.
(b) Assume a 100.0 g sample of ilmenite. From the percent composition data, a 100.0 g sample contains 31.63 g O, 31.56 g Ti, and 36.81 g Fe.

$$31.63 \text{ g O } \times \frac{1 \text{ mol O}}{16.00 \text{ g O}} = 1.977 \text{ mol O}$$

$$31.56 \text{ g Ti } \times \frac{1 \text{ mol Ti}}{47.87 \text{ g Ti}} = 0.6593 \text{ mol Ti}$$

$$36.81 \text{ g Fe } \times \frac{1 \text{ mol Fe}}{55.85 \text{ g Fe}} = 0.6591 \text{ mol Fe}$$

$Fe_{0.6591}Ti_{0.6593}O_{1.977}$; divide each subscript by the smallest, 0.6591.
$Fe_{0.6591 / 0.6591}Ti_{0.6593 / 0.6591}O_{1.977 / 0.6591}$
The empirical formula is $FeTiO_3$.

(c) Assume a 100.0 g sample of sodium thiosulfate. From the percent composition data, a 100.0 g sample contains 30.36 g O, 29.08 g Na, and 40.56 g S.

$$30.36 \text{ g O } \times \frac{1 \text{ mol O}}{16.00 \text{ g O}} = 1.897 \text{ mol O}$$

$$29.08 \text{ g Na } \times \frac{1 \text{ mol Na}}{22.99 \text{ g Na}} = 1.265 \text{ mol Na}$$

$$40.56 \text{ g S } \times \frac{1 \text{ mol S}}{32.07 \text{ g S}} = 1.265 \text{ mol S}$$

$Na_{1.265}S_{1.265}O_{1.897}$; divide each subscript by the smallest, 1.265.
$Na_{1.265 / 1.265}S_{1.265 / 1.265}O_{1.897 / 1.265}$
$NaSO_{1.5}$; multiply each subscript by 2 to obtain integers.
The empirical formula is $Na_2S_2O_3$.

3.90   Assume a 100.0 g sample. From the percent composition data, a 100.0 g sample contains 24.25 g F and 75.75 g Sn.

$$24.25 \text{ g F } \times \frac{1 \text{ mol F}}{19.00 \text{ g F}} = 1.276 \text{ mol F}$$

$$75.75 \text{ g Sn } \times \frac{1 \text{ mol Sn}}{118.7 \text{ g Sn}} = 0.6382 \text{ mol Sn}$$

$Sn_{0.6382}F_{1.276}$; divide each subscript by the smaller, 0.6382.
$Sn_{0.6382 / 0.6382}F_{1.276 / 0.6382}$
The empirical formula is SnF_2.

Formulas and Elemental Analysis (Section 3.8)

3.92   Assume a 100.0 g sample of liquid. From the percent composition data, a 100.0 g sample of liquid contains 5.57 g H, 28.01 g Cl, and 66.42 g C.

$$66.42 \text{ g C } \times \frac{1 \text{ mol C}}{12.01 \text{ g C}} = 5.530 \text{ mol C}$$

$$5.57 \text{ g H } \times \frac{1 \text{ mol H}}{1.01 \text{ g H}} = 5.51 \text{ mol H}$$

$$28.01 \text{ g Cl } \times \frac{1 \text{ mol Cl}}{35.45 \text{ g Cl}} = 0.7901 \text{ mol Cl}$$

$C_{5.530}H_{5.51}Cl_{0.7901}$; divide each subscript by the smallest, 0.7901.

$C_{5.530\,/\,0.7901}H_{5.51\,/\,0.7901}Cl_{0.7901\,/\,0.7901}$

C_7H_7Cl

The empirical formula is C_7H_7Cl, 126.59.

Because the molecular weight equals the empirical formula weight, the empirical formula is also the molecular formula.

3.94   Mass of toluene sample = 45.62 mg = 0.045 62 g; mass of CO_2 = 152.5 mg = 0.1525 g; mass of H_2O = 35.67 mg = 0.035 67 g

$$0.1525 \text{ g } CO_2 \text{ x } \frac{1 \text{ mol } CO_2}{44.01 \text{ g } CO_2} \text{ x } \frac{1 \text{ mol C}}{1 \text{ mol } CO_2} = 0.003\ 465 \text{ mol C}$$

$$\text{mass C} = 0.003\ 465 \text{ mol C x } \frac{12.011 \text{ g C}}{1 \text{ mol C}} = 0.041\ 62 \text{ g C}$$

$$0.035\ 67 \text{ g } H_2O \text{ x } \frac{1 \text{ mol } H_2O}{18.02 \text{ g } H_2O} \text{ x } \frac{2 \text{ mol H}}{1 \text{ mol } H_2O} = 0.003\ 959 \text{ mol H}$$

$$\text{mass H} = 0.003\ 959 \text{ mol H x } \frac{1.008 \text{ g H}}{1 \text{ mol H}} = 0.003\ 991 \text{ g H}$$

The (mass C + mass H) = 0.041 62 g + 0.003 991 g = 0.045 61 g. The calculated mass of (C + H) essentially equals the mass of the toluene sample, this means that toluene contains only C and H and no other elements.

$C_{0.003\ 465}H_{0.003\ 959}$; divide each subscript by the smaller, 0.003 465.

$C_{0.003\ 465\,/\,0.003\ 465}H_{0.003\ 959\,/\,0.003\ 465}$

$CH_{1.14}$; multiply each subscript by 7 to obtain integers.

The empirical formula is C_7H_8.

3.96   Let X equal the molecular weight of cytochrome c.

$$0.0043 = \frac{55.847 \text{ u}}{X}; \quad X = \frac{55.847 \text{ u}}{0.0043} = 13,000 \text{ u}$$

3.98   Let X equal the molecular weight of disilane.

$$0.9028 = \frac{2 \text{ x } 28.09 \text{ u}}{X}; \quad X = \frac{2 \text{ x } 28.09 \text{ u}}{0.9028} = 62.23 \text{ u}$$

62.23 – 2(Si atomic weight) = 62.23 – 2(28.09) = 6.05

6.05 is the total mass of H atoms.

$$6.05 \text{ u x } \frac{1 \text{ H atom}}{1.01 \text{ u}} = 6 \text{ H atoms}; \text{ Disilane is } Si_2H_6.$$

3.100 $C_{12}Br_{10}$, 943.2; $C_{12}Br_{10}O$, 959.2; 17.33 mg = 0.017 33 g

$$\text{For } C_{12}Br_{10}, \text{ \% C} = \frac{12 \text{ x } 12.01 \text{ g C}}{943.2 \text{ g}} \text{ x } 100\% = 15.28\%$$

$$\text{For } C_{12}Br_{10}O, \text{ \% C} = \frac{12 \text{ x } 12.01 \text{ g C}}{959.2 \text{ g}} \text{ x } 100\% = 15.03\%$$

Calculate the mass of C in 17.33 mg of CO_2.

$$0.017\ 33\ \text{g CO}_2 \times \frac{1\ \text{mol CO}_2}{44.01\ \text{g CO}_2} \times \frac{1\ \text{mol C}}{1\ \text{mol CO}_2} \times \frac{12.01\ \text{g C}}{1\ \text{mol C}} = 0.004\ 729\ \text{g} = 4.729\ \text{mg C}$$

Calculate the %C in the 31.472 mg sample.

$$\%\text{C} = \frac{4.729\ \text{mg C}}{31.472\ \text{mg}} \times 100\% = 15.03\%$$

Decabrom is $C_{12}Br_{10}O$.

Mass Spectrometry (Section 3.9)

3.102   A neutral molecule will travel in a straight, undeflected, path in a mass spectrometer. Ionization is necessary as electric and magnetic fields will only exert a force on a charged species, not a neutral molecule. Ions of different masses are then accelerated by an electric field and passed between the poles of a strong magnet, which deflects them through a curved, evacuated pipe. The radius of deflection of a charged ion, M^+, as it passes between the magnet poles depends on its mass, with lighter ions deflected more strongly than heavier ones.

3.104   Mass of the sample is 70.042 11.
For C_5H_{10}, mass = $5(12.000\ 000) + 10(1.007\ 825) = 70.078\ 250$
For C_4H_6O, mass = $4(12.000\ 000) + 6(1.007\ 825) + 1(15.994\ 915) = 70.041\ 865$
For $C_3H_6N_2$, mass = $3(12.000\ 000) + 6(1.007\ 825) + 2(14.003\ 074) = 70.053\ 098$
The sample is C_4H_6O.

3.106   mass of CO_2 = 169.2 mg = 0.1692 g; mass of H_2O = 34.6 mg = 0.0346 g

$$0.1692\ \text{g CO}_2 \times \frac{1\ \text{mol CO}_2}{44.01\ \text{g CO}_2} \times \frac{1\ \text{mol C}}{1\ \text{mol CO}_2} = 0.003\ 845\ \text{mol C}$$

$$0.0346\ \text{g H}_2\text{O} \times \frac{1\ \text{mol H}_2\text{O}}{18.02\ \text{g H}_2\text{O}} \times \frac{2\ \text{mol H}}{1\ \text{mol H}_2\text{O}} = 0.003\ 84\ \text{mol H}$$

$C_{0.003\ 845}H_{0.003\ 84}$; divide each subscript by the smaller, 0.003 84.
$C_{0.003\ 845\ /\ 0.003\ 84}H_{0.003\ 84\ /\ 0.003\ 84}$
The empirical formula is CH, 13.0.
The peak with the largest mass in the mass spectrum occurs near a mass of 78, which would be the molecular weight of the compound.
78/13.0 = 6;   molecular formula = $C_{(6 \times 1)}H_{(6 \times 1)} = C_6H_6$

Multiconcept Problems

3.108   High-resolution mass spectrometry is capable of measuring the mass of molecules with a particular isotopic composition.

3.110   The combustion reaction is: $2\ C_8H_{18} + 25\ O_2 \rightarrow 16\ CO_2 + 18\ H_2O$
C_8H_{18}, 114.23; CO_2, 44.01

$$\text{pounds } CO_2 = 1.00 \text{ gal } \times \frac{3.7854 \text{ L}}{1 \text{ gal}} \times \frac{1000 \text{ mL}}{1 \text{ L}} \times \frac{0.703 \text{ g } C_8H_{18}}{1 \text{ mL}} \times \frac{1 \text{ mol } C_8H_{18}}{114.23 \text{ g } C_8H_{18}} \times$$

$$\frac{16 \text{ mol } CO_2}{2 \text{ mol } C_8H_{18}} \times \frac{44.01 \text{ g } CO_2}{1 \text{ mol } CO_2} \times \frac{1 \text{ lb}}{453.59 \text{ g}} = 18.1 \text{ lb } CO_2$$

3.112 $CaCO_3$, 100.09

$$\% \text{ Ca} = \frac{40.08 \text{ g Ca}}{100.09 \text{ g}} \times 100\% = 40.04\%$$

$$\% \text{ C} = \frac{12.01 \text{ g C}}{100.09 \text{ g}} \times 100\% = 12.00\%$$

$$\% \text{ O} = \frac{3 \times 16.00 \text{ g O}}{100.09 \text{ g}} \times 100\% = 47.96\%$$

Because the mass %'s for the pulverized rock are different from the mass %'s for pure $CaCO_3$ calculated here, the pulverized rock cannot be pure $CaCO_3$.

3.114 FeO, 71.85; Fe_2O_3, 159.7
Let X equal the mass of FeO and Y the mass of Fe_2O_3 in the 10.0 g mixture. Therefore, X + Y = 10.0 g.

$$\text{mol Fe} = 7.43 \text{ g} \times \frac{1 \text{ mol Fe}}{55.85 \text{ g Fe}} = 0.133 \text{ mol Fe}$$

$$\text{mol FeO} + 2 \times \text{mol } Fe_2O_3 = 0.133 \text{ mol Fe}$$

$$X \times \frac{1 \text{ mol FeO}}{71.85 \text{ g FeO}} + 2 \times \left(Y \times \frac{1 \text{ mol } Fe_2O_3}{159.7 \text{ g } Fe_2O_3} \right) = 0.133 \text{ mol Fe}$$

Rearrange to get X = 10.0 g − Y and then substitute it into the equation above to solve for Y.

$$(10.0 \text{ g} - Y) \times \frac{1 \text{ mol FeO}}{71.85 \text{ g FeO}} + 2 \times \left(Y \times \frac{1 \text{ mol } Fe_2O_3}{159.7 \text{ g } Fe_2O_3} \right) = 0.133 \text{ mol Fe}$$

$$\frac{10.0 \text{ mol}}{71.85} - \frac{Y \text{ mol}}{71.85 \text{ g}} + \frac{2 \text{ Y mol}}{159.7 \text{ g}} = 0.133 \text{ mol}$$

$$-\frac{Y \text{ mol}}{71.85 \text{ g}} + \frac{2 \text{ Y mol}}{159.7 \text{ g}} = 0.133 \text{ mol} - \frac{10.0 \text{ mol}}{71.85} = -0.0062 \text{ mol}$$

$$\frac{(-\text{Y mol})(159.7 \text{ g}) + (2 \text{ Y mol})(71.85 \text{ g})}{(71.85 \text{ g})(159.7 \text{ g})} = -0.0062 \text{ mol}$$

$$\frac{-16.0 \text{ Y mol}}{11474 \text{ g}} = -0.0062 \text{ mol}; \quad \frac{16.0 \text{ Y}}{11474 \text{ g}} = 0.0062$$

Y = (0.0062)(11474 g)/16.0 = 4.44 g = 4.4 g Fe_2O_3
X = 10.0 g − Y = 10.0 g − 4.4 g = 5.6 g FeO

3.116 $C_6H_{12}O_6 + 6 O_2 \rightarrow 6 CO_2 + 6 H_2O$; $C_6H_{12}O_6$, 180.16; CO_2, 44.01

$$66.3 \text{ g } C_6H_{12}O_6 \times \frac{1 \text{ mol } C_6H_{12}O_6}{180.16 \text{ g } C_6H_{12}O_6} \times \frac{6 \text{ mol } CO_2}{1 \text{ mol } C_6H_{12}O_6} \times \frac{44.01 \text{ g } CO_2}{1 \text{ mol } CO_2} = 97.2 \text{ g } CO_2$$

$$66.3 \text{ g } C_6H_{12}O_6 \text{ x } \frac{1 \text{ mol } C_6H_{12}O_6}{180.16 \text{ g } C_6H_{12}O_6} \text{ x } \frac{6 \text{ mol } CO_2}{1 \text{ mol } C_6H_{12}O_6} \text{ x } \frac{25.4 \text{ L } CO_2}{1 \text{ mol } CO_2} = 56.1 \text{ L } CO_2$$

3.118   Mass of added Cl = mass of XCl_5 – mass of XCl_3 = 13.233 g – 8.729 g = 4.504 g

mass of Cl in XCl_5 = 5 Cl's x $\dfrac{4.504 \text{ g}}{2 \text{ Cl's}}$ = 11.26 g Cl

mass of X in XCl_5 = 13.233 g – 11.26 g = 1.973 g X

11.26 g Cl x $\dfrac{1 \text{ mol Cl}}{35.45 \text{ g Cl}}$ = 0.3176 mol Cl

0.3176 mol Cl x $\dfrac{1 \text{ mol X}}{5 \text{ mol Cl}}$ = 0.063 52 mol X

molar mass of X = $\dfrac{1.973 \text{ g X}}{0.063 \ 52 \text{ mol X}}$ = 31.1 g/mol; atomic weight =31.1, X = P

3.120 NH_4NO_3, 80.04; $(NH_4)_2HPO_4$, 132.06

Assume you have a 100.0 g sample of the mixture.

Let X = grams of NH_4NO_3 and (100.0 – X) = grams of $(NH_4)_2HPO_4$.

Both compounds contain 2 nitrogen atoms per formula unit.

Because the mass % N in the sample is 30.43%, the 100.0 g sample contains 30.43 g N.

mol NH_4NO_3 = (X) x $\dfrac{1 \text{ mol } NH_4NO_3}{80.04 \text{ g}}$

mol $(NH_4)_2HPO_4$ = (100.0 – X) x $\dfrac{1 \text{ mol } (NH_4)_2HPO_4}{132.06 \text{ g}}$

mass N = $\left(\left((X) \text{ x } \dfrac{1 \text{ mol } NH_4NO_3}{80.04 \text{ g}} \right) + \left((100.0 - X) \text{ x } \dfrac{1 \text{ mol } (NH_4)_2HPO_4}{132.06 \text{ g}} \right) \right)$ x

$\left(\dfrac{2 \text{ mol N}}{1 \text{ mol ammonium cmpds}} \right)$ x $\left(\dfrac{14.0067 \text{ g N}}{1 \text{ mol N}} \right)$ = 30.43 g

Solve for X.

$\left(\dfrac{X}{80.04} + \dfrac{100.0 - X}{132.06} \right) (2)(14.0067) = 30.43$

$\left(\dfrac{X}{80.04} + \dfrac{100.0 - X}{132.06} \right) = 1.08627$

$\dfrac{(132.06)(X) + (100.0 - X)(80.04)}{(80.04)(132.06)} = 1.08627$

(132.06)(X) + (100.0 – X)(80.04) = (1.08627)(80.04)(132.06)

132.06X + 8004 – 80.04X = 11481.96

132.06X – 80.04X = 11481.96 – 8004

52.02X = 3477.96

X = $\dfrac{3477.96}{52.02}$ = 66.86 g NH_4NO_3

(100.0 – X) = (100.0 – 66.86) = 33.14 g $(NH_4)_2HPO_4$

$$\frac{mass_{NH_4NO_3}}{mass_{(NH_4)_2HPO_4}} = \frac{66.86 \text{ g}}{33.14 \text{ g}} = 2.018$$

The mass ratio of NH_4NO_3 to $(NH_4)_2HPO_4$ in the mixture is 2 to 1.

3.122 (a) 56.0 mL = 0.0560 L

$$mol \ X_2 = (0.0560 \text{ L } X_2)\left(\frac{1 \text{ mol}}{22.41 \text{ L}}\right) = 0.00250 \text{ mol } X_2$$

mass X_2 = 1.12 g MX_2 – 0.720 g MX = 0.40 g X_2

$$molar \ mass \ X_2 = \frac{0.40 \text{ g}}{0.00250 \text{ mol}} = 160 \text{ g/mol}$$

atomic weight of X = 160/2 = 80;   X is Br.

(b) $mol \ MX = 0.00250 \text{ mol } X_2 \times \dfrac{2 \text{ mol MX}}{1 \text{ mol } X_2} = 0.00500 \text{ mol MX}$

mass of X in MX = $0.00500 \text{ mol MX} \times \dfrac{1 \text{ mol X}}{1 \text{ mol MX}} \times \dfrac{80 \text{ g X}}{1 \text{ mol X}} = 0.40 \text{ g X}$

mass of M in MX = 0.720 g MX – 0.40 g X = 0.32 g M

$molar \ mass \ M = \dfrac{0.32 \text{ g}}{0.00500 \text{ mol}} = 64 \text{ g/mol}$

atomic weight of X = 64;   M is Cu.

3.124 (a) $0.138 \text{ g } H_2O \times \dfrac{1 \text{ mol } H_2O}{18.0 \text{ g } H_2O} \times \dfrac{2 \text{ mol H}}{1 \text{ mol } H_2O} = 0.0153 \text{ mol H}$

$1.617 \text{ g } CO_2 \times \dfrac{1 \text{ mol } CO_2}{44.0 \text{ g } CO_2} \times \dfrac{1 \text{ mol C}}{1 \text{ mol } CO_2} = 0.0368 \text{ mol C}$

$0.0153 \text{ mol H} \times \dfrac{1.01 \text{ g H}}{1 \text{ mol H}} = 0.016 \text{ g H}$

$0.0368 \text{ mol C} \times \dfrac{12.0 \text{ g C}}{1 \text{ mol C}} = 0.442 \text{ g C}$

1.0 g total – (0.016 g H + 0.442 g C) = 0.542 g Cl

$0.542 \text{ g Cl} \times \dfrac{1 \text{ mol Cl}}{35.5 \text{ g Cl}} = 0.0153 \text{ mol Cl}$

$C_{0.0368}H_{0.0153}Cl_{0.0153}$; divide each subscript by the smallest, 0.0153.
$C_{0.0368/0.0153}H_{0.0153/0.0153}Cl_{0.0153/0.0153}$
$C_{2.5}HCl$
Multiply each subscript by 2 to get all integer subscripts.
$C_{(2 \times 2.5)}H_{(2 \times 1)}Cl_{(2 \times 1)}$
The empirical formula is $C_5H_2Cl_2$, 133.0.
(b) 326.26/133.0 = 2.45 = 2.5;   molecular formula = $C_{(2.5 \times 5)}H_{(2.5 \times 2)}Cl_{(2.5 \times 2)} = C_{12}H_5Cl_5$
(c) No, because Cl does not form a useful oxide, so both O and Cl would have to be obtained by mass difference but that is impossible. You can only determine one element by mass difference.

4

Reactions in Aqueous Solution

4.1 $C_{12}H_{22}O_{11}$, 342.3;   355 mL = 0.355 L

$$43.0 \text{ g } C_{12}H_{22}O_{11} \times \frac{1 \text{ mol } C_{12}H_{22}O_{11}}{342.3 \text{ g } C_{12}H_{22}O_{11}} = 0.126 \text{ mol } C_6H_{12}O_6$$

$$\text{molarity} = \frac{0.126 \text{ mol}}{0.355 \text{ L}} = 0.355 \text{ M}$$

4.2 $C_6H_{12}O_6$, 180.2; NaCl, 58.4; 250.0 mL = 0.2500 L

$$13.252 \text{ g } C_6H_{12}O_6 \times \frac{1 \text{ mol } C_6H_{12}O_6}{180.2 \text{ g } C_6H_{12}O_6} = 0.073\ 54 \text{ mol } C_6H_{12}O_6$$

$$C_6H_{12}O_6 \text{ molarity} = \frac{0.073\ 54 \text{ mol}}{0.2500 \text{ L}} = 0.2942 \text{ M}$$

$$0.686 \text{ g NaCl} \times \frac{1 \text{ mol NaCl}}{58.4 \text{ g NaCl}} = 0.0117 \text{ mol NaCl}$$

$$\text{NaCl molarity} = \frac{0.0117 \text{ mol}}{0.2500 \text{ L}} = 0.0470 \text{ M}$$

4.3   (a) 125 mL = 0.125 L; (0.20 mol/L)(0.125 L) = 0.025 mol $NaHCO_3$

4.4   (a) $C_{27}H_{46}O$, 386.7; 750 mL = 0.750 L

$$\frac{0.0050 \text{ mol } C_{27}H_{46}O}{L} \times 0.750 \text{ L} \times \frac{386.7 \text{ g } C_{27}H_{46}O}{1 \text{ mol } C_{27}H_{46}O} = 1 \text{ g } C_{27}H_{46}O$$

(b) $25 \text{ mg } C_{27}H_{46}O \times \dfrac{1 \times 10^{-3} \text{ g}}{1 \text{ mg}} \times \dfrac{1 \text{ mol } C_{27}H_{46}O}{386.7 \text{ g } C_{27}H_{46}O}$

$$\times \frac{1 \text{ L}}{0.005 \text{ mol } C_{27}H_{46}O} \times \frac{1 \text{ mL}}{1 \times 10^{-3} \text{ L}} = 13 \text{ mL blood}$$

4.5 $M_i \times V_i = M_f \times V_f$; $M_f = \dfrac{M_i \times V_i}{V_f} = \dfrac{3.50 \text{ M} \times 75.0 \text{ mL}}{400.0 \text{ mL}} = 0.656 \text{ M}$

4.6 $M_i \times V_i = M_f \times V_f$; $V_i = \dfrac{M_f \times V_f}{M_i} = \dfrac{0.500 \text{ M} \times 250.0 \text{ mL}}{18.0 \text{ M}} = 6.94 \text{ mL}$

4.7 $FeBr_3$ contains 3 Br^- ions. The molar concentration of Br^- ions = 3 x 0.225 M = 0.675 M.

4.8   (a) A_2Y is the strongest electrolyte because all the molecules have dissociated into ions. A_2X is the weakest because only 1 out of 5 molecules have dissociated into ions.
(b) In a 0.350 M solution of A_2Y, the molar concentration of A ions = 2 x 0.350 = 0.700 M.
In a 0.350 M solution of A_2Y, the molar concentration of Y ions = 1 x 0.350 = 0.350 M.
(c) There are 5 A_2X molecules initially in the solution. Only one of them is ionized. The percent ionization = (1/5) x 100% = 20%

4.9   (a) Ionic equation:
$2 \text{ Ag}^+(aq) + 2 \text{ NO}_3^-(aq) + 2 \text{ Na}^+(aq) + \text{CrO}_4^{2-}(aq) \rightarrow \text{Ag}_2\text{CrO}_4(s) + 2 \text{ Na}^+(aq) + 2 \text{ NO}_3^-(aq)$
Delete spectator ions from the ionic equation to get the net ionic equation.
Net ionic equation: $2 \text{ Ag}^+(aq) + \text{CrO}_4^{2-}(aq) \rightarrow \text{Ag}_2\text{CrO}_4(s)$

4.10   Spectator ions: Ca^{2+} and NO_3^-
Net ionic equation: $\text{Ag}^+(aq) + \text{Cl}^-(aq) \rightarrow \text{AgCl}(s)$
Molecular equation: $2 \text{ AgNO}_3(aq) + \text{CaCl}_2(aq) \rightarrow 2 \text{ AgCl}(s) + \text{Ca(NO}_3)_2(aq)$

4.11   Ionic equation:
$2 \text{ Ag}^+(aq) + 2 \text{ ClO}_4^-(aq) + \text{Ca}^{2+}(aq) + 2 \text{ Br}^-(aq) \rightarrow 2 \text{ AgBr}(s) + \text{Ca}^{2+}(aq) + 2 \text{ ClO}_4^-(aq)$
Delete spectator ions from the ionic equation and reduce coefficients to get the net ionic equation.
Net ionic equation: $\text{Ag}^+(aq) + \text{Br}^-(aq) \rightarrow \text{AgBr}(s)$
A precipitation reaction occurs.

4.12 $3 \text{ CaCl}_2(aq) + 2 \text{ Na}_3\text{PO}_4(aq) \rightarrow \text{Ca}_3(\text{PO}_4)_2(s) + 6 \text{ NaCl}(aq)$
Ionic equation:
$3 \text{ Ca}^{2+}(aq) + 6 \text{ Cl}^-(aq) + 6 \text{ Na}^+(aq) + 2 \text{ PO}_4^{3-}(aq) \rightarrow \text{Ca}_3(\text{PO}_4)_2(s) + 6 \text{ Na}^+(aq) + 6 \text{ Cl}^-(aq)$
Delete spectator ions from the ionic equation to get the net ionic equation.
Net ionic equation: $3 \text{ Ca}^{2+}(aq) + 2 \text{ PO}_4^{3-}(aq) \rightarrow \text{Ca}_3(\text{PO}_4)_2(s)$

4.13   A precipitate results from the reaction. The precipitate contains cations and anions in a 3:2 ratio. The precipitate is $Mg_3(PO_4)_2$.

4.14   The precipitate is $PbBr_2$.

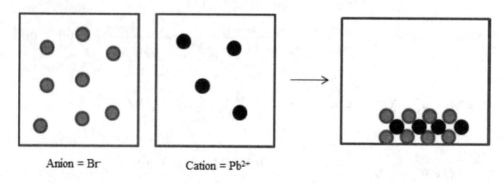

Anion = Br⁻   Cation = Pb²⁺

4.15   (a) HI, hydroiodic acid   (b) $HBrO_2$, bromous acid

4.16   (a) H_3PO_3 (b) H_2Se

4.17   Ionic equation:
$Ca^{2+}(aq) + 2\ OH^-(aq) + 2\ CH_3CO_2H(aq) \rightarrow Ca^{2+}(aq) + 2\ CH_3CO_2^-(aq) + 2\ H_2O(l)$
Delete spectator ions from the ionic equation and reduce coefficients to get the net ionic equation.
Net ionic equation: $CH_3CO_2H(aq) + OH^-(aq) \rightarrow CH_3CO_2^-(aq) + H_2O(l)$

4.18 $Mg^{2+}(aq) + 2\ OH^-(aq) + 2\ H^+(aq) + 2\ Cl^-(aq) \rightarrow Mg^{2+}(aq) + 2\ Cl^-(aq) + H_2O(l)$
$H^+(aq) + OH^-(aq) \rightarrow H_2O(l)$

4.19   50.0 mL = 0.0500 L
(0.100 mol/L)(0.0500 L) = 5.00 x 10^{-3} mol NaOH

5.00×10^{-3} mol NaOH x $\dfrac{1\ \text{mol}\ H_2SO_4}{2\ \text{mol NaOH}}$ = 2.50 x 10^{-3} mol H_2SO_4

2.50×10^{-3} mol H_2SO_4 x $\dfrac{1\ L}{0.250\ \text{mol}\ H_2SO_4}$ x $\dfrac{1000\ \text{mL}}{1\ L}$ = 10.0 mL

4.20   25.0 mL = 0.0250 L;   68.5 mL = 0.0685 L
From the reaction stoichiometry, moles KOH = moles HNO_3
(0.150 mol/L)(0.0250 L) = 3.75 x 10^{-3} mol KOH = 3.75 x 10^{-3} mol HNO_3

molarity = $\dfrac{3.75 \times 10^{-3}\ \text{mol}}{0.0685\ L}$ = 0.0547 M

4.21   25.0 mL = 0.0250 L;   94.7 mL = 0.0947 L
From the reaction stoichiometry, moles NaOH = moles CH_3CO_2H
(0.200 mol/L)(0.0947 L) = 0.018 94 mol NaOH = 0.018 94 mol CH_3CO_2H

molarity = $\dfrac{0.018\ 94\ \text{mol}}{0.0250\ L}$ = 0.758 M

4.22   vol H^+ solution = 8 OH^- x $\dfrac{1\ H^+}{1\ OH^-}$ x $\dfrac{100\ \text{mL}}{12\ H^+}$ = 66.7 mL

4.23   (a) $SnCl_4$:   Cl –1, Sn +4   (b) CrO_3:   O –2, Cr +6
(c) $VOCl_3$:   O –2, Cl –1, V +5   (d) V_2O_3:   O –2, V +3
(e) HNO_3:   O –2, H +1, N +5   (f) $FeSO_4$:   O –2, S +6, Fe +2

4.24   (a) +2, ClO, chlorine monoxide; +3, Cl_2O_3, dichlorine trioxide; +6, ClO_3, chlorine trioxide; +7, Cl_2O_7, dichlorine heptoxide
(b) Cl_2O_7 cannot react with O_2 because Cl has its maximum oxidation number, +7, and cannot be further oxidized.

4.25 $4 NH_3(g) + 5 O_2(g) \rightarrow 4 NO(g) + 6 H_2O(l)$

The N in NH_3 is oxidized (its oxidation number increases from -3 to $+2$). NH_3 is the reducing agent.

Each O in O_2 is reduced (its oxidation number decreases from 0 to -2). O_2 is the oxidizing agent.

4.26   (a) C in C_2H_5OH oxidized and Cr in $K_2Cr_2O_7$ gets reduced.

(b) $K_2Cr_2O_7$ is the oxidizing agent; C_2H_5OH is the reducing agent.

4.27   Mg is below Ca in the activity series; therefore NO REACTION.

4.28   "Any element higher in the activity series will react with the ion of any element lower in the activity series."

$A + D^+ \rightarrow A^+ + D$; therefore A is a stronger reducing agent than D.

$B^+ + D \rightarrow B + D^+$; therefore D is a stronger reducing agent than B.

$C^+ + D \rightarrow C + D^+$; therefore D is a stronger reducing agent than C.

$B + C^+ \rightarrow B^+ + C$; therefore B is a stronger reducing agent than C.

The net result is A(strongest) > D > B > C(weakest).

4.29   31.50 mL = 0.031 50 L;   10.00 mL = 0.010 00 L

$$0.031\ 50\ L \times \frac{0.105\ mol\ BrO_3^-}{1\ L} \times \frac{6\ mol\ Fe^{2+}}{1\ mol\ BrO_3^-} = 1.98 \times 10^{-2}\ mol\ Fe^{2+}$$

$$molarity = \frac{1.98 \times 10^{-2}\ mol\ Fe^{2+}}{0.010\ 00\ L} = 1.98\ M\ Fe^{2+}\ solution$$

4.30   14.92 mL = 0.014 92 L

$$0.014\ 92\ L \times \frac{0.100\ mol\ Cr_2O_7^{-2}}{1\ L} \times \frac{6\ mol\ Fe^{2+}}{1\ mol\ Cr_2O_7^{-2}} \times \frac{55.85\ g\ Fe^{2+}}{1\ mol\ Fe^{2+}} \times \frac{1000\ mg}{1\ g} = 50.0\ mg\ Fe^{2+}$$

4.31   (a) Sodium chloride, sodium citrate, and potassium dihydrogen phosphate are strong electrolytes since they are soluble ionic compounds; citric acid and vitamin B3 are weak electrolytes due the presence of the carboxylic acid (CO_2H) group; and fructose is a nonelectrolyte. (b) Sodium chloride, potassium dihydrogen phosphate, and sodium citrate replenish electrolytes with important biological functions.

4.32   150 mg = 0.15 g;   35 mg = 0.035 g;   360 mL = 0.36 L

$$molarity = \frac{\left(0.15\ g\ Na^+ \times \dfrac{1\ mol\ Na^+}{23\ g\ Na^+}\right)}{0.36\ L} = 0.018\ M$$

$$molarity = \frac{\left(0.35\ g\ K^+ \times \dfrac{1\ mol\ K^+}{39\ g\ K^+}\right)}{0.36\ L} = 0.0025\ M$$

4.33   NaCl, 58.4; 0.500 L = 500 mL

$$0.416 \text{ mg Na}^+/\text{mL} \times \frac{1 \times 10^{-3} \text{ g}}{1 \text{ mg}} \times 500 \text{ mL} \times \frac{1 \text{ mol Na}^+}{23.0 \text{ g Na}^+}$$

$$\times \frac{1 \text{ mol NaCl}}{1 \text{ mol Na}^+} \times \frac{58.4 \text{ g NaCl}}{1 \text{ mol NaCl}} = 0.528 \text{ g NaCl}$$

4.34   (a) Ba^{2+} (b) $3 \text{ Ba}^{2+}(aq) + 2 \text{ PO}_4^{3-}(aq) \rightarrow \text{Ba}_3(\text{PO}_4)_2(s)$

4.35   AgCl, 143.3; 172 mg = 0.172 g; 100.0 mL = 0.1000 L

$$0.172 \text{ g AgCl} \times \frac{1 \text{ mol AgCl}}{143.3 \text{ g AgCl}} \times \frac{1 \text{ mol Cl}^-}{1 \text{ mol AgCl}} = 0.001\,20 \text{ mol Cl}^-$$

$$\text{Cl}^- \text{ molarity} = \frac{0.001\,20 \text{ mol Cl}^-}{0.1000 \text{ L}} = 0.0120 \text{ M}$$

4.36   35.6 mL = 0.0356 L; 25.0 mL = 0.0250 L
(0.0356 L)(0.0400 mol/L) = 0.001 42 mol NaOH

$$0.001\,42 \text{ mol NaOH} \times \frac{1 \text{ mol H}_3\text{C}_6\text{H}_5\text{O}_7}{3 \text{ mol NaOH}} = 4.75 \times 10^{-4} \text{ mol H}_3\text{C}_6\text{H}_5\text{O}_7$$

$$\text{H}_3\text{C}_6\text{H}_5\text{O}_7 \text{ molarity} = \frac{4.75 \times 10^{-4} \text{ mol}}{0.0250 \text{ L}} = 0.0190 \text{ M}$$

Conceptual Problems

4.38   (a) $2 \text{ Na}^+(aq) + \text{CO}_3^{2-}(aq)$ does not form a precipitate. This is represented by box (1).
        (b) $\text{Ba}^{2+}(aq) + \text{CrO}_4^{2-}(aq) \rightarrow \text{BaCrO}_4(s)$. This is represented by box (2).
        (c) $2 \text{ Ag}^+(aq) + \text{SO}_4^{2-}(aq) \rightarrow \text{Ag}_2\text{SO}_4(s)$. This is represented by box (3).

4.40   HY is the strongest acid because it is completely dissociated.
        HX is the weakest acid because it is the least dissociated.

4.42   The concentration in the buret is three times that in the flask. The NaOCl concentration
        is 0.040 M. Because the I^- concentration in the buret is three times the OCl^-
        concentration in the flask and the reaction requires 2 I^- ions per OCl^- ion, 2/3 or 67% of
        the I^- solution from the buret must be added to the flask to react with all of the OCl^-.

4.44   (a) $\text{Sr}^+ + \text{At} \rightarrow \text{Sr} + \text{At}^+$ No reaction.
        (b) $\text{Si} + \text{At}^+ \rightarrow \text{Si}^+ + \text{At}$ Reaction would occur.
        (c) $\text{Sr} + \text{Si}^+ \rightarrow \text{Sr}^+ + \text{Si}$ Reaction would occur.

Section Problems
Molarity (Section 4.1)

4.46 (a) 35.0 mL = 0.0350 L; $\dfrac{1.200 \text{ mol HNO}_3}{L}$ x 0.0350 L = 0.0420 mol HNO$_3$

(b) 175 mL = 0.175 L; $\dfrac{0.67 \text{ mol C}_6\text{H}_{12}\text{O}_6}{L}$ x 0.175 L = 0.12 mol C$_6$H$_{12}$O$_6$

4.48 BaCl$_2$, 208.2

15.0 g BaCl$_2$ x $\dfrac{1 \text{ mol BaCl}_2}{208.2 \text{ g BaCl}_2}$ = 0.0720 mol BaCl$_2$

0.0720 mol x $\dfrac{1.0 \text{ L}}{0.45 \text{ mol}}$ = 0.16 L; 0.16 L = 160 mL

4.50 NaCl, 58.4; 400 mg = 0.400 g; 100 mL = 0.100 L

0.400 g NaCl x $\dfrac{1 \text{ mol NaCl}}{58.4 \text{ g NaCl}}$ = 0.006 85 mol NaCl

molarity = $\dfrac{0.006 \ 85 \text{ mol}}{0.100 \text{ L}}$ = 0.0685 M

4.52 3.045 g Cu x $\dfrac{1 \text{ mol Cu}}{63.546 \text{ g Cu}}$ = 0.047 92 mol Cu; 50.0 mL = 0.0500 L

Cu(NO$_3$)$_2$ molarity = $\dfrac{0.047 \ 92 \text{ mol}}{0.0500 \text{ L}}$ = 0.958 M

4.54 (a) NaOH, 40.0; 500.0 mL = 0.5000 L

$\dfrac{1.25 \text{ mol NaOH}}{L}$ x 0.500 L x $\dfrac{40.0 \text{ g NaOH}}{1 \text{ mol NaOH}}$ = 25.0 g NaOH

(b) C$_6$H$_{12}$O$_6$, 180.2

$\dfrac{0.250 \text{ mol C}_6\text{H}_{12}\text{O}_6}{L}$ x 1.50 L x $\dfrac{180.2 \text{ g C}_6\text{H}_{12}\text{O}_6}{1 \text{ mol C}_6\text{H}_{12}\text{O}_6}$ = 67.6 g C$_6$H$_{12}$O$_6$

4.56 CaF$_2$, 78.07; 250 mL = 0.250 L
(0.250 L)(0.100 mol/L) = 0.0250 mol F$^-$ ions

0.0250 mol F$^-$ ions x $\dfrac{1 \text{ mol CaF}_2}{2 \text{ mol F}^-}$ x $\dfrac{78.07 \text{ g CaF}_2}{1 \text{ mol CaF}_2}$ = 0.976 g CaF$_2$

Place 0.976 g of CaF$_2$ in a 250 mL volumetric flask and fill it to the mark with water.

Dilutions (Section 4.2)

4.58 M$_f$ x V$_f$ = M$_i$ x V$_i$; $\quad$ M$_f$ = $\dfrac{M_i \text{ x } V_i}{V_f}$ = $\dfrac{12.0 \text{ M x } 35.7 \text{ mL}}{250.0 \text{ mL}}$ = 1.71 M HCl

4.60 250 mL = 0.250 L

(0.250 L)(0.100 mol/L) = 0.0250 mol Cl⁻ ions

$$Cl^- \text{ molarity} = 3.00 \text{ M CaCl}_2 \times \frac{2 \text{ mol Cl}^-}{1 \text{ mol CaCl}_2} = 6.00 \text{ M Cl}^-$$

$$M_f \times V_f = M_i \times V_i; \quad V_i = \frac{M_f \times V_f}{M_i} = \frac{0.100 \text{ M} \times 250 \text{ mL}}{6.00 \text{ M}} = 4.17 \text{ mL}$$

Using a 10 mL graduated cylinder, measure out 4.17 mL of 3.00 M CaCl₂ solution. Pour the 4.17 mL into a 250 mL volumetric flask and fill it to the mark with water.

Electrolytes (Section 4.3)

4.62 (a) strong electrolyte, bright (b) nonelectrolyte, dark (c) weak electrolyte, dim

4.64 Ba(OH)₂ is soluble in aqueous solution, dissociates into Ba²⁺(aq) and 2 OH⁻(aq), and conducts electricity. In aqueous solution, H₂SO₄ dissociates into H⁺(aq) and HSO₄⁻(aq). H₂SO₄ solutions conduct electricity. When equal molar solutions of Ba(OH)₂ and H₂SO₄ are mixed, the insoluble BaSO₄ is formed along with two H₂O. In water, BaSO₄ does not produce any appreciable amount of ions and the mixture does not conduct electricity.

4.66 (a) HBr, strong electrolyte   (b) HF, weak electrolyte
(c) NaClO₄, strong electrolyte   (d) (NH₄)₂CO₃, strong electrolyte
(e) NH₃, weak electrolyte   (f) C₂H₅OH, nonelectrolyte

4.68 (a) K₂CO₃ contains 3 ions (2 K⁺ and 1 CO₃²⁻).
The molar concentration of ions = 3 x 0.750 M = 2.25 M.
(b) AlCl₃ contains 4 ions (1 Al³⁺ and 3 Cl⁻).
The molar concentration of ions = 4 x 0.355 M = 1.42 M.

4.70 NaCl, 58.4; KCl, 74.6; CaCl₂, 111.0; 500 mL = 0.500 L

$$4.30 \text{ g NaCl} \times \frac{1 \text{ mol NaCl}}{58.4 \text{ g NaCl}} = 0.0736 \text{ mol NaCl}$$

$$0.150 \text{ g KCl} \times \frac{1 \text{ mol KCl}}{74.6 \text{ g KCl}} = 0.002 \ 01 \text{ mol KCl}$$

$$0.165 \text{ g CaCl}_2 \times \frac{1 \text{ mol CaCl}_2}{111.0 \text{ g CaCl}_2} = 0.001 \ 49 \text{ mol CaCl}_2$$

0.0736 mol + 0.002 01 mol + 2(0.001 49 mol) = 0.0786 mol Cl⁻

$$Na^+ \text{ molarity} = \frac{0.0736 \text{ mol}}{0.500 \text{ L}} = 0.147 \text{ M}$$

$$Ca^{2+} \text{ molarity} = \frac{0.001 \ 49 \text{ mol}}{0.500 \text{ L}} = 0.002 \ 98 \text{ M}$$

$$K^+ \text{ molarity} = \frac{0.002 \ 01 \text{ mol}}{0.500 \text{ L}} = 0.004 \ 02 \text{ M}$$

$$\text{Cl}^- \text{ molarity} = \frac{0.0786 \text{ mol}}{0.500 \text{ L}} = 0.157 \text{ M}$$

Net Ionic Equations and Aqueous Reactions (Sections 4.4–4.5)

4.72 (a) precipitation   (b) redox   (c) acid-base neutralization

4.74 (a) Ionic equation:
$Hg^{2+}(aq) + 2 NO_3^-(aq) + 2 Na^+(aq) + 2 I^-(aq) \rightarrow 2 Na^+(aq) + 2 NO_3^-(aq) + HgI_2(s)$
Delete spectator ions from the ionic equation to get the net ionic equation.
Net ionic equation: $Hg^{2+}(aq) + 2 I^-(aq) \rightarrow HgI_2(s)$

(b) $2 HgO(s) \xrightarrow{\text{Heat}} 2 Hg(l) + O_2(g)$
(c) Ionic equation:
$H_3PO_4(aq) + 3 K^+(aq) + 3 OH^-(aq) \rightarrow 3 K^+(aq) + PO_4^{3-}(aq) + 3 H_2O(l)$
Delete spectator ions from the ionic equation to get the net ionic equation.
Net ionic equation: $H_3PO_4(aq) + 3 OH^-(aq) \rightarrow PO_4^{3-}(aq) + 3 H_2O(l)$

Precipitation Reactions and Solubility Guidelines (Section 4.6)

4.76 (a) $PbSO_4$, insoluble   (b) $Ba(NO_3)_2$, soluble
(c) $SnCO_3$, insoluble   (d) $(NH_4)_3PO_4$, soluble

4.78 (a) No precipitate will form.   (b) $Fe^{2+}(aq) + 2 OH^-(aq) \rightarrow Fe(OH)_2(s)$
(c) No precipitate will form.   (d) No precipitate will form.

4.80 (a) 0.10 M $LiNO_3$ will not form a precipitate.
(b) $BaSO_4(s)$ will precipitate.   (c) $AgCl(s)$ will precipitate.

4.82 (a) $Pb(NO_3)_2(aq) + Na_2SO_4(aq) \rightarrow PbSO_4(s) + 2 NaNO_3(aq)$
(b) $3 MgCl_2(aq) + 2 K_3PO_4(aq) \rightarrow Mg_3(PO_4)_2(s) + 6 KCl(aq)$
(c) $ZnSO_4(aq) + Na_2CrO_4(aq) \rightarrow ZnCrO_4(s) + Na_2SO_4(aq)$

4.84 $Ag^+(aq) + NO_3^-(aq) + H^+(aq) + Cl^-(aq) \rightarrow AgCl(s) + H^+(aq) + NO_3^-(aq)$

$$\text{mol Cl}^- = 30.0 \text{ mL} \times \frac{1 \times 10^{-3} \text{ L}}{1 \text{ mL}} \times 0.150 \text{ mol/L} = 0.00450 \text{ mol}$$

$$\text{mol Ag}^+ = 25.0 \text{ mL} \times \frac{1 \times 10^{-3} \text{ L}}{1 \text{ mL}} \times 0.200 \text{ mol/L} = 0.00500 \text{ mol}$$

Because the mole ratio between Ag^+ and Cl^- is 1 to 1, Cl^- is the limiting reactant because it is the smaller mol quantity. 0.00450 mol of AgCl is produced.

AgCl, 143.3;   mass AgCl $= 0.00450$ mol AgCl $\times \dfrac{143.3 \text{ g AgCl}}{1 \text{ mol AgCl}} = 0.645$ g AgCl

4.86 Add HCl(aq); it will selectively precipitate AgCl(s).

4.88 Ag^+ is eliminated because it would have precipitated as $AgCl(s)$; Ba^{2+} is eliminated because it would have precipitated as $BaSO_4(s)$. The solution might contain Cs^+ and/or NH_4^+. Neither of these will precipitate with OH^-, SO_4^{2-}, or Cl^-.

4.90   (a) Add HCl to precipitate Hg_2Cl_2. $Hg_2^{2+}(aq) + 2Cl^-(aq) \rightarrow Hg_2Cl_2(s)$
        (b) Add H_2SO_4 to precipitate $PbSO_4$. $Pb^{2+}(aq) + SO_4^{2-}(aq) \rightarrow PbSO_4(s)$
        (c) Add Na_2CO_3 to precipitate $CaCO_3$. $Ca^{2+}(aq) + CO_3^{2-}(aq) \rightarrow CaCO_3(s)$
        (d) Add Na_2SO_4 to precipitate $BaSO_4$. $Ba^{2+}(aq) + SO_4^{2-}(aq) \rightarrow BaSO_4(s)$

4.92   100.0 mL = 0.1000 L and 50.0 mL = 0.0500 L
        mol Na_2SO_4 = (0.1000 L)(0.100 mol/L) = 0.0100 mol Na_2SO_4
        mol SO_4^{2-} = mol Na_2SO_4 = 0.0100 mol SO_4^{2-}

$$\text{mol } Na^+ = 0.0100 \text{ mol } Na_2SO_4 \times \frac{2 \text{ mol } Na^+}{1 \text{ mol } Na_2SO_4} = 0.0200 \text{ mol } Na^+$$

mol $ZnCl_2$ = (0.0500 L)(0.300 mol/L) = 0.0150 mol $ZnCl_2$
mol Zn^{2+} = mol $ZnCl_2$ = 0.0150 mol Zn^{2+}

$$\text{mol } Cl^- = 0.0150 \text{ mol } ZnCl_2 \times \frac{2 \text{ mol } Cl^-}{1 \text{ mol } ZnCl_2} = 0.0300 \text{ mol } Cl^-$$

mol $Ba(CN)_2$ = (0.1000 L)(0.200 mol/L) = 0.0200 mol $Ba(CN)_2$
mol Ba^{2+} = mol $Ba(CN)_2$ = 0.0200 mol Ba^{2+}

$$\text{mol } CN^- = 0.0200 \text{ mol } Ba(CN)_2 \times \frac{2 \text{ mol } CN^-}{1 \text{ mol } Ba(CN)_2} = 0.0400 \text{ mol } CN^-$$

The following two reactions will take place to form precipitates.
$Zn^{2+}(aq) + 2 CN^-(aq) \rightarrow Zn(CN)_2(s)$
$Ba^{2+}(aq) + SO_4^{2-}(aq) \rightarrow BaSO_4(s)$

$$\text{For } Zn^{2+}, \text{ mol } CN^- \text{ needed} = 0.0150 \text{ mol } Zn^{2+} \times \frac{2 \text{ mol } CN^-}{1 \text{ mol } Zn^{2+}} = 0.0300 \text{ mol } CN^- \text{ needed}$$

CN^- is in excess, so Zn^{2+} is the limiting reactant and is totally consumed.
mol CN^- remaining after reaction = 0.0400 mol – 0.0300 mol = 0.0100 mol CN^-
For Ba^{2+}, mol SO_4^{2-} needed = mol Ba^{2+} = 0.0200 mol SO_4^{2-} needed
Ba^{2+} is in excess, so SO_4^{2-} is the limiting reactant and is totally consumed.
mol Ba^{2+} remaining after reaction = 0.0200 mol – 0.0100 mol = 0.0100 mol Ba^{2+}
total volume = 0.1000 L + 0.0500 L + 0.1000 L = 0.2500 L
$[Zn^{2+}] = [SO_4^{2-}] = 0$

$$[Na^+] = \frac{0.0200 \text{ mol}}{0.2500 \text{ L}} = 0.0800 \text{ M}$$

$$[Cl^-] = \frac{0.0300 \text{ mol}}{0.2500 \text{ L}} = 0.120 \text{ M}$$

$$[CN^-] = \frac{0.0100 \text{ mol}}{0.2500 \text{ L}} = 0.0400 \text{ M}$$

$$[Ba^{2+}] = \frac{0.0100 \text{ mol}}{0.2500 \text{ L}} = 0.0400 \text{ M}$$

Acids, Bases, and Neutralization Reactions (Section 4.7)

4.94   Add the solution to an active metal, such as magnesium. Bubbles of H_2 gas indicate the presence of an acid.

4.96   (a) $2 H^+(aq) + 2 ClO_4^-(aq) + Ca^{2+}(aq) + 2 OH^-(aq) \rightarrow Ca^{2+}(aq) + 2 ClO_4^-(aq) + 2 H_2O(l)$
        (b) $CH_3CO_2H(aq) + Na^+(aq) + OH^-(aq) \rightarrow CH_3CO_2^-(aq) + Na^+(aq) + H_2O(l)$

4.98   (a) $LiOH(aq) + HI(aq) \rightarrow LiI(aq) + H_2O(l)$
        Ionic equation: $Li^+(aq) + OH^-(aq) + H^+(aq) + I^-(aq) \rightarrow Li^+(aq) + I^-(aq) + H_2O(l)$
        Delete spectator ions from the ionic equation to get the net ionic equation.
        Net ionic equation: $H^+(aq) + OH^-(aq) \rightarrow H_2O(l)$
        (b) $2 HBr(aq) + Ca(OH)_2(aq) \rightarrow CaBr_2(aq) + 2 H_2O(l)$
        Ionic equation:
        $2 H^+(aq) + 2 Br^-(aq) + Ca^{2+}(aq) + 2 OH^-(aq) \rightarrow Ca^{2+}(aq) + 2 Br^-(aq) + 2 H_2O(l)$
        Delete spectator ions from the ionic equation to get the net ionic equation.
        Net ionic equation: $H^+(aq) + OH^-(aq) \rightarrow H_2O(l)$

Solution Stoichiometry and Titration (Sections 4.8–4.9)

4.100 $2 HBr(aq) + K_2CO_3(aq) \rightarrow 2 KBr(aq) + CO_2(g) + H_2O(l)$
        K_2CO_3, 138.2; 450 mL = 0.450 L

$$\frac{0.500 \text{ mol HBr}}{L} \times 0.450 \text{ L} = 0.225 \text{ mol HBr}$$

$$0.225 \text{ mol HBr} \times \frac{1 \text{ mol } K_2CO_3}{2 \text{ mol HBr}} \times \frac{138.2 \text{ g } K_2CO_3}{1 \text{ mol } K_2CO_3} = 15.5 \text{ g } K_2CO_3$$

4.102 $H_2C_2O_4$, 90.0

$$3.225 \text{ g } H_2C_2O_4 \times \frac{1 \text{ mol } H_2C_2O_4}{90.0 \text{ g } H_2C_2O_4} \times \frac{2 \text{ mol KMnO}_4}{5 \text{ mol } H_2C_2O_4} = 0.0143 \text{ mol KMnO}_4$$

$$0.0143 \text{ mol} \times \frac{1 \text{ L}}{0.250 \text{ mol}} = 0.0572 \text{ L} = 57.2 \text{ mL}$$

4.104   (a) $LiOH(aq) + HBr(aq) \rightarrow LiBr(aq) + H_2O(l)$
        25.0 mL = 0.0250 L and 75.0 mL = 0.0750 L
        (0.240 mol/L LiOH)(0.0250 L) = 0.006 00 mol LiOH
        (0.200 mol/L HBr)(0.0750 L) = 0.0150 mol HBr
        HBr and LiOH react in a one to one mole ratio.
        mol HBr left over = 0.0150 mol HBr – 0.006 00 mol LiOH = 0.0090 mol HBr
        $KOH(aq) + HBr(aq) \rightarrow KBr(aq) + H_2O(l)$
        HBr and KOH react in a one to one mole ratio.
        0.0090 mol KOH will neutralize the solution.

$$\text{volume} = \frac{\text{mol}}{M} = \frac{0.0090 \text{ mol KOH}}{1.00 \text{ mol/L}} = 0.0090 \text{ L} = 9.0 \text{ mL KOH}$$

(b) $HCl(aq) + NaOH(aq) \rightarrow NaCl(aq) + H_2O(l)$
45.0 mL = 0.0450 L and 10.0 mL = 0.0100 L
(0.300 mol/L HCl)(0.0450 L) = 0.0135 mol HCl
(0.250 mol/L NaOH)(0.0100 L) = 0.002 50 mol NaOH
HCl and NaOH react in a one to one mole ratio.
mol HCl left over = 0.0135 mol HCl – 0.002 50 mol NaOH = 0.0110 mol HCl
$KOH(aq) + HCl(aq) \rightarrow KCl(aq) + H_2O(l)$
HCl and KOH react in a one to one mole ratio.
0.0110 mol KOH will neutralize the solution.
$$\text{volume} = \frac{\text{mol}}{\text{M}} = \frac{0.0110 \text{ mol KOH}}{1.00 \text{ mol/L}} = 0.0110 \text{ L} = 11.0 \text{ mL KOH}$$

4.106 (a) $HBr(aq) + KOH(aq) \rightarrow KBr(aq) + H_2O(l)$
50.0 mL = 0.0500 L and 30.0 mL = 0.0300 L
(0.100 mol/L HBr)(0.0500 L) = 0.005 00 mol HBr
(0.200 mol/L KOH)(0.0300 L) = 0.006 00 mol KOH
HBr and KOH react in a one to one mole ratio. The resulting solution is basic because there is an excess of KOH.

(b) $2 HCl(aq) + Ba(OH)_2(aq) \rightarrow BaCl_2(aq) + 2 H_2O(l))$
100.0 mL = 0.1000 L and 75 mL = 0.0750 L
(0.0750 mol/L HCl)(0.1000 L) = 0.007 50 mol HCl
(0.100 mol/L Ba(OH)$_2$)(0.0750 L) = 0.007 50 mol Ba(OH)$_2$
$$0.007\ 50 \text{ mol HCl} \times \frac{1 \text{ mol Ba(OH)}_2}{2 \text{ mol HCl}} = 0.003\ 75 \text{ mol Ba(OH)}_2 \text{ needed}$$
The resulting solution is basic because there is an excess of Ba(OH)$_2$.

Oxidation Numbers (Section 4.10)

4.108 (a) NO_2 — O –2, N +4   (b) SO_3 — O –2, S +6
(c) $COCl_2$ — O –2, Cl –1, C +4   (d) CH_2Cl_2 — Cl –1, H +1, C 0
(e) $KClO_3$ — O –2, K +1, Cl +5   (f) HNO_3 — O –2, H +1, N +5

4.110 (a) ClO_3^- — O –2, Cl +5   (b) SO_3^{2-} — O –2, S +4
(c) $C_2O_4^{2-}$ — O –2, C +3   (d) NO_2^- — O –2, N +3
(e) BrO^- — O –2, Br +1   (f) AsO_4^{3-} — O –2, As +5

4.112 (a) +1, N_2O, nitrous oxide; +2, NO, nitric oxide; +4, N_2O_4, dinitrogen tetroxide; +5, N_2O_5, dinitrogen pentoxide
(b) N_2O_5 cannot react with O_2 because N has its maximum oxidation number, +5, and cannot be further oxidized.

Redox Reactions (Section 4.11)

4.114 The best reducing agents are at the bottom left of the periodic table. The best oxidizing agents are at the top right of the periodic table (excluding the noble gases).

4.116 (a) An oxidizing agent gains electrons.
(b) A reducing agent loses electrons.
(c) A substance undergoing oxidation loses electrons.
(d) A substance undergoing reduction gains electrons.

4.118 (a) $Ca(s) + Sn^{2+}(aq) \rightarrow Ca^{2+}(aq) + Sn(s)$
$Ca(s)$ is oxidized (oxidation number increases from 0 to +2).
$Sn^{2+}(aq)$ is reduced (oxidation number decreases from +2 to 0).
(b) $ICl(s) + H_2O(l) \rightarrow HCl(aq) + HOI(aq)$
No oxidation numbers change. The reaction is not a redox reaction.

Activity Series (Section 4.12)

4.120 (a) Zn is below Na^+; therefore no reaction.
(b) Pt is below H^+; therefore no reaction.
(c) Au is below Ag^+; therefore no reaction.
(d) Ag is above Au^{3+}; the reaction is $Au^{3+}(aq) + 3\ Ag(s) \rightarrow 3\ Ag^+(aq) + Au(s)$.

4.122 (a) "Any element higher in the activity series will react with the ion of any element lower in the activity series."
$A + B^+ \rightarrow A^+ + B$; therefore A is higher than B.
$C^+ + D \rightarrow$ no reaction; therefore C is higher than D.
$B + D^+ \rightarrow B^+ + D$; therefore B is higher than D.
$B + C^+ \rightarrow B^+ + C$; therefore B is higher than C.
The net result is $A > B > C > D$.
(b) (1) C is below A^+; therefore no reaction.
(2) D is below A^+; therefore no reaction.

Redox Titrations (Section 4.13)

4.124 $I_2(aq) + 2\ S_2O_3{}^{2-}(aq) \rightarrow S_4O_6{}^{2-}(aq) + 2\ I^-(aq);\quad$ 35.20 mL = 0.032 50 L

$$0.035\ 20\ L \times \frac{0.150\ mol\ S_2O_3{}^{2-}}{L} \times \frac{1\ mol\ I_2}{2\ mol\ S_2O_3{}^{2-}} \times \frac{253.8\ g\ I_2}{1\ mol\ I_2} = 0.670\ g\ I_2$$

4.126 46.99 mL = 0.046 99 L; 50.00 mL = 0.050 00 L
$(0.2004\ mol/L\ Cr_2O_7{}^{2-})(0.046\ 99\ L) = 0.009\ 417\ mol\ Cr_2O_7{}^{2-}$

$$0.009\ 417\ mol\ Cr_2O_7{}^{2-} \times \frac{6\ mol\ Fe^{2+}}{1\ mol\ Cr_2O_7{}^{2-}} = 0.056\ 50\ mol\ Fe^{2+}$$

$$[Fe^{2+}] = \frac{0.056\ 50\ mol}{0.050\ 00\ L} = 1.130\ M$$

4.128 $3\ H_3AsO_3(aq) + BrO_3{}^-(aq) \rightarrow Br^-(aq) + 3\ H_3AsO_4(aq)$
22.35 mL = 0.022 35 L and 50.00 mL = 0.050 00 L

$$0.022\ 35\ L \times \frac{0.100\ mol\ BrO_3^-}{L} \times \frac{3\ mol\ H_3AsO_3}{1\ mol\ BrO_3^-} = 6.70 \times 10^{-3}\ mol\ H_3AsO_3$$

$$molarity = \frac{6.70 \times 10^{-3}\ mol}{0.050\ 00\ L} = 0.134\ M\ As(III)$$

4.130 $2\ Fe^{3+}(aq) + Sn^{2+}(aq) \rightarrow 2\ Fe^{2+}(aq) + Sn^{4+}(aq);\quad 13.28\ mL = 0.013\ 28\ L$

$$0.013\ 28\ L \times \frac{0.1015\ mol\ Sn^{2+}}{L} \times \frac{2\ mol\ Fe^{3+}}{1\ mol\ Sn^{2+}} \times \frac{55.845\ g\ Fe^{3+}}{1\ mol\ Fe^{3+}} = 0.1506\ g\ Fe^{3+}$$

$$mass\ \%\ Fe = \frac{0.1506\ g}{0.1875\ g} \times 100\% = 80.32\%$$

4.132 $C_2H_5OH(aq) + 2\ Cr_2O_7^{2-}(aq) + 16\ H^+(aq) \rightarrow 2\ CO_2(g) + 4\ Cr^{3+}(aq) + 11\ H_2O(l)$
$C_2H_5OH,\ 46.1;\ 8.76\ mL = 0.008\ 76\ L$

$$0.008\ 76\ L \times \frac{0.049\ 88\ mol\ Cr_2O_7^{2-}}{L} \times \frac{1\ mol\ C_2H_5OH}{2\ mol\ Cr_2O_7^{2-}} \times \frac{46.1\ g\ C_2H_5OH}{1\ mol\ C_2H_5OH}$$

$$= 0.0101\ g\ C_2H_5OH$$

$$mass\ \%\ C_2H_5OH = \frac{0.0101\ g}{10.002\ g} \times 100\% = 0.101\%$$

Multiconcept Problems

4.134 Let X equal the mass of benzoic acid and Y the mass of gallic acid in the 1.00 g mixture.
Therefore, $X + Y = 1.00$ g.
Because both acids contain only one acidic hydrogen, there is a 1 to 1 mol ratio between each acid and NaOH in the acid-base titration.
In the titration, mol benzoic acid + mol gallic acid = mol NaOH.

$$Therefore,\ X \times \frac{1\ mol\ BA}{122\ g\ BA} + Y \times \frac{1\ mol\ GA}{170\ g\ GA} = mol\ NaOH$$

$$mol\ NaOH = 14.7\ mL \times \frac{1 \times 10^{-3}\ L}{1\ mL} \times \frac{0.500\ mol\ NaOH}{1\ L} = 0.00735\ mol\ NaOH$$

We have two unknowns, X and Y, and two equations.
$X + Y = 1.00$ g

$$X \times \frac{1\ mol\ BA}{122\ g\ BA} + Y \times \frac{1\ mol\ GA}{170\ g\ GA} = 0.00735\ mol\ NaOH$$

Rearrange to get $X = 1.00$ g $- Y$ and then substitute it into the equation above to solve for Y.

$$(1.00\ g - Y) \times \frac{1\ mol\ BA}{122\ g\ BA} + Y \times \frac{1\ mol\ GA}{170\ g\ GA} = 0.00735\ mol\ NaOH$$

$$\frac{1\ mol}{122} - \frac{Y\ mol}{122\ g} + \frac{Y\ mol}{170\ g} = 0.00735\ mol$$

$$-\frac{Y\ mol}{122\ g} + \frac{Y\ mol}{170\ g} = 0.00735\ mol - \frac{1\ mol}{122} = -8.47 \times 10^{-4}\ mol$$

$$\frac{(-\,Y\ mol)(170\ g)\ +\ (Y\ mol)(122\ g)}{(170\ g)(122\ g)} = -\,8.47\ x\ 10^{-4}\ mol$$

$$\frac{-\,48\ Y\ mol}{20740\ g} = -\,8.47\ x\ 10^{-4}\ mol; \qquad \frac{48\ Y}{20740\ g} = 8.47\ x\ 10^{-4}$$

$$Y = \frac{(20740\ g)(8.47\ x\ 10^{-4})}{48} = 0.366\ g$$

X = 1.00 g – 0.366 g = 0.634 g

In the 1.00 g mixture there is 0.63 g of benzoic acid and 0.37 g of gallic acid.

4.136　54.91 mL = 0.054 91 L

$$0.054\ 91\ L\ x\ \frac{0.1018\ mol\ Ce^{4+}}{L}\ x\ \frac{1\ mol\ Fe^{2+}}{1\ mol\ Ce^{4+}}\ x\ \frac{55.85\ g\ Fe^{2+}}{1\ mol\ Fe^{2+}} = 0.3122\ g\ Fe^{2+}$$

$$mass\ \%\ Fe = \frac{0.3122\ g}{1.2284\ g}\ x\ 100\% = 25.41\%$$

4.138　10.49 mL = 0.010 49 L

$(0.100\ mol/L\ S_2O_3^{2-})(0.010\ 49\ L) = 0.001\ 049\ mol\ S_2O_3^{2-}$

$$0.001\ 049\ mol\ S_2O_3^{2-}\ x\ \frac{1\ mol\ I_3^{-}}{2\ mol\ S_2O_3^{2-}}\ x\ \frac{2\ mol\ Cu^{2+}}{1\ mol\ I_3^{-}} = 0.001\ 049\ mol\ Cu^{2-}$$

$$0.001\ 049\ mol\ Cu^{2-}\ x\ \frac{63.55\ g\ Cu}{1\ mol\ Cu^{2+}} = 0.066\ 66\ g\ Cu$$

$$mass\ \%\ Cu = \frac{0.066\ 66\ g\ Cu}{14.98\ g\ sample}\ x\ 100\% = 0.4450\ \%\ Cu$$

4.140　$MgF_2(s) \rightleftarrows Mg^{2+}(aq)\ +\ 2\ F^{-}(aq)$

　　　　　　　　　x　　　　　　2x

$[Mg^{2+}] = x = 2.6\ x\ 10^{-4}\ M$ and $[F^{-}] = 2x = 2(2.6\ x\ 10^{-4}\ M) = 5.2\ x\ 10^{-4}\ M$ in a saturated solution.

$K_{sp} = [Mg^{2+}][F^{-}]^2 = (2.6\ x\ 10^{-4}\ M)(5.2\ x\ 10^{-4}\ M)^2 = 7.0\ x\ 10^{-11}$

4.142　CuO, 79.55;　Cu_2O, 143.09

Let X equal the mass of CuO and Y the mass of Cu_2O in the 10.50 g mixture. Therefore, X + Y = 10.50 g.

$$mol\ Cu = 8.66\ g\ x\ \frac{1\ mol\ Cu}{63.546\ g\ Cu} = 0.1363\ mol\ Cu$$

$mol\ CuO\ +\ 2\ x\ mol\ Cu_2O\ =\ 0.1363\ mol\ Cu$

$$X\ x\ \frac{1\ mol\ CuO}{79.55\ g\ CuO}\ +\ 2\ x\ \left(Y\ x\ \frac{1\ mol\ Cu_2O}{143.09\ g\ Cu_2O}\right) = 0.1363\ mol\ Cu$$

Rearrange to get X = 10.50 g – Y and then substitute it into the equation above to solve for Y.

$$(10.50 \text{ g} - Y) \times \frac{1 \text{ mol CuO}}{79.55 \text{ g CuO}} + 2 \times \left(Y \times \frac{1 \text{ mol Cu}_2O}{143.09 \text{ g Cu}_2O} \right) = 0.1363 \text{ mol Cu}$$

$$\frac{10.50 \text{ mol}}{79.55} - \frac{Y \text{ mol}}{79.55 \text{ g}} + \frac{2 \text{ Y mol}}{143.09 \text{ g}} = 0.1363 \text{ mol}$$

$$-\frac{Y \text{ mol}}{79.55 \text{ g}} + \frac{2 \text{ Y mol}}{143.09 \text{ g}} = 0.1363 \text{ mol} - \frac{10.50 \text{ mol}}{79.55} = 0.0043 \text{ mol}$$

$$\frac{(- Y \text{ mol})(143.09 \text{ g}) + (2 \text{ Y mol})(79.55 \text{ g})}{(79.55 \text{ g})(143.09 \text{ g})} = 0.0043 \text{ mol}$$

$$\frac{16.01 \text{ Y mol}}{11383 \text{ g}} = 0.0043 \text{ mol}; \quad \frac{16.01 \text{ Y}}{11383 \text{ g}} = 0.0043$$

$Y = (0.0043)(11383 \text{ g})/16.01 = 3.06 \text{ g Cu}_2O$

$X = 10.50 \text{ g} - Y = 10.50 \text{ g} - 3.06 \text{ g} = 7.44 \text{ g CuO}$

4.144 Mass S in MS = 1.504 g MS − 1.000 g M = 0.504 g S

$$0.504 \text{ g S} \times \frac{1 \text{ mol S}}{32.06 \text{ g S}} = 0.0157 \text{ mol S}$$

$$0.0157 \text{ mol S} \times \frac{1 \text{ mol M}}{1 \text{ mol S}} = 0.0157 \text{ mol M}$$

$$\text{molar mass M} = \frac{1.000 \text{ g M}}{0.0157 \text{ mol M}} = 63.69 \text{ g/mol; M is Cu}$$

4.146 48.39 mL = 0.048 39 L

$(0.1116 \text{ mol/L MnO}_4^-)(0.048 39 \text{ L}) = 5.400 \times 10^{-3} \text{ mol MnO}_4^-$

$$5.400 \times 10^{-3} \text{ mol MnO}_4^- \times \frac{5 \text{ mol Fe}^{2+}}{1 \text{ mol MnO}_4^-} \times \frac{55.84 \text{ g Fe}^{2+}}{1 \text{ mol Fe}^{2+}} = 1.508 \text{ g Fe}^{2+}$$

$$\text{mass \% Fe} = \frac{1.508 \text{ g Fe}^{2+}}{2.368 \text{ g}} \times 100\% = 63.68\%$$

4.148 Let SA stand for salicylic acid.

$$\text{mol C in 1.00 g of SA} = 2.23 \text{ g CO}_2 \times \frac{1 \text{ mol CO}_2}{44.01 \text{ g CO}_2} \times \frac{1 \text{ mol C}}{1 \text{ mol CO}_2} = 0.0507 \text{ mol C}$$

$$\text{mass C} = 0.0507 \text{ mol C} \times \frac{12.011 \text{ g C}}{1 \text{ mol C}} = 0.609 \text{ g C}$$

$$\text{mol H in 1.00 g of SA} = 0.39 \text{ g H}_2O \times \frac{1 \text{ mol H}_2O}{18.02 \text{ g H}_2O} \times \frac{2 \text{ mol H}}{1 \text{ mol H}_2O} = 0.043 \text{ mol H}$$

$$\text{mass H} = 0.043 \text{ mol H} \times \frac{1.008 \text{ g H}}{1 \text{ mol H}} = 0.043 \text{ g H}$$

mass O = 1.00 g − mass C − mass H = 1.00 − 0.609 g − 0.043 g = 0.35 g O

$$\text{mol O in 1.00 g of} = 0.35 \text{ g N} \times \frac{1 \text{ mol O}}{16.00 \text{ g O}} = 0.022 \text{ mol O}$$

Determine empirical formula.

$C_{0.0507}H_{0.043}O_{0.022}$; divide each subscript by the smallest, 0.022.

$C_{0.0507/0.022}H_{0.043/0.022}O_{0.022/0.022}$

$C_{2.3}H_2O$, multiply each subscript by 3 to get integers.

The empirical formula is $C_7H_6O_3$. The empirical formula mass = 138.12 g/mol.

Because salicylic acid has only one acidic hydrogen, there is a 1 to 1 mol ratio between salicylic acid and NaOH in the acid-base titration.

$$\text{mol SA in 1.00 g SA} = 72.4 \text{ mL} \times \frac{1 \times 10^{-3} \text{ L}}{1 \text{ mL}} \times \frac{0.100 \text{ mol NaOH}}{1 \text{ L}} \times \frac{1 \text{ mol SA}}{1 \text{ mol NaOH}} =$$

$$0.00724 \text{ mol SA}$$

$$\text{SA molar mass} = \frac{1.00 \text{ g}}{0.00724 \text{ mol}} = 138 \text{ g/mol}$$

Because the empirical formula mass and the molar mass are the same, the empirical formula is the molecular formula for salicylic acid.

4.150   100.00 mL = 0.100 00 L; 71.02 mL = 0.071 02 L

$$\text{mol H}_2\text{SO}_4 = \frac{0.1083 \text{ mol H}_2\text{SO}_4}{\text{L}} \times 0.100 \text{ 00 L} = 0.010 \text{ 83 mol H}_2\text{SO}_4$$

$$\text{mol NaOH} = \frac{0.1241 \text{ mol NaOH}}{\text{L}} \times 0.071 \text{ 02 L} = 0.008 \text{ 814 mol NaOH}$$

$$\text{H}_2\text{SO}_4 + 2 \text{ NaOH} \rightarrow \text{Na}_2\text{SO}_4 + 2 \text{ H}_2\text{O}$$

$$\text{mol H}_2\text{SO}_4 \text{ reacted with NaOH} = 0.008 \text{ 814 mol NaOH} \times \frac{1 \text{ mol H}_2\text{SO}_4}{2 \text{ mol NaOH}} = 0.004 \text{ 407 mol H}_2\text{SO}_4$$

mol H_2SO_4 reacted with MCO_3 = 0.010 83 mol – 0.004 407 mol = 0.006 423 mol H_2SO_4

mol H_2SO_4 reacted with MCO_3 = mol CO_3^{2-} in MCO_3 = mol CO_2 produced = 0.006 423 mol CO_2

(a) CO_3^{2-}, 60.01; $0.006 \text{ 423 mol CO}_3^{2-} \times \dfrac{60.01 \text{ g CO}_3^{2-}}{1 \text{ mol CO}_3^{2-}} = 0.3854 \text{ g CO}_3^{2-}$

mass of M = 1.268 g – 0.3854 g = 0.8826 g M

$$\text{molar mass of M} = \frac{0.8826 \text{ g}}{0.006 \text{ 423 mol}} = 137.4 \text{ g/mol}; \text{M is Ba}$$

(b) $0.006 \text{ 423 mol CO}_2 \times \dfrac{44.01 \text{ g CO}_2}{1 \text{ mol CO}_2} \times \dfrac{1 \text{ L}}{1.799 \text{ g}} = 0.1571 \text{ L CO}_2$

4.152   NaOH, 40.00; $Ba(OH)_2$, 171.34

Let X equal the mass of NaOH and Y the mass of $Ba(OH)_2$ in the 10.0 g mixture.
Therefore, X + Y = 10.0 g.

$$\text{mol HCl} = 108.9 \text{ mL} \times \frac{1 \times 10^{-3} \text{ L}}{1 \text{ mL}} \times \frac{1.50 \text{ mol HCl}}{1 \text{ L}} = 0.163 \text{ mol HCl}$$

mol NaOH + 2 x mol $Ba(OH)_2$ = 0.163 mol HCl

$$X \times \frac{1 \text{ mol NaOH}}{40.00 \text{ g NaOH}} + 2 \times \left(Y \times \frac{1 \text{ mol Ba(OH)}_2}{171.34 \text{ g Ba(OH)}_2} \right) = 0.163 \text{ mol HCl}$$

Rearrange to get $X = 10.0 \text{ g} - Y$ and then substitute it into the equation above to solve for Y.

$$(10.0 \text{ g} - Y) \times \frac{1 \text{ mol NaOH}}{40.00 \text{ g NaOH}} + 2 \times \left(Y \times \frac{1 \text{ mol Ba(OH)}_2}{171.34 \text{ g Ba(OH)}_2} \right) = 0.163 \text{ mol HCl}$$

$$\frac{10.00 \text{ mol}}{40.00} - \frac{Y \text{ mol}}{40.00 \text{ g}} + \frac{2 Y \text{ mol}}{171.34 \text{ g}} = 0.163 \text{ mol}$$

$$-\frac{Y \text{ mol}}{40.00 \text{ g}} + \frac{2 Y \text{ mol}}{171.34 \text{ g}} = 0.163 \text{ mol} - \frac{10.00 \text{ mol}}{40.00} = -0.087 \text{ mol}$$

$$\frac{(-Y \text{ mol})(171.34 \text{ g}) + (2 Y \text{ mol})(40.00 \text{ g})}{(40.00 \text{ g})(171.34 \text{ g})} = -0.087 \text{ mol}$$

$$\frac{-91.34 \text{ Y mol}}{6853.6 \text{ g}} = -0.087 \text{ mol}; \quad \frac{91.34 \text{ Y}}{6853.6 \text{ g}} = 0.087$$

$Y = (0.087)(6853.6 \text{ g})/91.34 = 6.5 \text{ g Ba(OH)}_2$
$X = 10.0 \text{ g} - Y = 10.0 \text{ g} - 6.5 \text{ g} = 3.5 \text{ g NaOH}$

4.154 (a) $Cr^{2+}(aq) + Cr_2O_7^{2-}(aq) \rightarrow Cr^{3+}(aq)$
$[Cr^{2+}(aq) \rightarrow Cr^{3+}(aq) + e^-] \times 6$　　　　　(oxidation half reaction)

$Cr_2O_7^{2-}(aq) \rightarrow Cr^{3+}(aq)$
$Cr_2O_7^{2-}(aq) \rightarrow 2 Cr^{3+}(aq)$
$Cr_2O_7^{2-}(aq) \rightarrow 2 Cr^{3+}(aq) + 7 H_2O(l)$
$14 H^+(aq) + Cr_2O_7^{2-}(aq) \rightarrow 2 Cr^{3+}(aq) + 7 H_2O(l)$
$6 e^- + 14 H^+(aq) + Cr_2O_7^{2-}(aq) \rightarrow 2 Cr^{3+}(aq) + 7 H_2O(l)$　　(reduction half reaction)

Combine the two half reactions.
$14 H^+(aq) + Cr_2O_7^{2-}(aq) + 6 Cr^{2+}(aq) \rightarrow 8 Cr^{3+}(aq) + 7 H_2O(l)$

(b) total volume = 100.0 ml + 20.0 mL = 120.0 mL = 0.1200 L
Initial moles:

$$0.120 \, \frac{\text{mol Cr(NO}_3)_2}{1 \text{ L}} \times 0.1000 \text{ L} = 0.0120 \text{ mol Cr(NO}_3)_2$$

$$0.500 \, \frac{\text{mol HNO}_3}{1 \text{ L}} \times 0.1000 \text{ L} = 0.0500 \text{ mol HNO}_3$$

$$0.250 \, \frac{\text{mol K}_2\text{Cr}_2\text{O}_7}{1 \text{ L}} \times 0.0200 \text{ L} = 0.005\,00 \text{ mol K}_2\text{Cr}_2\text{O}_7$$

Check for the limiting reactant. 0.0120 mol of Cr^{2+} requires $(0.0120)/6 = 0.00200$ mol $Cr_2O_7^{2-}$ and $(14/6)(0.0120) = 0.0280$ mol H^+. Both are in excess of the required amounts, so Cr^{2+} is the limiting reactant.

$$14\ H^+(aq)\ +\ Cr_2O_7^{2-}(aq)\ +\ 6\ Cr^{2+}(aq)\ \rightarrow\ 8\ Cr^{3+}(aq)\ +\ 7\ H_2O(l)$$

Initial moles	0.0500	0.00500	0.0120	0
Change	−14x	−x	−6x	+8x

Because Cr^{2+} is the limiting reactant, $6x = 0.0120$ and $x = 0.00200$

Final moles	0.0220	0.00300	0	0.0160

$$\text{mol } K^+ = 0.00500 \text{ mol } K_2Cr_2O_7 \times \frac{2 \text{ mol } K^+}{1 \text{ mol } K_2Cr_2O_7} = 0.0100 \text{ mol } K^+$$

$$\text{mol } NO_3^- = 0.0120 \text{ mol } Cr(NO_3)_2 \times \frac{2 \text{ mol } NO_3^-}{1 \text{ mol } Cr(NO_3)_2}$$

$$+ 0.0500 \text{ mol } HNO_3 \times \frac{1 \text{ mol } NO_3^-}{1 \text{ mol } HNO_3} = 0.0740 \text{ mol } NO_3^-$$

$$\text{mol } H^+ = 0.0220 \text{ mol}; \quad \text{mol } Cr_2O_7^{2-} = 0.00300 \text{ mol}; \quad \text{mol } Cr^{3+} = 0.01600 \text{ mol}$$

Check for charge neutrality.

Total moles of +charge = $0.0100 + 0.0220 + 3 \times (0.01600) = 0.0800$ mol +charge

Total moles of −charge = $0.0740 + 2 \times (0.00300) = 0.0800$ mol −charge

The charges balance and there is electrical neutrality in the solution after the reaction.

$$K^+ \text{ molarity} = \frac{0.0100 \text{ mol } K^+}{0.1200 \text{ L}} = 0.0833 \text{ M}$$

$$NO_3^- \text{ molarity} = \frac{0.0740 \text{ mol } NO_3^-}{0.1200 \text{ L}} = 0.617 \text{ M}$$

$$H^+ \text{ molarity} = \frac{0.0220 \text{ mol } H^+}{0.1200 \text{ L}} = 0.183 \text{ M}$$

$$Cr_2O_7^{2-} \text{ molarity} = \frac{0.00300 \text{ mol } Cr_2O_7^{2-}}{0.1200 \text{ L}} = 0.0250 \text{ M}$$

$$Cr^{3+} \text{ molarity} = \frac{0.0160 \text{ mol } Cr^{3+}}{0.1200 \text{ L}} = 0.133 \text{ M}$$

4.156 (a) (1) $Cu(s)\ \rightarrow\ Cu^{2+}(aq)$

$[Cu(s)\ \rightarrow\ Cu^{2+}(aq)\ +\ 2\ e^-] \times 3$ (oxidation half reaction)

$NO_3^-(aq)\ \rightarrow\ NO(g)$

$NO_3^-(aq)\ \rightarrow\ NO(g)\ +\ 2\ H_2O(l)$

$4\ H^+(aq)\ +\ NO_3^-(aq)\ \rightarrow\ NO(g)\ +\ 2\ H_2O(l)$

$[3\ e^-\ +\ 4\ H^+(aq)\ +\ NO_3^-(aq)\ \rightarrow\ NO(g)\ +\ 2\ H_2O(l)] \times 2$ (reduction half reaction)

Combine the two half reactions.

$3\ Cu(s)\ +\ 8\ H^+(aq)\ +\ 2\ NO_3^-(aq)\ \rightarrow\ 3\ Cu^{2+}(aq)\ +\ 2\ NO(g)\ +\ 4\ H_2O(l)$

(2) $Cu^{2+}(aq)\ +\ SCN^-(aq)\ \rightarrow\ CuSCN(s)$

$[e^-\ +\ Cu^{2+}(aq)\ +\ SCN^-(aq)\ \rightarrow\ CuSCN(s)] \times 2$ (reduction half reaction)

$HSO_3^-(aq) \rightarrow HSO_4^-(aq)$

$H_2O(l) + HSO_3^-(aq) \rightarrow HSO_4^-(aq)$

$H_2O(l) + HSO_3^-(aq) \rightarrow HSO_4^-(aq) + 2\,H^+(aq)$

$H_2O(l) + HSO_3^-(aq) \rightarrow HSO_4^-(aq) + 2\,H^+(aq) + 2\,e^-$

(oxidation half reaction)

Combine the two half reactions.

$2\,Cu^{2+}(aq) + 2\,SCN^-(aq) + H_2O(l) + HSO_3^-(aq) \rightarrow$
$$2\,CuSCN(s) + HSO_4^-(aq) + 2\,H^+(aq)$$

(3) $Cu^+(aq) \rightarrow Cu^{2+}(aq)$

$[Cu^+(aq) \rightarrow Cu^{2+}(aq) + e^-] \times 10$ (oxidation half reaction)

$IO_3^-(aq) \rightarrow I_2(aq)$

$2\,IO_3^-(aq) \rightarrow I_2(aq)$

$2\,IO_3^-(aq) \rightarrow I_2(aq) + 6\,H_2O(l)$

$12\,H^+(aq) + 2\,IO_3^-(aq) \rightarrow I_2(aq) + 6\,H_2O(l)$

$10\,e^- + 12\,H^+(aq) + 2\,IO_3^-(aq) \rightarrow I_2(aq) + 6\,H_2O(l)$ (reduction half reaction)

Combine the two half reactions.

$10\,Cu^+(aq) + 12\,H^+(aq) + 2\,IO_3^-(aq) \rightarrow 10\,Cu^{2+}(aq) + I_2(aq) + 6\,H_2O(l)$

(4) $I_2(aq) \rightarrow I^-(aq)$

$I_2(aq) \rightarrow 2\,I^-(aq)$

$2\,e^- + I_2(aq) \rightarrow 2\,I^-(aq)$ (reduction half reaction)

$S_2O_3^{2-}(aq) \rightarrow S_4O_6^{2-}(aq)$

$2\,S_2O_3^{2-}(aq) \rightarrow S_4O_6^{2-}(aq)$

$2\,S_2O_3^{2-}(aq) \rightarrow S_4O_6^{2-}(aq) + 2\,e^-$ (oxidation half reaction)

Combine the two half reactions.

$I_2(aq) + 2\,S_2O_3^{2-}(aq) \rightarrow 2\,I^-(aq) + S_4O_6^{2-}(aq)$

(5) $2\,ZnNH_4PO_4 \rightarrow Zn_2P_2O_7 + H_2O + 2\,NH_3$

(b) $10.82\,mL = 0.01082\,L$

mol $S_2O_3^{2-} = (0.1220\,mol/L)(0.01082\,L) = 0.00132\,mol\,S_2O_3^{2-}$

$$mol\,I_2 = 0.00132\,mol\,S_2O_3^{2-} \times \frac{1\,mol\,I_2}{2\,mol\,S_2O_3^{2-}} = 6.60 \times 10^{-4}\,mol\,I_2$$

$$mol\,Cu^+ = 6.60 \times 10^{-4}\,mol\,I_2 \times \frac{10\,mol\,Cu^+}{1\,mol\,I_2} = 6.60 \times 10^{-3}\,mol\,Cu^+\,(Cu)$$

g Cu $= (6.60 \times 10^{-3}\,mol)(63.546\,g/mol) = 0.419\,g\,Cu$

$$mass\,\%\,Cu\,in\,brass = \frac{0.419\,g\,Cu}{0.544\,g\,brass} \times 100\% = 77.1\%\,Cu$$

(c) $Zn_2P_2O_7$, 304.72

mass % Zn in $Zn_2P_2O_7 = \dfrac{2 \times 65.39\ g}{304.72\ g} \times 100\% = 42.92\%$

mass of Zn in $Zn_2P_2O_7 = (0.4292)(0.246\ g) = 0.106\ g\ Zn$

mass % Zn in brass $= \dfrac{0.106\ g\ Zn}{0.544\ g\ brass} \times 100\% = 19.5\%\ Zn$

4.158   (a) $H_3MO_3(aq) \rightarrow H_3MO_4(aq)$
$H_3MO_3(aq) + H_2O(l) \rightarrow H_3MO_4(aq)$
$H_3MO_3(aq) + H_2O(l) \rightarrow H_3MO_4(aq) + 2\ H^+(aq)$
$[H_3MO_3(aq) + H_2O(l) \rightarrow H_3MO_4(aq) + 2\ H^+(aq) + 2\ e^-] \times 5$ (oxidation half reaction)

$MnO_4^-(aq) \rightarrow Mn^{2+}(aq)$
$MnO_4^-(aq) \rightarrow Mn^{2+}(aq) + 4\ H_2O(l)$
$MnO_4^-(aq) + 8\ H^+(aq) \rightarrow Mn^{2+}(aq) + 4\ H_2O(l)$
$[MnO_4^-(aq) + 8\ H^+(aq) + 5\ e^- \rightarrow Mn^{2+}(aq) + 4\ H_2O(l)] \times 2$ (reduction half reaction)

Combine the two half reactions.
$5\ H_3MO_3(aq) + 5\ H_2O(l) + 2\ MnO_4^-(aq) + 16\ H^+(aq) \rightarrow$
$\qquad\qquad\qquad 5\ H_3MO_4(aq) + 10\ H^+(aq) + 2\ Mn^{2+}(aq) + 8\ H_2O(l)$
$5\ H_3MO_3(aq) + 2\ MnO_4^-(aq) + 6\ H^+(aq) \rightarrow 5\ H_3MO_4(aq) + 2\ Mn^{2+}(aq) + 3\ H_2O(l)$

(b) 10.7 mL = 0.0107 L
mol $MnO_4^- = (0.0107\ L)(0.100\ mol/L) = 1.07 \times 10^{-3}$ mol MnO_4^-

mol $H_3MO_3 = 1.07 \times 10^{-3}$ mol $MnO_4^- \times \dfrac{5\ mol\ H_3MO_3}{2\ mol\ MnO_4^-} = 2.67 \times 10^{-3}$ mol H_3MO_3

mol $M_2O_3 = 2.67 \times 10^{-3}$ mol $H_3MO_3 \times \dfrac{1\ mol\ M_2O_3}{2\ mol\ H_3MO_3} = 1.34 \times 10^{-3}$ mol M_2O_3

mol M in $M_2O_3 = 1.34 \times 10^{-3}$ mol $M_2O_3 \times \dfrac{2\ mol\ M}{1\ mol\ M_2O_3} = 2.68 \times 10^{-3}$ mol M

(c) M molar mass $= \dfrac{0.200\ g}{2.68 \times 10^{-3}\ mol} = 74.6$ g/mol;   M atomic weight = 74.6

M is As.

5

Periodicity and the Electronic Structure of Atoms

5.1 $\nu = 102.5$ MHz $= 102.5 \times 10^6$ Hz $= 102.5 \times 10^6$ s^{-1}

$$\lambda = \frac{c}{\nu} = \frac{3.00 \times 10^8 \text{ m/s}}{102.5 \times 10^6 \text{ s}^{-1}} = 2.93 \text{ m}$$

5.2 The wave with the shorter wavelength (b) has the higher frequency. The wave with the larger amplitude (b) represents the more intense beam of light. The wave with the shorter wavelength (b) represents blue light. The wave with the longer wavelength (a) represents red light.

5.3 IR, $\lambda = 1.55 \times 10^{-6}$ m

$$E = \frac{hc}{\lambda} = (6.626 \times 10^{-34} \text{ J} \cdot \text{s})\left(\frac{3.00 \times 10^8 \text{ m/s}}{1.55 \times 10^{-6} \text{ m}}\right)(6.022 \times 10^{23} / \text{mol})$$

$E = 7.72 \times 10^4$ J/mol $= 77.2$ kJ/mol

UV, $\lambda = 250$ nm $= 250 \times 10^{-9}$ m

$$E = \frac{hc}{\lambda} = (6.626 \times 10^{-34} \text{ J} \cdot \text{s})\left(\frac{3.00 \times 10^8 \text{ m/s}}{250 \times 10^{-9} \text{ m}}\right)(6.022 \times 10^{23} / \text{mol})$$

$E = 4.79 \times 10^5$ J/mol $= 479$ kJ/mol

X ray, $\lambda = 5.49$ nm $= 5.49 \times 10^{-9}$ m

$$E = \frac{hc}{\lambda} = (6.626 \times 10^{-34} \text{ J} \cdot \text{s})\left(\frac{3.00 \times 10^8 \text{ m/s}}{5.49 \times 10^{-9} \text{ m}}\right)(6.022 \times 10^{23} / \text{mol})$$

$E = 2.18 \times 10^7$ J/mol $= 2.18 \times 10^4$ kJ/mol

5.4 $$E = \frac{hc}{\lambda} = (6.626 \times 10^{-34} \text{ J} \cdot \text{s})\left(\frac{3.00 \times 10^8 \text{ m/s}}{2.3 \times 10^{-3} \text{ m}}\right)(6.022 \times 10^{23} / \text{mol})$$

$E = 52$ J/mol $= 0.052$ kJ/mol

$$74 \text{ kJ} \times \frac{1 \text{ mol photons}}{0.052 \text{ kJ}} = 1.4 \times 10^3 \text{ mol photons}$$

5.5 $$E = \frac{hc}{\lambda} = (6.626 \times 10^{-34} \text{ J} \cdot \text{s})\left(\frac{3.00 \times 10^8 \text{ m/s}}{390 \times 10^{-9} \text{ m}}\right)(6.022 \times 10^{23} / \text{mol})$$

$E = 3.07 \times 10^5$ J/mol $= 307$ kJ/mol

The energy is less than the work function. Electrons will not be ejected.

5.6   (a) Ag is predicted to have the higher work function because Rb is further left on the periodic table and holds its electrons less tightly.
(b) Rb because lower energies correspond to longer wavelength.

5.7 $m = 2$; $R_\infty = 1.097 \times 10^{-2} \text{ nm}^{-1}$

$$\frac{1}{\lambda} = R_\infty \left[\frac{1}{m^2} - \frac{1}{n^2} \right]; \quad \frac{1}{\lambda} = R_\infty \left[\frac{1}{2^2} - \frac{1}{7^2} \right]; \quad \frac{1}{\lambda} = 2.519 \times 10^{-3} \text{ nm}^{-1}; \quad \lambda = 397.0 \text{ nm}$$

$$E = \frac{hc}{\lambda} = (6.626 \times 10^{-34} \text{ J} \cdot \text{s}) \left(\frac{3.00 \times 10^8 \text{ m/s}}{397.0 \times 10^{-9} \text{ m}} \right) (6.022 \times 10^{23} / \text{mol})$$

$E = 3.015 \times 10^5 \text{ J/mol} = 301.5 \text{ kJ/mol}$

5.8 $m = 3$; $R_\infty = 1.097 \times 10^{-2} \text{ nm}^{-1}$

(a) $\frac{1}{\lambda} = R_\infty \left[\frac{1}{m^2} - \frac{1}{n^2} \right]; \quad \frac{1}{\lambda} = R_\infty \left[\frac{1}{3^2} - \frac{1}{4^2} \right]; \quad \frac{1}{\lambda} = 5.333 \times 10^{-4} \text{ nm}^{-1}; \quad \lambda = 1875 \text{ nm}$

(b) $\frac{1}{\lambda} = R_\infty \left[\frac{1}{m^2} - \frac{1}{n^2} \right]; \quad \frac{1}{\lambda} = R_\infty \left[\frac{1}{3^2} - \frac{1}{\infty^2} \right]; \quad \frac{1}{\lambda} = 1.219 \times 10^{-3} \text{ nm}^{-1}; \quad \lambda = 820.4 \text{ nm}$

5.9 $\lambda = \dfrac{h}{mv} = \dfrac{6.626 \times 10^{-34} \text{ kg m}^2 \text{ s}^{-1}}{(1150 \text{ kg})(24.6 \text{ m/s})} = 2.34 \times 10^{-38} \text{ m}$

5.10 $\lambda = 1\text{nm}/10 = 1 \times 10^{-10} \text{ m}$

$\lambda = \dfrac{h}{mv}; \quad v = \dfrac{h}{m\lambda} = \dfrac{6.626 \times 10^{-34} \text{ kg m}^2 \text{ s}^{-1}}{(9.1 \times 10^{-31} \text{ kg})(1 \times 10^{-10} \text{ m})} = 7 \times 10^6 \text{ m/s}$

5.11   4f

5.12

n	l	m_l	Orbital	No. of Orbitals
5	0	0	5s	1
	1	–1, 0, +1	5p	3
	2	–2, –1, 0, +1, +2	5d	5
	3	–3, –2, –1, 0, +1, +2, +3	5f	7
	4	–4, –3, –2, –1, 0, +1, +2, +3, +4	5g	9

There are 25 possible orbitals in the fifth shell.

5.13 $n = 4, l = 0$, 4s

5.14   The g orbitals have four nodal planes.

5.15   Ti, $1s^2 2s^2 2p^6 3s^2 3p^6 4s^2 3d^2$ or $[Ar] 4s^2 3d^2$
[Ar] $\underline{\uparrow\downarrow}$ $\underline{\uparrow}$ $\underline{\uparrow}$ __ __ __
    4s   3d

5.16   (a) 43 electrons = Tc   (b) 28 electrons = Ni

5.17   Si (valence shell of n = 3) and Sn (valence shell of n = 5) are in the same group (4A).
Atoms get larger as you go down a group, therefore, Sn is larger than Si. Cs (valence shell
of n = 6) is in group 1A. atoms get smaller as you go across a period. Cs is below and to
the left of Si and Sn, therefore Cs is larger than both Si and Sn.
smallest Si < Sn < Cs largest

5.18   Iodine has the largest atomic radius of the three halogens.   C–I would be the longest bond
length.

5.19   (a) 5%   (b) 20%

5.20   (b) kinetic energy of high speed electrons

5.21   (a) The fluorescent bulb does not emit all the wavelengths of light that would be emitted
from a white light source. Notice that there are dark regions between the colored peaks.
(b) Fluorescent light does appear as "white light" because its line spectrum has
contributions from all the colors (blue, green, yellow, orange, and red).

5.22   (a) $[Xe] 6s^2 4f^{14} 5d^{10}$
(b) [Xe] $\underline{\uparrow\downarrow}$ $\underline{\uparrow\downarrow}$ $\underline{\uparrow\downarrow}$ $\underline{\uparrow\downarrow}$ $\underline{\uparrow\downarrow}$ $\underline{\uparrow\downarrow}$ $\underline{\uparrow\downarrow}$ $\underline{\uparrow\downarrow}$ $\underline{\uparrow\downarrow}$ $\underline{\uparrow\downarrow}$ $\underline{\uparrow\downarrow}$ $\underline{\uparrow\downarrow}$ $\underline{\uparrow\downarrow}$
    6s   4f   5d
(c) There are no unpaired electrons.

5.23   (a) $[Xe] 6s^1 4f^{14} 5d^{10} 6p^1$
(b) [Xe] $\underline{\uparrow}$ $\underline{\uparrow\downarrow}$ $\underline{\uparrow\downarrow}$ $\underline{\uparrow\downarrow}$ $\underline{\uparrow\downarrow}$ $\underline{\uparrow\downarrow}$ $\underline{\uparrow\downarrow}$ $\underline{\uparrow\downarrow}$ $\underline{\uparrow\downarrow}$ $\underline{\uparrow\downarrow}$ $\underline{\uparrow\downarrow}$ $\underline{\uparrow\downarrow}$ $\underline{\uparrow\downarrow}$ $\underline{\uparrow}$ __ __
    6s   4f   5d   6p
(c) There are 2 unpaired electrons.

5.24   (a) 7d, n = 7, l = 2, m_l = –2, –1, 0, 1, 2
(b) 6p, n = 6, l = 1, m_l = –1, 0, 1

(c) $E = \dfrac{hc}{\lambda} = (6.626 \times 10^{-34}\ J\cdot s)\left(\dfrac{3.00 \times 10^8\ m/s}{434.7 \times 10^{-9}\ m}\right)(6.022 \times 10^{23}\ /mol)$

$E = 2.75 \times 10^5\ J/mol = 275\ kJ/mol$

5.25   The shortest wavelength corresponds to the highest energy, therefore, 126.8 nm
corresponds to 8p → 6s; 140.2 nm corresponds to 7p → 6s; and 185.0 nm corresponds to
6p → 6s.

Conceptual Problems

5.26   The wave with the larger amplitude (a) has the greater intensity. The wave with the shorter wavelength (a) has the higher energy radiation. The wave with the shorter wavelength (a) represents yellow light. The wave with the longer wavelength (b) represents infrared radiation.

5.28   (a) $3p_y$ $n = 3, l = 1$ (b) $4d_{z^2}$ $n = 4, l = 2$

5.30   The green element, molybdenum, has an anomalous electron configuration. Its predicted electron configuration is $[Ar]\ 5s^2\ 4d^4$. Its anomalous electron configuration is $[Ar]\ 5s^1\ 4d^5$ because of the resulting half-filled d-orbitals.

5.32   There are 34 total electrons in the atom, so there are also 34 protons in the nucleus. The atom is selenium (Se)

Se, $[Ar]$ $\underset{4s}{\uparrow\downarrow}$ $\underset{3d}{\uparrow\downarrow\ \ \uparrow\downarrow\ \ \uparrow\downarrow\ \ \uparrow\downarrow\ \ \uparrow\downarrow}$ $\underset{4p}{\uparrow\downarrow\ \ \uparrow\ \ \uparrow}$

Section Problems
Wave Properties of Radiant Energy (Section 5.1)

5.34   Violet has the higher frequency and energy. Red has the higher wavelength.

5.36 1.15×10^{-7} m $= 115 \times 10^{-9}$ m $= 115$ nm $=$ UV
        2.0×10^{-6} m $= 2000 \times 10^{-9}$ m $= 2000$ nm $=$ IR
        The visible region is (380 to 780 nm) is completely within this range. The ultraviolet and infrared regions are partially in this range.

5.38 $\lambda = \dfrac{c}{\nu} = \dfrac{3.00 \times 10^8 \text{ m/s}}{5.5 \times 10^{15} \text{ s}^{-1}} = 5.5 \times 10^{-8}$ m

5.40   (a) $\lambda = \dfrac{c}{\nu} = \dfrac{3.00 \times 10^8 \text{ m/s}}{825 \times 10^6 \text{ s}^{-1}} \times \dfrac{1 \text{ cm}}{1 \times 10^{-2} \text{ m}} = 36.4$ cm

        (b) $\lambda = \dfrac{c}{\nu} = \dfrac{3.00 \times 10^8 \text{ m/s}}{875 \times 10^6 \text{ s}^{-1}} \times \dfrac{1 \text{ cm}}{1 \times 10^{-2} \text{ m}} = 34.3$ cm

Particlelike Properties of Radiant Energy (Section 5.2)

5.42   (a) $\nu = 99.5$ MHz $= 99.5 \times 10^6$ s^{-1}
        $E = h\nu = (6.626 \times 10^{-34} \text{ J·s})(99.5 \times 10^6 \text{ s}^{-1})(6.022 \times 10^{23} \text{ /mol})$
        $E = 3.97 \times 10^{-2}$ J/mol $= 3.97 \times 10^{-5}$ kJ/mol
        $\nu = 1150$ kHz $= 1150 \times 10^3$ s^{-1}
        $E = h\nu = (6.626 \times 10^{-34} \text{ J·s})(1150 \times 10^3 \text{ s}^{-1})(6.022 \times 10^{23} \text{ /mol})$
        $E = 4.589 \times 10^{-4}$ J/mol $= 4.589 \times 10^{-7}$ kJ/mol

The FM radio wave (99.5 MHz) has the higher energy.

(b) $\lambda = 3.44 \times 10^{-9}$ m

$$E = \frac{hc}{\lambda} = (6.626 \times 10^{-34} \text{ J·s})\left(\frac{3.00 \times 10^8 \text{ m/s}}{3.44 \times 10^{-9} \text{ m}}\right)(6.022 \times 10^{23}/\text{mol})$$

$E = 3.48 \times 10^7$ J/mol $= 3.48 \times 10^4$ kJ/mol

$\lambda = 6.71 \times 10^{-2}$ m

$$E = \frac{hc}{\lambda} = (6.626 \times 10^{-34} \text{ J·s})\left(\frac{3.00 \times 10^8 \text{ m/s}}{6.71 \times 10^{-2} \text{ m}}\right)(6.022 \times 10^{23}/\text{mol})$$

$E = 1.78$ J/mol $= 1.78 \times 10^{-3}$ kJ/mol

The X ray ($\lambda = 3.44 \times 10^{-9}$ m) has the higher energy.

5.44   (a) $E = 90.5$ kJ/mol $\times \dfrac{1000 \text{ J}}{1 \text{ kJ}} \times \dfrac{1 \text{ mol}}{6.02 \times 10^{23}} = 1.50 \times 10^{-19}$ J

$$\nu = \frac{E}{h} = \frac{1.50 \times 10^{-19} \text{ J}}{6.626 \times 10^{-34} \text{ J·s}} = 2.27 \times 10^{14} \text{ s}^{-1}$$

$$\lambda = \frac{c}{\nu} = \frac{3.00 \times 10^8 \text{ m/s}}{2.27 \times 10^{14} \text{ s}^{-1}} = 1.32 \times 10^{-6} \text{ m} = 1320 \times 10^{-9} \text{ m} = 1320 \text{ nm, near IR}$$

(b) $E = 8.05 \times 10^{-4}$ kJ/mol $\times \dfrac{1000 \text{ J}}{1 \text{ kJ}} \times \dfrac{1 \text{ mol}}{6.02 \times 10^{23}} = 1.34 \times 10^{-24}$ J

$$\nu = \frac{E}{h} = \frac{1.34 \times 10^{-24} \text{ J}}{6.626 \times 10^{-34} \text{ J·s}} = 2.02 \times 10^9 \text{ s}^{-1}$$

$$\lambda = \frac{c}{\nu} = \frac{3.00 \times 10^8 \text{ m/s}}{2.02 \times 10^9 \text{ s}^{-1}} = 0.149 \text{ m, radio wave}$$

(c) $E = 1.83 \times 10^3$ kJ/mol $\times \dfrac{1000 \text{ J}}{1 \text{ kJ}} \times \dfrac{1 \text{ mol}}{6.02 \times 10^{23}} = 3.04 \times 10^{-18}$ J

$$\nu = \frac{E}{h} = \frac{3.04 \times 10^{-18} \text{ J}}{6.626 \times 10^{-34} \text{ J·s}} = 4.59 \times 10^{15} \text{ s}^{-1}$$

$$\lambda = \frac{c}{\nu} = \frac{3.00 \times 10^8 \text{ m/s}}{4.59 \times 10^{15} \text{ s}^{-1}} = 6.54 \times 10^{-8} \text{ m} = 65.4 \times 10^{-9} \text{ m} = 65.4 \text{ nm, UV}$$

5.46   (a) $\lambda = \dfrac{c}{\nu} = \dfrac{3.00 \times 10^8 \text{ m/s}}{2.9 \times 10^{18} \text{ s}^{-1}} = 1.0 \times 10^{-10}$ m

(b) $E = h\nu = (6.626 \times 10^{-34} \text{ J·s})(2.9 \times 10^{18} \text{ s}^{-1})\left(\dfrac{1 \text{ kJ}}{1000 \text{ J}}\right)(6.022 \times 10^{23}/\text{mol})$

$E = 1.2 \times 10^6$ kJ/mol

(c) X rays

5.48 $\nu = 9{,}192{,}631{,}770 \text{ s}^{-1} = 9.19263 \times 10^9 \text{ s}^{-1}$

$E = h\nu = (6.626 \times 10^{-34} \text{ J}\cdot\text{s})(9.19263 \times 10^9 \text{ s}^{-1})\left(\dfrac{1 \text{ kJ}}{1000 \text{ J}}\right)(6.022 \times 10^{23}/\text{mol}) = 3.668 \times 10^{-3} \text{ kJ/mol}$

5.50   (a) $\lambda = \dfrac{c}{\nu} = \dfrac{3.00 \times 10^8 \text{ m/s}}{3.85 \times 10^{14} \text{ s}^{-1}} = 7.79 \times 10^{-7} \text{ m} = 779 \times 10^{-9} \text{ m} = 779 \text{ nm}$

$E = h\nu = (6.626 \times 10^{-34} \text{ J}\cdot\text{s})(3.85 \times 10^{14} \text{ s}^{-1}) = 2.55 \times 10^{-19} \text{ J}$

(b) $\lambda = \dfrac{c}{\nu} = \dfrac{3.00 \times 10^8 \text{ m/s}}{4.62 \times 10^{14} \text{ s}^{-1}} = 6.49 \times 10^{-7} \text{ m} = 649 \times 10^{-9} \text{ m} = 649 \text{ nm}$

$E = h\nu = (6.626 \times 10^{-34} \text{ J}\cdot\text{s})(4.62 \times 10^{14} \text{ s}^{-1}) = 3.06 \times 10^{-19} \text{ J}$

(c) $\lambda = \dfrac{c}{\nu} = \dfrac{3.00 \times 10^8 \text{ m/s}}{7.41 \times 10^{14} \text{ s}^{-1}} = 4.05 \times 10^{-7} \text{ m} = 405 \times 10^{-9} \text{ m} = 405 \text{ nm}$

$E = h\nu = (6.626 \times 10^{-34} \text{ J}\cdot\text{s})(7.41 \times 10^{14} \text{ s}^{-1}) = 4.91 \times 10^{-19} \text{ J}$

5.52   Both (c) & (d) are below the threshold energy and no electrons would be ejected. (b) would eject the least number of electrons.

5.54 $E = (436 \text{ kJ/mol})\left(\dfrac{1000 \text{ J}}{1 \text{ kJ}}\right)\left(\dfrac{1 \text{ mol}}{6.022 \times 10^{23} \text{ photon}}\right) = 7.24 \times 10^{-19} \text{ J/photon}$

$\nu = \dfrac{E}{h} = \dfrac{7.24 \times 10^{-19} \text{ J}}{6.626 \times 10^{-34} \text{ J}\cdot\text{s}} = 1.09 \times 10^{15} \text{ s}^{-1} = 1.09 \times 10^{15} \text{ Hz}$

Atomic Line Spectra and Quantized Energy (Section 5.3)

5.56   The deuterium lamp produces a continuous emission spectrum.

5.58   For $n = 3$; $\lambda = 656.3 \text{ nm} = 656.3 \times 10^{-9} \text{ m}$

$E = \dfrac{hc}{\lambda} = (6.626 \times 10^{-34} \text{ J}\cdot\text{s})\left(\dfrac{2.998 \times 10^8 \text{ m/s}}{656.3 \times 10^{-9} \text{ m}}\right)\left(\dfrac{1 \text{ kJ}}{1000 \text{ J}}\right)(6.022 \times 10^{23}/\text{mol})$

$E = 182.3 \text{ kJ/mol}$

For $n = 4$; $\lambda = 486.1 \text{ nm} = 486.1 \times 10^{-9} \text{ m}$

$E = \dfrac{hc}{\lambda} = (6.626 \times 10^{-34} \text{ J}\cdot\text{s})\left(\dfrac{2.998 \times 10^8 \text{ m/s}}{486.1 \times 10^{-9} \text{ m}}\right)\left(\dfrac{1 \text{ kJ}}{1000 \text{ J}}\right)(6.022 \times 10^{23}/\text{mol})$

$E = 246.1 \text{ kJ/mol}$

For $n = 5$; $\lambda = 434.0 \text{ nm} = 434.0 \times 10^{-9} \text{ m}$

$E = \dfrac{hc}{\lambda} = (6.626 \times 10^{-34} \text{ J}\cdot\text{s})\left(\dfrac{2.998 \times 10^8 \text{ m/s}}{434.0 \times 10^{-9} \text{ m}}\right)\left(\dfrac{1 \text{ kJ}}{1000 \text{ J}}\right)(6.022 \times 10^{23}/\text{mol})$

$E = 275.6 \text{ kJ/mol}$

5.60 $m = 1, n = \infty$; $R_\infty = 1.097 \times 10^{-2}$ nm^{-1}

$$\frac{1}{\lambda} = R_\infty \left[\frac{1}{m^2} - \frac{1}{n^2} \right]; \quad \frac{1}{\lambda} = R_\infty \left[\frac{1}{1^2} - \frac{1}{\infty^2} \right] = R_\infty = 1.097 \times 10^{-2} \text{ nm}^{-1}; \quad \lambda = 91.16 \text{ nm}$$

$$E = \frac{hc}{\lambda} = (6.626 \times 10^{-34} \text{ J·s}) \left(\frac{2.998 \times 10^8 \text{ m/s}}{91.16 \times 10^{-9} \text{ m}} \right) \left(\frac{1 \text{ kJ}}{1000 \text{ J}} \right) (6.022 \times 10^{23}/\text{mol})$$

$$E = 1312 \text{ kJ/mol}$$

5.62 $m = 4, n = 5$; $R_\infty = 1.097 \times 10^{-2}$ nm^{-1}

$$\frac{1}{\lambda} = R_\infty \left[\frac{1}{m^2} - \frac{1}{n^2} \right]; \quad \frac{1}{\lambda} = R_\infty \left[\frac{1}{4^2} - \frac{1}{5^2} \right] = 2.468 \times 10^{-4} \text{ nm}^{-1}; \quad \lambda = 4051 \text{ nm}$$

$$E = \frac{hc}{\lambda} = (6.626 \times 10^{-34} \text{ J·s}) \left(\frac{2.998 \times 10^8 \text{ m/s}}{4051 \times 10^{-9} \text{ m}} \right) \left(\frac{1 \text{ kJ}}{1000 \text{ J}} \right) (6.022 \times 10^{23}/\text{mol})$$

$$E = 29.55 \text{ kJ/mol, IR}$$

$m = 4, n = 6$; $R_\infty = 1.097 \times 10^{-2}$ nm^{-1}

$$\frac{1}{\lambda} = R_\infty \left[\frac{1}{m^2} - \frac{1}{n^2} \right]; \quad \frac{1}{\lambda} = R_\infty \left[\frac{1}{4^2} - \frac{1}{6^2} \right] = 3.809 \times 10^{-4} \text{ nm}^{-1}; \quad \lambda = 2625 \text{ nm}$$

$$E = \frac{hc}{\lambda} = (6.626 \times 10^{-34} \text{ J·s}) \left(\frac{2.998 \times 10^8 \text{ m/s}}{2625 \times 10^{-9} \text{ m}} \right) \left(\frac{1 \text{ kJ}}{1000 \text{ J}} \right) (6.022 \times 10^{23}/\text{mol})$$

$$E = 45.60 \text{ kJ/mol, IR}$$

5.64 $m = 2$; $R_\infty = 1.097 \times 10^{-2}$ nm^{-1}

$$\frac{1}{\lambda} = R_\infty \left[\frac{1}{m^2} - \frac{1}{n^2} \right]; \quad \frac{1}{\lambda} = R_\infty \left[\frac{1}{2^2} - \frac{1}{6^2} \right] = 2.438 \times 10^{-3} \text{ nm}^{-1}$$

$\lambda = 410.2 \text{ nm} = 410.2 \times 10^{-9} \text{ m}$

$$E = \frac{hc}{\lambda} = (6.626 \times 10^{-34} \text{ J·s}) \left(\frac{2.998 \times 10^8 \text{ m/s}}{410.2 \times 10^{-9} \text{ m}} \right) \left(\frac{1 \text{ kJ}}{1000 \text{ J}} \right) (6.022 \times 10^{23}/\text{mol})$$

$$E = 291.6 \text{ kJ/mol}$$

Wavelike Properties of Particles (Section 5.4)

5.66 $\lambda = \dfrac{h}{mv} = \dfrac{6.626 \times 10^{-34} \text{ kg m}^2 \text{ s}^{-1}}{(9.11 \times 10^{-31} \text{ kg})(0.99 \times 3.00 \times 10^8 \text{ m/s})} = 2.45 \times 10^{-12} \text{ m, } \gamma \text{ ray}$

5.68 $156 \text{ km/h} = 156 \times 10^3 \text{ m}/3600 \text{ s} = 43.3 \text{ m/s}$; $145 \text{ g} = 0.145 \text{ kg}$

$\lambda = \dfrac{h}{mv} = \dfrac{6.626 \times 10^{-34} \text{ kg m}^2 \text{ s}^{-1}}{(0.145 \text{ kg})(43.3 \text{ m/s})} = 1.06 \times 10^{-34} \text{ m}$

The wavelength is too small, compared to the object, to observe.

5.70   145 g = 0.145 kg; 0.500 nm = 0.500 x 10^{-9} m

$$v = \frac{h}{m\lambda} = \frac{6.626 \times 10^{-34} \text{ kg m}^2 \text{ s}^{-1}}{(0.145 \text{ kg})(0.500 \times 10^{-9} \text{ m})} = 9.14 \times 10^{-24} \text{ m/s}$$

Orbitals and Quantum Mechanics (Sections 5.5–5.8)

5.72   0.68 g = 0.68 x 10^{-3} kg

$$(\Delta x)(\Delta mv) \geq \frac{h}{4\pi}$$

$$\Delta x \geq \frac{h}{4\pi(\Delta mv)} = \frac{6.626 \times 10^{-34} \text{ kg m}^2 \text{ s}^{-1}}{4\pi(0.68 \times 10^{-3} \text{ kg})(0.1 \text{ m/s})} = 8 \times 10^{-31} \text{ m}$$

5.74   The Heisenberg uncertainty principle states that one can never know both the position and
        the velocity of an electron beyond a certain level of precision. This means we cannot
        think of electrons circling the nucleus in specific orbital paths, but we can think of
        electrons as being found in certain three-dimensional regions of space around the nucleus,
        called orbitals.

5.76   n is the principal quantum number. The size and energy level of an orbital depends on n.
        l is the angular-momentum quantum number. l defines the three-dimensional shape of an
        orbital. m_l is the magnetic quantum number. m_l defines the spatial orientation of an
        orbital. m_s is the spin quantum number. m_s indicates the spin of the electron and can
        have either of two values, +½ or –½.

5.78   (a) 4s n = 4; l = 0; m_l = 0; m_s = ±½
        (b) 3p n = 3; l = 1; m_l = –1, 0, +1; m_s = ±½
        (c) 5f n = 5; l = 3; m_l = –3, –2, –1, 0, +1, +2, +3; m_s = ±½
        (d) 5d n = 5; l = 2; m_l = –2, –1, 0, +1, +2; m_s = ±½

5.80   Co $1s^2\ 2s^2\ 2p^6\ 3s^2\ 3p^6\ 4s^2\ 3d^7$
        (a) is not allowed because for l = 0, m_l = 0 only.
        (b) is not allowed because n = 4 and l = 2 is for a 4d orbital.
        (c) is allowed because n = 3 and l = 1 is for a 3p orbital.

5.82   For n = 5, the maximum number of electrons will occur when the 5g orbital is filled:
        [Rn] $7s^2\ 5f^{14}\ 6d^{10}\ 7p^6\ 8s^2\ 5g^{18}$ = 138 electrons

5.84 λ = 330 nm = 330 x 10^{-9} m

$$E = \frac{hc}{\lambda} = (6.626 \times 10^{-34} \text{ J}\cdot\text{s})\left(\frac{3.00 \times 10^8 \text{ m/s}}{330 \times 10^{-9} \text{ m}}\right)\left(\frac{1 \text{ kJ}}{1000 \text{ J}}\right)(6.022 \times 10^{23}/\text{mol})$$

        E = 363 kJ/mol

5.86 C $1s^2\ 2s^2\ 2p^2$

n	l	m_l	m_s
1	0	0	$+\frac{1}{2}$
1	0	0	$-\frac{1}{2}$
2	0	0	$+\frac{1}{2}$
2	0	0	$-\frac{1}{2}$
2	1	-1	$+\frac{1}{2}$
2	1	0	$+\frac{1}{2}$

5.88 Sr $[Kr]\ 5s^2$

n	l	m_l	m_s
5	0	0	$+\frac{1}{2}$
5	0	0	$-\frac{1}{2}$

5.90 A 4s orbital has three nodal surfaces.

4s orbital    nodes are white

regions of maximum electron probability are black

5.92 The different colors in the two lobes of the p orbitals indicate different phases. The different phases play a role in bonding.

Orbital Energy Levels in Multielectron Atoms (Section 5.9)

5.94 Part of the electron-nucleus attraction is canceled by the electron-electron repulsion, an effect we describe by saying that the electrons are shielded from the nucleus by the other electrons. The net nuclear charge actually felt by an electron is called the effective nuclear charge, Z_{eff}, and is often substantially lower than the actual nuclear charge, Z_{actual}.
$$Z_{eff} = Z_{actual} - \text{electron shielding}$$

5.96 4s > 4d > 4f

5.98 The number of elements in successive periods of the periodic table increases by the progression 2, 8, 18, 32 because the principal quantum number n increases by 1 from one period to the next. As the principal quantum number increases, the number of orbitals in a shell increases. The progression of elements parallels the number of electrons in a particular shell.

5.100 (a) 5d   (b) 4s   (c) 6s

5.102 (a) 3d after 4s   (b) 4p after 3d   (c) 6d after 5f   (d) 6s after 5p

Electron Configurations (Sections 5.10 –5.12)

5.104 (a) Ti, $Z = 22$ $1s^2 2s^2 2p^6 3s^2 3p^6 4s^2 3d^2$

(b) Ru, $Z = 44$ $1s^2 2s^2 2p^6 3s^2 3p^6 4s^2 3d^{10} 4p^6 5s^2 4d^6$

(c) Sn, $Z = 50$ $1s^2 2s^2 2p^6 3s^2 3p^6 4s^2 3d^{10} 4p^6 5s^2 4d^{10} 5p^2$

(d) Sr, $Z = 38$ $1s^2 2s^2 2p^6 3s^2 3p^6 4s^2 3d^{10} 4p^6 5s^2$

(e) Se, $Z = 34$ $1s^2 2s^2 2p^6 3s^2 3p^6 4s^2 3d^{10} 4p^4$

5.106 (a) Rb, $Z = 37$ [Kr] $\uparrow$ 5s

(b) W, $Z = 74$ [Xe] $\uparrow\downarrow$ (6s) $\uparrow\downarrow$ $\uparrow\downarrow$ $\uparrow\downarrow$ $\uparrow\downarrow$ $\uparrow\downarrow$ $\uparrow\downarrow$ $\uparrow\downarrow$ (4f) $\uparrow$ $\uparrow$ $\uparrow$ $\uparrow$ _ (5d)

(c) Ge, $Z = 32$ [Ar] $\uparrow\downarrow$ (4s) $\uparrow\downarrow$ $\uparrow\downarrow$ $\uparrow\downarrow$ $\uparrow\downarrow$ $\uparrow\downarrow$ (3d) $\uparrow$ $\uparrow$ _ (4p)

(d) Zr, $Z = 40$ [Kr] $\uparrow\downarrow$ (5s) $\uparrow$ $\uparrow$ _ _ _ (4d)

5.108 (a) O $1s^2 2s^2 2p^4$ $\uparrow\downarrow$ $\uparrow$ $\uparrow$ (2p)   2 unpaired e⁻

(b) Si $1s^2 2s^2 2p^6 3s^2 3p^2$ $\uparrow$ $\uparrow$ _ (3p)   2 unpaired e⁻

(c) K   [Ar] $4s^1$ 1 unpaired e⁻

(d) As   [Ar] $4s^2 3d^{10} 4p^3$ $\uparrow$ $\uparrow$ $\uparrow$ (4p)   3 unpaired e⁻

5.110 (a) Ra   [Rn] $7s^2$ [Rn] $\uparrow\downarrow$ (7s)

(b) Sc   [Ar] $4s^2 3d^1$ [Ar] $\uparrow\downarrow$ (4s) $\uparrow$ _ _ _ _ (3d)

(c) Lr   [Rn] $7s^2 5f^{14} 6d^1$ [Rn] $\uparrow\downarrow$ (7s) $\uparrow\downarrow$ $\uparrow\downarrow$ $\uparrow\downarrow$ $\uparrow\downarrow$ $\uparrow\downarrow$ $\uparrow\downarrow$ $\uparrow\downarrow$ (5f) $\uparrow$ _ _ _ _ (6d)

(d) B   [He] $2s^2 2p^1$ [He] $\uparrow\downarrow$ (2s) $\uparrow$ _ _ (2p)

(e) Te   [Kr] $5s^2 4d^{10} 5p^4$ [Kr] $\uparrow\downarrow$ (5s) $\uparrow\downarrow$ $\uparrow\downarrow$ $\uparrow\downarrow$ $\uparrow\downarrow$ $\uparrow\downarrow$ (4d) $\uparrow\downarrow$ $\uparrow$ $\uparrow$ (5p)

5.112 Order of orbital filling:

$1s \rightarrow 2s \rightarrow 2p \rightarrow 3s \rightarrow 3p \rightarrow 4s \rightarrow 3d \rightarrow 4p \rightarrow 5s \rightarrow 4d \rightarrow 5p \rightarrow 6s \rightarrow 4f \rightarrow 5d \rightarrow 6p \rightarrow 7s \rightarrow 5f \rightarrow 6d \rightarrow 7p \rightarrow 8s \rightarrow 5g$
$Z = 121$

5.114 Na⁺ $1s^2 2s^2 2p^6$

5.116 $Z = 116$ [Rn] $7s^2 5f^{14} 6d^{10} 7p^4$

Electron Configurations and Periodic Properties (Section 5.13)

5.118 Atomic radii increase down a group because the electron shells are farther away from the nucleus.

5.120 $F < O < S$

5.122 (a) K, lower in group 1A
(b) Ta, lower in group 5B
(c) V, farther to the left in same period
(d) Ba, four periods lower and only one group to the right

Multiconcept Problems

5.124 $m = 2$; $n = 3$; $R = 1.097 \times 10^{-2}$ nm^{-1}

$$\frac{1}{\lambda} = Z^2 R \left[\frac{1}{m^2} - \frac{1}{n^2} \right]; \quad \frac{1}{\lambda} = (2^2) R \left[\frac{1}{2^2} - \frac{1}{3^2} \right] = 6.094 \times 10^{-3} \text{ nm}^{-1}$$

$\lambda = 164$ nm

5.126 (a) row 1 $n = 1$, $l = 0$ 1s   2 elements
                          $l = 1$ 1p   6 elements
                          $l = 2$ 1d   10 elements
     row 2 $n = 2$, $l = 0$ 2s   2 elements
                          $l = 1$ 2p   6 elements
                          $l = 2$ 2d   10 elements
                          $l = 3$ 2f   14 elements
There would be 50 elements in the first two rows.
(b) There would be 18 elements in the first row [see (a) above]. The fifth element in the second row would have atomic number = 23.
(c) $Z = 12$

↑↓		↑↓	↑↓	↑↓		↑	↑	↑	↑	—
1s			1p					1d		

5.128 La ([Xe] $6s^2$ $5d^1$) is directly below Y ([Kr] $5s^2$ $4d^1$) in the periodic table. Both have similar valence electron configurations, but for La the valence electrons are one shell farther out leading to its larger radius.

Although Hf ([Xe] $6s^2$ $4f^{14}$ $5d^2$) is directly below Zr ([Kr] $5s^2$ $4d^2$) in the periodic table, Zr and Hf have almost identical atomic radii because the 4f electrons in Hf are not effective in shielding the valence electrons. The valence electrons in Hf are drawn in closer to the nucleus by the higher Z_{eff}.

5.130 75 W = 75 J/s; 550 nm = 550×10^{-9} m; $(0.05)(75 \text{ J/s}) = 3.75$ J/s

$$E = \frac{hc}{\lambda} = (6.626 \times 10^{-34} \text{ J·s}) \left(\frac{3.00 \times 10^8 \text{ m/s}}{550 \times 10^{-9} \text{ m}} \right) = 3.61 \times 10^{-19} \text{ J/photon}$$

$$\text{number of photons} = \frac{3.75 \text{ J/s}}{3.61 \text{ x } 10^{-19} \text{ J/photon}} = 1.0 \text{ x } 10^{19} \text{ photons/s}$$

5.132 $48.2 \text{ nm} = 48.2 \text{ x } 10^{-9} \text{ m}$

$$E(\text{photon}) = 6.626 \text{ x } 10^{-34} \text{ J·s x } \frac{3.00 \text{ x } 10^8 \text{ m/s}}{48.2 \text{ x } 10^{-9} \text{ m}} \text{ x } \frac{1 \text{ kJ}}{1000 \text{ J}} \text{ x } \frac{6.022 \text{ x } 10^{23}}{\text{mol}} = 2.48 \text{ x } 10^3 \text{ kJ/mol}$$

$$E_K = E(\text{electron}) = \tfrac{1}{2}(9.109 \text{ x } 10^{-31} \text{ kg})(2.371 \text{ x } 10^6 \text{ m/s})^2 \left(\frac{1 \text{ kJ}}{1000 \text{ J}}\right)\left(\frac{6.022 \text{ x } 10^{23}}{\text{mol}}\right)$$

$$E_K = 1.54 \text{ x } 10^3 \text{ kJ/mol}$$

$$E(\text{photon}) = E_i + E_K; \quad E_i = E(\text{photon}) - E_K = (2.48 \text{ x } 10^3) - (1.54 \text{ x } 10^3) = 940 \text{ kJ/mol}$$

5.134 Substitute the equation for the orbit radius, r, into the equation for the energy level, E, to

get $E = \dfrac{-Ze^2}{2\left(\dfrac{n^2 a_o}{Z}\right)} = \dfrac{-Z^2 e^2}{2a_o n^2}$

Let E_1 be the energy of an electron in a lower orbit and E_2 the energy of an electron in a higher orbit. The difference between the two energy levels is

$$\Delta E = E_2 - E_1 = \frac{-Z^2 e^2}{2a_o n_2^2} - \frac{-Z^2 e^2}{2a_o n_1^2} = \frac{-Z^2 e^2}{2a_o n_2^2} + \frac{Z^2 e^2}{2a_o n_1^2} = \frac{Z^2 e^2}{2a_o n_1^2} - \frac{Z^2 e^2}{2a_o n_2^2}$$

$$\Delta E = \frac{Z^2 e^2}{2a_o}\left[\frac{1}{n_1^2} - \frac{1}{n_2^2}\right]$$

Because Z, e, and a_o are constants, this equation shows that ΔE is proportional to

$\left[\dfrac{1}{n_1^2} - \dfrac{1}{n_2^2}\right]$ where n_1 and n_2 are integers with $n_2 > n_1$.

This is similar to the Balmer-Rydberg equation where $1/\lambda$ or ν for the emission spectra of

atoms is proportional to $\left[\dfrac{1}{m^2} - \dfrac{1}{n^2}\right]$ where m and n are integers with n > m.

5.136 (a) $E = h\nu$; $\nu = \dfrac{E}{h} = \dfrac{7.21 \text{ x } 10^{-19} \text{ J}}{6.626 \text{ x } 10^{-34} \text{ J·s}} = 1.09 \text{ x } 10^{15} \text{ s}^{-1}$

(b) $E(\text{photon}) = E_i + E_K$; from (a), $E_i = 7.21 \text{ x } 10^{-19} \text{ J}$

$$E(\text{photon}) = \frac{hc}{\lambda} = (6.626 \text{ x } 10^{-34} \text{ J·s})\left(\frac{3.00 \text{ x } 10^8 \text{ m/s}}{2.50 \text{ x } 10^{-7} \text{ m}}\right) = 7.95 \text{ x } 10^{-19} \text{ J}$$

$E_K = E(\text{photon}) - E_i = (7.95 \text{ x } 10^{-19} \text{ J}) - (7.21 \text{ x } 10^{-19} \text{ J}) = 7.4 \text{ x } 10^{-20} \text{ J}$
Calculate the electron velocity from the kinetic energy, E_K.
$E_K = 7.4 \text{ x } 10^{-20} \text{ J} = 7.4 \text{ x } 10^{-20} \text{ kg·m}^2/\text{s}^2 = \tfrac{1}{2}mv^2 = \tfrac{1}{2}(9.109 \text{ x } 10^{-31} \text{ kg})v^2$

$$v = \sqrt{\frac{2 \times (7.4 \times 10^{-20}\ kg \cdot m^2/s^2)}{9.109 \times 10^{-31}\ kg}} = 4.0 \times 10^5\ m/s$$

$$\text{de Broglie wavelength} = \frac{h}{mv} = \frac{6.626 \times 10^{-34}\ kg \cdot m^2/s}{(9.109 \times 10^{-31}\ kg)(4.0 \times 10^5\ m/s)} = 1.8 \times 10^{-9}\ m = 1.8\ nm$$

5.138 (a) 5f subshell: $n = 5$, $l = 3$, $m_l = -3, -2, -1, 0, +1, +2, +3$

3d subshell: $n = 3$, $l = 2$, $m_l = -2, -1, 0, +1, +2$

(b) In the H atom the subshells in a particular energy level are all degenerate, i.e., all have the same energy. Therefore, you only need to consider the principal quantum number, n, to calculate the wavelength emitted for an electron that drops from the 5f to the 3d subshell.

$m = 3$, $n = 5$; $R_\infty = 1.097 \times 10^{-2}\ nm^{-1}$

$$\frac{1}{\lambda} = R_\infty \left[\frac{1}{m^2} - \frac{1}{n^2} \right]; \quad \frac{1}{\lambda} = R_\infty \left[\frac{1}{3^2} - \frac{1}{5^2} \right]; \quad \frac{1}{\lambda} = 7.801 \times 10^{-4}\ nm^{-1}; \quad \lambda = 1282\ nm$$

(c) $m = 3$, $n = \infty$; $R_\infty = 1.097 \times 10^{-2}\ nm^{-1}$

$$\frac{1}{\lambda} = R_\infty \left[\frac{1}{m^2} - \frac{1}{n^2} \right]; \quad \frac{1}{\lambda} = R_\infty \left[\frac{1}{3^2} - \frac{1}{\infty^2} \right]; \quad \frac{1}{\lambda} = R_\infty \left[\frac{1}{3^2} \right] = 1.219 \times 10^{-3}\ nm^{-1}; \quad \lambda = 820.4\ nm$$

$$E = (6.626 \times 10^{-34}\ J \cdot s) \left(\frac{3.00 \times 10^8\ m/s}{820.4 \times 10^{-9}\ m} \right) (6.022 \times 10^{23}/mol) = 1.46 \times 10^5\ J/mol = 146\ kJ/mol$$

6 Ionic Compounds: Periodic Trends and Bonding Theory

6.1   (a) Ra^{2+} [Rn]   (b) Ni^{2+} [Ar] $3d^8$ (c) N^{3-} [Ne]

6.2   (b) Ti^{4+}, Ca^{2+}, and Cl^- are isoelectronic. They all have the electron configuration of Ar.
       (c) Na^+, Mg^{2+}, and Al^{3+} are isoelectronic. They all have the electron configuration of Ne.

6.3   As the positive charge of an element increases, Z_{eff} increases and the size decreases. Fe is larger than both Fe^{2+} and Fe^{3+}.

6.4 K^+, Ca^{2+}, and Cl^- are isoelectronic. The Z_{eff} for $Ca^{2+} > K^+ > Cl^-$.
       In terms of size, $Cl^- > K^+ > Ca^{2+}$. Cl^- is yellow, K^+ is green, and Ca^{2+} is red.

6.5   Ionization energy generally increases from left to right across a row of the periodic table and decreases from top to bottom down a group.   Rb < Se < O

6.6   (c) < (b) < (a) < (d)

6.7   Be $1s^2 2s^2$;   C $1s^2 2s^2 2p^2$;   N $1s^2 2s^2 2p^3$
       Be would have the larger third ionization energy because this electron would come from the 1s orbital.

6.8   The first 3 electrons are relatively easy to remove compared to the fourth and fifth electrons. The atom is Al.

6.9   Ge $1s^2 2s^2 2p^6 3s^2 3p^6 4s^2 3d^{10} 4p^2$
       As $1s^2 2s^2 2p^6 3s^2 3p^6 4s^2 3d^{10} 4p^3$
       Br $1s^2 2s^2 2p^6 3s^2 3p^6 4s^2 3d^{10} 4p^5$
       As has the least favorable E_{ea} because you are adding an electron to a half-filled set of 4p orbitals, pairing one pair of electrons and increasing repulsion. Br has the most favorable E_{ea} because on adding an electron you are achieving an inert gas configuration. Adding an electron to Ge goes into an empty 4p orbital which is favorable (but not as fovorable as Br) and E_{ea} for Ge is between that for As and Br.
       least favorable As < Ge < Br most favorable.

6.10   The least favorable E_{ea} is for Kr (red) because it is a noble gas with a filled set of 4p orbitals. The most favorable E_{ea} is for Ge (blue) because the 4p orbitals would become half filled. In addition, Z_{eff} is larger for Ge than it is for K (green).

6.11   Sr in $SrCO_3$ is Sr^{2+}, $1s^2 2s^2 2p^6 3s^2 3p^6 4s^2 3d^{10} 4p^6$

6.12 Group 6A elements will gain 2 electrons. The ion charge will be 2–.

6.13
$Mg(s) \rightarrow Mg(g)$	+147.7	kJ/mol
$Mg(g) \rightarrow Mg^+(g) + e^-$	+737.7	kJ/mol
$Mg^+(g) \rightarrow Mg^{2+}(g) + e^-$	+1450.7	kJ/mol
$F_2(g) \rightarrow 2\ F(g)$	+158	kJ/mol
$2[F(g) + e^- \rightarrow F^-(g)]$	2(–328)	kJ/mol
$Mg^{2+}(g) + 2\ F^-(g) \rightarrow MgF_2(s)$	–2957	kJ/mol

$$\text{Sum} = -1119 \quad \text{kJ/mol for } Mg(s)\ +\ F_2(g)\ \rightarrow\ MgF_2(s)$$

6.14
$Li(s) \rightarrow Li(g)$	+161	kJ/mol
$Li(s) \rightarrow Li(g) + e^-$	+520	kJ/mol
$\frac{1}{2}[Cl_2(g) \rightarrow 2\ Cl(g)]$	+243/2	kJ/mol
$Cl(g) + e^- \rightarrow Cl^-(g)$	–349	kJ/mol
$Li^+(g) + Cl^-(g) \rightarrow LiCl(s)$	–U	kJ/mol

$$\text{Sum} = -409 \text{ kJ/mol for } Li(s)\ +\ \tfrac{1}{2}\,Cl_2(g) \rightarrow LiCl(s)$$

electrostatic attraction $= -U = -409 - 161 - 520 - 243/2 + 349 = -863$ kJ/mol

6.15 Sr, Cu and Mg are 2+ cations. Cs is a large 1+ cation. I is a large 1– anion. F is a small 1– anion. O is a small 2– anion. Mg^{2+} is smaller than either Sr^{+2} or Cu^{2+}. Smaller ions with higher charges result in larger lattice energies. MgO has the largest lattice energy of the group.

6.16 The anions are larger than the cations. Cl^- is larger than O^{2-} because it is below it in the periodic table. Therefore, (a) is NaCl and (b) is MgO. Because of the higher ion charge and shorter cation–anion distance, MgO has the larger lattice energy.

6.17 (a), (c) and (d)

6.18 In ionic liquids the cation has an irregular shape and one or both of the ions are large and bulky to disperse charges over a large volume. Both factors minimize the crystal lattice energy, making the solid less stable and favoring the liquid.

6.19 (a) Iodide ions are larger than bromide ions. Tetraheptylammonium bromide corresponds to picture (ii) and tetraheptylammonium iodide corresponds to picture (i).
(b) Tetraheptylammonium bromide has the larger lattice energy because bromide ions are smaller than iodide ions.
(c) Tetraheptylammonium bromide has melting point of 88 °C and tetraheptylammonium iodide has a melting point of 39 °C.

6.20 (a) F^-, $1s^2\,2s^2\,2p^6$
Se^{2-}, $1s^2\,2s^2\,2p^6\,3s^2\,3p^6\,4s^2\,3d^{10}\,4p^6$
O^{2-}, $1s^2\,2s^2\,2p^6$
Br^-, $1s^2\,2s^2\,2p^6\,3s^2\,3p^6\,4s^2\,3d^{10}\,4p^6$
(b) F^- and O^{2-} are isoelectronic; Se^{2-} and Br^- are isoelectronic.
(c) The larger Br^-.

Conceptual Problems

6.22   The first sphere gets larger on going from reactant to product. This is consistent with it being a nonmetal gaining an electron and becoming an anion. The second sphere gets smaller on going from reactant to product. This is consistent with it being a metal losing an electron and becoming a cation.

6.24   Ca (red) would have the largest third ionization energy of the three because the electron being removed is from a filled valence shell. For Al (green) and Kr (blue), the electron being removed is from a partially filled valence shell. The third ionization energy for Kr would be larger than that for Al because the electron being removed from Kr is coming out of a 4p orbital while the electron being removed from Al makes Al isoelectronic with Ne. In addition, Z_{eff} is larger for Kr than for Al. The ease of losing its third electron is Al < Kr < Ca.

6.26   (a) $MgSO_4$ (b) Li_2CO_3 (c) $FeCl_2$ (d) $Ca_3(PO_4)_2$

6.28   (a) I_2 (b) Na   (c) NaCl   (d) Cl_2

6.30   All the ions in both drawings are singly charged, so only the size of the ions is important. The ions in drawing (b) are smaller and closer together, so (b) has the larger lattice energy.

6.32

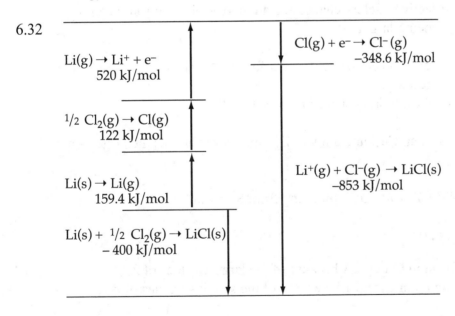

Section Problems
Electron Configuration of Ions (Section 6.1)

6.34   A covalent bond results when two atoms share several (usually two) of their electrons. An ionic bond results from a complete transfer of one or more electrons from one atom to another.

6.36   A molecule is the unit of matter that results when two or more atoms are joined by covalent bonds. An ion results when an atom gains or loses electrons.

6.38   (a) Be^{2+}, 4 protons and 2 electrons   (b) Rb^+, 37 protons and 36 electrons
        (c) Se^{2-}, 34 protons and 36 electrons   (d) Au^{3+}, 79 protons and 76 electrons

6.40   (a) La^{3+},   [Xe]   (b) Ag^+, $[Kr]\,4d^{10}$ (c) Sn^{2+}, $[Kr]\,5s^2\,4d^{10}$

6.42 Ca^{2+}, [Ar]; Ti^{2+}, $[Ar]\,3d^2$

6.44   The neutral atom contains 12 e^- and is Mg. The ion is Mg^{2+}.

6.46 Cr^{2+} $[Ar]\,3d^4$

↑ ↑ ↑ ↑ __
3d

        Fe^{2+} $[Ar]\,3d^6$

↑↓ ↑ ↑ ↑ ↑
3d

Ionic Radii (Section 6.2)

6.48   (a) S^{2-}; decrease in effective nuclear charge and an increase in electron-electron repulsions leads to the anion being larger.
        (b) Ca; in Ca^{2+} electrons are removed from a larger valence shell and there is an increase in effective nuclear charge leading to the smaller cation.
        (c) O^{2-}; decrease in effective nuclear charge and an increase in electron-electron repulsions leads to O^{2-} being larger.

6.50 Sr^{2+}, Se^{2-}, Br^-, and Rb^+ are isoelectronic. The Z_{eff} for $Sr^{2+} >$ $Rb^+ >$ $Br^- >$ Se^{2-}. The smallest ion has the largest Z_{eff}.
        Ions arranged from smallest to largest are $Sr^{2+} <$ $Rb^+ <$ $Br^- <$ Se^{2-} .

6.52 Cu^+ has more electrons than Cu^{2+} and a lower Z_{eff}; therefore Cu^+ has the larger ionic radius.

6.54 $S^{2-} > Ca^{2+} > Sc^{3+} > Ti^{4+}$, Z_{eff} increases on going from S^{2-} to Ti^{4+}.

Ionization Energy (Section 6.3)

6.56   The largest E_{i1} are found in Group 8A because of the largest values of Z_{eff}.
        The smallest E_{i1} are found in Group 1A because of the smallest values of Z_{eff}.

6.58   Using Figure 6.4 as a reference:

	Lowest E_{i1}	Highest E_{i1}
(a)	K	Li
(b)	B	Cl
(c)	Ca	Cl

Higher Ionization Energies (Section 6.4)

6.60 (a) K   [Ar] $4s^1$ Ca   [Ar] $4s^2$
Ca has the smaller second ionization energy because it is easier to remove the second 4s valence electron in Ca than it is to remove the second electron in K from the filled 3p orbitals.
(b) Ca   [Ar] $4s^2$ Ga   [Ar] $4s^2 3d^{10} 4p^1$
Ca has the larger third ionization energy because it is more difficult to remove the third electron in Ca from the filled 3p orbitals than it is to remove the third electron (second 4s valence electron) from Ga.

6.62 (a) $1s^2 2s^2 2p^6 3s^2 3p^3$ is P   (b) $1s^2 2s^2 2p^6 3s^2 3p^6$ is Ar   (c) $1s^2 2s^2 2p^6 3s^2 3p^6 4s^2$ is Ca
Ar has the highest E_{i2}. Ar has a higher Z_{eff} than P. The 4s electrons in Ca are easier to remove than any 3p electrons.
Ar has the lowest E_{i7}. It is difficult to remove 3p electrons from Ca, and it is difficult to remove 2p electrons from P.

6.64 The likely second row element is boron because it has three valence electrons. The large fourth ionization energy is from an inner shell 1s electron.

Electron Affinity (Section 6.5)

6.66 The relationship between the electron affinity of a univalent cation and the ionization energy of the neutral atom is that they have the same magnitude but opposite signs.

6.68 Na^+ has a more negative electron affinity than either Na or Cl because of its positive charge.

6.70 Energy is usually released when an electron is added to a neutral atom but absorbed when an electron is removed from a neutral atom because of the positive Z_{eff}.

6.72 The electron-electron repulsion is large and Z_{eff} is low.

Octet Rule (Section 6.6)

6.74 (a) [Ne], N^{3-} (b) [Ar], Ca^{2+} (c) [Ar], S^{2-} (d) [Kr], Br^-

6.76 (a) Mg^{2+} and Cl^-, $MgCl_2$, magnesium chloride
(b) Ca^{2+} and O^{2-}, CaO, calcium oxide
(c) Li^+ and N^{3-}, Li_3N, lithium nitride
(d) Al^{3+} and O^{2-}, Al_2O_3, aluminum oxide

6.78 (a) Because X reacts by losing electrons, it is likely to be a metal.
(b) Because Y reacts by gaining electrons, it is likely to be a nonmetal.
(c) X_2Y_3
(d) X is likely to be in group 3A and Y is likely to be in group 6A.

Formation of Ionic Compounds (Section 6.7)

6.80 $Li \rightarrow Li^+ + e^-$ +520 kJ/mol

         $Br + e^- \rightarrow Br^-$ <u>−325 kJ/mol</u>

                           +195 kJ/mol

The total energy = +195 kJ/mol, which is unfavorable because it is positive.

6.82 $Li(s) \rightarrow Li(g)$ +159.4 kJ/mol

         $Li(s) \rightarrow Li(g) + e^-$ +520   kJ/mol

         ½ $[Br_2(l) \rightarrow Br_2(g)]$ +15.4 kJ/mol

         ½ $[Br_2(g) \rightarrow 2 Br(g)]$ +112   kJ/mol

         $Br(g) + e^- \rightarrow Br^-(g)$ −325   kJ/mol

         $Li^+(g) + Br^-(g) \rightarrow LiBr(s)$ <u>−807   kJ/mol</u>

                        Sum = −325   kJ/mol for $Li(s) + ½ Br_2(l) \rightarrow LiBr(s)$

6.84 $Na(s) \rightarrow Na(g)$ +107.3   kJ/mol

         $Na(g) \rightarrow Na^+(g) + e^-$ +495.8   kJ/mol

         ½ $[H_2(g) \rightarrow 2 H(g)]$ ½(+435.9) kJ/mol

         $H(g) + e^- \rightarrow H^-(g)$ −72.8   kJ/mol

         $Na^+(g) + H^-(g) \rightarrow NaH(s)$ <u>−U</u>

                    Sum =   −60 kJ/mol for $Na(s) + ½ H_2(g) \rightarrow NaH(s)$

      $−U = −60 − 107.3 − 495.8 − 435.9/2 + 72.8 = −808$ kJ/mol;   U = 808 kJ/mol

6.86 $Cs(s) \rightarrow Cs(g)$ +76.1 kJ/mol

         $Cs(g) \rightarrow Cs^+(g) + e^-$ +375.7 kJ/mol

         ½ $[F_2(g) \rightarrow 2 F(g)]$ +79   kJ/mol

         $F(g) + e^- \rightarrow F^-(g)$ −328   kJ/mol

         $Cs^+(g) + F^-(g) \rightarrow CsF(s)$ <u>−740   kJ/mol</u>

                    Sum = −537   kJ/mol for $Cs(s) + ½ F_2(g) \rightarrow CsF(s)$

6.88 $Ca(s) \rightarrow Ca(g)$ +178.2   kJ/mol

         $Ca(g) \rightarrow Ca^+(g) + e^-$ +589.8   kJ/mol

         ½$[Cl_2(g) \rightarrow 2 Cl(g)]$ +121.5   kJ/mol

         $Cl(g) + e^- \rightarrow Cl^-(g)$ −348.6   kJ/mol

         $Ca^+(g) + Cl^-(g) \rightarrow CaCl(s)$ <u>−717   kJ/mol</u>

                    Sum = −176   kJ/mol for $Ca(s) + ½ Cl_2(g) \rightarrow CaCl(s)$

6.90

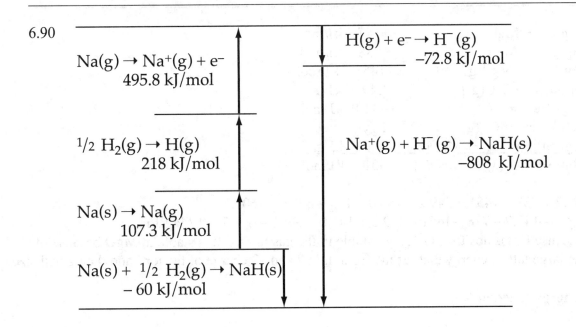

$Na(g) \rightarrow Na^+(g) + e^-$
495.8 kJ/mol

$H(g) + e^- \rightarrow H^-(g)$
–72.8 kJ/mol

½ $H_2(g) \rightarrow H(g)$
218 kJ/mol

$Na^+(g) + H^-(g) \rightarrow NaH(s)$
–808 kJ/mol

$Na(s) \rightarrow Na(g)$
107.3 kJ/mol

$Na(s) + $ ½ $H_2(g) \rightarrow NaH(s)$
– 60 kJ/mol

6.92
$Mg(s) \rightarrow Mg(g)$	+147.7 kJ/mol
$Mg(g) \rightarrow Mg^+(g) + e^-$	+737.7 kJ/mol
½ $F_2(g) \rightarrow F(g)$	+79 kJ/mol
$F(g) + e^- \rightarrow F^-(g)$	–328 kJ/mol
$Mg^+(g) + F^-(g) \rightarrow MgF(s)$	<u>–930</u> kJ/mol

Sum = –294 kJ/mol for $Mg(s) + $ ½ $F_2(g) \rightarrow MgF(s)$

$Mg(s) \rightarrow Mg(g)$	+147.7 kJ/mol
$Mg(g) \rightarrow Mg^+(g) + e^-$	+737.7 kJ/mol
$Mg^+(g) \rightarrow Mg^{2+}(g) + e^-$	+1450.7 kJ/mol
$F_2(g) \rightarrow 2 F(g)$	+158 kJ/mol
$2[F(g) + e^- \rightarrow F^-(g)]$	2(–328) kJ/mol
$Mg^{2+}(g) + 2 F^-(g) \rightarrow MgF_2(s)$	<u>–2952</u> kJ/mol

Sum = –1114 kJ/mol for $Mg(s) + F_2(g) \rightarrow MgF_2(s)$

In the reaction of magnesium with fluorine, MgF_2 will form because the overall energy for the formation of MgF_2 is much more negative than for the formation of MgF.

6.94
$Na(s) \rightarrow Na(g)$	+107.3 kJ/mol
$Na(g) + e^- \rightarrow Na^-(g)$	–52.9 kJ/mol
½$[Cl_2(g) \rightarrow 2 Cl(g)]$	+122 kJ/mol
$Cl(g) \rightarrow Cl^+(g) + e^-$	+1251 kJ/mol
$Na^-(g) + Cl^+(g) \rightarrow ClNa(s)$	<u>–787</u> kJ/mol

Sum = +640 kJ/mol for $Na(s) + $ ½ $Cl_2(g) \rightarrow Cl^+Na^-(s)$

The formation of Cl^+Na^- from its elements is not favored because the net energy change is positive whereas it is negative for the formation of Na^+Cl^-.

6.96 $Mg(s) \rightarrow Mg(g)$ +147.7 kJ/mol
       $Mg(g) \rightarrow Mg^+(g) + e^-$ +738   kJ/mol
       $Mg^+(g) \rightarrow Mg^{2+}(g) + e^-$ +1451 kJ/mol
       $\frac{1}{2}[O_2(g) \rightarrow 2\,O(g)]$ +249.2 kJ/mol
       $O(g) + e^- \rightarrow O^-(g)$ −141.0 kJ/mol
       $O^-(g) + e^- \rightarrow O^{2-}(g)$ E_{ea2}
       $\underline{Mg^{2+}(g) + O^{2-}(g) \rightarrow MgO(s)}$ −3791 kJ/mol
       $Mg(s) + \frac{1}{2}O_2(g) \rightarrow MgO(s)$ −601.7 kJ/mol

   $147.7 + 738 + 1451 + 249.2 - 141.0 + E_{ea2} - 3791 = -601.7$
   $E_{ea2} = -147.7 - 738 - 1451 - 249.2 + 141.0 + 3791 - 601.7 = +744$ kJ/mol
   Because E_{ea2} is positive, O^{2-} is not stable in the gas phase. It is stable in MgO because of the large lattice energy that results from the +2 and −2 charge of the ions and their small size.

Lattice Energy (Section 6.8)

6.98 $MgCl_2 > LiCl > KCl > KBr$

Multiconcept Problems

6.100 When moving diagonally down and right on the periodic table, the increase in atomic radius caused by going to a larger shell is offset by a decrease caused by a higher Z_{eff}. Thus, there is little net change in the charge density.

6.102 $Cr(s) \rightarrow Cr(g)$ +397 kJ/mol
       $Cr(g) \rightarrow Cr^+(g)$ +652 kJ/mol
       $Cr^+(g) \rightarrow Cr^{2+}(g)$ +1588 kJ/mol
       $Cr^{2+}(g) \rightarrow Cr^{3+}(g)$ +2882 kJ/mol
       $\frac{1}{2}(I_2(s) \rightarrow I_2(g))$ +62/2 kJ/mol
       $\frac{1}{2}(I_2(g) \rightarrow 2\,I(g))$ +151/2 kJ/mol
       $I(g) + e^- \rightarrow I^-(g)$ −295 kJ/mol
       $Cl_2(g) \rightarrow 2\,Cl(g)$ +243 kJ/mol
       $2(Cl(g) + e^- \rightarrow Cl^-(g))$ 2(−349) kJ/mol
       $\underline{Cr^{3+}(g) + 2\,Cl^-(g) + I^-(g) \rightarrow CrCl_2I(s)}$ −U
       $Cr(s) + Cl_2(g) + \frac{1}{2}I_2(g) \rightarrow CrCl_2I(s)$ − 420 kJ/mol

   $-U = -420 - 397 - 652 - 1588 - 2882 - 62/2 - 151/2 + 295 - 243 + 2(349) = -5295.5$ kJ/mol
   $U = 5295$ kJ/mol

6.104 (a) Fe   [Ar] $4s^2 3d^6$
          Fe^{2+} [Ar] $3d^6$
          Fe^{3+} [Ar] $3d^5$
       (b) A 3d electron is removed on going from Fe^{2+} to Fe^{3+}. For the 3d electron, $n = 3$ and $l = 2$.
       (c) E(J/photon) = 2952 kJ/mol x $\dfrac{1\ \text{mol photons}}{6.022 \times 10^{23}\ \text{photons}}$ x $\dfrac{1000\ \text{J}}{1\ \text{kJ}}$ = 4.90×10^{-18} J/photon

$$E = \frac{hc}{\lambda}$$

$$\lambda = \frac{hc}{E} = \frac{(6.626 \times 10^{-34} \text{ J} \cdot \text{s})(3.00 \times 10^{8} \text{ m/s})}{4.90 \times 10^{-18} \text{ J}} = 4.06 \times 10^{-8} \text{ m} = 40.6 \times 10^{-9} \text{ m} = 40.6 \text{ nm}$$

(d) Ru is directly below Fe in the periodic table and the two metals have similar electron configurations. The electron removed from Ru to go from Ru^{2+} to Ru^{3+} is a 4d electron. The electron with the higher principal quantum number, n = 4, is farther from the nucleus, less tightly held, and requires less energy to remove.

7

Covalent Bonding and Electron-Dot Structures

7.1 Only (a) is labeled incorrectly. The bonds in $SiCl_4$ are polar covalent.

7.2 H is positively polarized (blue). O is negatively polarized (red). This is consistent with the electronegativity values for O (3.5) and H (2.1). The more negatively polarized atom should be the one with the larger electronegativity.

7.3 $$\mu = Q \times r = (1.60 \times 10^{-19} \text{ C})(228 \times 10^{-12} \text{ m})\left(\frac{1 \text{ D}}{3.336 \times 10^{-30} \text{ C} \cdot \text{m}}\right) = 10.94 \text{ D}$$

% ionic character for AgCl $= \dfrac{6.08 \text{ D}}{10.94 \text{ D}} \times 100\% = 55.6\%$

7.4 Li–H $\quad \Delta EN = EN(H) - EN(Li) = 2.1 - 1.0 = 1.1$
H–F $\quad \Delta EN = EN(F) - EN(H) = 4.0 - 2.1 = 1.9$
Based on the larger ΔEN value, you would predict that HF would have the higher % ionic character. The calculations below show the opposite is in fact the case.

LiH; $\quad \mu = Q \times r = (1.60 \times 10^{-19} \text{ C})(160 \times 10^{-12} \text{ m})\left(\dfrac{1 \text{ D}}{3.336 \times 10^{-30} \text{ C} \cdot \text{m}}\right) = 7.67 \text{ D}$

% ionic character for LiH $= \dfrac{6.00 \text{ D}}{7.67 \text{ D}} \times 100\% = 78.2\%$

HF; $\quad \mu = Q \times r = (1.60 \times 10^{-19} \text{ C})(92 \times 10^{-12} \text{ m})\left(\dfrac{1 \text{ D}}{3.336 \times 10^{-30} \text{ C} \cdot \text{m}}\right) = 4.41 \text{ D}$

% ionic character for HF $= \dfrac{1.83 \text{ D}}{4.41 \text{ D}} \times 100\% = 41.4\%$

7.5 (c) $H \overset{..}{\underset{..}{:S:}} H$

7.6 (a) OF_2 (b) $SiCl_4$

7.7 (a) is incorrect, the least electronegative atom (P) should be placed in the center.
(b) correct
(c) is incorrect, oxygen does not have a complete octet.
(d) is incorrect, there are 2 too many valence electrons.

7.8 (a) 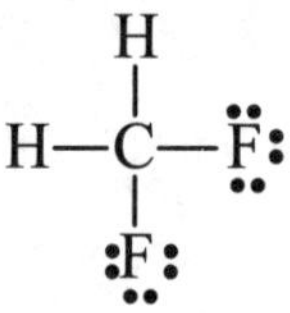 H—C—F with H on top, F below, and lone pairs on F atoms

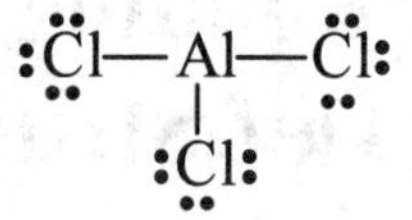

(b) :Cl—Al—Cl: with :Cl: below

(c) [:O: / :O—Cl—O: / :O:]⁻ perchlorate structure

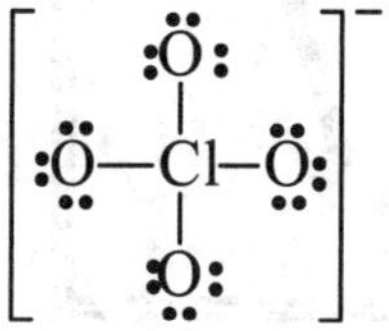

(d) 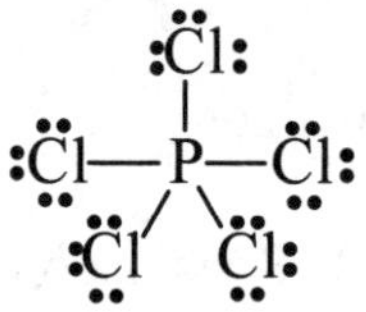 :Cl—P—Cl: with :Cl: above and :Cl: :Cl: below

(e) 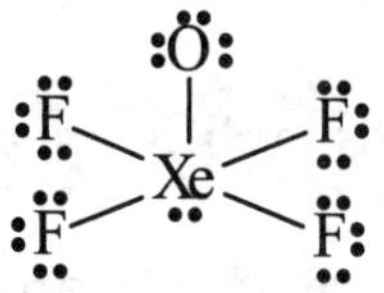 :F: Xe :F: with :O: on top and :F: :F: arrangement

(e) 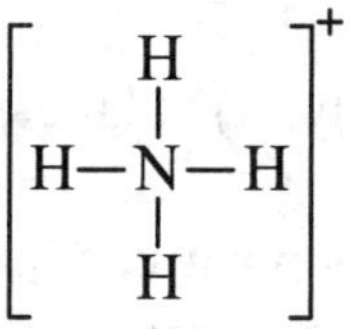 [H—N—H with H on top and H on bottom]⁺

7.9 (d) $:C{\equiv}O:$

7.10 (e) Both (b) and (c) are correct.

7.11 (c) is the correct structure because it has 19 valence electrons, the two oxygens have complete octets, and the odd electron is on the chlorine.

7.12 O_2^- is a radical and the presence of an unpaired electron leads to a highly reactive species.

7.13 (a) 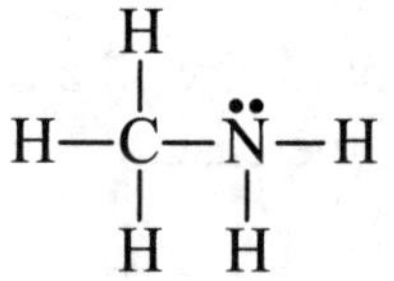 H—C—N—H structure with H atoms

(b) H—C=C—H with H atoms

(c) H—O—O—H with lone pairs

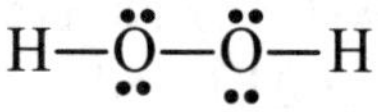

(d) 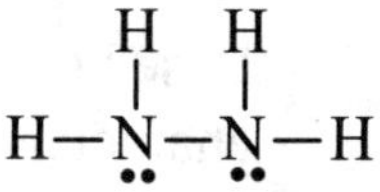 H—N—N—H with H atoms and lone pairs

7.14 H—C—C—O—H structure with H atoms   and   H—C—O—C—H structure with H atoms

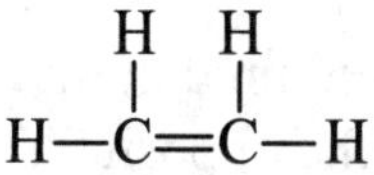

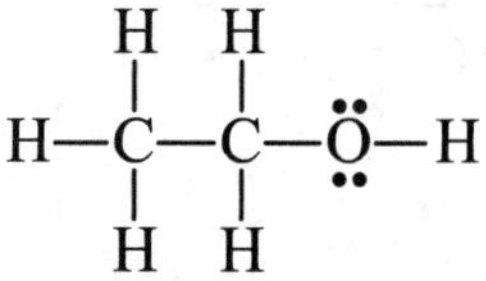

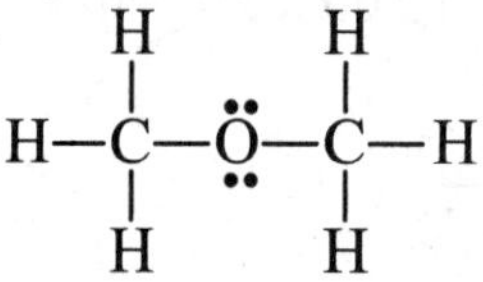

7.15 Molecular formula: $C_4H_5N_3O$

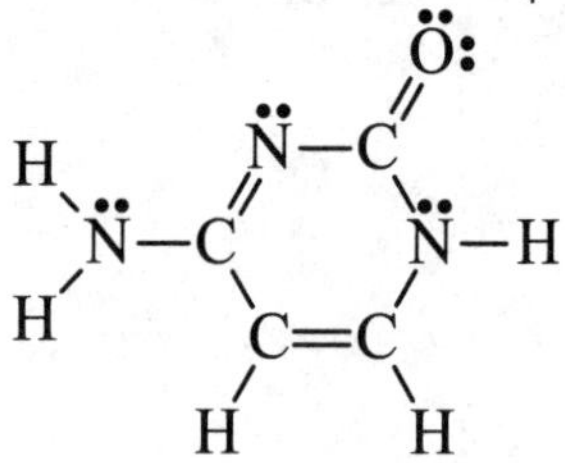

ring structure with O, N, C, H atoms

7.16 (a) 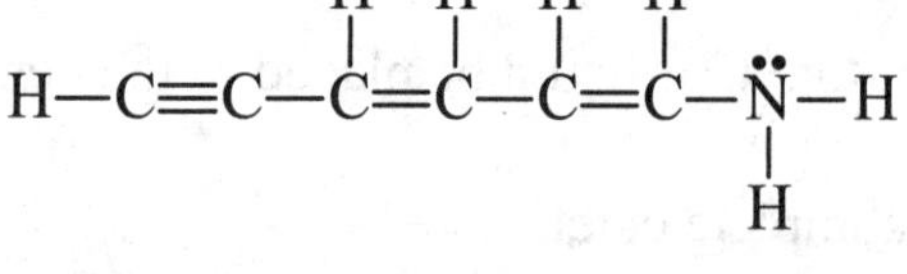 H—C≡C—C=C—C=C—N—H structure with H atoms

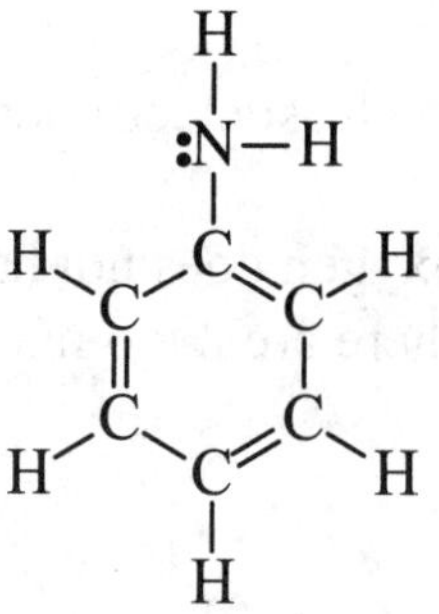

N—H aromatic ring structure with C and H atoms

(b)

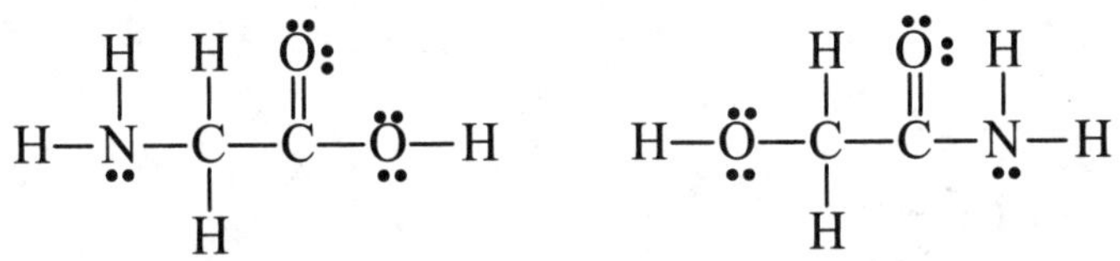

7.17 (a)

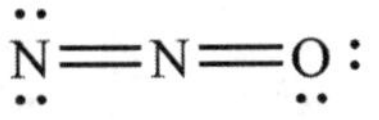

(b)

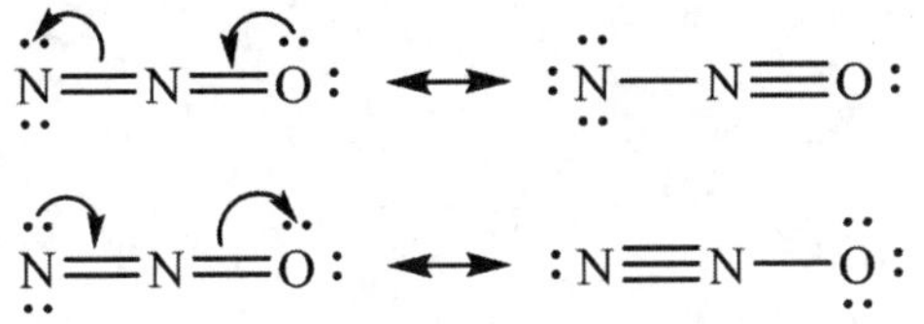

7.18 (a)

(b)

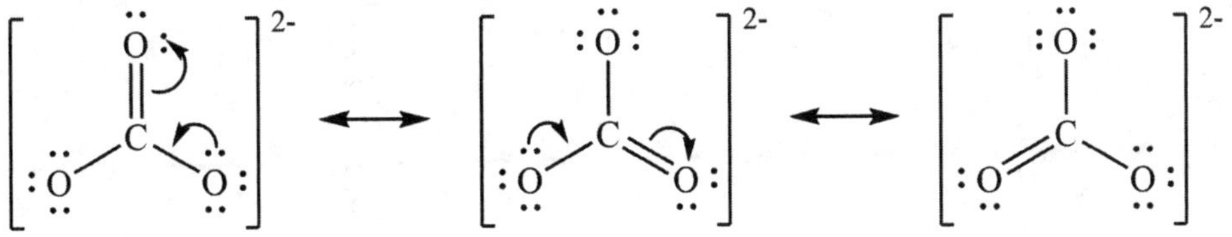

(c)

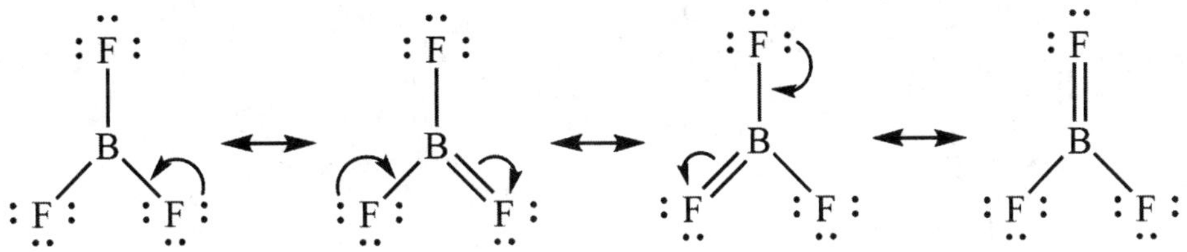

7.19 (a)

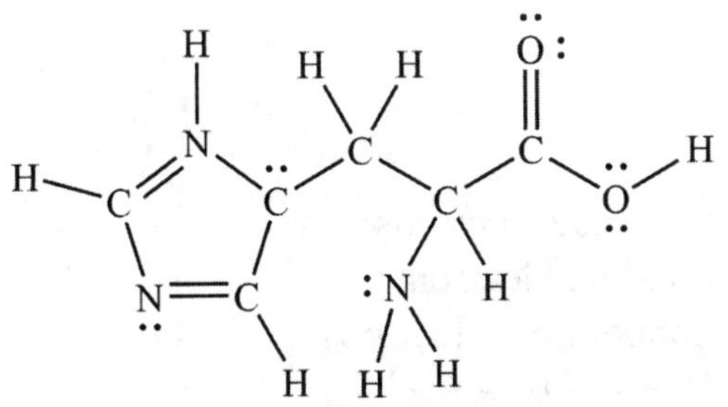

This is a valid resonance structure.

(b)

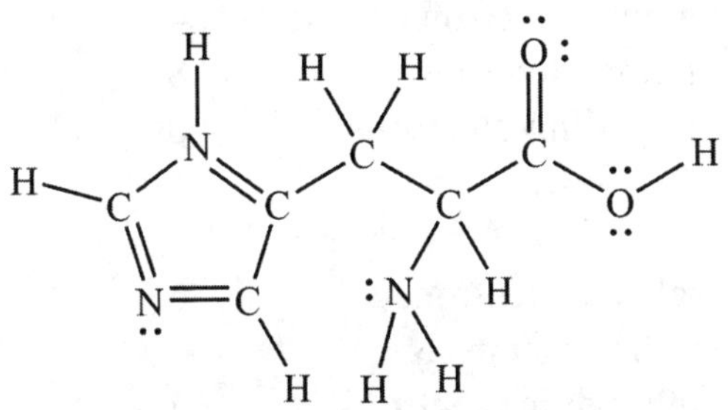

This is not a valid resonance structure. The N with two double bonds has 10 electrons, not 8.

7.20 (a)

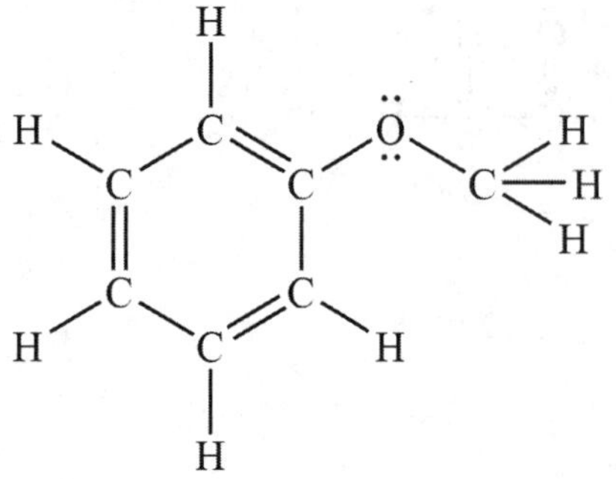

(b)

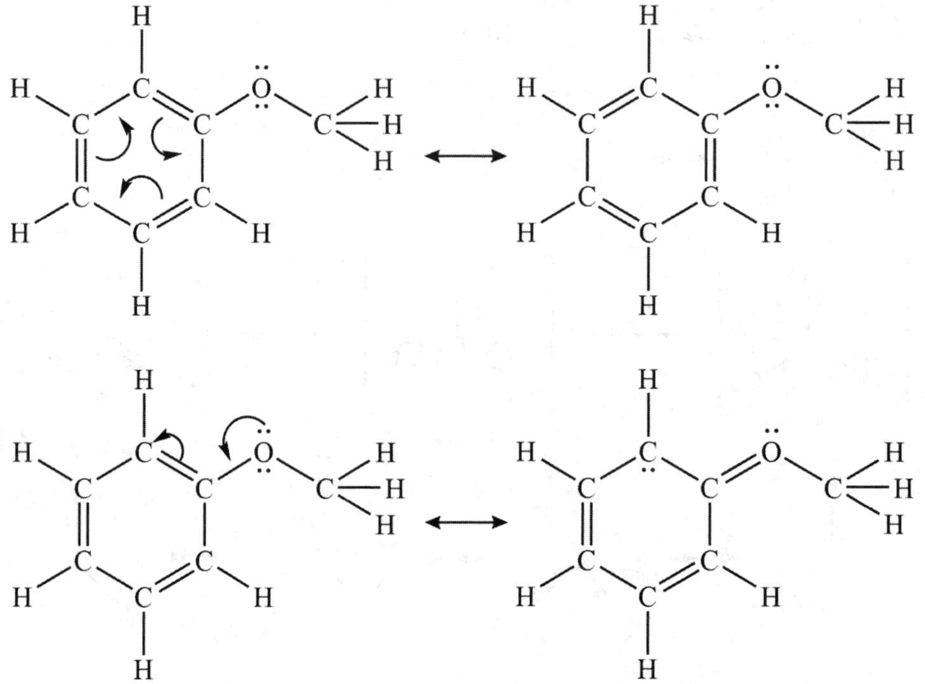

7.21

$$\left[:\ddot{N}=C=\ddot{O}: \right]^{-}$$

For nitrogen:	Isolated nitrogen valence electrons	5
	Bound nitrogen bonding electrons	4
	Bound nitrogen nonbonding electrons	4
	Formal charge = $5 - \frac{1}{2}(4) - 4 = -1$	

For carbon:	Isolated carbon valence electrons	4
	Bound carbon bonding electrons	8
	Bound carbon nonbonding electrons	0
	Formal charge = $4 - \frac{1}{2}(8) - 0 = 0$	

For oxygen:	Isolated oxygen valence electrons	6
	Bound oxygen bonding electrons	4
	Bound oxygen nonbonding electrons	4
	Formal charge = $6 - \frac{1}{2}(4) - 4 = 0$	

7.22 (a)

$$\left[\ddot{N}\!=\!C\!=\!\ddot{O}\right]^{-} \longleftrightarrow \left[:N\!\equiv\!C\!-\!\ddot{\ddot{O}}:\right]^{-}$$

$$\left[\ddot{N}\!=\!C\!=\!\ddot{O}\right]^{-} \longleftrightarrow \left[:\ddot{N}\!-\!C\!\equiv\!O:\right]^{-}$$

(b) $\left[\ddot{\ddot{N}}\!=\!C\!=\!\ddot{\ddot{O}}\right]^{-}$

For nitrogen:	Isolated nitrogen valence electrons	5
	Bound nitrogen bonding electrons	4
	Bound nitrogen nonbonding electrons	4
	Formal charge = 5 – ½(4) – 4 = –1	

For carbon:	Isolated carbon valence electrons	4
	Bound carbon bonding electrons	8
	Bound carbon nonbonding electrons	0
	Formal charge = 4 – ½(8) – 0 = 0	

For oxygen:	Isolated oxygen valence electrons	6
	Bound oxygen bonding electrons	4
	Bound oxygen nonbonding electrons	4
	Formal charge = 6 – ½(4) – 4 = 0	

$\left[:N\!\equiv\!C\!-\!\ddot{\ddot{O}}:\right]^{-}$

For nitrogen:	Isolated nitrogen valence electrons	5
	Bound nitrogen bonding electrons	6
	Bound nitrogen nonbonding electrons	2
	Formal charge = 5 – ½(6) – 2 = 0	

For carbon:	Isolated carbon valence electrons	4
	Bound carbon bonding electrons	8
	Bound carbon nonbonding electrons	0
	Formal charge = 4 – ½(8) – 0 = 0	

For oxygen:	Isolated oxygen valence electrons	6
	Bound oxygen bonding electrons	2
	Bound oxygen nonbonding electrons	6
	Formal charge = 6 – ½(2) – 6 = –1	

$\left[:\ddot{N}\!-\!C\!\equiv\!O:\right]^{-}$

For nitrogen:	Isolated nitrogen valence electrons	5
	Bound nitrogen bonding electrons	2
	Bound nitrogen nonbonding electrons	6
	Formal charge = 5 – ½(2) – 6 = –2	

For carbon:

Isolated carbon valence electrons	4
Bound carbon bonding electrons	8
Bound carbon nonbonding electrons	0

Formal charge = $4 - \frac{1}{2}(8) - 0 = 0$

For oxygen:

Isolated oxygen valence electrons	6
Bound oxygen bonding electrons	6
Bound oxygen nonbonding electrons	2

Formal charge = $6 - \frac{1}{2}(6) - 2 = +1$

The first two structures make the largest contribution to the resonance hybrid because the −1 formal charge is on either of the electronegative N or O. The third structure has a +1 formal charge on O and does not significantly contribute to the resonance hybrid.

(c) Carbon–nitrogen because of the triple bond contribution to the resonance hybrid.

7.23

All atoms in this structure have 0 formal charge.

For top oxygen:

Isolated oxygen valence electrons	6
Bound oxygen bonding electrons	2
Bound oxygen nonbonding electrons	6

Formal charge = $6 - \frac{1}{2}(2) - 6 = -1$

For right oxygen:

Isolated oxygen valence electrons	6
Bound oxygen bonding electrons	6
Bound oxygen nonbonding electrons	2

Formal charge = $6 - \frac{1}{2}(6) - 2 = +1$

All the other atoms in this structure have 0 formal charge.

The structure without formal charges makes a larger contribution to the resonance hybrid because energy is required to separate + and − charges. Thus, the actual electronic structure of acetic acid is closer to that of the more favorable, lower energy structure.

7.24

All atoms in structures 1 and 2 have 0 formal charge.

In structure 3:

For carbon:	Isolated carbon valence electrons	4
	Bound carbon bonding electrons	6
	Bound carbon nonbonding electrons	2
	Formal charge = $4 - \frac{1}{2}(6) - 2 = -1$	

For oxygen:	Isolated oxygen valence electrons	6
	Bound oxygen bonding electrons	6
	Bound oxygen nonbonding electrons	2
	Formal charge = $6 - \frac{1}{2}(6) - 2 = +1$	

All the other atoms in structure 3 have 0 formal charge.

The structures 1 and 2 without formal charges make the larger contribution to the resonance hybrid because energy is required to separate + and – charges.

7.25   (a) $\Delta EN(P=O) = 1.4$ and $\Delta EN(P=S) = 0.4$. Both bonds are polar covalent, but the phosphorus-oxygen bond is more polar.
(b) For the reaction between the organophosphate insecticide and the enzyme to occur, the phosphorus atom must bear a positive charge. Greater positive charge leads to increased rate of reaction and increased toxicity of the insecticide. Since oxygen is more electronegative than sulfur it has greater ability to pull shared electrons toward itself. Therefore, the molecule with the P=O bond is more toxic due to the greater positive charge on phosphorus.

7.26   For the reaction between an organophosphate insecticide and the enzyme to occur, the phosphorus atom must bear a positive charge. Greater positive charge leads to increased rate of reaction and increased toxicity of the insecticide. The more electronegative the X group, the more positive the phosphorus.   (a) Cl   (b) CF_3

7.27   Phosphorus is in the third row of the periodic table and can utilize d orbitals to hold extra electrons, and therefore form more than four bonds.

7.28 $C_{11}H_{19}N_2PSO_3$

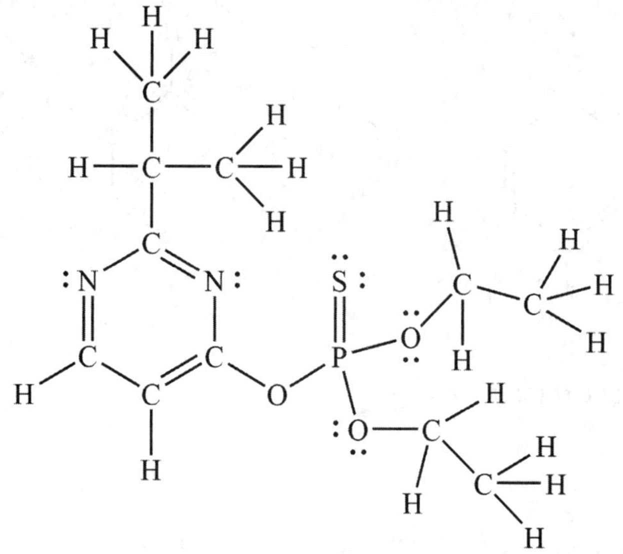

7.29

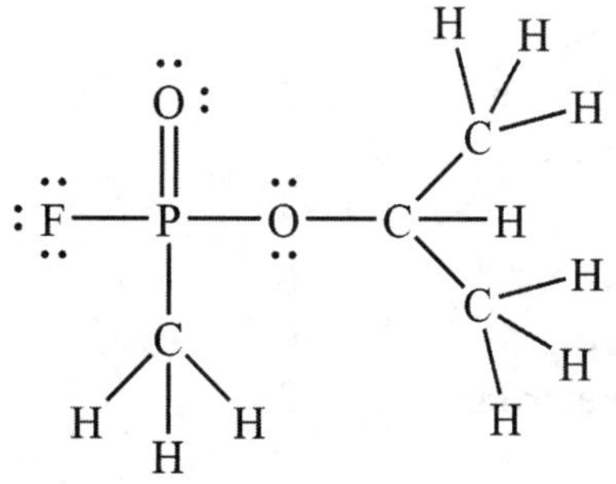

For phosphorus:	Isolated phosphorus valence electrons	5
	Bound phosphorus bonding electrons	10
	Bound phosphorus nonbonding electrons	0
	Formal charge = 5 – ½(10) – 0 = 0	

For top oxygen:	Isolated oxygen valence electrons	6
	Bound oxygen bonding electrons	4
	Bound oxygen nonbonding electrons	4
	Formal charge = 6 – ½(4) – 4 = 0	

For right oxygen:	Isolated oxygen valence electrons	6
	Bound oxygen bonding electrons	4
	Bound oxygen nonbonding electrons	4
	Formal charge = 6 – ½(4) – 4 = 0	

For fluorine:	Isolated fluorine valence electrons	7
	Bound fluorine bonding electrons	2
	Bound fluorine nonbonding electrons	6
	Formal charge = 7 – ½(2) – 6 = 0	

For carbon: Isolated carbon valence electrons   4
               Bound carbon bonding electrons   8
               Bound carbon nonbonding electrons   0
               Formal charge = 4 – ½(8) – 0 = 0

7.30 (a)

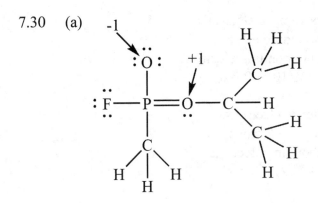

For fluorine: Isolated fluorine valence electrons   7
                 Bound fluorine bonding electrons   2
                 Bound fluorine nonbonding electrons   6
                 Formal charge = 7 – ½(2) – 6 = 0

For phosphorus: Isolated phosphorus valence electrons   5
                   Bound phosphorus bonding electrons   10
                   Bound phosphorus nonbonding electrons   0
                   Formal charge = 5 – ½(10) – 0 = 0

For top oxygen: Isolated oxygen valence electrons   6
                   Bound oxygen bonding electrons   2
                   Bound oxygen nonbonding electrons   6
                   Formal charge = 6 – ½(2) – 6 = –1

For right oxygen: Isolated oxygen valence electrons   6
                     Bound oxygen bonding electrons   6
                     Bound oxygen nonbonding electrons   2
                     Formal charge = 6 – ½(6) – 2 = +1

(b)

For fluorine: Isolated fluorine valence electrons 7
      Bound fluorine bonding electrons   4
      Bound fluorine nonbonding electrons 4
      Formal charge = $7 - \frac{1}{2}(4) - 4 = +1$

For phosphorus: Isolated phosphorus valence electrons 5
      Bound phosphorus bonding electrons 10
      Bound phosphorus nonbonding electrons 0
      Formal charge = $5 - \frac{1}{2}(10) - 0 = 0$

For top oxygen: Isolated oxygen valence electrons 6
      Bound oxygen bonding electrons   2
      Bound oxygen nonbonding electrons 6
      Formal charge = $6 - \frac{1}{2}(2) - 6 = -1$

For right oxygen: Isolated oxygen valence electrons 6
      Bound oxygen bonding electrons   4
      Bound oxygen nonbonding electrons 4
      Formal charge = $6 - \frac{1}{2}(4) - 4 = 0$

The problem 7.29 structure without formal charges makes the largest contribution to the resonance hybrid because energy is required to separate + and − charges.

Conceptual Problems

7.32 C–D is the stronger bond. A–B is the longer bond.

7.34 (a) fluoroethane (b) ethane (c) ethanol (d) acetaldehyde

7.36 (a) $C_8H_9NO_2$

7.38

Section Problems
Covalent Bonds (Sections 7.1)

7.40   (a) ionic   (b) nonpolar covalent   (c) covalent

Strengths of Covalent Bonds (Section 7.2)

7.42   C–F   450 kJ/mol $\Delta EN = EN(F) - EN(C) = 4.0 - 2.5 = 1.5$
        N–F   270 kJ/mol $\Delta EN = EN(F) - EN(N) = 4.0 - 3.0 = 1.0$
        O–F   180 kJ/mol $\Delta EN = EN(F) - EN(O) = 4.0 - 3.5 = 0.5$
        F–F   159 kJ/mol $\Delta EN = EN(F) - EN(F) = 4.0 - 4.0 = 0$
        In general, increased bond polarity leads to increased bond strength.

7.44   N–H $\Delta EN = EN(N) - EN(H) = 3.0 - 2.1 = 0.9$
        O–H $\Delta EN = EN(O) - EN(H) = 3.5 - 2.1 = 1.4$
        S–H $\Delta EN = EN(S) - EN(H) = 2.5 - 2.1 = 0.4$
        In general, increased bond polarity leads to increased bond strength. The most polar bond
        is the O–H bond and should be the strongest of the three.

Polar Covalent Bonds: Electronegativity (Section 7.3)

7.46   Electronegativity increases from left to right across a period and decreases down a group.

7.48   K < Li < Mg < Pb < C < Br

7.50   (a) HF   fluorine   EN = 4.0
                      hydrogen   EN = 2.1
                                      $\Delta EN = 1.9$ HF is polar covalent.

        (b) HI   iodine   EN = 2.5
                      hydrogen   EN = 2.1
                                      $\Delta EN = 0.4$ HI is polar covalent.

(c) $PdCl_2$ chlorine   EN = 3.0
                  palladium   EN = 2.2
                             ΔEN = 0.8 $PdCl_2$ is polar covalent.

(d) BBr_3 bromine   EN = 2.8
                  boron   EN = 2.0
                             ΔEN = 0.8 BBr_3 is polar covalent.

(e) NaOH $Na^+ - OH^-$ is ionic
      OH^- oxygen   EN = 3.5
                  hydrogen   EN = 2.1
                             ΔEN = 1.4 OH^- is polar covalent.

(f) CH_3Li lithium   EN = 1.0
                  carbon   EN = 2.5
                             ΔEN = 1.5 CH_3Li is polar covalent.

7.52 (a) $\overset{\delta-}{C} - \overset{\delta+}{H}$ $\overset{\delta+}{C} - \overset{\delta-}{Cl}$ (b) $\overset{\delta-}{Si} - \overset{\delta+}{Li}$ $\overset{\delta+}{Si} - \overset{\delta-}{Cl}$

     (c) $N - Cl$ $\overset{\delta-}{N} - \overset{\delta+}{Mg}$

7.54 (a) MgO, $BaCl_2$ (b) P_4 (c) $CdBr_2$, BrF_3, NF_3, $POCl_3$, $LiBr$

7.56 (a) CCl_4 chlorine   EN = 3.0
                  carbon   EN = 2.5
                            ΔEN = 0.5

     (b) $BaCl_2$ chlorine   EN = 3.0
                  barium   EN = 0.9
                            ΔEN = 2.1

     (c) $TiCl_3$ chlorine   EN = 3.0
                  titanium   EN = 1.5
                            ΔEN = 1.5

     (d) Cl_2O oxygen   EN = 3.5
                  chlorine   EN = 3.0
                            ΔEN = 0.5

Increasing ionic character: $CCl_4 \sim ClO_2 < TiCl_3 < BaCl_2$

7.58 (a) $MgBr_2$ (b) PBr_3

7.60 $\mu = Q \times r = (1.60 \times 10^{-19} \text{ C})(213.9 \times 10^{-12} \text{ m})\left(\dfrac{1 \text{ D}}{3.336 \times 10^{-30} \text{ C} \cdot \text{m}}\right) = 10.26 \text{ D}$

% ionic character for BrCl = $\dfrac{0.518 \text{ D}}{10.26 \text{ D}} \times 100\% = 5.05\%$

A Comparison of Ionic and Covalent Compounds (Section 7.4)

7.62   (b) and (c) are ionic compounds; (a) is a covalent compound and is most likely a gas at room temperature.

Electron-Dot Structures and Resonance (Sections 7.5–7.7)

7.64   The octet rule states that main-group elements tend to react so that they attain a noble gas electron configuration with filled s and p sublevels (8 electrons) in their valence electron shells. The transition metals are characterized by partially filled d orbitals that can be used to expand their valence shell beyond the normal octet of electrons.

7.66   (a)
:Br:
:Br—C—Br:
:Br:

(b) :Cl—N—Cl:
:Cl:

(c)
H H
| |
H—C—C—Cl:
| |
H H

(d)
$\left[\begin{array}{c} :F: \\ | \\ :F—B—F: \\ | \\ :F: \end{array}\right]^{-}$

(e) $\left[:\ddot{O}—\ddot{O}:\right]^{2-}$

(f) $\left[:N≡O:\right]^{+}$

7.68   (c) is the correct electron-dot for XeF_5^+ because it accounts for the required number of bonding (10) and lone pair (32) electrons.

7.70
:O   O:
‖   ‖
H—O—C—C—O—H

7.72   (a) The anion has 32 valence electrons. Each Cl has seven valence electrons (28 total). The minus one charge on the anion accounts for one valence electron. This leaves three valence electrons for X. X is Al.

(b) The cation has eight valence electrons. Each H has one valence electron (4 total). X is left with four valence electrons. Since this is a cation, one valence electron was removed from X. X has five valence electrons. X is P.

7.74   (a)
O:
‖   H
|
:Cl—C—O—C—H
|
H

(b)
H
|
H—C—C≡C—H
|
H

7.76 $H-\ddot{N}=\ddot{N}-H$

N$_2$H$_2$ has the stronger N–N bond because of the higher bond order.

7.78 (a)

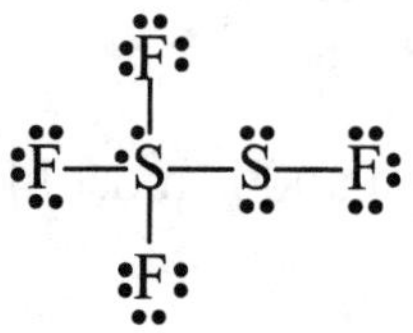

(b)

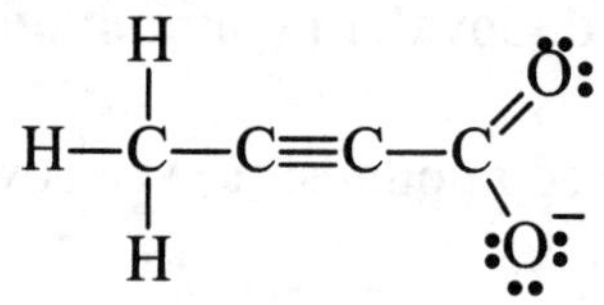

Electron-Dot Structures for Molecules with Second-Row Elements (Section 7.8)

7.80

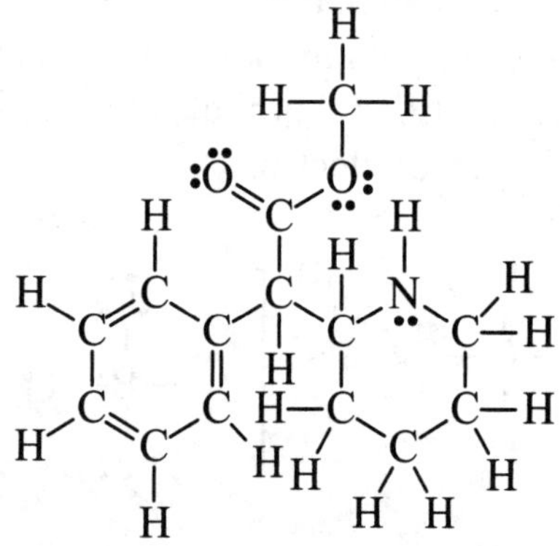

7.82

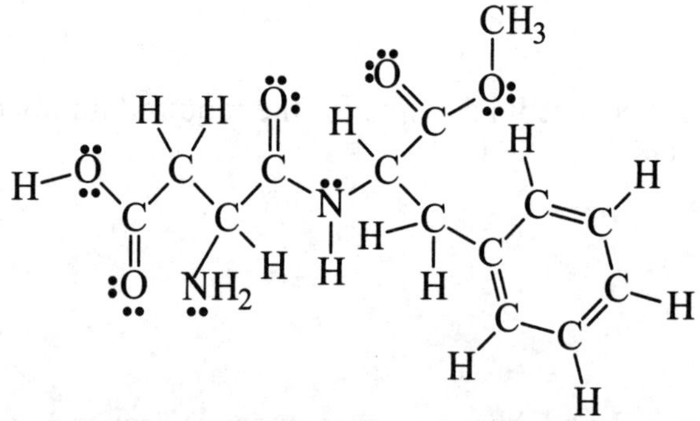

Resonance (Section 7.9)

7.84 (a)

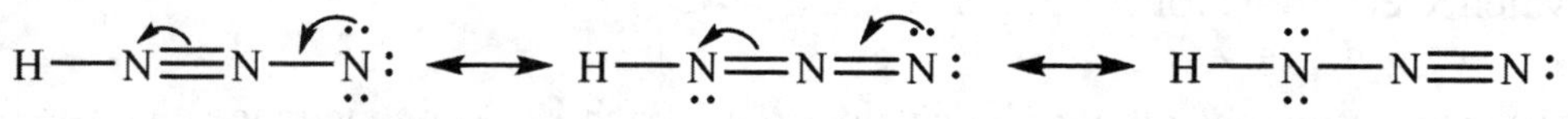

(b)

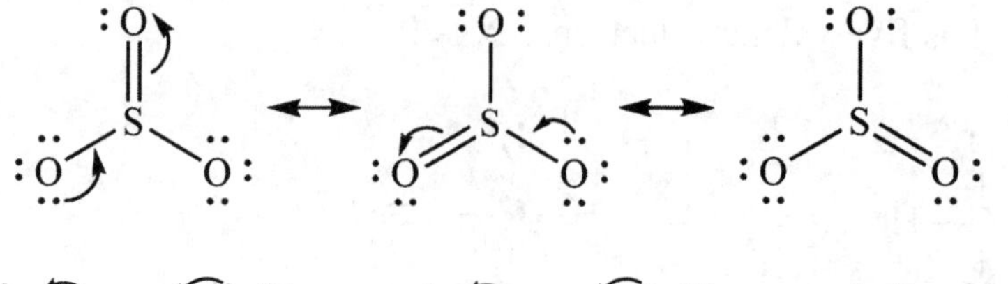

(c)

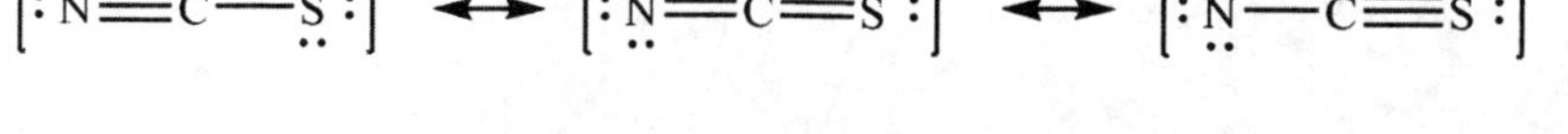

7.86   (a) yes   (b) no   (c) yes   (d) yes

7.88

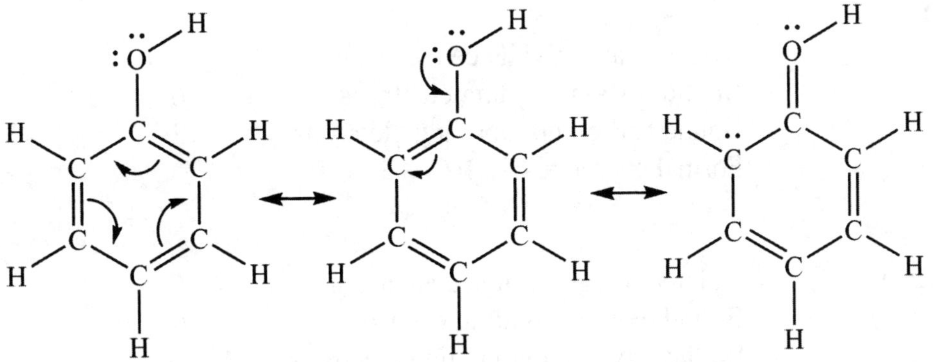

7.90 (a)

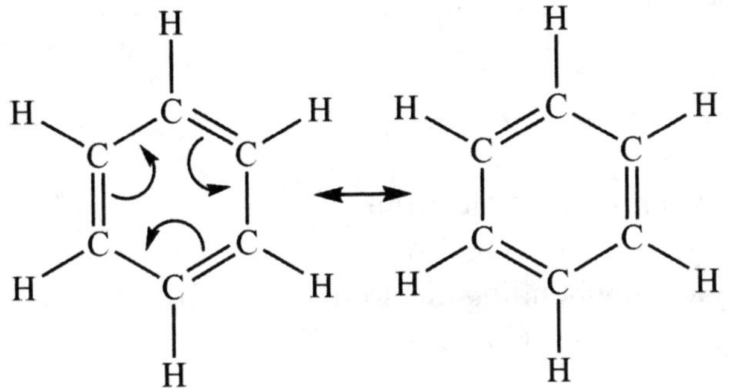

(b) ii)

7.92   (a)

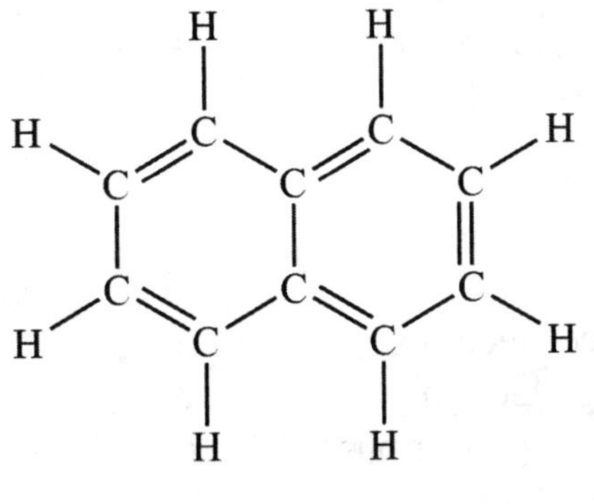

(b)

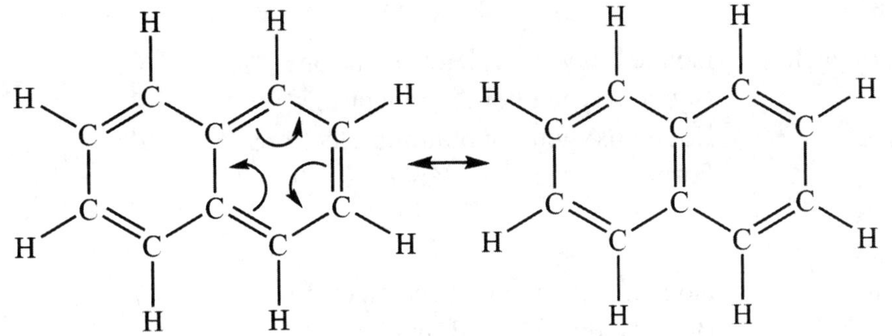

Formal Charges (Section 7.10)

7.94   :C≡O:

For carbon:	Isolated carbon valence electrons	4
	Bound carbon bonding electrons	6
	Bound carbon nonbonding electrons	2
	Formal charge = $4 - \frac{1}{2}(6) - 2 = -1$	

For oxygen:	Isolated oxygen valence electrons	6
	Bound oxygen bonding electrons	6
	Bound oxygen nonbonding electrons	2
	Formal charge = $6 - \frac{1}{2}(6) - 2 = +1$	

7.96 $\left[:\ddot{O}-\ddot{C}l-\ddot{O}: \right]^{-}$

For both oxygens:	Isolated oxygen valence electrons	6
	Bound oxygen bonding electrons	2
	Bound oxygen nonbonding electrons	6
	Formal charge = $6 - \frac{1}{2}(2) - 6 = -1$	

For chlorine:	Isolated chlorine valence electrons	7
	Bound chlorine bonding electrons	4
	Bound chlorine nonbonding electrons	4
	Formal charge = $7 - \frac{1}{2}(4) - 4 = +1$	

$\left[:\ddot{O}-\ddot{C}l=\ddot{O} \right]^{-}$

For left oxygen:	Isolated oxygen valence electrons	6
	Bound oxygen bonding electrons	2
	Bound oxygen nonbonding electrons	6
	Formal charge = $6 - \frac{1}{2}(2) - 6 = -1$	

For right oxygen:	Isolated oxygen valence electrons	6
	Bound oxygen bonding electrons	4
	Bound oxygen nonbonding electrons	4
	Formal charge = $6 - \frac{1}{2}(4) - 4 = 0$	

For chlorine:	Isolated chlorine valence electrons	7
	Bound chlorine bonding electrons	6
	Bound chlorine nonbonding electrons	4
	Formal charge = $7 - \frac{1}{2}(6) - 4 = 0$	

7.98   (a)   H
              \
               C=N=N̈:
              /
            H

For hydrogen:	Isolated hydrogen valence electrons	1
	Bound hydrogen bonding electrons	2
	Bound hydrogen nonbonding electrons	0
	Formal charge = $1 - \frac{1}{2}(2) - 0 = 0$

For nitrogen: (central)	Isolated nitrogen valence electrons	5
	Bound nitrogen bonding electrons	8
	Bound nitrogen nonbonding electrons	0
	Formal charge = $5 - \frac{1}{2}(8) - 0 = +1$

For nitrogen: (terminal)	Isolated nitrogen valence electrons	5
	Bound nitrogen bonding electrons	4
	Bound nitrogen nonbonding electrons	4
	Formal charge = $5 - \frac{1}{2}(4) - 4 = -1$

For carbon:	Isolated carbon valence electrons	4
	Bound carbon bonding electrons	8
	Bound carbon nonbonding electrons	0
	Formal charge = $4 - \frac{1}{2}(8) - 0 = 0$

(b)   H
      \
       C—N̈=N̈:
      /
    H

For hydrogen:	Isolated hydrogen valence electrons	1
	Bound hydrogen bonding electrons	2
	Bound hydrogen nonbonding electrons	0
	Formal charge = $1 - \frac{1}{2}(2) - 0 = 0$

For nitrogen: (central)	Isolated nitrogen valence electrons	5
	Bound nitrogen bonding electrons	6
	Bound nitrogen nonbonding electrons	2
	Formal charge = $5 - \frac{1}{2}(6) - 2 = 0$

For nitrogen: (terminal)	Isolated nitrogen valence electrons	5
	Bound nitrogen bonding electrons	4
	Bound nitrogen nonbonding electrons	4
	Formal charge = $5 - \frac{1}{2}(4) - 4 = -1$

For carbon:	Isolated carbon valence electrons	4
	Bound carbon bonding electrons	6
	Bound carbon nonbonding electrons	0
	Formal charge = $4 - \frac{1}{2}(6) - 0 = +1$

Structure (a) is more important because of the octet of electrons around carbon.

7.100

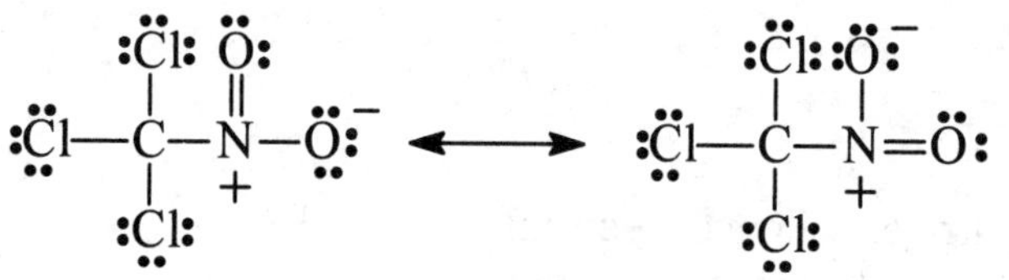

For chlorine:	Isolated chlorine valence electrons	7
	Bound chlorine bonding electrons	2
	Bound chlorine nonbonding electrons	6
	Formal charge = $7 - \frac{1}{2}(2) - 6 = 0$	

For carbon:	Isolated carbon valence electrons	4
	Bound carbon bonding electrons	8
	Bound carbon nonbonding electrons	0
	Formal charge = $4 - \frac{1}{2}(8) - 0 = 0$	

For nitrogen:	Isolated nitrogen valence electrons	5
	Bound nitrogen bonding electrons	8
	Bound nitrogen nonbonding electrons	0
	Formal charge = $5 - \frac{1}{2}(8) - 0 = +1$	

For oxygen: (double bonded)	Isolated oxygen valence electrons	6
	Bound oxygen bonding electrons	4
	Bound oxygen nonbonding electrons	4
	Formal charge = $6 - \frac{1}{2}(4) - 4 = 0$	

For oxygen: (single bonded)	Isolated oxygen valence electrons	6
	Bound oxygen bonding electrons	2
	Bound oxygen nonbonding electrons	6
	Formal charge = $6 - \frac{1}{2}(2) - 6 = -1$	

7.102

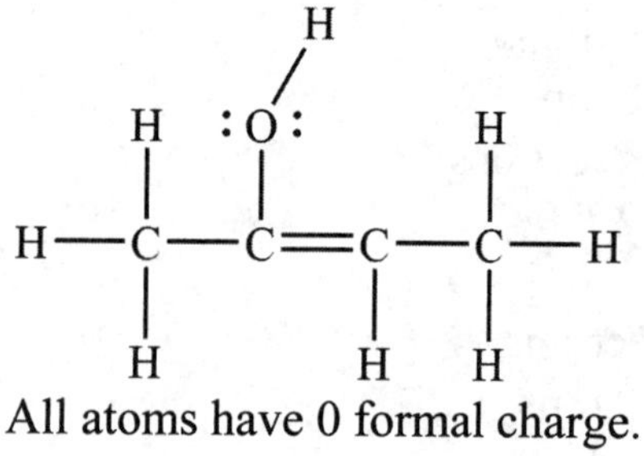

All atoms have 0 formal charge.

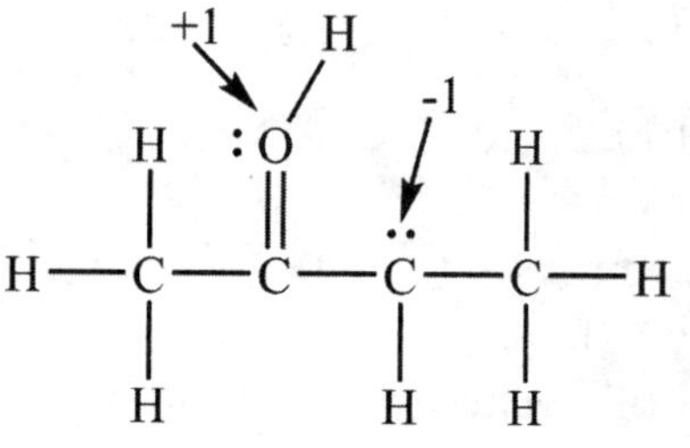

For oxygen: Isolated oxygen valence electrons   6
                Bound oxygen bonding electrons   6
                Bound oxygen nonbonding electrons   2
                Formal charge = $6 - \frac{1}{2}(6) - 2 = +1$

For carbon: Isolated carbon valence electrons   4
                Bound carbon bonding electrons   6
                Bound carbon nonbonding electrons   2
                Formal charge = $4 - \frac{1}{2}(6) - 2 = -1$

All other atoms have 0 formal charge.
The original structure is the larger contributor.

7.104 (a) reactants

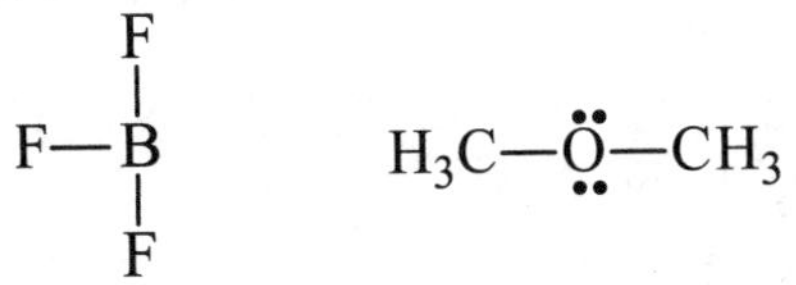

$$F-B \begin{matrix} F \\ | \\ | \\ F \end{matrix} \qquad H_3C-\overset{..}{\underset{..}{O}}-CH_3$$

For boron: Isolated boron valence electrons   3
                Bound boron bonding electrons   6
                Bound boron nonbonding electrons   0
                Formal charge = $3 - \frac{1}{2}(6) - 0 = 0$

For oxygen: Isolated oxygen valence electrons   6
                Bound oxygen bonding electrons   4
                Bound oxygen nonbonding electrons   4
                Formal charge = $6 - \frac{1}{2}(4) - 4 = 0$

product

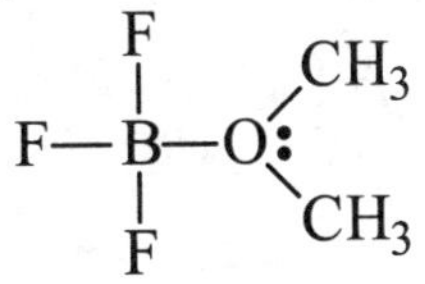

$$F-B \begin{matrix} F \\ | \\ | \\ F \end{matrix} -O\overset{CH_3}{\underset{CH_3}{:}}$$

For boron: Isolated boron valence electrons   3
                Bound boron bonding electrons   8
                Bound boron nonbonding electrons   0
                Formal charge = $3 - \frac{1}{2}(8) - 0 = -1$

For oxygen: Isolated oxygen valence electrons   6
                Bound oxygen bonding electrons   6
                Bound oxygen nonbonding electrons   2
                Formal charge = $6 - \frac{1}{2}(6) - 2 = +1$

7.106

$$H-\overset{\overset{\displaystyle H}{|}}{\underset{\underset{\displaystyle H}{|}}{C}}-\ddot{N}=C=\ddot{O}: \quad \longleftrightarrow \quad H-\overset{\overset{\displaystyle H}{|}}{\underset{\underset{\displaystyle H}{|}}{C}}-\overset{+}{N}\equiv C-\overset{\displaystyle -}{\underset{\displaystyle \cdot\cdot}{\ddot{O}}}:$$

For $\quad H-\overset{\overset{\displaystyle H}{|}}{\underset{\underset{\displaystyle H}{|}}{C}}-\ddot{N}=C=\ddot{O}:$

For hydrogen:	Isolated hydrogen valence electrons	1
	Bound hydrogen bonding electrons	2
	Bound hydrogen nonbonding electrons	0
	Formal charge = $1 - \frac{1}{2}(2) - 0 = 0$	

For carbon: (left)	Isolated carbon valence electrons	4
	Bound carbon bonding electrons	8
	Bound carbon nonbonding electrons	0
	Formal charge = $4 - \frac{1}{2}(8) - 0 = 0$	

For nitrogen:	Isolated nitrogen valence electrons	5
	Bound nitrogen bonding electrons	6
	Bound nitrogen nonbonding electrons	2
	Formal charge = $5 - \frac{1}{2}(6) - 2 = 0$	

For carbon: (right)	Isolated carbon valence electrons	4
	Bound carbon bonding electrons	8
	Bound carbon nonbonding electrons	0
	Formal charge = $4 - \frac{1}{2}(8) - 0 = 0$	

For oxygen:	Isolated oxygen valence electrons	6
	Bound oxygen bonding electrons	4
	Bound oxygen nonbonding electrons	4
	Formal charge = $6 - \frac{1}{2}(4) - 4 = 0$	

For $\quad H-\overset{\overset{\displaystyle H}{|}}{\underset{\underset{\displaystyle H}{|}}{C}}-\overset{+}{N}\equiv C-\overset{\displaystyle -}{\underset{\displaystyle \cdot\cdot}{\ddot{O}}}:$

For hydrogen:	Isolated hydrogen valence electrons	1
	Bound hydrogen bonding electrons	2
	Bound hydrogen nonbonding electrons	0
	Formal charge = $1 - \frac{1}{2}(2) - 0 = 0$	

For carbon: (left)	Isolated carbon valence electrons	4
	Bound carbon bonding electrons	8
	Bound carbon nonbonding electrons	0
	Formal charge = $4 - \frac{1}{2}(8) - 0 = 0$	

For nitrogen:
Isolated nitrogen valence electrons   5
Bound nitrogen bonding electrons   8
Bound nitrogen nonbonding electrons   0
Formal charge = $5 - \frac{1}{2}(8) - 0 = +1$

For carbon:
(right)
Isolated carbon valence electrons   4
Bound carbon bonding electrons   8
Bound carbon nonbonding electrons   0
Formal charge = $4 - \frac{1}{2}(8) - 0 = 0$

For oxygen:
Isolated oxygen valence electrons   6
Bound oxygen bonding electrons   2
Bound oxygen nonbonding electrons   6
Formal charge = $6 - \frac{1}{2}(2) - 6 = -1$

Multiconcept Problems

7.108 (a)   (b)   (c)

(d)   (e)   (f)

(g)   (h)

Structures (a) – (d) make more important contributions to the resonance hybrid because of only –1 and 0 formal charges on the oxygens. A +1 formal charge is unlikely.

7.110 (a) Assume a 100.0 g sample. From the percent composition data, a 100.0 g sample contains 47.5 g S and 52.5 g Cl.

$$69.6 \text{ g S} \times \frac{1 \text{ mol S}}{32.065 \text{ g S}} = 2.17 \text{ mol S}; \quad 30.4 \text{ g N} \times \frac{1 \text{ mol N}}{14.007 \text{ g N}} = 2.17 \text{ mol N}$$

Because the two mole quantities are the same, the empirical formula is SN.
The empirical formula weight = 47.07

$$\text{Multiple} = \frac{\text{molecular weight}}{\text{empirical formula weight}} = \frac{184.3}{47.07} = 4.00$$

The molecular formula is S_4N_4.

(b)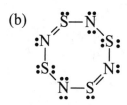

For sulfur: (1 lone pair)	Isolated sulfur valence electrons	6
	Bound sulfur bonding electrons	6
	Bound sulfur nonbonding electrons	2
	Formal charge = 6 – ½(6) – 2 = +1	

For sulfur: (2 lone pairs)	Isolated sulfur valence electrons	6
	Bound sulfur bonding electrons	4
	Bound sulfur nonbonding electrons	4
	Formal charge = 6 – ½(4) – 4 = 0	

For nitrogen: (1 lone pair)	Isolated nitrogen valence electrons	5
	Bound nitrogen bonding electrons	6
	Bound nitrogen nonbonding electrons	2
	Formal charge = 5 – ½(6) – 2 = 0	

For nitrogen: (2 lone pairs)	Isolated nitrogen valence electrons	5
	Bound nitrogen bonding electrons	4
	Bound nitrogen nonbonding electrons	4
	Formal charge = 5 – ½(4) – 4 = –1	

For sulfur: (1 lone pair)	Isolated sulfur valence electrons	6
	Bound sulfur bonding electrons	6
	Bound sulfur nonbonding electrons	2
	Formal charge = 6 – ½(6) – 2 = +1	

For sulfur: (2 lone pairs)	Isolated sulfur valence electrons	6
	Bound sulfur bonding electrons	6
	Bound sulfur nonbonding electrons	4
	Formal charge = 6 – ½(6) – 4 = –1	

For nitrogen:	Isolated nitrogen valence electrons	5
	Bound nitrogen bonding electrons	6
	Bound nitrogen nonbonding electrons	2
	Formal charge = 5 – ½(6) – 2 = 0	

(c) Multiple $= \dfrac{\text{molecular weight}}{\text{empirical formula weight}} = \dfrac{92.2}{47.07} = 1.96 = 2$

The molecular formula is S_2N_2. Two possible structures are shown here.

$$\ddot{N}=\ddot{S}$$
$$\mid \qquad \mid$$
$$:\ddot{S}-\ddot{N}:$$

$$\ddot{N}-\ddot{N}:$$
$$\parallel \qquad \mid$$
$$\ddot{S}-\ddot{S}:$$

7.112 (a) (4 orbitals)(3 electrons) = 12 outer-shell electrons
(b) 3 electrons
(c) $1s^3\, 2s^3\, 2p^6$; $:\!\ddot{\ddot{X}}\!:$
(d) 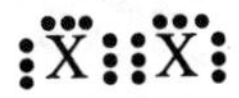 $:\!\overset{\bullet\bullet\bullet}{X}\!:\!:\!\overset{\bullet\bullet\bullet}{X}\!:$

8

Covalent Compounds: Bonding Theories and Molecular Structure

8.1 ICl_4^-, 6 charge clouds, octahedral charge cloud arrangement, 4 bonding pairs, 2 lone pairs, square planar molecular geometry

8.2 (a) 4 charge clouds, tetrahedral charge cloud arrangement, tetrahedral molecular geometry
(b) 5 charge clouds, trigonal bipyramidal charge cloud arrangement, seesaw molecular geometry.

8.3

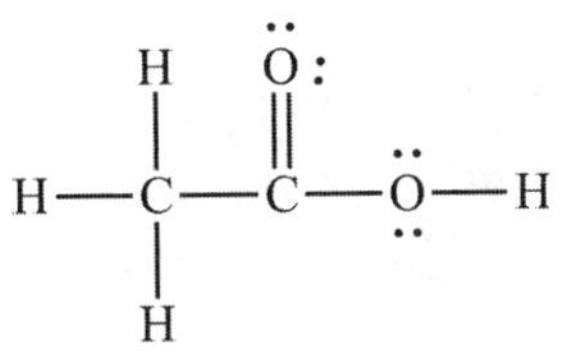

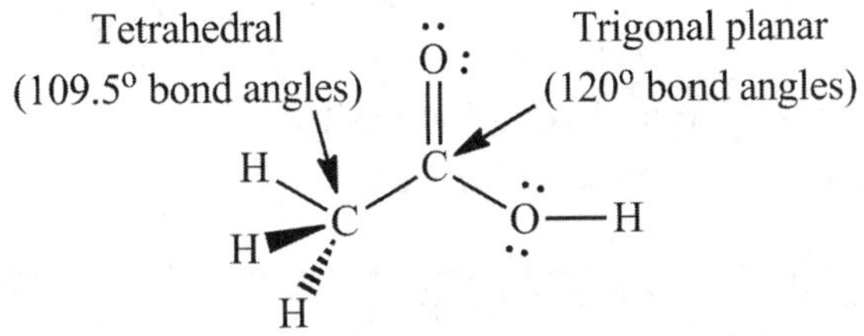

8.4

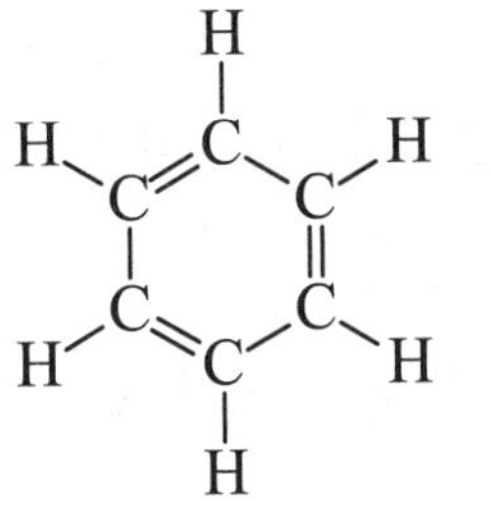

The bond angle around every carbon is 120° (trigonal planar). Benzene is a planar hexagon.

8.5 CH_2Cl_2; The C is sp^3 hybridized. The C–H bonds are formed by the overlap of one singly occupied sp^3 orbital on C with a singly occupied H 1s orbital. The C–Cl bonds are formed by the overlap of one singly occupied sp^3 orbital on C with a singly occupied Cl 2p orbital.

8.6

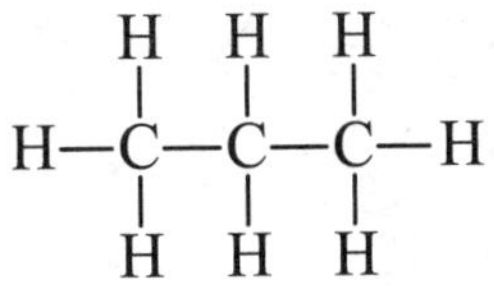

Each C is sp^3 hybridized. The C–C bonds are formed by the overlap of one singly occupied sp^3 hybrid orbital from each C. The C–H bonds are formed by the overlap of one singly occupied sp^3 orbital on C with a singly occupied H 1s orbital.

8.7

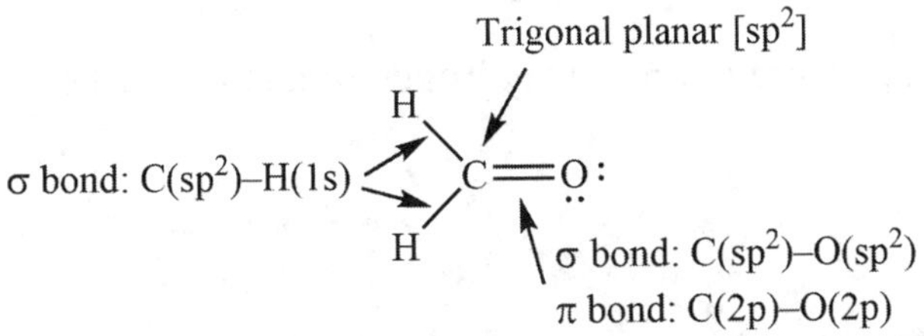

8.8

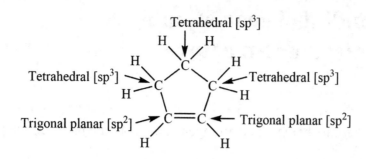

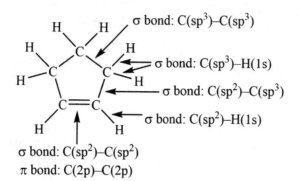

8.9   Both σ_1 and σ_2 $O(sp^2)–C(sp)$
π_1 $O(2p_x)–C(2p_x)$
π_2 $O(2p_y)–C(2p_y)$

8.10

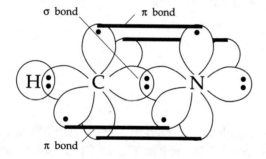

In HCN the carbon is sp hybridized.

8.11   Only (c) BrF_3 has dipole moment.

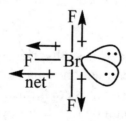

8.12   (a) CF_4 (b) CH_2F_2 (c) CHF_3 (d) CH_3F

8.13   (a) is the incorrect depiction of hydrogen bonding. A hydrogen bonded to carbon cannot hydrogen bond.

8.14 Hydrogen bonds are shown as dashed lines.

Because of the three hydrogen bonds, DNA regions that are high in G–C pairs would have the higher melting point.

8.15 Cl_2 dispersion
CCl_4 dispersion
CH_3F dipole-dipole, dispersion
HF   hydrogen bonding, dipole-dipole, dispersion
(a) is false; (b), (c) & (d) are true

8.16 H_2S dipole-dipole, dispersion
CH_3OH hydrogen bonding, dipole-dipole, dispersion
C_2H_6 dispersion
Ar   dispersion
$Ar < C_2H_6 < H_2S < CH_3OH$

8.17 For He_2^+ σ^*_{1s} $\uparrow$
               σ_{1s} $\uparrow\downarrow$

$$He_2^+ \text{ Bond order} = \frac{\left(\begin{array}{c}\text{number of}\\\text{bonding electrons}\end{array}\right) - \left(\begin{array}{c}\text{number of}\\\text{antibonding electrons}\end{array}\right)}{2} = \frac{2-1}{2} = 1/2$$

He_2^+ should be stable with a bond order of 1/2.

8.18   The bond order in He_2^{2+} is 1, which is greater than the bond order of 1/2 in He_2^+; therefore He_2^{2+} is predicted to have a stronger bond and be a more stable species

8.19   For B_2

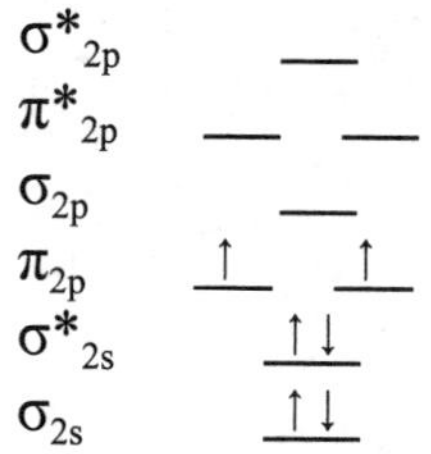

σ^*_{2p} ⎯

π^*_{2p} ⎯ ⎯

σ_{2p} ⎯

π_{2p} ↑   ↑

σ^*_{2s} ↑↓

σ_{2s} ↑↓

B_2 Bond order $= \dfrac{\left(\begin{array}{c}\text{number of}\\\text{bonding electrons}\end{array}\right) - \left(\begin{array}{c}\text{number of}\\\text{antibonding electrons}\end{array}\right)}{2} = \dfrac{4-2}{2} = 1$

B_2 is paramagnetic because it has two unpaired electrons in the π_{2p} molecular orbitals.

8.20   The bond orders are: $O_2^{2-} = 1$, $O_2^- = 1.5$, $O_2 = 2$, $O_2^+ = 2.5$, $O_2^{2+} = 3$.
The order from weakest to strongest bond is: $O_2^{2-} < O_2^- < O_2 < O_2^+ < O_2^{2+}$.
The order from shortest to longest bond is: $O_2^{2+} < O_2^+ < O_2 < O_2^- < O_2^{2-}$.

8.21

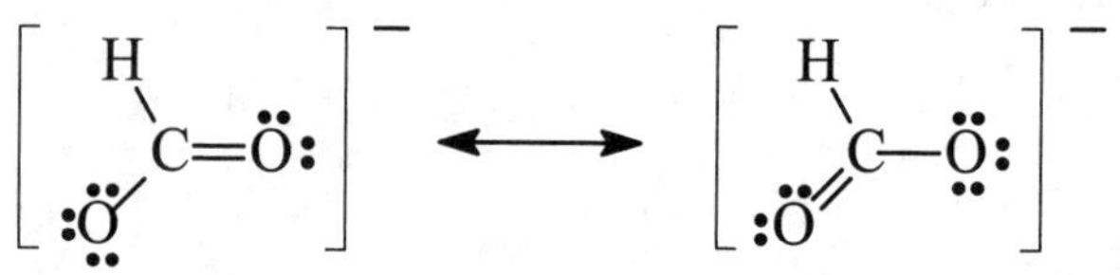

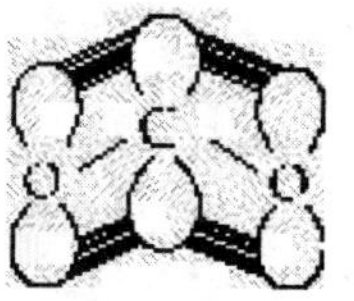

8.22

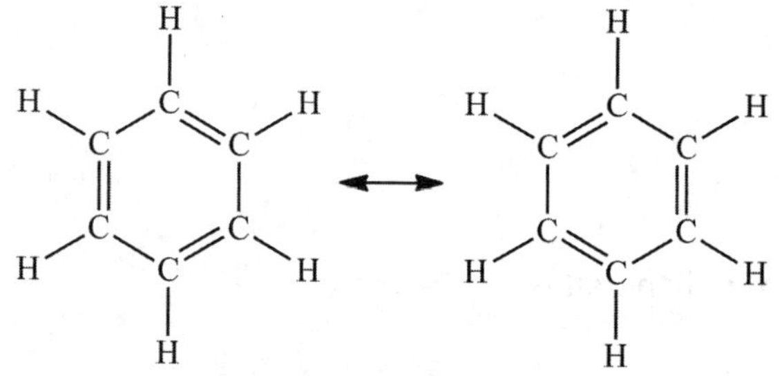

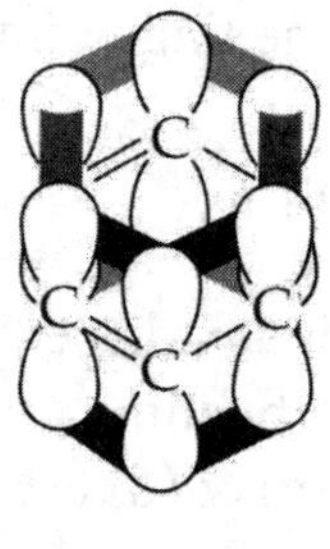

8.23   (a) $C_a \sim 109°$; $C_b \sim 109°$; $C_c \sim 120°$
(b) $C_a(sp^3)$; $C_b(sp^3)$; $C_c(sp^2)$
(c) σ bond: $C(sp^2)$–$C(sp^2)$; π bond: $C(2p)$–$C(2p)$

8.24   Only interactions (1) and (2) are hydrogen bonds. For interaction (3), a H bonded to a C cannot hydrogen bond. For interaction (4), a H cannot hydrogen bond with another H.

8.25

(a) 8 lone pairs, 4 pi bonds   (b) dispersion and dipole-dipole
(c) solid, large molecules with dipole-dipole and dispersion forces are usually solid.

8.26   The carbons in Ring 1 are all sp^2 hybridized and has a π molecular orbital that delocalizes electron density. The atoms in Ring 2 are all sp^3 hybridized with no delocalization.

Conceptual Problems

8.28   (a) trigonal bipyramidal   (b) tetrahedral
(c) square pyramidal (4 ligands in the horizontal plane, including one hidden)

8.30   (a) sp^2 (b) sp   (c) sp^3

8.32   (a) $C_8H_9NO_2$
(b), (c), and (d)

8.34   The electronegative O atoms are electron rich (red), while the rest of the molecule is electron poor (blue).

8.36   The N atom is electron rich (red) because of its high electronegativity. The C and H atoms are electron poor (blue) because they are less electronegative.

Section Problems
The VSEPR Model (Section 8.1)

8.38   From data in Table 8.1:
(a) trigonal planar   (b) trigonal bipyramidal   (c) linear   (d) octahedral

8.40   From data in Table 8.1:

    (a) tetrahedral, 4   (b) octahedral, 6   (c) bent, 3 or 4

    (d) linear, 2 or 5   (e) square pyramidal, 6   (f) trigonal pyramidal, 4

8.42

	Number of Bonded Atoms	Number of Lone Pairs	Shape
(a) H_2Se	2	2	bent
(b) $TiCl_4$	4	0	tetrahedral
(c) O_3	2	1	bent
(d) GaH_3	3	0	trigonal planar

8.44

	Number of Bonded Atoms	Number of Lone Pairs	Shape
(a) SbF_5	5	0	trigonal bipyramidal
(b) IF_4^+	4	1	see saw
(c) SeO_3^{2-}	3	1	trigonal pyramidal
(d) CrO_4^{2-}	4	0	tetrahedral

8.46

	Number of Bonded Atoms	Number of Lone Pairs	Shape
(a) PO_4^{3-}	4	0	tetrahedral
(b) MnO_4^-	4	0	tetrahedral
(c) SO_4^{2-}	4	0	tetrahedral
(d) SO_3^{2-}	3	1	trigonal pyramidal
(e) ClO_4^-	4	0	tetrahedral
(f) SCN^-	2	0	linear

            (C is the central atom)

8.48   (a) In SF_2 the sulfur is bound to two fluorines and contains two lone pairs of electrons. SF_2 is bent and the F–S–F bond angle is approximately $109°$.

(b) In N_2H_2 each nitrogen is bound to the other nitrogen and one hydrogen. Each nitrogen has one lone pair of electrons. The H–N–N bond angle is approximately $120°$.

(c) In KrF_4 the krypton is bound to four fluorines and contains two lone pairs of electrons. KrF_4 is square planar, and the F–Kr–F bond angle is $90°$.

(d) In NOCl the nitrogen is bound to one oxygen and one chlorine and contains one lone pair of electrons. NOCl is bent, and the Cl–N–O bond angle is approximately $120°$.

8.50

$H - C_a - H$	$\sim 120°$
$H - C_a - C_b$	$\sim 120°$
$C_a - C_b - C_c$	$\sim 120°$
$C_b - C_c - N$	$180°$
$C_a - C_b - H$	$\sim 120°$
$H - C_b - C_c$	$\sim 120°$

8.52

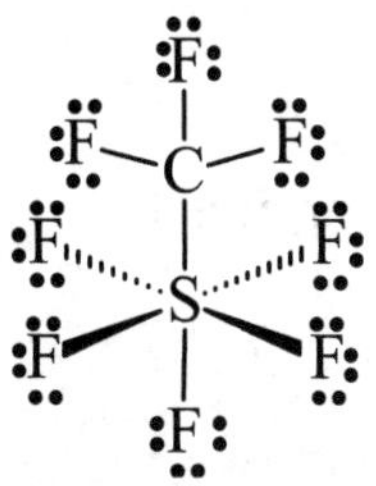

The bond angles are 109.5° around carbon and 90° around S.

8.54 All six carbons in cyclohexane are bonded to two other carbons and two hydrogens (i.e., four charge clouds). The geometry about each carbon is tetrahedral with a C–C–C bond angle of approximately 109°. Because the geometry about each carbon is tetrahedral, the cyclohexane ring cannot be flat.

8.56

	Number of Bonded Atoms	Number of Lone Pairs	Shape
(a) BF_3	3	0	trigonal planar (120°)
PF_3	3	1	trigonal pyramidal (~107°)

PF_3 has the smaller F–X–F angles.

(b) PCl_4^+	4	0	tetrahedral (109.5°)
ICl_2^-	2	3	linear (180°)

PCl_4^+ has the smaller Cl–X–Cl angles.

(c) CCl_3^-	3	1	trigonal pyramidal (~107°)
PCl_6^-	6	0	octahedral (90°)

PCl_6^- has the smaller Cl–X–Cl angles.

Valence Bond Theory and Hybridization (Sections 8.2–8.4)

8.58 In a π bond, the shared electrons occupy a region above and below a line connecting the two nuclei. A σ bond has its shared electrons located along the axis between the two nuclei.

8.60 See Table 8.2. (a) sp (b) sp^2 (c) sp^3

8.62 (a) sp^2 (b) sp^2 (c) sp^3 (d) sp^2

8.64

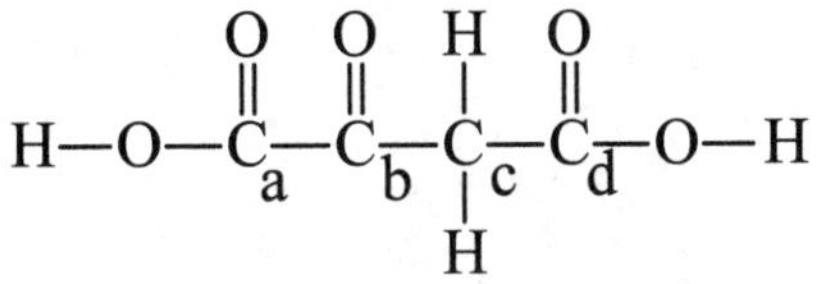

Carbons a, b, and d are sp^2 hybridized and carbon c is sp^3 hybridized.

The bond angles around carbons a, b, and d are ~120°. The bond angles around carbon c are ~109°. The terminal H–O–C bond angles are ~109°.

8.66

In Cl_2CO the carbon is sp2 hybridized.

8.68

(a) There are 34 σ bonds and 4 π bonds.
(b) and (c) Each C with four single bonds is sp³ hybridized with bond angles of 109.5°.
Each C with a double bond is sp² hybridized with bond angles of 120°.
(d) The nitrogen is sp³ hybridized.

8.70 Both the B and N are sp² hybridized. All bond angles are ~120°. The overall geometry of the molecule is planar.

8.72

21 σ bonds
5 π bonds
Each C with a double bond is sp² hybridized.
The –CH₃ carbon is sp³ hybridized.

8.74 (a)

(b) There are 6 σ bonds and 3 π bonds.
(c) see figure
(d) The shortest bond is the C–N triple bond.

8.76

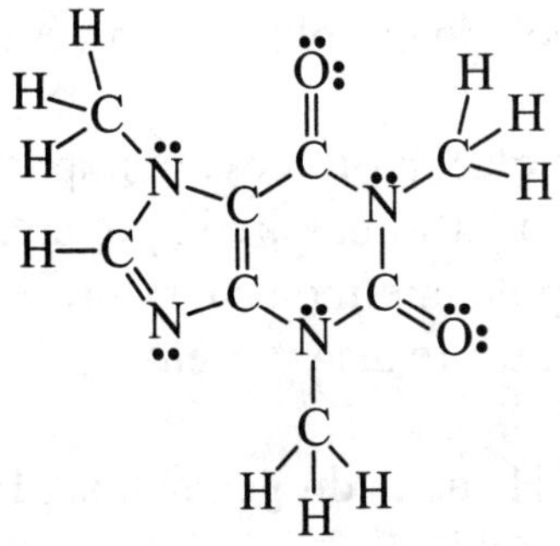

Each C with four single bonds is sp^3 hybridized.
Each C with a double bond is sp^2 hybridized.

Dipole Moments (Section 8.5)

8.78   (a)

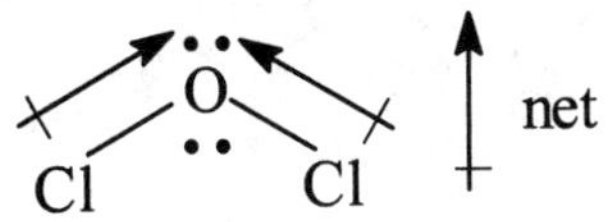

net

(b)

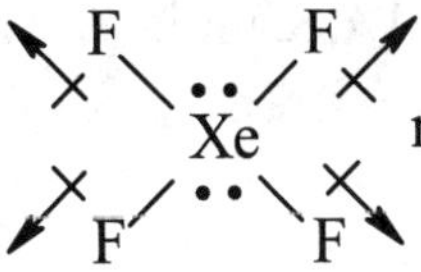

net dipole moment = 0

(c)

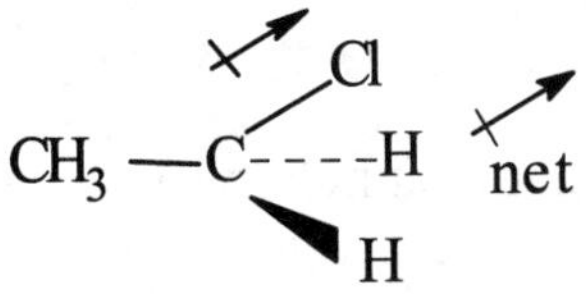

net

(d)

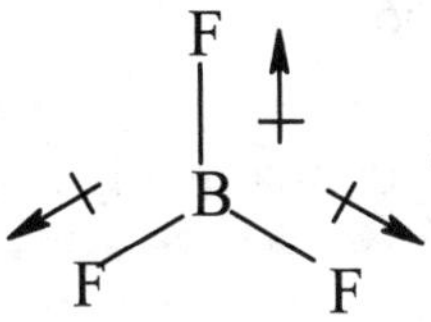

net dipole moment = 0

8.80

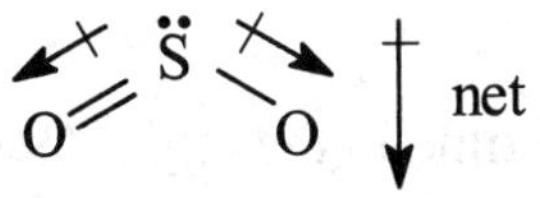

net

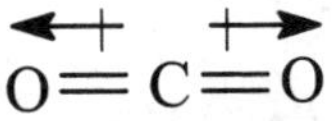

net dipole moment = 0

SO_2 is bent and the individual bond dipole moments add to give the molecule a net dipole moment.

CO_2 is linear and the individual bond dipole moments point in opposite directions to cancel each other out. CO_2 has no net dipole moment.

8.82   (a)

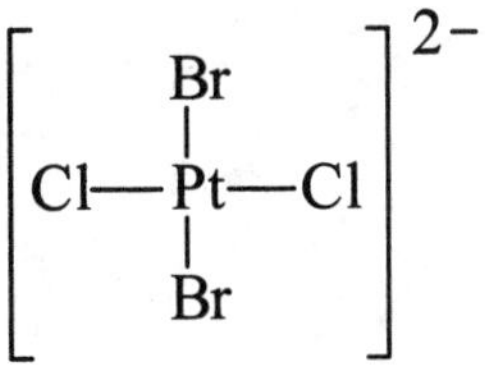

(b)

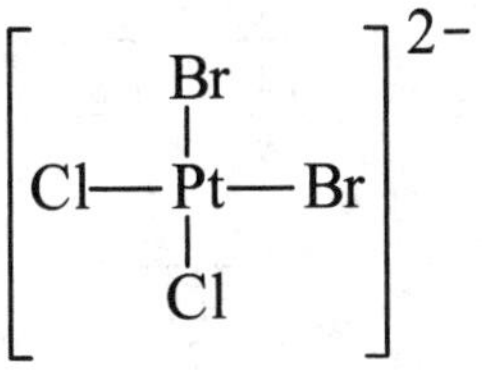

8.84   If a molecule has polar covalent bonds, the molecular shape (and location of lone pairs of electrons) determines whether the bond dipoles cancel and thus whether the molecule has a dipole moment.

Intermolecular Forces (Section 8.6)

8.86   Dipole-dipole forces arise between molecules that have permanent dipole moments.
London dispersion forces arise between molecules as a result of induced temporary dipoles.

8.88   (a) $CHCl_3$ has a permanent dipole moment. Dipole-dipole intermolecular forces are important. London dispersion forces are also present.
(b) O_2 has no dipole moment. London dispersion intermolecular forces are important.
(c) Polyethylene, C_nH_{2n+2}. London dispersion intermolecular forces are important.
(d) CH_3OH has a permanent dipole moment. Dipole-dipole intermolecular forces and hydrogen bonding are important. London dispersion forces are also present.

8.90   For CH_3OH and CH_4, dispersion forces are small. CH_3OH can hydrogen bond; CH_4 cannot. This accounts for the large difference in boiling points.
For 1-decanol and decane, dispersion forces are comparable and relatively large along the C–H chain. 1-decanol can hydrogen bond; decane cannot. This accounts for the 57 °C higher boiling point for 1-decanol.

8.92

8.94   Illustrations (ii) and (iii) depict the hydrogen bonding that occurs between methylamine and water.

Molecular Orbital Theory (Sections 8.7– 8.9)

8.96   Electrons in a bonding molecular orbital spend most of their time in the region between the two nuclei, helping to bond the atoms together. Electrons in an antibonding molecular orbital cannot occupy the central region between the nuclei and cannot contribute to bonding.

8.98

$$\text{Bond order} = \frac{\left(\begin{array}{c}\text{number of}\\\text{bonding electrons}\end{array}\right) - \left(\begin{array}{c}\text{number of}\\\text{antibonding electrons}\end{array}\right)}{2}$$

O_2^+ bond order $= \dfrac{8-3}{2} = 2.5$ O_2 bond order $= \dfrac{8-4}{2} = 2$

O_2^- bond order $= \dfrac{8-5}{2} = 1.5$

All are stable with bond orders between 1.5 and 2.5. All have unpaired electrons.

8.100

$$
\begin{array}{c|cc|cc}
\sigma^*_{2p} & \underline{\quad} & & \underline{\quad} & \\
\pi^*_{2p} & \underline{\quad} & \underline{\quad} & \underline{\quad} & \underline{\quad} \\
\sigma_{2p} & \underline{\quad} & & \underline{\uparrow} & \\
\pi_{2p} & \underline{\uparrow\downarrow} & \underline{\uparrow\downarrow} & \underline{\uparrow\downarrow} & \underline{\uparrow\downarrow} \\
\sigma^*_{2s} & \underline{\uparrow\downarrow} & & \underline{\uparrow\downarrow} & \\
\sigma_{2s} & \underline{\uparrow\downarrow} & & \underline{\uparrow\downarrow} & \\
& C_2 & & C_2^- &
\end{array}
$$

$$
\text{Bond order} = \frac{\left(\begin{array}{c}\text{number of} \\ \text{bonding electrons}\end{array}\right) - \left(\begin{array}{c}\text{number of} \\ \text{antibonding electrons}\end{array}\right)}{2}
$$

(a) C_2 bond order $= \dfrac{6-2}{2} = 2$

(b) Add one electron because it will go into a bonding molecular orbital.

(c) C_2^- bond order $= \dfrac{7-2}{2} = 2.5$

8.102

(a) C_2^{2-}
diamagnetic

(b) C_2^{2+}
paramagnetic

(c) F_2^-
paramagnetic

(d) Cl_2
diamagnetic

(e) Li_2^+
paramagnetic

8.104 Li_2

$$
\begin{array}{c|c}
\sigma^*_{2s} & \underline{\quad} \\
\sigma_{2s} & \underline{\uparrow\downarrow} \\
\sigma^*_{1s} & \underline{\uparrow\downarrow} \\
\sigma_{1s} & \underline{\uparrow\downarrow}
\end{array}
$$

Li_2 Bond order $= \dfrac{\left(\begin{array}{c}\text{number of} \\ \text{bonding electrons}\end{array}\right) - \left(\begin{array}{c}\text{number of} \\ \text{antibonding electrons}\end{array}\right)}{2} = \dfrac{4-2}{2} = 1$

The bond order for Li_2 is 1, and the molecule is likely to be stable.

8.106 (a)

$$
\begin{array}{c|cc|cc}
\sigma^*_{3p} & \underline{\quad} & & \underline{\quad} & \\
\pi^*_{3p} & \underline{\uparrow} & \underline{\uparrow} & \underline{\uparrow\downarrow} & \underline{\uparrow\downarrow} \\
\pi_{3p} & \underline{\uparrow\downarrow} & \underline{\uparrow\downarrow} & \underline{\uparrow\downarrow} & \underline{\uparrow\downarrow} \\
\sigma_{3p} & \underline{\uparrow\downarrow} & & \underline{\uparrow\downarrow} & \\
\sigma^*_{3s} & \underline{\uparrow\downarrow} & & \underline{\uparrow\downarrow} & \\
\sigma_{3s} & \underline{\uparrow\downarrow} & & \underline{\uparrow\downarrow} & \\
& S_2 & & S_2^{2-} &
\end{array}
$$

(b) S_2 would be paramagnetic with two unpaired electrons in the π^*_{3p} MOs.

(c)   Bond order $= \dfrac{\left(\begin{array}{c}\text{number of} \\ \text{bonding electrons}\end{array}\right) - \left(\begin{array}{c}\text{number of} \\ \text{antibonding electrons}\end{array}\right)}{2}$

S_2 bond order $= \dfrac{8 - 4}{2} = 2$

(d) S_2^{2-} bond order $= \dfrac{8 - 6}{2} = 1$

The two added electrons go into the antibonding π^*_{3p} MOs, the bond order drops from 2 to 1, and the bond length in S_2^{2-} should be longer than the bond length in S_2.

8.108   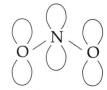   p orbitals in NO_2^-

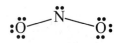 NO_2^- showing only the σ bonds (N is sp^2 hybridized)

   delocalized MO model for π bonding in NO_2^-

Multiconcept Problems

8.110   (a) H—C≡C—H   (b) H—N̈=N̈—H   (c)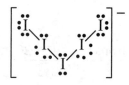

8.112

$$\left[\begin{array}{c} \ddot{\underset{\cdot\cdot}{I}} \diagdown \ddot{\underset{\cdot\cdot}{I}} \\ \ddot{\underset{\cdot\cdot}{I}} \diagdown \diagup \ddot{\underset{\cdot\cdot}{I}} \\ \ddot{\underset{\cdot\cdot}{I}} \end{array}\right]^{-}$$

8.114   (a)

H \ / Cl C=C H / \ Cl	Cl \ / Cl C=C H / \ H	Cl \ / H C=C H / \ Cl
polar	polar	nonpolar

(b) All three molecules are planar. The first two structures are polar because they both have an unsymmetrical distribution of atoms about the center of the molecule (the middle of the double bond), and bond polarities do not cancel. Structure 3 is nonpolar because the H's and Cl's, respectively, are symmetrically distributed about the center of the molecule, both being opposite each other. In this arrangement, bond polarities cancel.

(c) 200 nm = 200 x 10^{-9} m

$$E = \frac{hc}{\lambda} = \frac{(6.626 \text{ x } 10^{-34} \text{ J·s})(3.00 \text{ x } 10^{8} \text{ m/s})}{200 \text{ x } 10^{-9} \text{ m}} \ (6.022 \text{ x } 10^{23} \text{ /mol})$$

E = 5.99 x 10^{5} J/mol = **599 kJ/mol**

(d)

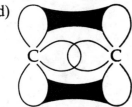

The π bond must be broken **before rotation can occur.**

9

Thermochemistry: Chemical Energy

9.1 $\Delta V = (4.3 \text{ L} - 8.6 \text{ L}) = -4.3 \text{ L}$

$w = -P\Delta V = -(44 \text{ atm})(-4.3 \text{ L}) = +189.2 \text{ L} \cdot \text{atm}$

$w = (189.2 \text{ L} \cdot \text{atm})(101 \dfrac{\text{J}}{\text{L} \cdot \text{atm}}) = +1.9 \times 10^4 \text{ J}$

The positive sign for the work indicates that the surroundings do work on the system. Work energy flows into the system.

9.2 $w = -P\Delta V = -(2.5 \text{ atm})(3 \text{ L} - 2 \text{ L}) = -2.5 \text{ L} \cdot \text{atm}$

$w = (-2.5 \text{ L} \cdot \text{atm}) \left(101 \dfrac{\text{J}}{\text{L} \cdot \text{atm}} \right) = -252.5 \text{ J} = -250 \text{ J} = -0.25 \text{ kJ}$

The negative sign indicates that the expanding system loses work energy and does work on the surroundings.

9.3 $P\Delta V = (1.00 \text{ atm})(-24.4 \text{ L}) = -24.4 \text{ L} \cdot \text{atm}$

$P\Delta V = (-24.4 \text{ L} \cdot \text{atm})(101 \dfrac{\text{J}}{\text{L} \cdot \text{atm}}) = -2464 \text{ J} = -2.46 \text{ kJ}$

$\Delta E = \Delta H - P\Delta V = -484 \text{ kJ} - (-2.46 \text{ kJ}) = -481.5 \text{ kJ} = -482 \text{ kJ}$

9.4   (a) $w = -P\Delta V$ is positive and $P\Delta V$ is negative for this reaction because the system volume is decreased at constant pressure.
(b) $P\Delta V$ is small compared to ΔE.
$\Delta H = \Delta E + P\Delta V$; ΔH is negative. Its value is slightly more negative than ΔE.

9.5 $2 \text{ H}_2\text{O(l)} \rightarrow 2 \text{ H}_2\text{(g)} + \text{O}_2\text{(g)}$ $\Delta H° = +571.6 \text{ kJ}$

$\text{heat} = 10.0 \text{ g H}_2\text{O} \times \dfrac{1 \text{ mol H}_2\text{O}}{18.02 \text{ g H}_2\text{O}} \times \dfrac{571.6 \text{ kJ}}{2 \text{ mol H}_2\text{O}} = 158.6 \text{ kJ} = 159 \text{ kJ}$

159 kJ are absorbed.

9.6 C_3H_8, 44.09

$1.8 \times 10^6 \text{ kJ} \times \dfrac{1 \text{ mol C}_3\text{H}_8}{2044 \text{ kJ}} \times \dfrac{44.09 \text{ g C}_3\text{H}_8}{1 \text{ mol C}_3\text{H}_8} \times \dfrac{1 \text{ kg}}{1000 \text{ g}} = 39 \text{ kg C}_3\text{H}_8$

9.7   The temperature of the water increased. This means that the reaction between HCl and NaOH produced heat. A reaction that produces heat is exothermic. $\Delta H°$ is negative.

9.8 LiF, 25.9; NH_4NO_3, 80.04

Only LiF and NH_4NO_3 will produce endothermic reactions and lower the water temperature.

$$10.0 \text{ g LiF} \times \frac{1 \text{ mol LiF}}{25.9 \text{ g LiF}} \times \frac{5.5 \text{ kJ}}{1 \text{ mol LiF}} = 2.12 \text{ kJ}$$

$$10.0 \text{ g } NH_4NO_3 \times \frac{1 \text{ mol } NH_4NO_3}{80.0 \text{ g } NH_4NO_3} \times \frac{25.7 \text{ kJ}}{1 \text{ mol } NH_4NO_3} = 3.21 \text{ kJ}$$

10.0 g of NH_4NO_3 would result in the largest temperature decrease.

9.9 $q = $ (specific heat capacity) x m x ΔT

$$\text{specific heat capacity} = \frac{q}{m \times \Delta T} = \frac{97.2 \text{ J}}{(75.0 \text{ g})(10.0 \,^{\circ}C)} = 0.130 \text{ J/(g} \cdot \,^{\circ}C)$$

$$\text{molar heat capacity} = 0.130 \text{ J/(g} \cdot \,^{\circ}C) \times \frac{207.2 \text{ g Pb}}{1 \text{ mol Pb}} = 26.9 \text{ J/(mol} \cdot \,^{\circ}C)$$

9.10 $$\text{heat capacity} = \frac{1650 \text{ J}}{2.00 \,^{\circ}C} = 825 \text{ J/}^{\circ}C$$

9.11 25.0 mL = 0.0250 L and 50.0 mL = 0.0500 L

mol H_2SO_4 = (1.00 mol/L)(0.0250 L) = 0.0250 mol H_2SO_4

mol NaOH = (1.00 mol/L)(0.0500 L) = 0.0500 mol NaOH

NaOH and H_2SO_4 are present in a 2:1 mol ratio. This matches the stoichiometric ratio in the balanced equation.

$q = $ (specific heat) x m x ΔT

m = (25.0 mL + 50.0 mL)(1.00 g/mL) = 75.0 g

$$q = (4.18 \frac{J}{g \cdot \,^{\circ}C})(75.0 \text{ g})(33.9 \,^{\circ}C - 25.0 \,^{\circ}C) = 2790 \text{ J}$$

$$\text{mol } H_2SO_4 = 0.0250 \text{ L} \times 1.00 \frac{mol}{L} \, H_2SO_4 = 0.0250 \text{ mol } H_2SO_4$$

$$\text{Heat evolved per mole of } H_2SO_4 = \frac{2.79 \times 10^3 \text{ J}}{0.0250 \text{ mol } H_2SO_4} = 1.12 \times 10^5 \text{ J/mol } H_2SO_4$$

Because the reaction evolves heat, the sign for ΔH is negative.

$$\Delta H = -1.12 \times 10^5 \text{ J/mol} \times \frac{1 \text{ kJ}}{1000 \text{ J}} = -1.12 \times 10^2 \text{ kJ/mol}$$

9.12 NH_4NO_3, 80.04, 150.0 mL of H_2O = 150.0 g of H_2O

$q = $ (specific heat) x m x ΔT

m = 15.0 g + 150.0 g = 165.0 g

$$q = (4.18 \frac{J}{g \cdot \,^{\circ}C})(165.0 \text{ g})(17.5 \,^{\circ}C - 24.5 \,^{\circ}C) = -4828 \text{ J}$$

$\Delta H = -q = -(-4828 \text{ J}) = 4828 \text{ J} = 4.83 \text{ kJ}$

$$15.0 \text{ g NH}_4\text{NO}_3 \ \times \ \frac{1 \text{ mol NH}_4\text{NO}_3}{80.0 \text{ g NH}_4\text{NO}_3} = 0.188 \text{ mol NH}_4\text{NO}_3$$

$$\Delta H^\circ = \frac{4.83 \text{ kJ}}{0.188 \text{ mol}} = 25.7 \text{ kJ/mol}$$

9.13 C_7H_8, 92.14

$$q_{reaction} = -q_{calorimeter} = -(C \times \Delta T) = -\left[\left(\frac{3.21 \text{ kJ}}{^\circ C}\right)(13.0\,^\circ C)\right] = -41.7 \text{ kJ}$$

$$1 \text{ g } C_7H_8 \ \times \ \frac{1 \text{ mol } C_7H_8}{92.14 \text{ g } C_7H_8} = 0.010\ 85 \text{ mol } C_7H_8$$

$$\Delta E = \frac{-41.7 \text{ kJ}}{0.010\ 85 \text{ mol}} = -3843 \text{ kJ/mol}$$

9.14 C_8H_{18}, 114.2

$$1.00 \text{ g } C_8H_{18} \ \times \ \frac{1 \text{ mol } C_8H_{18}}{114.2 \text{ g } C_8H_{18}} = 0.008\ 757 \text{mol } C_8H_{18}$$

$$\Delta E = \frac{q_{reaction}}{\text{mol}} ; \quad q_{reaction} = \Delta E \times \text{mol} = (-5468 \text{ kJ/mol})(0.008\ 757\text{mol}) = -47.9 \text{ kJ}$$

$$q_{reaction} = -q_{calorimeter} = -(C \times \Delta T)$$

$$\frac{q_{calorimeter}}{\Delta T} = C = \frac{47.9 \text{ kJ}}{(37.15\,^\circ C - 25.67\,^\circ C)} = 4.17 \text{ kJ/}^\circ C$$

9.15

$C(s) + O_2(g) \rightarrow CO_2(g)$	$\Delta H^\circ = -393.5$ kJ
$H_2O(g) \rightarrow 1/2 \ O_2(g) + H_2(g)$	$\Delta H^\circ = +483.6/2$ kJ
$CO2(g) \rightarrow 1/2 \ O_2(g) + CO(g)$	$\Delta H^\circ = +566.0/2$ kJ
sum $C(s) + H_2O(g) \rightarrow CO(g) + H_2(g)$	$\Delta H^\circ = +131.3$ kJ

9.16   (a) $A + 2 \ B \rightarrow D$; $\Delta H^\circ = -100$ kJ $+ (-50$ kJ$) = -150$ kJ
(b) The red arrow corresponds to step 1: $A + B \rightarrow C$
The green arrow corresponds to step 2: $C + B \rightarrow D$
The blue arrow corresponds to the overall reaction.
(c) The top energy level represents $A + 2 \ B$.
The middle energy level represents $C + B$.
The bottom energy level represents D.

9.17 $4 \ NH_3(g) + 5 \ O_2(g) \rightarrow 4 \ NO(g) + 6 \ H_2O(g)$
$\Delta H^\circ_{rxn} = [4 \ \Delta H^\circ_f (NO) + 6 \ \Delta H^\circ_f (H_2O)] - [4 \ \Delta H^\circ_f (NH_3)]$
$\Delta H^\circ_{rxn} = [(4 \text{ mol})(91.3 \text{ kJ/mol}) + (6 \text{ mol})(-241.8 \text{ kJ/mol})] - [(4 \text{ mol})(-46.1 \text{ kJ/mol})]$
$\Delta H^\circ_{rxn} = -901.2$ kJ

9.18 $2 C_8H_{18}(l) + 25 O_2(g) \rightarrow 8 CO_2(g) + 9 H_2O(l)$
$\Delta H°_{rxn} = \Delta H°_c = -5220$ kJ
$\Delta H°_{rxn} = [8 \Delta H°_f(CO_2) + 9 \Delta H°_f(H_2O)] - [2 \Delta H°_f(C_8H_{18})]$
-5220 kJ $= [(8$ mol$)(-393.5$ kJ/mol$) + (9$ mol$)(-285.8$ kJ/mol$)] - [(2$ mol$)(\Delta H°_f(C_8H_{18}))]$
Solve for $\Delta H°_f(C_8H_{18})$
-5220 kJ $= -5720.2$ kJ $- (2$ mol$)(\Delta H°_f(C_8H_{18}))$
$+500$ kJ $= -(2$ mol$)(\Delta H°_f(C_8H_{18}))$
$\Delta H°_f(C_8H_{18}) = \dfrac{500 \text{ kJ}}{-2 \text{ mol}} = -250$ kJ/mol

9.19 $H_2C{=}CH_2(g) + H_2O(g) \rightarrow C_2H_5OH(g)$
$\Delta H°_{rxn} = D$ (Reactant bonds) $- D$ (Product bonds)
$\Delta H°_{rxn} = (D_{C=C} + 4 D_{C-H} + 2 D_{O-H}) - (D_{C-C} + D_{C-O} + 5 D_{C-H} + D_{O-H})$
$\Delta H°_{rxn} = [(1$ mol$)(728$ kJ/mol$) + (4$ mol$)(410$ kJ/mol$) + (2$ mol$)(460$ kJ/mol$)]$
$- [(1$ mol$)(350$ kJ/mol$) + (1$ mol$)(350$ kJ/mol$) + (5$ mol$)(410$ kJ/mol$) + (1$ mol$)(460$ kJ/mol$)]$
$\Delta H°_{rxn} = +78$ kJ

9.20 $\Delta H°_{rxn} = D$ (Reactant bonds) $- D$ (Product bonds)
(a) $2 C_6H_6(g) + 15 O_2(g) \rightarrow 12 CO_2(g) + 6 H_2O(g)$
$\Delta H°_{rxn} = (12 D_{C-C} + 12 D_{C-H} + 15 D_{O=O}) - (24 D_{C-O} + 12 D_{O-H}) = -6339$ kJ
$\Delta H°_{rxn} = [(12$ mol$)(D_{C-C}) + (12$ mol$)(410$ kJ/mol$) + (15$ mol$)(498$ kJ/mol$)]$
$\qquad - [(24$ mol$)(804$ kJ/mol$) + (12$ mol$)(460$ kJ/mol$)] = -6339$ kJ
$(12$ mol$)(D_{C-C}) = -6339$ kJ $+ [(24$ mol$)(804$ kJ/mol$) + (12$ mol$)(460$ kJ/mol$)]$
$\qquad\qquad - [(12$ mol$)(410$ kJ/mol$) + (15$ mol$)(498$ kJ/mol$)]$
$(12$ mol$)(D_{C-C}) = 6087$ kJ
$D_{C-C} = \dfrac{6087 \text{ kJ}}{12 \text{ mol}} = 507$ kJ/mol for C– C bond in benzene

9.21 $\Delta S°$ is negative because the reaction decreases the number of moles of gaseous molecules.

9.22   The reaction proceeds from a solid and a gas (reactants) to all gas (product). Randomness increases, so $\Delta S°$ is positive.

9.23   The reaction involves only bond making, so it is exothermic and ΔH is negative. There are more reactant atoms than product molecules. The randomness of the system decreases on going from reactant to product, therefore ΔS is negative. Because the reaction is spontaneous, ΔG is negative.

9.24   (a) $2 A_2 + B_2 \rightarrow 2 A_2B$
(b) Because the reaction is exothermic, ΔH is negative. There are more reactant molecules than product molecules. The randomness of the system decreases on going from reactant to product, therefore ΔS is negative.
(c) Because $\Delta G = \Delta H - T\Delta S$, a reaction with both ΔH and ΔS negative is favored at low temperatures where the negative ΔH term is larger than the positive $- T\Delta S$, and ΔG is negative.

9.25 $\Delta G^\circ = \Delta H^\circ - T\Delta S^\circ = (-92.2 \text{ kJ}) - (298 \text{ K})(-0.199 \text{ kJ/K}) = -32.9 \text{ kJ}$
Because ΔG° is negative, the reaction is spontaneous.
Set $\Delta G^\circ = 0$ and solve for T.

$$\Delta G^\circ = 0 = \Delta H^\circ - T\Delta S^\circ; \quad T = \frac{\Delta H^\circ}{\Delta S^\circ} = \frac{-92.2 \text{ kJ}}{-0.199 \text{ kJ/K}} = 463 \text{ K} = 190 \text{ °C}$$

9.26 $\Delta G^\circ = \Delta H^\circ - T\Delta S^\circ$
Set $\Delta G^\circ = 0$ and solve for T.

$$\Delta G^\circ = 0 = \Delta H^\circ - T\Delta S^\circ; \quad T = \frac{\Delta H^\circ}{\Delta S^\circ} = \frac{+75.0 \text{ kJ}}{+0.231 \text{ kJ/K}} = 325 \text{ K} = 52 \text{ °C}$$

The reaction is at equilibrium at 325 K (52 °C). The temperature should be increased to make the reaction spontaneous.

9.27 Ethanol is produced by fermenting corn. Biodiesel is produced by reaction of vegetable oil and methyl alcohol.

9.28 Biofuels are renewable and almost carbon neutral. Burning biofuels does not contribute to an increase in CO_2 in the atmosphere. Gasoline is a fossil fuel and not renewable. Burning gasoline contributes to an increase in CO_2 in the atmosphere.

9.29 (a) $C_2H_6O(l) + 3 O_2(g) \rightarrow 2 CO_2(g) + 3 H_2O(l)$
(b) $\Delta H^\circ_c = [2 \Delta H^\circ_f(CO_2) + 3 \Delta H^\circ_f(H_2O)] - \Delta H^\circ_f(C_2H_6O)$
$\Delta H^\circ_c = [(2 \text{ mol})(-393.5 \text{ kJ/mol}) + (3 \text{ mol})(-285.8 \text{ kJ/mol})] - [(1 \text{ mol})(-277.7 \text{ kJ/mol})]$
$\Delta H^\circ_c = -1367 \text{ kJ}$

9.30 $\Delta H^\circ_{rxn} = D \text{ (Reactant bonds)} - D \text{ (Product bonds)}$
$C_2H_6O(l) + 3 O_2(g) \rightarrow 2 CO_2(g) + 3 H_2O(l)$
$\Delta H^\circ_c = (D_{C-C} + 5 D_{C-H} + D_{C-O} + D_{O-H} + 3 D_{O=O}) - (4 D_{C=O} + 6 D_{O-H})$
$\Delta H^\circ_c = [(1 \text{ mol})(350 \text{ kJ/mol}) + (5 \text{ mol})(410 \text{ kJ/mol}) + (1 \text{ mol})(350 \text{ kJ/mol})$
$\qquad\qquad + (1 \text{ mol})(460 \text{ kJ/mol}) + (3 \text{ mol})(498 \text{ kJ/mol})]$
$\qquad\qquad\qquad - [(4 \text{ mol})(804 \text{ kJ/mol}) + (6 \text{ mol})(460 \text{ kJ/mol})] = -1272 \text{ kJ/mol}$

9.31 (a) $\Delta V = (49.0 \text{ L} - 73.5 \text{ L}) = -24.5 \text{ L}$
$P\Delta V = (1.00 \text{ atm})(-24.5 \text{ L}) = -24.5 \text{ L} \cdot \text{atm}$

$$P\Delta V = (-24.5 \text{ L} \cdot \text{atm})(101 \frac{\text{J}}{\text{L} \cdot \text{atm}}) = -2475 \text{ J} = -2.47 \text{ kJ}$$

$w = -P\Delta V = +2.47 \text{ kJ}$
Work energy is transferred to the system.
(b) $\Delta E = \Delta H - P\Delta V = -1367 \text{ kJ} - (-2.47 \text{ kJ}) = -1365 \text{ kJ/mol}$

9.32 $C_{19}H_{38}O_2$,

$$q_{reaction} = -q_{calorimeter} = -(C \times \Delta T) = -\left[\left(\frac{2.50 \text{ kJ}}{\text{°C}}\right)(5.29 \text{ °C})\right] = -13.2 \text{ kJ}$$

$$\text{combustion energy} = \frac{-13.2 \text{ kJ}}{0.350 \text{ g } C_{19}H_{38}O_2} = -37.7 \text{ kJ/g}$$

9.33 $C_{19}H_{38}O_2$, 298.5

$$\text{heat} = 39.9 \text{ kg } C_{19}H_{38}O_2 \text{ x } \frac{1000 \text{ g}}{1 \text{ kg}} \text{ x } \frac{1 \text{ mol } C_{19}H_{38}O_2}{298.5 \text{ g } C_{19}H_{38}O_2} \text{ x}$$

$$\frac{-11,236 \text{ kJ}}{1 \text{ mol } C_{19}H_{38}O_2} \text{ x } \frac{1 \text{ MJ}}{1000 \text{ kJ}} = -1500 \text{ MJ}$$

heat (octane) = 1530 MJ

Conceptual Problems

9.34 $\Delta H > 0$ and $w < 0$

9.36   (a)   (b)

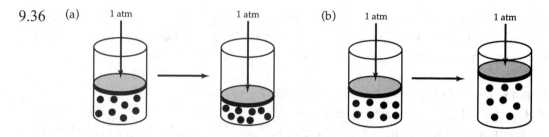

9.38   (a) Diagram 1   (b) $\Delta H° = +30$ kJ

9.40   (a) $2 \text{ AB}_2 \rightarrow A_2 + 2 \text{ B}_2$
       (b) Because the reaction is exothermic, ΔH is negative. Because the number of molecules increases going from reactants to products, ΔS is positive.
       (c) When ΔH is negative and ΔS is positive, the reaction is likely to be spontaneous at all temperatures.

9.42   The change is the spontaneous conversion of a liquid to a gas. ΔG is negative because the change is spontaneous. The conversion of a liquid to a gas is endothermic, therefore ΔH is positive. ΔS is positive because randomness increases when a liquid is converted to a gas.

Section Problems
Heat, Work, and Energy (Sections 9.1–9.3)

9.44   Heat is the energy transferred from one object to another as the result of a temperature difference between them. Temperature is a measure of the kinetic energy of molecular motion. Energy is the capacity to do work or supply heat. Work is defined as the distance moved times the force that opposes the motion ($w = d$ x F).
       Kinetic energy is the energy of motion. Potential energy is stored energy.

9.46   Car: $E_K = \frac{1}{2}(1400 \text{ kg})\left(\dfrac{115 \times 10^3 \text{ m}}{3600 \text{ s}}\right)^2 = 7.1 \times 10^5 \text{ J}$

Truck: $E_K = \frac{1}{2}(12,000 \text{ kg})\left(\dfrac{38 \times 10^3 \text{ m}}{3600 \text{ s}}\right)^2 = 6.7 \times 10^5 \text{ J}$

The car has more kinetic energy.

9.48   (a) and (b) are state functions; (c) is not.

9.50 $w = - P\Delta V = - (3.6 \text{ atm})(3.4 \text{ L} - 3.2 \text{ L}) = - 0.72 \text{ L} \cdot \text{atm}$

$w = (- 0.72 \text{ L} \cdot \text{atm})\left(\dfrac{101 \text{ J}}{1 \text{ L} \cdot \text{atm}}\right) = -72.7 \text{ J} = -70 \text{ J};$ The energy change is negative.

9.52   (a) 100 W = 100 J/s

$50 \text{ Cal} \times \dfrac{1000 \text{ cal}}{1 \text{ Cal}} \times \dfrac{4.184 \text{ J}}{1 \text{ cal}} \times \dfrac{1 \text{ s}}{100 \text{ J}} \times \dfrac{1 \text{ min}}{60 \text{ s}} = 35 \text{ min}$

(b) 23 W = 23 J/s

$50 \text{ Cal} \times \dfrac{1000 \text{ cal}}{1 \text{ Cal}} \times \dfrac{4.184 \text{ J}}{1 \text{ cal}} \times \dfrac{1 \text{ s}}{23 \text{ J}} \times \dfrac{1 \text{ min}}{60 \text{ s}} = 150 \text{ min}$

9.54   (a) $w = -P\Delta V = - (0.975 \text{ atm})(18.0 \text{ L} - 12.0 \text{ L}) = -5.85 \text{ L} \cdot \text{atm}$

$w = (-5.85 \text{ L} \cdot \text{atm})\left(\dfrac{101.325 \text{ J}}{1 \text{ L} \cdot \text{atm}}\right) = -593 \text{ J}$

(b) $\Delta V = 6.0 \text{ L} = 6000 \text{ mL} = 6000 \text{ cm}^3;$ $r = \dfrac{17.0 \text{ cm}}{2} = 8.50 \text{ cm}$

$V = \pi r^2 h;$ $h = \dfrac{V}{\pi r^2} = \dfrac{6000 \text{ cm}^3}{\pi (8.50 \text{ cm})^2} = 26.4 \text{ cm}$

9.56 $\Delta H = \Delta E + P\Delta V$
$\Delta H = -5.67 \text{ kJ} = -7.20 \text{ kJ} + P\Delta V$
$P\Delta V = +1.53 \text{ kJ}$
$w = -P\Delta V = -1.53 \text{ kJ}$

$w = -P\Delta V = -1.53 \text{ kJ} \times \dfrac{1000 \text{ J}}{1 \text{ kJ}} \times \dfrac{1}{\left(\dfrac{101 \text{ J}}{\text{L} \cdot \text{atm}}\right)} = -15.1 \text{ L} \cdot \text{atm}$

$P\Delta V = 15.1 \text{ L} \cdot \text{atm} = (1 \text{ atm})(\Delta V)$
$\Delta V = 15.1 \text{ L}$

Energy and Enthalpy (Section 9.4)

9.58 $\Delta E = q_v$ is the heat change associated with a reaction at constant volume. Since $\Delta V = 0$, no PV work is done.

$\Delta H = q_p$ is the heat change associated with a reaction at constant pressure. Since $\Delta V \neq 0$, PV work can also be done.

9.60 $\Delta V = 448$ L and assume P = 1.00 atm

$w = -P\Delta V = -(1.00 \text{ atm})(448 \text{ L}) = -448 \text{ L} \cdot \text{atm}$

$w = -(448 \text{ L} \cdot \text{atm})(101 \dfrac{J}{L \cdot atm}) = -4.52 \times 10^4 \text{ J}$

$w = -4.52 \times 10^4 \text{ J} \times \dfrac{1 \text{ kJ}}{1000 \text{ J}} = -45.2 \text{ kJ}$

9.62 $P\Delta V = -7.6 \text{ J}$ (from Problem 9.51)

$\Delta H = \Delta E + P\Delta V$

$\Delta E = \Delta H - P\Delta V = -0.31 \text{ kJ} - (-7.6 \times 10^{-3} \text{ kJ}) = -0.30 \text{ kJ}$

9.64 $\Delta H = \Delta E + P\Delta V = 44.0 \text{ kJ} + (1.00 \text{ atm})(14.0 \text{ L})\left(\dfrac{101.325 \text{ J}}{1 \text{ L} \cdot \text{atm}} \times \dfrac{1 \text{ kJ}}{1000 \text{ J}} \right) = 45.4 \text{ kJ}$

Thermochemical Equations for Chemical and Physical Changes (Sections 9.5–9.6)

9.66   (a) heat is transferred to the system, endothermic, $\Delta H° > 0$
(b) heat is transferred to the surroundings, exothermic, $\Delta H° < 0$
(c) heat is transferred to the system, endothermic, $\Delta H° > 0$

9.68 $C_4H_{10}O$, 74.12; mass of $C_4H_{10}O = (0.7138 \text{ g/mL})(100 \text{ mL}) = 71.38 \text{ g}$

mol $C_4H_{10}O = 71.38 \text{ g} \times \dfrac{1 \text{ mol}}{74.12 \text{ g}} = 0.9630 \text{ mol}$

$q = n \times \Delta H_{vap} = 0.9630 \text{ mol} \times 26.5 \text{ kJ/mol} = 25.5 \text{ kJ}$

9.70   Al, 26.98

mol Al $= 5.00 \text{ g} \times \dfrac{1 \text{ mol}}{26.98 \text{ g}} = 0.1853 \text{ mol}$

$q = n \times \Delta H° = 0.1853 \text{ mol Al} \times \dfrac{-1408.4 \text{ kJ}}{2 \text{ mol Al}} = -131 \text{ kJ}$; 131 kJ is released.

9.72   (a) C_3H_8, 44.10; $\Delta H° = -2220 \text{ kJ/mol } C_3H_8$

$15.5 \text{ g} \times \dfrac{1 \text{ mol } C_3H_8}{44.10 \text{ g } C_3H_8} \times \dfrac{-2220 \text{ kJ}}{1 \text{ mol } C_3H_8} = -780. \text{ kJ}$

780. kJ of heat is evolved.

(b) $Ba(OH)_2 \cdot 8 \text{ H}_2O$, 315.5; $\Delta H° = +80.3 \text{ kJ/mol } Ba(OH)_2 \cdot 8 \text{ H}_2O$

$4.88 \text{ g} \times \dfrac{1 \text{ mol } Ba(OH)_2 \cdot 8 \text{ H}_2O}{315.5 \text{ g } Ba(OH)_2 \cdot 8 \text{ H}_2O} \times \dfrac{80.3 \text{ kJ}}{1 \text{ mol } Ba(OH)_2 \cdot 8 \text{ H}_2O} = +1.24 \text{ kJ}$

1.24 kJ of heat is absorbed.

9.74 Fe_2O_3, 159.7;   mol $Fe_2O_3 = 2.50$ g x $\dfrac{1 \text{ mol}}{159.7 \text{ g}} = 0.015\ 65$ mol

$q = n \times \Delta H° = 0.015\ 65$ mol Fe_2O_3 x $\dfrac{-24.8 \text{ kJ}}{1 \text{ mol } Fe_2O_3} = -0.388$ kJ; 0.388 kJ is evolved.

Because ΔH is negative, the reaction is exothermic.

Calorimetry and Heat Capacity (Section 9.7)

9.76   Heat capacity is the amount of heat required to raise the temperature of a substance a given amount. Specific heat is the amount of heat necessary to raise the temperature of exactly 1 g of a substance by exactly 1 °C.

9.78   Na, 22.99

specific heat = 28.2 $\dfrac{J}{\text{mol}\cdot°C}$ x $\dfrac{1 \text{ mol}}{22.99 \text{ g}} = 1.23$ J/(g·°C)

9.80 $q =$ (specific heat) x m x $\Delta T = (4.18 \dfrac{J}{g\cdot°C})(350 \text{ g})(3 °C - 25 °C) = -3.2 \times 10^4$ J

$q = -3.2 \times 10^4$ J x $\dfrac{1 \text{ kJ}}{1000 \text{ J}} = -32$ kJ

9.82 NH_4NO_3, 80.04;   assume 125 mL = 125 g H_2O

50.0 g NH_4NO_3 x $\dfrac{1 \text{ mol } NH_4NO_3}{80.04 \text{ g } NH_4NO_3} = 0.625$ mol NH_4NO_3

$q_p = \Delta H \times n = (+25.7 \text{ kJ/mol})(0.625 \text{ mol}) = 16.1$ kJ = 16,100 J
$q_{soln} = -q_p = -16,100$ J
$q_{soln} =$ (specific heat) x m x ΔT

$\Delta T = \dfrac{q_{soln}}{(\text{specific heat}) \times m} = \dfrac{-16,100 \text{ J}}{\left(4.18 \dfrac{J}{g\cdot°C}\right)(50 \text{ g} + 125 \text{ g})} = -22.0 °C$

$\Delta T = -22.0 °C = T_{final} - T_{initial} = T_{final} - 25.0 °C;$ $T_{final} = -22.0 °C + 25.0 °C = 3.0 °C$

9.84   Mass of solution = 50.0 g + 1.045 g = 51.0 g
$q =$ (specific heat) x m x ΔT

$q = \left(4.18 \dfrac{J}{g\cdot°C}\right)(51.0 \text{ g})(32.3 °C - 25.0 °C) = 1.56 \times 10^3$ J = 1.56 kJ

CaO, 56.08;   mol CaO = 1.045 g x $\dfrac{1 \text{ mol}}{56.08 \text{ g}} = 0.018\ 63$ mol

Heat evolved per mole of CaO = $\dfrac{1.56 \text{ kJ}}{0.018\ 63 \text{ mol}} = 83.7$ kJ/mol CaO

Because the reaction evolves heat, the sign for ΔH is negative. $\Delta H = -83.7$ kJ

9.86 C_6H_6, 78.1

$$q_{reaction} = -q_{calorimeter} = -(C \times \Delta T) = -\left[\left(\frac{2.46 \text{ kJ}}{{}^\circ C}\right)(3.45\,{}^\circ C)\right] = -8.49 \text{ kJ}$$

$$\Delta E = \frac{-8.49 \text{ kJ}}{0.187 \text{ g } C_6H_6} = -45.4 \text{ kJ/g}$$

$$\Delta E = -45.4 \text{ kJ/g} \times \frac{78.1 \text{ g } C_6H_6}{1 \text{ mol } C_6H_6} = -3545 \text{ kJ/mol}$$

9.88 $Mg(s) + 2 HCl(aq) \rightarrow MgCl_2(aq) + H_2(g)$

$$\text{mol Mg} = 1.50 \text{ g} \times \frac{1 \text{ mol}}{24.3 \text{ g}} = 0.0617 \text{ mol Mg}$$

$$\text{mol HCl} = 0.200 \text{ L} \times 6.00\,\frac{\text{mol}}{L} = 1.20 \text{ mol HCl}$$

There is an excess of HCl. Mg is the limiting reactant.

$$q = \left(4.18\,\frac{J}{g \cdot {}^\circ C}\right)(200 \text{ g})(42.9\,{}^\circ C - 25.0\,{}^\circ C) + \left(776\,\frac{J}{{}^\circ C}\right)(42.9\,{}^\circ C - 25.0\,{}^\circ C) = 2.89 \times 10^4 \text{ J}$$

$$q = 2.89 \times 10^4 \text{ J} \times \frac{1 \text{ kJ}}{1000 \text{ J}} = 28.9 \text{ kJ}$$

$$\text{Heat evolved per mole of Mg} = \frac{28.9 \text{ kJ}}{0.0617 \text{ mol}} = 468 \text{ kJ/mol}$$

Because the reaction evolves heat, the sign for ΔH is negative. $\Delta H = -468 \text{ kJ}$

9.90   There is a large excess of NaOH.   5.00 mL = 0.005 00 L
mol citric acid = (0.005 00 L)(0.64 mol/L) = 0.0032 mol citric acid

$$q_{H_2O} = (51.6 \text{ g})[4.0 \text{ J/(g} \cdot {}^\circ C)](27.9\,{}^\circ C - 26.0\,{}^\circ C) = 392 \text{ J}$$

$$q_{rxn} = -q_{H_2O} = -392 \text{ J}$$

$$\Delta H = -\frac{392 \text{ J}}{0.0032 \text{ mol}} \times \frac{1 \text{ kJ}}{1000 \text{ J}} = -123 \text{ kJ/mol} = -120 \text{ kJ/mol citric acid}$$

Hess's Law (Sections 9.8)

9.92   The standard state of an element is its most stable form at 1 atm and the specified temperature, usually 25 °C.

Because enthalpy is a state function, ΔH is the same regardless of the path taken between reactants and products. Thus, the sum of the enthalpy changes for the individual steps in a reaction is equal to the overall enthalpy change for the entire reaction, a relationship known as Hess's law. Hess's law works because of the law of the conservation of energy.

9.94 $CH_4(g) + Cl_2(g) \rightarrow CH_3Cl(g) + HCl(g)$ $\Delta H°_1 = -98.3$ kJ
$\underline{CH_3Cl(g) + Cl_2(g) \rightarrow CH_2Cl_2(g) + HCl(g)}$ $\Delta H°_2 = -104$ kJ
Sum $CH_4(g) + 2 Cl_2(g) \rightarrow CH_2Cl_2(g) + 2 HCl(g)$
$\Delta H° = \Delta H°_1 + \Delta H°_2 = -202$ kJ

9.96 $2 NO(g) + O_2(g) \rightarrow 2 NO_2(g)$ $\Delta H°_1 = 2(-58.1$ kJ$)$
$\underline{2 NO_2(g) \rightarrow N_2O_4(g)}$ $\Delta H°_2 = -55.3$ kJ
Sum $2 NO(g) + O_2(g) \rightarrow N_2O_4(g)$
$\Delta H° = \Delta H°_1 + \Delta H°_2 = -171.5$ kJ

Heats of Formation (Sections 9.9)

9.98 A compound's standard heat of formation is the amount of heat associated with the formation of 1 mole of a compound from its elements (in their standard states).

9.100 (a) Cl_2, gas   (b) Hg, liquid   (c) CO_2, gas   (d) Ga, solid

9.102 (a) $2 Fe(s) + 3/2 O_2(g) \rightarrow Fe_2O_3(s)$
(b) $12 C(s) + 11 H_2(g) + 11/2 O_2(g) \rightarrow C_{12}H_{22}O_{11}(s)$
(c) $U(s) + 3 F_2(g) \rightarrow UF_6(s)$

9.104 $S(s) + O_2(g) \rightarrow SO_2(g)$ $\Delta H°_1 = -296.8$ kJ
$\underline{SO_2 + ½ O_2(g) \rightarrow SO_3(g)}$ $\Delta H°_2 = -98.9$ kJ
Sum $S(s) + 3/2 O_2(g) \rightarrow SO_3(g)$ $\Delta H°_3 = \Delta H°_1 + \Delta H°_2$
$\Delta H°_f = \Delta H°_3 = -296.8$ kJ $+ (-98.9$ kJ$) = -395.7$ kJ/mol

9.106 $SO_3(g) + H_2O(l) \rightarrow H_2SO_4(aq)$ $\Delta H°_1 = -227.8$ kJ
$H_2(g) + ½ O_2(g) \rightarrow H_2O(l)$ $\Delta H°_2 = \Delta H°_f = -285.8$ kJ
$\underline{S(s) + 3/2 O_2(g) \rightarrow SO_3(g)}$ $\Delta H°_3 = \Delta H°_f = -395.7$
Sum $S(s) + H_2(g) + 2 O_2(g) \rightarrow H_2SO_4(aq)$ $\Delta H°_f (H_2SO_4) = ?$
$\Delta H°_f (H_2SO_4) = \Delta H°_1 + \Delta H°_2 + \Delta H°_3 = -909.3$ kJ

9.108 $C_8H_8(l) + 10 O_2(g) \rightarrow 8 CO_2(g) + 4 H_2O(l)$
$\Delta H°_{rxn} = \Delta H°_c = -4395$ kJ
$\Delta H°_{rxn} = [8 \Delta H°_f(CO_2) + 4 \Delta H°_f(H_2O)] - \Delta H°_f(C_8H_8)$
-4395 kJ $= [(8$ mol$)(-393.5$ kJ/mol$) + (4$ mol$)(-285.8$ kJ/mol$)] - [(1$ mol$)(\Delta H°_f(C_8H_8))]$
Solve for $\Delta H°_f(C_8H_8)$
-4395 kJ $= -4291.2$ kJ $- (1$ mol$)(\Delta H°_f(C_8H_8))$
-103.8 kJ $= -(1$ mol$)(\Delta H°_f(C_8H_8))$
$\Delta H°_f(C_8H_8) = \dfrac{-103.8 \text{ kJ}}{-1 \text{ mol}} = +103.8$ kJ/mol $= +104$ kJ/mol

9.110 $\Delta H°_{rxn} = \Delta H°_f(MTBE) - [\Delta H°_f(2\text{-Methylpropene}) + \Delta H°_f(CH_3OH)]$
-57.5 kJ $= -313.6$ kJ $- [(1$ mol$)(\Delta H°_f(2\text{-Methylpropene})) + (-239.2$ kJ$)]$
Solve for $\Delta H°_f(2\text{-Methylpropene})$.

-16.9 kJ $= (1$ mol$)(\Delta H^\circ_f(2\text{-Methylpropene}))$
$\Delta H^\circ_f(2\text{-Methylpropene}) = -16.9$ kJ/mol

9.112 $CaCO_3(s) \rightarrow CaO(s) + CO_2(g)$
$\Delta H^\circ_{rxn} = [\Delta H^\circ_f(CaO) + \Delta H^\circ_f(CO_2)] - \Delta H^\circ_f(CaCO_3)$
$\Delta H^\circ_{rxn} = [(1$ mol$)(-634.9$ kJ/mol$) + (1$ mol$)(-393.5$ kJ/mol$)] - [(1$ mol$)(-1207.6$ kJ/mol$)]$
$\Delta H^\circ_{rxn} = + 179.2$ kJ

9.114 $CaCO_3(s) \rightarrow CaO(s) + CO_2(g)$
$\Delta H^\circ_{rxn} = [\Delta H^\circ_f(CaO) + \Delta H^\circ_f(CO_2)] - \Delta H^\circ_f(CaCO_3)$
$\Delta H^\circ_{rxn} = [(1$ mol$)(-634.9$ kJ/mol$) + (1$ mol$)(-393.5$ kJ/mol$)] - [(1$ mol$)(-1207.6$ kJ/mol$)]$
$\Delta H^\circ_{rxn} = + 179.2$ kJ

9.116   (a) $C(s) + CO_2(g) \rightarrow 2\ CO(g)$
$\Delta H^\circ_{rxn} = [2\ \Delta H^\circ_f(CO)] - \Delta H^\circ_f(CO_2)$
$\Delta H^\circ_{rxn} = [(2$ mol$)(-110.5$ kJ/mol$)] - [(1$ mol$)(-393.5$ kJ/mol$)] = +172.5$ kJ
(b) $2\ H_2O_2(aq) \rightarrow 2\ H_2O(l) + O_2(g)$
$\Delta H^\circ_{rxn} = [2\ \Delta H^\circ_f(H_2O)] - [2\ \Delta H^\circ_f(H_2O_2)]$
$\Delta H^\circ_{rxn} = [(2$ mol$)(-285.8$ kJ/mol$)] - [(2$ mol$)(-191.2$ kJ/mol$)] = -189.2$ kJ
(c) $Fe_2O_3(s) + 3\ CO(g) \rightarrow 2\ Fe(s) + 3\ CO_2(g)$
$\Delta H^\circ_{rxn} = [3\ \Delta H^\circ_f(CO_2)] - [\Delta H^\circ_f(Fe_2O_3) + 3\ \Delta H^\circ_f(CO)]$
$\Delta H^\circ_{rxn} = [(3$ mol$)(-393.5$ kJ/mol$)]$
$\quad\quad - [(1$ mol$)(-824.2$ kJ/mol$) + (3$ mol$)(-110.5$ kJ/mol$)] = -24.8$ kJ

Bond Dissociation Energies (Section 9.10)

9.118 $\Delta H^\circ_{rxn} = D$ (Reactant bonds) $- D$ (Product bonds)
$C_2H_6(g) + 7/2\ O_2(g) \rightarrow 2\ CO_2(g) + 3\ H_2O(g)$
$\Delta H^\circ_{rxn} = (1\ D_{C-C} + 6\ D_{C-H} + 7/2\ D_{O=O}) - (4\ D_{C=O} + 6\ D_{O-H})$
$\Delta H^\circ_{rxn} = [(1$ mol$)(350$ kJ/mol$) + (6$ mol$)(410$ kJ/mol$) + (7/2$ mol$)(498$ kJ/mol$)]$
$\quad\quad - [(4$ mol$)(804$ kJ/mol$) + (6$ mol$)(460$ kJ/mol$)] = -1423$ kJ

9.120 $\Delta H^\circ_{rxn} = D$ (Reactant bonds) $- D$ (Product bonds)
(a) $2\ CH_4(g) \rightarrow C_2H_6(g) + H_2(g)$
$\Delta H^\circ_{rxn} = (8\ D_{C-H}) - (D_{C-C} + 6\ D_{C-H} + D_{H-H})$
$\Delta H^\circ_{rxn} = [(8$ mol$)(410$ kJ/mol$)] - [(1$ mol$)(350$ kJ/mol$) + (6$ mol$)(410$ kJ/mol$)$
$\quad\quad + (1$ mol$)(436$ kJ/mol$)] = +34$ kJ

(b) $C_2H_6(g) + F_2(g) \rightarrow C_2H_5F(g) + HF(g)$
$\Delta H^\circ_{rxn} = (6\ D_{C-H} + D_{C-C} + D_{F-F}) - (5\ D_{C-H} + D_{C-C} + D_{C-F} + D_{H-F})$
$\Delta H^\circ_{rxn} = [(6$ mol$)(410$ kJ/mol$) + (1$ mol$)(350$ kJ/mol$) + (1$ mol$)(159$ kJ/mol$)]$
$\quad\quad - [(5$ mol$)(410$ kJ/mol$) + (1$ mol$)(350$ kJ/mol$) + (1$ mol$)(450$ kJ/mol$)$
$\quad\quad + (1$ mol$)(570$ kJ/mol$)] = -451$ kJ

(c) $N_2(g) + 3 H_2(g) \rightarrow 2 NH_3(g)$
The bond dissociation energy for N_2 is 945 kJ/mol.
$\Delta H°_{rxn} = (D_{N_2} + 3 D_{H-H}) - (6 D_{N-H})$
$\Delta H°_{rxn} = [(1\ mol)(945\ kJ/mol) + (3\ mol)(436\ kJ/mol)] - [(6\ mol)(390\ kJ/mol)] = -87\ kJ$

Free Energy and Entropy (Sections 9.11–9.12)

9.122 Entropy is a measure of molecular randomness.

9.124 A reaction can be spontaneous yet endothermic if ΔS is positive (more randomness) and the $T\Delta S$ term is larger than ΔH.

9.126 (a) positive   (more randomness)   (b) negative   (less randomness)

9.128 (a) zero   (equilibrium)   (b) zero   (equilibrium)
      (c) negative   (spontaneous)

9.130 ΔS is positive. The reaction increases the total number of molecules.

9.132 $\Delta G = \Delta H - T\Delta S$
    (a) $\Delta G = -48\ kJ - (400\ K)(135 \times 10^{-3}\ kJ/K) = -102\ kJ$
    $\Delta G < 0$, spontaneous; $\Delta H < 0$, exothermic.
    (b) $\Delta G = -48\ kJ - (400\ K)(-135 \times 10^{-3}\ kJ/K) = +6\ kJ$
    $\Delta G > 0$, nonspontaneous; $\Delta H < 0$, exothermic.
    (c) $\Delta G = +48\ kJ - (400\ K)(135 \times 10^{-3}\ kJ/K) = -6\ kJ$
    $\Delta G < 0$, spontaneous; $\Delta H > 0$, endothermic.
    (d) $\Delta G = +48\ kJ - (400\ K)(-135 \times 10^{-3}\ kJ/K) = +102\ kJ$
    $\Delta G > 0$, nonspontaneous; $\Delta H > 0$, endothermic.

9.134 $\Delta G = \Delta H - T\Delta S$;   Set $\Delta G = 0$ and solve for T (the crossover temperature).
    $T = \dfrac{\Delta H}{\Delta S} = \dfrac{-33\ kJ}{-0.058\ kJ/K} = 570\ K$

9.136 (a) $\Delta H < 0$ and $\Delta S > 0$; reaction is spontaneous at all temperatures.
    (b) $\Delta H < 0$ and $\Delta S < 0$; reaction has a crossover temperature.
    (c) $\Delta H > 0$ and $\Delta S > 0$; reaction has a crossover temperature.
    (d) $\Delta H > 0$ and $\Delta S < 0$; reaction is nonspontaneous at all temperatures.

9.138 $T = -114.1\ °C = 273.15 + (-114.1) = 159.0\ K$
    $\Delta G_{fus} = \Delta H_{fus} - T\Delta S_{fus}$; $\Delta G = 0$ at the melting point temperature.
    Set $\Delta G = 0$ and solve for ΔS_{fus}.
    $\Delta G = 0 = \Delta H_{fus} - T\Delta S_{fus}$
    $\Delta S_{fus} = \dfrac{\Delta H_{fus}}{T} = \dfrac{5.02\ kJ/mol}{159.0\ K} = 0.0316\ kJ/(K \cdot mol) = 31.6\ J/(K \cdot mol)$

9.140 $\Delta G = \Delta H - T\Delta S$; at equilibrium $\Delta G = 0$. Set $\Delta G = 0$ and solve for T.

$$\Delta G = 0 = \Delta H - T\Delta S; \qquad T = \frac{\Delta H}{\Delta S} = \frac{30.91 \text{ kJ/mol}}{93.2 \text{ x } 10^{-3} \text{ kJ/(K} \cdot \text{mol)}} = 332 \text{ K} = 59 \text{ }^{\circ}\text{C}$$

9.142 $HgS(s) + O_2(g) \rightarrow Hg(l) + SO_2(g)$
(a) $\Delta H^{\circ}_{rxn} = \Delta H^{\circ}_f(SO_2) - \Delta H^{\circ}_f(HgS)$
$\Delta H^{\circ}_{rxn} = [(1 \text{ mol})(-296.8 \text{ kJ/mol})] - [(1 \text{ mol})(-58.2 \text{ kJ/mol})] = -238.6 \text{ kJ}$
(b) and (c) Because $\Delta H < 0$ and $\Delta S > 0$, the reaction is spontaneous at all temperatures.

Multiconcept Problems

9.144 C_2H_4, 28.05;   HCl, 36.46
$w = -P\Delta V = -(1.00 \text{ atm})(-71.5 \text{ L}) = 71.5 \text{ L} \cdot \text{atm}$

$$w = (71.5 \text{ L} \cdot \text{atm})\left(\frac{101 \text{ J}}{1 \text{ L} \cdot \text{atm}}\right) = 7222 \text{ J} = 7.22 \text{ kJ}$$

$$89.5 \text{ g } C_2H_4 \text{ x } \frac{1 \text{ mol } C_2H_4}{28.05 \text{ g } C_2H_4} = 3.19 \text{ mol } C_2H_4; \quad 125 \text{ g HCl x } \frac{1 \text{ mol HCl}}{36.46 \text{ g HCl}} = 3.43 \text{ mol HCl}$$

Because the reaction stoichiometry between C_2H_4 and HCl is 1:1, C_2H_4 is the limiting reactant.
$\Delta H^{\circ} = -72.3 \text{ kJ/mol } C_2H_4$
$q = (-72.3 \text{ kJ/mol})(3.19 \text{ mol}) = -231 \text{ kJ}$
$\Delta E = \Delta H - P\Delta V = -231 \text{ kJ} - (-7.22 \text{ kJ}) = -224 \text{ kJ}$

9.146 (a) $\Delta S_{total} = \Delta S_{system} + \Delta S_{surr}$ and $\Delta S_{surr} = -\Delta H/T$
$\Delta S_{total} = \Delta S_{system} + (-\Delta H/T) = \Delta S_{system} - \Delta H/T$
$\Delta S_{system} = \Delta S_{total} + \Delta H/T$
$\Delta G = \Delta H - T\Delta S$ (substitute ΔS_{system} for ΔS in this equation)
$\Delta G = \Delta H - T(\Delta S_{total} + \Delta H/T) = -T\Delta S_{total}$
$\Delta G = -T\Delta S_{total}$ For a spontaneous reaction, if $\Delta S_{total} > 0$ then $\Delta G < 0$.
(b) $\Delta G^{\circ} = \Delta H^{\circ} - T\Delta S^{\circ}$

$\Delta H^{\circ} = \Delta G^{\circ} + T\Delta S^{\circ}$

$$\Delta S_{surr} = -\frac{\Delta H^{\circ}}{T} = -\frac{[\Delta G^{\circ} + T\Delta S^{\circ}]}{T} = -\frac{[2879 \text{ x } 10^3 \text{ J/mol} + (298 \text{ K})(-262 \text{ J/(K} \cdot \text{mol)})]}{298 \text{ K}}$$

$\Delta S_{surr} = -9399 \text{ J/(K} \cdot \text{mol)}$

9.148 $NaNO_3$, 84.99; KF, 58.10

For $NaNO_3(s) \rightarrow NaNO_3(aq)$, $q = 20.4 \text{ kJ/mol x } \dfrac{1 \text{ mol } NaNO_3}{84.99 \text{ g } NaNO_3} = 0.240 \text{ kJ/g}$

For $KF(s) \rightarrow KF(aq)$, $q = -17.7 \text{ kJ/mol x } \dfrac{1 \text{ mol KF}}{58.10 \text{ g KF}} = -0.305 \text{ kJ/g}$

$q_{soln} = (110.0 \text{ g})[4.18 \text{ J/(g} \cdot {}^{\circ}\text{C)}](2.22 \text{ }^{\circ}\text{C}) = 1021 \text{ J} = 1.02 \text{ kJ}$

$q_{rxn} = -q_{soln} = -1.02 \text{ kJ}$

Let X = mass of $NaNO_3$ and Y = mass of KF

X + Y = 10.0 g, so Y = 10.0 g – X

q_{rxn} = –1.02 kJ = X(0.240 kJ/g) + Y(– 0.305 kJ/g)   (substitute for Y and solve for X)

–1.02 kJ = X(0.240 kJ/g) + (10.0 g – X)(– 0.305 kJ/g)

–1.02 kJ = (0.240 kJ)X – 3.05 kJ + (0.305 kJ)X

2.03 kJ = (0.545 kJ)X

$X = \dfrac{2.03 \text{ kJ}}{0.545 \text{ kJ}} = 3.72$ g $NaNO_3$

Y = 10.0 g – X = 10.0 g – 3.72 g = 6.28 g KF = 6.3 g KF

9.150 (a) Each S has 2 bonding pairs and 2 lone pairs of electrons. Each S is sp^3 hybridized and the geometry around each S is bent.

(b) $\Delta H = D$ (reactant bonds) – D (product bonds) = $(8\ D_{S-S}) - (4\ D_{S=S}) = +237$ kJ

$\Delta H = [(8 \text{ mol})(225 \text{ kJ/mol}) - [(4 \text{ mol})(D_{S=S})] = +237$ kJ

$- (4 \text{ mol})(D_{S=S}) = 237 \text{ kJ} - 1800 \text{ kJ} = -1563$ kJ

$D_{S=S} = (1563 \text{ kJ})/(4 \text{ mol}) = 391$ kJ/mol

(c)

σ^*_{3p}	—	
π^*_{3p}	↑	↑
π_{3p}	↑↓	↑↓
σ_{3p}	↑↓	
σ^*_{3s}	↑↓	
σ_{3s}	↑↓	

S_2

S_2 should be paramagnetic with two unpaired electrons in the π^*_{3p} MOs.

9.152 (a) (1) $2\ CH_3CO_2H(l) + Na_2CO_3(s) \rightarrow 2\ CH_3CO_2Na(aq) + CO_2(g) + H_2O(l)$

(2) $CH_3CO_2H(l) + NaHCO_3(s) \rightarrow CH_3CO_2Na(aq) + CO_2(g) + H_2O(l)$

(b) CH_3CO_2H, 60.05; Na_2CO_3, 105.99; $NaHCO_3$, 84.01

$1 \text{ gal} \times \dfrac{3.7854 \text{ L}}{1 \text{ gal}} \times \dfrac{1 \text{ mL}}{1 \times 10^{-3} \text{ L}} \times \dfrac{1.049 \text{ g } CH_3CO_2H}{1 \text{ mL}} = 3971 \text{ g } CH_3CO_2H$

$3971 \text{ g } CH_3CO_2H \times \dfrac{1 \text{ mol } CH_3CO_2H}{60.05 \text{ g } CH_3CO_2H} = 66.13 \text{ mol } CH_3CO_2H$

For reaction (1)

$66.13 \text{ mol } CH_3CO_2H \times \dfrac{1 \text{ mol } Na_2CO_3}{2 \text{ mol } CH_3CO_2H} \times \dfrac{105.99 \text{ g } Na_2CO_3}{1 \text{ mol } Na_2CO_3} \times \dfrac{1 \text{ kg}}{1000 \text{ g}} = 3.505 \text{ kg } Na_2CO_3$

For reaction (2)

$66.13 \text{ mol } CH_3CO_2H \times \dfrac{1 \text{ mol } NaHCO_3}{1 \text{ mol } CH_3CO_2H} \times \dfrac{84.01 \text{ g } NaHCO_3}{1 \text{ mol } NaHCO_3} \times \dfrac{1 \text{ kg}}{1000 \text{ g}} = 5.556 \text{ kg } NaHCO_3$

(c) $2\ CH_3CO_2H(l) + Na_2CO_3(s) \rightarrow 2\ CH_3CO_2Na(aq) + CO_2(g) + H_2O(l)$

$\Delta H°_{rxn} = [2\ \Delta H°_f(CH_3CO_2Na) + \Delta H°_f(CO_2) + \Delta H°_f(H_2O)]$

$- [2\ \Delta H°_f(CH_3CO_2H) + \Delta H°_f(Na_2CO_3)]$

$\Delta H^\circ_{rxn} = [(2 \text{ mol})(-726.1 \text{ kJ/mol}) + (1 \text{ mol})(-393.5 \text{ kJ/mol}) + (1 \text{ mol})(-285.8 \text{ kJ/mol})]$
$\quad - [(2 \text{ mol})(-484.5 \text{ kJ/mol}) + (1 \text{ mol})(-1130.7 \text{ kJ/mol})]$

$\Delta H^\circ_{rxn} = -31.8 \text{ kJ for 2 mol } CH_3CO_2H$

$\text{Heat} = -\dfrac{31.8 \text{ kJ}}{2 \text{ mol } CH_3CO_2H} \times 66.13 \text{ mol } CH_3CO_2H = -1050 \text{ kJ (liberated)}$

$CH_3CO_2H(l) + NaHCO_3(s) \rightarrow CH_3CO_2Na(aq) + CO_2(g) + H_2O(l)$

$\Delta H^\circ_{rxn} = [\Delta H^\circ_f(CH_3CO_2Na) + \Delta H^\circ_f(CO_2) + \Delta H^\circ_f(H_2O)]$
$\quad - [\Delta H^\circ_f(CH_3CO_2H) + \Delta H^\circ_f(NaHCO_3)]$

$\Delta H^\circ_{rxn} = [(1 \text{ mol})(-726.1 \text{ kJ/mol}) + (1 \text{ mol})(-393.5 \text{ kJ/mol}) + (1 \text{ mol})(-285.8 \text{ kJ/mol})]$
$\quad - [(1 \text{ mol})(-484.5 \text{ kJ/mol}) + (1 \text{ mol})(-950.8 \text{ kJ/mol})]$

$\Delta H^\circ_{rxn} = +29.9 \text{ kJ for 1 mol } CH_3CO_2H$

$q = \dfrac{29.9 \text{ kJ}}{1 \text{ mol } CH_3CO_2H} \times 66.13 \text{ mol } CH_3CO_2H = +1980 \text{ kJ (absorbed)}$

9.154 (a)

Each N is sp^3 hybridized and the geometry about each N is trigonal pyramidal.

(b)

$2/4 \ NH_3(g) + 3/4 \ N_2O(g) \rightarrow N_2(g) + 3/4 \ H_2O(l)$	$\Delta H^\circ_1 = \dfrac{-1011.2 \text{ kJ}}{4}$
$1/4 \ H_2O(l) + 1/4 \ N_2H_4(l) \rightarrow 1/8 \ O_2(g) + 1/2 \ NH_3(g)$	$\Delta H^\circ_2 = \dfrac{+286 \text{ kJ}}{8}$
$3/4 \ N_2H_4(l) + 3/4 \ H_2O(l) \rightarrow 3/4 \ N_2O(g) + 9/4 \ H_2(g)$	$\Delta H^\circ_3 = \dfrac{(3)(+317 \text{ kJ})}{4}$
$\underline{9/4 \ H_2(g) + 9/8 \ O_2(g) \rightarrow 9/4 \ H_2O(l)}$	$\Delta H^\circ_4 = \dfrac{(9)(-285.8 \text{ kJ})}{4}$
Sum $\quad N_2H_4(l) + O_2(g) \rightarrow N_2(g) + 2 \ H_2O(l)$	$\Delta H^\circ = -622 \text{ kJ}$

(c) N_2H_4, 32.045

$\text{mol } N_2H_4 = 100.0 \text{ g } N_2H_4 \times \dfrac{1 \text{ mol } N_2H_4}{32.045 \text{ g } N_2H_4} = 3.12 \text{ mol } N_2H_4$

$q = (3.12 \text{ mol } N_2H_4)(622 \text{ kJ/mol}) = 1940 \text{ kJ}$

10

Gases: Their Properties and Behavior

10.1 914 mbar = 914 x 10⁻³ bar

$$\text{pressure} = 914 \times 10^{-3} \text{ bar } \times \frac{760 \text{ mm Hg}}{1.013\ 25 \text{ bar}} = 686 \text{ mm Hg}$$

10.2 $\text{pressure} = \left[1 \text{ atm} + \left(702 \text{ ft } \times \frac{1 \text{ atm}}{33 \text{ ft}} \right) \right] \times \frac{1.013\ 25 \times 10^5 \text{Pa}}{1 \text{atm}} = 2.26 \times 10^6 \text{ Pa}$

10.3 The pressure in the flask is less than 0.975 atm because the liquid level is higher on the side connected to the flask. The 24.7 cm of Hg is the difference between the two pressures.

$$\text{Pressure difference} = 24.7 \text{ cm Hg} \times \frac{1.00 \text{ atm}}{76.0 \text{ cm Hg}} = 0.325 \text{ atm}$$

Pressure in flask = 0.975 atm − 0.325 atm = 0.650 atm

10.4 (a)

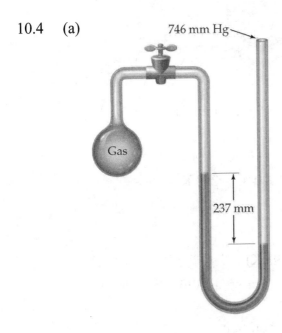

(b) (13.6 g/mL)/(0.822 g/mL) = 16.5

$$P_{\text{bulb}} = 746 \text{ mm Hg} - \left(237 \text{ mm min oil } \times \frac{1 \text{ mm Hg}}{16.5 \text{ mm min oil}} \right) = 732 \text{ mm Hg}$$

10.5 Assume an initial volume of 1.00 L.
First consider the volume change resulting from a change in the number of moles with the pressure and temperature constant.

$$\frac{V_i}{n_i} = \frac{V_f}{n_f}; \quad V_f = \frac{V_i\, n_f}{n_i} = \frac{(1.00\ L)(0.225\ mol)}{0.3\ mol} = 0.75\ L$$

Now consider the volume change from 0.75 L as a result of a change in temperature with the number of moles and the pressure constant.

$$\frac{V_i}{T_i} = \frac{V_f}{T_f}; \quad V_f = \frac{V_i\, T_f}{T_i} = \frac{(0.75\ L)(200\ K)}{300\ K} = 0.5\ L$$

The volume would be cut in half as a result of the decrease in the number of moles of gas and a temperature decrease.

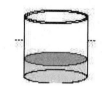

10.6   1 atm   (a)   (b)

10.7 $n = \dfrac{PV}{RT} = \dfrac{(1.00\ atm)(1.00 \times 10^5\ L)}{\left(0.082\ 06\ \dfrac{L \cdot atm}{K \cdot mol}\right)(273.15\ K)} = 4.46 \times 10^3\ mol\ CH_4$

10.8 C_3H_8, 44.10;   V = 350 mL = 0.350 L;   T = 20 °C = 293 K

$$n = 3.2\ g \times \frac{1\ mol\ C_3H_8}{44.10\ g\ C_3H_8} = 0.073\ mol\ C_3H_8$$

$$P = \frac{nRT}{V} = \frac{(0.073\ mol)\left(0.082\ 06\ \dfrac{L \cdot atm}{K \cdot mol}\right)(293\ K)}{0.350\ L} = 5.0\ atm$$

10.9   The volume and number of moles of gas remain constant.

$$\frac{nR}{V} = \frac{P_i}{T_i} = \frac{P_f}{T_f}; \quad T_f = \frac{P_f\, T_i}{P_i} = \frac{(2.37\ atm)(273\ K)}{2.15\ atm} = 301\ K = 28\ °C$$

10.10 $\dfrac{P_i V_i}{T_i} = \dfrac{P_f V_f}{T_f}$;   25.0 °C = 298 K; $V_f = \dfrac{P_i V_i T_f}{T_i P_f} = \dfrac{(745\ mm\ Hg)(45.0\ L)(225\ K)}{(298\ K)(178\ mm\ Hg)} = 142\ L$

10.11 $CaCO_3(s) + 2 HCl(aq) \rightarrow CaCl_2(aq) + CO_2(g) + H_2O(l)$
$CaCO_3$, 100.1

$$\text{mole } CO_2 = 33.7 \text{ g } CaCO_3 \times \frac{1 \text{ mol } CaCO_3}{100.1 \text{ g } CaCO_3} \times \frac{1 \text{ mol } CO_2}{1 \text{ mol } CaCO_3} = 0.337 \text{ mol } CO_2$$

$$V = \frac{nRT}{P} = \frac{(0.337 \text{ mol})\left(0.082\ 06\ \dfrac{L \cdot atm}{K \cdot mol}\right)(273 \text{ K})}{1.00 \text{ atm}} = 7.55 \text{ L}$$

10.12 $3 H_2(g) + N_2(g) \rightarrow 2 NH_3(g)$

$$V_{H_2} = 500.0 \text{ L } NH_3 \times \frac{3 \text{ mol } H_2}{2 \text{ mol } NH_3} = 750.0 \text{ L } H_2$$

$$V_{N_2} = 500.0 \text{ L } NH_3 \times \frac{1 \text{ mol } N_2}{2 \text{ mol } NH_3} = 250.0 \text{ L } N_2$$

10.13 $n = \dfrac{PV}{RT} = \dfrac{(1.00 \text{ atm})(1.00 \text{ L})}{\left(0.082\ 06\ \dfrac{L \cdot atm}{K \cdot mol}\right)(273 \text{ K})} = 0.0446 \text{ mol}$

$$\text{molar mass} = \frac{1.52 \text{ g}}{0.0446 \text{ mol}} = 34.1 \text{ g/mol}; \quad \text{molecular mass} = 34.1$$

$Na_2S(aq) + 2 HCl(aq) \rightarrow H_2S(g) + 2 NaCl(aq)$
The foul-smelling gas is H_2S, hydrogen sulfide.

10.14 (a) CO_2, 44.0

$$d = \frac{PM}{RT} = \frac{(1 \text{ atm})(44.0 \text{ g/mol})}{\left(0.082\ 06\ \dfrac{L \cdot atm}{K \cdot mol}\right)(298 \text{ K})} = 1.80 \text{ g/L}$$

(b) Carbon dioxide has a higher molar mass (44.0 g/mol) than the major components of air N_2 (28.0 g/mol) and O_2 (32.0 g/mol).

10.15 O_2, 32.0; N_2, 28.0

$$235.5 \text{ g } O_2 \times \frac{1 \text{ mol } O_2}{32.0 \text{ g } O_2} = 7.36 \text{ mol } O_2$$

$$366.8 \text{ g } N_2 \times \frac{1 \text{ mol } N_2}{28.0 \text{ g } N_2} = 13.1 \text{ mol } N_2$$

$n_{total} = 7.36 \text{ mol} + 13.1 \text{ mol} = 20.5 \text{ mol}$

$$X_{O_2} = \frac{7.36 \text{ mol}}{20.5 \text{ mol}} = 0.360; \quad X_{N_2} = \frac{13.1 \text{ mol}}{20.5 \text{ mol}} = 0.640$$

$P_{O_2} = X_{O_2} \cdot P_{total} = (0.360)(50.0 \text{ atm}) = 18.0 \text{ atm}$

$P_{N_2} = X_{N_2} \cdot P_{total} = (0.640)(50.0 \text{ atm}) = 32.0 \text{ atm}$

10.16 $P_{O_2} = P_{tot} \cdot X_{O_2}$

0.21 atm $= (8.38 \text{ atm}) \cdot X_{O_2}$

$X_{O_2} = \dfrac{0.21 \text{ atm}}{8.38 \text{ atm}} = 0.025$

10.17 O_2, 32.0; Kr, 83.8

$\dfrac{\text{rate } O_2}{\text{rate Kr}} = \sqrt{\dfrac{M_{Kr}}{M_{O_2}}} = \sqrt{\dfrac{83.8}{32.0}} = 1.62$

O_2 diffuses 1.62 times faster than Kr.

10.18 SO_2, 64.06

$\dfrac{\text{rate}_1}{\text{rate}_2} = 1.414 = \sqrt{\dfrac{M_{SO_2}}{M_x}}; \quad \sqrt{M_x} = \dfrac{\sqrt{M_x}}{1.414} = \dfrac{\sqrt{64.06}}{1.414} = 5.660$

$M_x = (5.660)^2 = 32.00;$ The unknown gas could be O_2.

10.19 For diethyl ether, MAC $= \dfrac{15 \text{ mm Hg}}{760 \text{ mm Hg}}$ x 100% = 2.0%

10.20 (a) $P_{chloroform} = X_{chloroform} \cdot P_{total} = \left(\dfrac{0.77}{100}\right)(760 \text{mmHg}) = 5.9 \text{ mm Hg}$

(b) volume of chloroform $= (10.0 \text{ L})(0.0077) = 0.077 \text{ L}$

$CHCl_3$, 119.4

$PV = nRT$

$n = \dfrac{PV}{RT} = \dfrac{(1 \text{ atm})(0.077 \text{ L})}{\left(0.0821\dfrac{\text{L} \cdot \text{atm}}{\text{mol} \cdot \text{K}}\right)(273 \text{ K})} = 0.003 \text{ 44 mol}$

mass $CHCl_3 = 0.003 \text{ 44 mol} \times \dfrac{119.4 \text{ g chloroform}}{1 \text{ mol chloroform}} = 0.41 \text{ g chloroform}$

10.21 (a) $P_{N_2O} = 0.70 \times 50 \text{ atm} = 35 \text{ atm}$ and $P_{O_2} = 0.30 \times 50 \text{ atm} = 15 \text{ atm}$

(b) $PV = nRT$

$n_{N_2O} = \dfrac{PV}{RT} = \dfrac{(35 \text{ atm})(10.0 \text{ L})}{\left(0.0821\dfrac{\text{L} \cdot \text{atm}}{\text{mol} \cdot \text{K}}\right)(298 \text{ K})} = 14.3 \text{ mol } N_2O$

mass $N_2O = 14.3 \text{ mol} \times \dfrac{44.01 \text{ g } N_2O}{1 \text{ mol } N_2O} = 629 \text{ g } N_2O$

$PV = nRT$

$$n_{O_2} = \frac{PV}{RT} = \frac{(15 \text{ atm})(10.0 \text{ L})}{\left(0.0821\dfrac{\text{L} \cdot \text{atm}}{\text{mol} \cdot \text{K}}\right)(298 \text{ K})} = 6.13 \text{ mol O}_2$$

$$\text{mass O}_2 = 6.13 \text{ mol} \times \frac{32.00 \text{ g O}_2}{1 \text{ mol O}_2} = 196 \text{ g O}_2$$

10.22 $P_{\text{sevoflurane}} = X_{\text{sevoflurane}} \cdot P_{\text{total}} = (0.08)(760 \text{ mm Hg}) = 61 \text{ mm Hg}$

$P_{N_2O} = X_{N_2O} \cdot P_{\text{total}} = (0.62)(760 \text{ mm Hg}) = 471 \text{ mm Hg}$

$P_{O_2} = X_{O_2} \cdot P_{\text{total}} = (0.30)(760 \text{ mm Hg}) = 228 \text{ mm Hg}$

Conceptual Problems

10.24 The gas pressure in the bulb in mm Hg is equal to the difference in the height of the Hg in the two arms of the manometer.

10.26 (a) 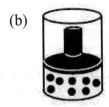 The volume of a gas is proportional to the kelvin temperature at constant pressure. As the temperature increases from 300 K to 450 K, the volume will increase by a factor of 1.5.

(b) 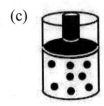 The volume of a gas is inversely proportional to pressure at constant temperature. As the pressure increases from 1 atm to 2 atm, the volume will decrease by a factor of 2.

(c) $PV = nRT$;   The amount of gas (n) is constant.

Therefore $nR = \dfrac{P_i V_i}{T_i} = \dfrac{P_f V_f}{T_f}$.

Assume $V_i = 1$ L and solve for V_f.

$$\frac{P_i V_i T_f}{T_i P_f} = \frac{(3 \text{ atm})(1 \text{ L})(200 \text{ K})}{(300 \text{ K})(2 \text{ atm})} = V_f = 1 \text{ L}$$

There is no change in volume.

10.28 The two gases should mix randomly and homogeneously and this is best represented by drawing (c).

10.30 The two gases will be equally distributed among the three flasks.

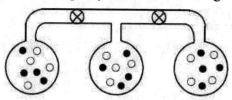

10.32 The picture on the right will be the same as that on the left, apart from random scrambling of the He and Ar atoms.

Section Problems
Gases and Gas Pressure (Section 10.1)

10.34 1.00 atm = 14.7 psi

$$1.00 \text{ mm Hg} \times \frac{1 \text{ atm}}{760 \text{ mm Hg}} \times \frac{14.7 \text{ psi}}{1 \text{ atm}} = 1.93 \times 10^{-2} \text{ psi}$$

10.36 A gas is a collection of atoms or molecules moving independently through a volume that is largely empty space. Collisions of randomly moving particles with the walls of their container exert a force per unit area that we perceive as pressure.

10.38 $P = 480 \text{ mm Hg} \times \dfrac{1.00 \text{ atm}}{760 \text{ mm Hg}} = 0.632 \text{ atm}$

$P = 480 \text{ mm Hg} \times \dfrac{101,325 \text{ Pa}}{760 \text{ mm Hg}} = 6.40 \times 10^4 \text{ Pa}$

10.40 $P_{flask} > 754.3 \text{ mm Hg};$ $P_{flask} = 754.3 \text{ mm Hg} + 176 \text{ mm Hg} = 930 \text{ mm Hg}$

10.42 $P_{flask} > 752.3 \text{ mm Hg}$ (see Figure 10.5); If the pressure in the flask can support a column of ethyl alcohol (d = 0.7893 g/mL) 55.1 cm high, then it can only support a column of Hg that is much shorter because of the higher density of Hg.

$$55.1 \text{ cm} \times \frac{0.7893 \text{ g/mL}}{13.546 \text{ g/mL}} = 3.21 \text{ cm Hg} = 32.1 \text{ mm Hg}$$

$P_{flask} = 752.3 \text{ mm Hg} + 32.1 \text{ mm Hg} = 784.4 \text{ mm Hg}$

$P_{flask} = 784.4 \text{ mm Hg} \times \dfrac{101,325 \text{ Pa}}{760 \text{ mm Hg}} = 1.046 \times 10^5 \text{ Pa}$

10.44 Average molecular mass of air = 28.96; CO_2, 44.01

$$P = 760 \text{ mm Hg} \times \frac{44.01 \text{ g/mol}}{28.96 \text{ g/mol}} = 1155 \text{ mm Hg}$$

10.46   % Volume
    N_2 78.08
    O_2 20.95
    Ar   0.93
    CO_2 0.037

The % volume for a particular gas is proportional to the number of molecules of that gas in a mixture of gases.

Average molecular mass of air

$= (0.7808)(\text{mol. mass } N_2) + (0.2095)(\text{mol. mass } O_2)$
$\qquad\qquad + (0.0093)(\text{at. mass Ar}) + (0.000\ 37)(\text{mol. mass } CO_2)$
$= (0.7808)(28.01) + (0.2095)(32.00)$
$\qquad\qquad + (0.0093)(39.95) + (0.000\ 37)(44.01) = 28.96$

The Gas Laws (Sections 10.2–10.3)

10.48 (a) $\dfrac{nR}{V} = \dfrac{P_i}{T_i} = \dfrac{P_f}{T_f}; \qquad \dfrac{P_i T_f}{T_i} = P_f$

Let $P_i = 1$ atm, $T_i = 100$ K, $T_f = 300$ K

$P_f = \dfrac{P_i T_f}{T_i} = \dfrac{(1\text{ atm})(300\text{ K})}{(100\text{ K})} = 3$ atm

The pressure would triple.

(b) $\dfrac{RT}{V} = \dfrac{P_i}{n_i} = \dfrac{P_f}{n_f}; \qquad \dfrac{P_i n_f}{n_i} = P_f$

Let $P_i = 1$ atm, $n_i = 3$ mol, $n_f = 1$ mol

$P_f = \dfrac{P_i n_f}{n_i} = \dfrac{(1\text{ atm})(1\text{ mol})}{(3\text{ mol})} = \dfrac{1}{3}$ atm

The pressure would be $\dfrac{1}{3}$ the initial pressure.

(c) $nRT = P_i V_i = P_f V_f; \qquad \dfrac{P_i V_i}{V_f} = P_f$

Let $P_i = 1$ atm, $V_i = 1$ L, $V_f = 1 - 0.45$ L $= 0.55$ L

$P_f = \dfrac{P_i V_i}{V_f} = \dfrac{(1\text{ atm})(1\text{ L})}{(0.55\text{ L})} = 1.8$ atm

The pressure would increase by 1.8 times.

(d) $nR = \dfrac{P_i V_i}{T_i} = \dfrac{P_f V_f}{T_f}; \qquad \dfrac{P_i V_i T_f}{T_i V_f} = P_f$

Let $P_i = 1$ atm, $V_i = 1$ L, $T_i = 200$ K, $V_f = 3$ L, $T_i = 100$ K

$$P_f = \frac{P_i V_i T_f}{T_i V_f} = \frac{(1\ \text{atm})(1\ \text{L})(100\ \text{K})}{(200\ \text{K})(3\ \text{L})} = 0.17\ \text{atm}$$

The pressure would be 0.17 times the initial pressure.

10.50 They all contain the same number of gas molecules.

10.52 n and T are constant; therefore $nRT = P_i V_i = P_f V_f$

$$V_f = \frac{P_i V_i}{P_f} = \frac{(150\ \text{atm})(49.0\ \text{L})}{(1.02\ \text{atm})} = 7210\ \text{L}$$

n and P are constant; therefore $\dfrac{nR}{P} = \dfrac{V_i}{T_i} = \dfrac{V_f}{T_f}$

$$V_f = \frac{V_i T_f}{T_i} = \frac{(49.0\ \text{L})(308\ \text{K})}{(293\ \text{K})} = 51.5\ \text{L}$$

10.54 $15.0\ \text{g}\ CO_2 \times \dfrac{1\ \text{mol}\ CO_2}{44.0\ \text{g}\ CO_2} = 0.341\ \text{mol}\ CO_2$

$$P = \frac{nRT}{V} = \frac{(0.341\ \text{mol})\left(0.082\ 06\ \dfrac{\text{L} \cdot \text{atm}}{\text{K} \cdot \text{mol}}\right)(300\ \text{K})}{(0.30\ \text{L})} = 27.98\ \text{atm}$$

$27.98\ \text{atm} \times \dfrac{760\ \text{mm Hg}}{1\ \text{atm}} = 2.1 \times 10^4\ \text{mm Hg}$

10.56 $\dfrac{1\ \text{H atom}}{\text{cm}^3} \times \dfrac{1\ \text{mol H}}{6.02 \times 10^{23}\ \text{atoms}} \times \dfrac{1000\ \text{cm}^3}{1\ \text{L}} = 1.7 \times 10^{-21}\ \text{mol H/L}$

$$P = \frac{nRT}{V} = \frac{(1.7 \times 10^{-21}\ \text{mol})\left(0.082\ 06\ \dfrac{\text{L} \cdot \text{atm}}{\text{K} \cdot \text{mol}}\right)(100\ \text{K})}{(1\ \text{L})} = 1.4 \times 10^{-20}\ \text{atm}$$

$P = 1.4 \times 10^{-20}\ \text{atm} \times \dfrac{760\ \text{mm Hg}}{1.0\ \text{atm}} = 1 \times 10^{-17}\ \text{mm Hg}$

10.58 $P = 13{,}800\ \text{kPa} \times \dfrac{1000\ \text{Pa}}{1\ \text{kPa}} \times \dfrac{1\ \text{atm}}{101{,}325\ \text{Pa}} = 136.2\ \text{atm}$

n and T are constant; therefore $nRT = P_i V_i = P_f V_f$

$$V_f = \frac{P_i V_i}{P_f} = \frac{(136.2\ \text{atm})(2.30\ \text{L})}{(1.25\ \text{atm})} = 250.6\ \text{L}$$

$250.6\ \text{L} \times \dfrac{1\ \text{balloon}}{1.5\ \text{L}} = 167\ \text{balloons}$

10.60 For steam, T = 123.0 °C = 396 K

$$n = \frac{PV}{RT} = \frac{(0.93 \text{ atm})(15.0 \text{ L})}{\left(0.082\ 06\ \dfrac{\text{L} \cdot \text{atm}}{\text{K} \cdot \text{mol}}\right)(396 \text{ K})} = 0.43 \text{ mol steam}$$

For ice, H_2O, 18.02; $n = 10.5 \text{ g} \times \dfrac{1 \text{ mol}}{18.02 \text{ g}} = 0.583 \text{ mol ice}$

Because the number of moles of ice is larger than the number of moles of steam, the ice contains more H_2O molecules.

10.62 The containers are identical. Both containers contain the same number of gas molecules. Weigh the containers. Because the molecular mass for O_2 is greater than the molecular mass for H_2, the heavier container contains O_2.

Gas Stoichiometry (Section 10.4)

10.64 room volume = 4.0 m x 5.0 m x 2.5 m x $\dfrac{1 \text{ L}}{10^{-3} \text{ m}^3}$ = 5.0 x 10⁴ L

$$n_{total} = \frac{PV}{RT} = \frac{(1.0 \text{ atm})(5.0 \times 10^4 \text{ L})}{\left(0.082\ 06\ \dfrac{\text{L} \cdot \text{atm}}{\text{K} \cdot \text{mol}}\right)(273 \text{ K})} = 2.23 \times 10^3 \text{ mol}$$

$n_{O_2} = (0.2095)n_{total} = (0.2095)(2.23 \times 10^3 \text{ mol}) = 467 \text{ mol } O_2$

mass O_2 = 467 mol x $\dfrac{32.0 \text{ g}}{1 \text{ mol}}$ = 1.5×10^4 g O_2

10.66 (a) CH_4, 16.04; $d = \dfrac{16.04 \text{ g}}{22.4 \text{ L}} = 0.716 \text{ g/L}$

(b) CO_2, 44.01; $d = \dfrac{44.01 \text{ g}}{22.4 \text{ L}} = 1.96 \text{ g/L}$

(c) O_2, 32.00; $d = \dfrac{32.00 \text{ g}}{22.4 \text{ L}} = 1.43 \text{ g/L}$

10.68 $n = \dfrac{PV}{RT} = \dfrac{\left(356 \text{ mm Hg} \times \dfrac{1.00 \text{ atm}}{760 \text{ mm Hg}}\right)(1.500 \text{ L})}{\left(0.082\ 06\ \dfrac{\text{L} \cdot \text{atm}}{\text{K} \cdot \text{mol}}\right)(295.5 \text{ K})} = 0.0290 \text{ mol}$

molar mass = $\dfrac{0.9847 \text{ g}}{0.0290 \text{ mol}}$ = 34.0 g/mol; molecular weight = 34.0

10.70 2 HgO(s) → 2 Hg(l) + O_2(g);   HgO, 216.59

10.57 g HgO x $\dfrac{1 \text{ mol HgO}}{216.59 \text{ g HgO}}$ x $\dfrac{1 \text{ mol } O_2}{2 \text{ mol HgO}}$ = 0.024 40 mol O_2

$$V = \frac{nRT}{P} = \frac{(0.024\ 40\ \text{mol})\left(0.082\ 06\ \dfrac{\text{L} \cdot \text{atm}}{\text{K} \cdot \text{mol}}\right)(273.15\ \text{K})}{1.000\ \text{atm}} = 0.5469\ \text{L}$$

10.72 $Zn(s) + 2\ HCl(aq) \rightarrow ZnCl_2(aq) + H_2(g)$

(a) $25.5\ \text{g Zn} \times \dfrac{1\ \text{mol Zn}}{65.39\ \text{g Zn}} \times \dfrac{1\ \text{mol } H_2}{1\ \text{mol Zn}} = 0.390\ \text{mol } H_2$

$$V = \frac{nRT}{P} = \frac{(0.390\ \text{mol})\left(0.082\ 06\ \dfrac{\text{L} \cdot \text{atm}}{\text{K} \cdot \text{mol}}\right)(288\ \text{K})}{\left(742\ \text{mm Hg} \times \dfrac{1.00\ \text{atm}}{760\ \text{mm Hg}}\right)} = 9.44\ \text{L}$$

(b) $n = \dfrac{PV}{RT} = \dfrac{\left(350\ \text{mm Hg} \times \dfrac{1.00\ \text{atm}}{760\ \text{mm Hg}}\right)(5.00\ \text{L})}{\left(0.082\ 06\ \dfrac{\text{L} \cdot \text{atm}}{\text{K} \cdot \text{mol}}\right)(303.15\ \text{K})} = 0.092\ 56\ \text{mol } H_2$

$0.092\ 56\ \text{mol } H_2 \times \dfrac{1\ \text{mol Zn}}{1\ \text{mol } H_2} \times \dfrac{65.39\ \text{g Zn}}{1\ \text{mol Zn}} = 6.05\ \text{g Zn}$

10.74 (a) $V_{24h} = (4.50\ \text{L/min})(60\ \text{min/h})(24\ \text{h/day}) = 6480\ \text{L}$
$V_{CO_2} = (0.034)V_{24h} = (0.034)(6480\ \text{L}) = 220\ \text{L}$

$$n = \frac{PV}{RT} = \frac{\left(735\ \text{mm Hg} \times \dfrac{1.00\ \text{atm}}{760\ \text{mm Hg}}\right)(220\ \text{L})}{\left(0.082\ 06\ \dfrac{\text{L} \cdot \text{atm}}{\text{K} \cdot \text{mol}}\right)(298\ \text{K})} = 8.70\ \text{mol } CO_2$$

$8.70\ \text{mol } CO_2 \times \dfrac{44.01\ \text{g } CO_2}{1\ \text{mol } CO_2} = 383\ \text{g} = 380\ \text{g } CO_2$

(b) $2\ Na_2O_2(s) + 2\ CO_2(g) \rightarrow 2\ Na_2CO_3(s) + O_2(g)$; Na_2O_2, 77.98
$3.65\ \text{kg} = 3650\ \text{g}$

$3650\ \text{g } Na_2O_2 \times \dfrac{1\ \text{mol } Na_2O_2}{77.98\ \text{g } Na_2O_2} \times \dfrac{2\ \text{mol } CO_2}{2\ \text{mol } Na_2O_2} \times \dfrac{1\ \text{day}}{8.70\ \text{mol } CO_2} = 5.4\ \text{days}$

10.76 He, 4.00; 365 mL = 0.365 L; 25 °C = 298 K

$$n_{He} = \frac{PV}{RT} = \frac{(7.8\ \text{atm})(0.365\ \text{L})}{\left(0.082\ 06\ \dfrac{\text{L} \cdot \text{atm}}{\text{K} \cdot \text{mol}}\right)(298\ \text{K})} = 0.116\ \text{mol He}$$

$\text{mass He} = 0.116\ \text{mol He} \times \dfrac{4.00\ \text{g He}}{1\ \text{mol He}} = 0.464\ \text{g He}$

10.78 S, 32.1

$$\text{mass S} = 1 \text{ kg} \times \frac{1000 \text{ g}}{1 \text{ kg}} \times 0.02 = 20 \text{ g S}$$

$$\text{mol S} = 20 \text{ g S} \times \frac{1 \text{ mol S}}{32.1 \text{ g S}} = 0.62 \text{ mol S} = 0.62 \text{ mol SO}_2$$

$$V_{SO_2} = \frac{nRT}{P} = \frac{(0.62 \text{ mol})\left(0.082\ 06\ \frac{\text{L} \cdot \text{atm}}{\text{K} \cdot \text{mol}}\right)(298 \text{ K})}{1.0 \text{ atm}} = 15 \text{ L}$$

Dalton's Law and Mole Fraction (Section 10.5)

10.80 Because of Avogadro's Law ($V \propto n$), the % volumes are also % moles.

	% mole
N_2	78.08
O_2	20.95
Ar	0.93
CO_2	0.038

In decimal form, % mole = mole fraction.

$$P_{N_2} = X_{N_2} \cdot P_{total} = (0.7808)(1.000 \text{ atm}) = 0.7808 \text{ atm}$$

$$P_{O_2} = X_{O_2} \cdot P_{total} = (0.2095)(1.000 \text{ atm}) = 0.2095 \text{ atm}$$

$$P_{Ar} = X_{Ar} \cdot P_{total} = (0.0093)(1.000 \text{ atm}) = 0.0093 \text{ atm}$$

$$P_{CO_2} = X_{CO_2} \cdot P_{total} = (0.000\ 38)(1.000 \text{ atm}) = 0.000\ 38 \text{ atm}$$

Pressures of the rest are negligible.

10.82 Assume a 100.0 g sample.   g CO_2 = 1.00 g and g O_2 = 99.0 g

$$\text{mol CO}_2 = 1.00 \text{ g CO}_2 \times \frac{1 \text{ mol CO}_2}{44.01 \text{ g CO}_2} = 0.0227 \text{ mol CO}_2$$

$$\text{mol O}_2 = 99.0 \text{ g O}_2 \times \frac{1 \text{ mol O}_2}{32.00 \text{ g O}_2} = 3.094 \text{ mol O}_2$$

$$n_{total} = 3.094 \text{ mol} + 0.0227 \text{ mol} = 3.117 \text{ mol}$$

$$X_{O_2} = \frac{3.094 \text{ mol}}{3.117 \text{ mol}} = 0.993; \quad X_{CO_2} = \frac{0.0227 \text{ mol}}{3.117 \text{ mol}} = 0.007\ 28$$

$$P_{O_2} = X_{O_2} \cdot P_{total} = (0.993)(0.977 \text{ atm}) = 0.970 \text{ atm}$$

$$P_{CO_2} = X_{CO_2} \cdot P_{total} = (0.007\ 28)(0.977 \text{ atm}) = 0.007\ 11 \text{ atm}$$

10.84 Assume a 100.0 g sample.

$$\text{g HCl} = (0.0500)(100.0 \text{ g}) = 5.00 \text{ g}; \quad 5.00 \text{ g HCl} \times \frac{1 \text{ mol HCl}}{36.5 \text{ g HCl}} = 0.137 \text{ mol HCl}$$

$$\text{g H}_2 = (0.0100)(100.0 \text{ g}) = 1.00 \text{ g}; \quad 1.00 \text{ g H}_2 \times \frac{1 \text{ mol H}_2}{2.016 \text{ g H}_2} = 0.496 \text{ mol H}_2$$

g Ne = (0.94)(100.0 g) = 94 g; $\quad$ 94 g Ne x $\dfrac{1 \text{ mol Ne}}{20.18 \text{ g Ne}}$ = 4.66 mol Ne

n_{total} = 0.137 + 0.496 + 4.66 = 5.3 mol

$X_{\text{HCl}} = \dfrac{0.137 \text{ mol}}{5.3 \text{ mol}} = 0.026; \quad X_{\text{H}_2} = \dfrac{0.496 \text{ mol}}{5.3 \text{ mol}} = 0.094; \quad X_{\text{Ne}} = \dfrac{4.66 \text{ mol}}{5.3 \text{ mol}} = 0.88$

10.86 (a) H_2, 2.016

$$P = \dfrac{nRT}{V} = \dfrac{\left(14.2 \text{ g x } \dfrac{1 \text{ mol}}{2.016 \text{ g}}\right)\left(0.082\ 06\ \dfrac{\text{L} \cdot \text{atm}}{\text{K} \cdot \text{mol}}\right)(290 \text{ K})}{(100.0 \text{ L})} = 1.68 \text{ atm}$$

(b) $$P = \dfrac{nRT}{V} = \dfrac{\left(36.7 \text{ g x } \dfrac{1 \text{ mol}}{39.95 \text{ g}}\right)\left(0.082\ 06\ \dfrac{\text{L} \cdot \text{atm}}{\text{K} \cdot \text{mol}}\right)(290 \text{ K})}{(100.0 \text{ L})} = 0.219 \text{ atm}$$

10.88 $P_{\text{total}} = P_{\text{H}_2} + P_{\text{H}_2\text{O}}; \quad P_{\text{H}_2} = P_{\text{total}} - P_{\text{H}_2\text{O}} = 747 \text{ mm Hg} - 23.8 \text{ mm Hg} = 723 \text{ mm Hg}$

$$n = \dfrac{PV}{RT} = \dfrac{\left(723 \text{ mm Hg x } \dfrac{1.00 \text{ atm}}{760 \text{ mm Hg}}\right)(3.557 \text{ L})}{\left(0.082\ 06\ \dfrac{\text{L} \cdot \text{atm}}{\text{K} \cdot \text{mol}}\right)(298 \text{ K})} = 0.1384 \text{ mol } H_2$$

0.1384 mol H_2 x $\dfrac{1 \text{ mol Mg}}{1 \text{ mol } H_2}$ x $\dfrac{24.3 \text{ g Mg}}{1 \text{ mol Mg}}$ = 3.36 g Mg

10.90 (a) average molecular mass for natural gas
$\qquad$ = (0.915)(16.04) + (0.085)(30.07) = 17.2

total moles of gas = 15.50 g x $\dfrac{1 \text{ mol gas}}{17.2 \text{ g gas}}$ = 0.901 mol gas

(b) $$P = \dfrac{(0.901 \text{ mol})\left(0.082\ 06\ \dfrac{\text{L} \cdot \text{atm}}{\text{K} \cdot \text{mol}}\right)(293 \text{ K})}{(15.00 \text{ L})} = 1.44 \text{ atm}$$

(c) $P_{\text{CH}_4} = X_{\text{CH}_4} \cdot P_{\text{total}} = (1.44 \text{ atm})(0.915) = 1.32 \text{ atm}$

$\quad P_{\text{C}_2\text{H}_6} = X_{\text{C}_2\text{H}_6} \cdot P_{\text{total}} = (1.44 \text{ atm})(0.085) = 0.12 \text{ atm}$

(d) $\Delta H_{\text{combustion}}(\text{CH}_4) = -890.3 \text{ kJ/mol}$ and $\Delta H_{\text{combustion}}(\text{C}_2\text{H}_6) = -1427.7 \text{ kJ/mol}$
Heat liberated = (0.915)(0.901 mol)(−890.3 kJ/mol)
$\qquad\qquad\qquad$ + (0.085)(0.901)(−1427.7 kJ/mol) = −843 kJ

Kinetic–Molecular Theory (Section 10.6)

10.92 The kinetic-molecular theory is based on the following assumptions:
1. A gas consists of tiny particles, either atoms or molecules, moving about at random.
2. The volume of the particles themselves is negligible compared with the total volume of the gas; most of the volume of a gas is empty space.

3. The gas particles act independently; there are no attractive or repulsive forces between particles.

4. Collisions of the gas particles, either with other particles or with the walls of the container, are elastic; that is, the total kinetic energy of the gas particles is constant at constant T.

5. The average kinetic energy of the gas particles is proportional to the Kelvin temperature of the sample.

10.94 $u = \sqrt{\dfrac{3\,RT}{M}} = \sqrt{\dfrac{3 \times 8.314\ \text{kg m}^2/(\text{s}^2\,\text{K mol}) \times 220\ \text{K}}{28.0 \times 10^{-3}\ \text{kg/mol}}} = 443\ \text{m/s}$

10.96 For Br_2: $u = \sqrt{\dfrac{3RT}{M}} = \sqrt{\dfrac{3 \times 8.314\ \text{kg m}^2/(\text{s}^2\,\text{K mol}) \times 293\ \text{K}}{159.8 \times 10^{-3}\ \text{kg/mol}}} = 214\ \text{m/s}$

For Xe: $u = 214\ \text{m/s} = \sqrt{\dfrac{3 \times 8.314\ \text{kg m}^2/(\text{s}^2\,\text{K mol}) \times T}{131.3 \times 10^{-3}\ \text{kg/mol}}}$

Square both sides of the equation and solve for T.

$45796\ \text{m}^2/\text{s}^2 = \dfrac{3 \times 8.314\ \text{kg m}^2/(\text{s}^2\,\text{K mol}) \times T}{131.3 \times 10^{-3}\ \text{kg/mol}}$

$T = 241\ \text{K} = -32\ ^\circ\text{C}$

10.98 For H_2, $u = \sqrt{\dfrac{3\,RT}{M}} = \sqrt{\dfrac{3 \times 8.314\ \text{kg m}^2/(\text{s}^2\,\text{K mol}) \times 150\ \text{K}}{2.02 \times 10^{-3}\ \text{kg/mol}}} = 1360\ \text{m/s}$

For He, $u = \sqrt{\dfrac{3 \times 8.314\ \text{kg m}^2/(\text{s}^2\,\text{K mol}) \times 648\ \text{K}}{4.00 \times 10^{-3}\ \text{kg/mol}}} = 2010\ \text{m/s}$

He at 375 °C has the higher average speed.

10.100 $u = 45\ \text{m/s} = \sqrt{\dfrac{3 \times 8.314\ \text{kg m}^2/(\text{s}^2\,\text{K mol}) \times T}{4.00 \times 10^{-3}\ \text{kg/mol}}}$

Square both sides of the equation and solve for T.

$2025\ \text{m}^2/\text{s}^2 = \dfrac{3 \times 8.314\ \text{kg m}^2/(\text{s}^2\,\text{K mol}) \times T}{4.00 \times 10^{-3}\ \text{kg/mol}}$

$T = 0.325\ \text{K} = -272.83\ ^\circ\text{C}$ (near absolute zero)

Graham's Law (Section 10.7)

10.102 Diffusion – The mixing of different gases by random molecular motion and with frequent collisions.

Effusion – The process in which gas molecules escape through a tiny hole in a membrane without collisions.

10.104 $\dfrac{\text{rate Xe}}{\text{rate Z}} = \sqrt{\dfrac{M_Z}{M_{Xe}}}; \quad \dfrac{1}{1.86} = \dfrac{\sqrt{M_Z}}{\sqrt{131.29}}; \quad \dfrac{\sqrt{131.29}}{1.86} = \sqrt{M_Z}; \quad \dfrac{131.29}{(1.86)^2} = M_Z$

$M_Z = 37.9$ g/mol; molecular weight = 37.9; The gas could be F_2.

10.106 HCl, 36.5; F_2, 38.0; Ar, 39.9

$\dfrac{\text{rate HCl}}{\text{rate Ar}} = \sqrt{\dfrac{M_{Ar}}{M_{HCl}}} = \sqrt{\dfrac{39.9}{36.5}} = 1.05; \quad \dfrac{\text{rate } F_2}{\text{rate Ar}} = \sqrt{\dfrac{M_{Ar}}{M_{F_2}}} = \sqrt{\dfrac{39.9}{38.0}} = 1.02$

The relative rates of diffusion are $HCl(1.05) > F_2(1.02) > Ar(1.00)$.

10.108 Both tanks contain the same number of gas particles (atoms or molecules) at the same temperature with the same average kinetic energy. Because O_2 is lighter than Kr, the average O_2 velocity is greater than the average Kr velocity.
(a) $Kr < O_2$ (b) $O_2 < Kr$ (c) $Kr < O_2$ (d) Both are the same.

Real Gases (Section 10.8)

10.110 (a) high (b) larger

10.112 $P = \dfrac{nRT}{V} = \dfrac{(0.500 \text{ mol})\left(0.082\ 06\ \dfrac{\text{L} \cdot \text{atm}}{\text{K} \cdot \text{mol}}\right)(300 \text{ K})}{(0.600 \text{ L})} = 20.5$ atm

$P = \dfrac{nRT}{V - nb} - \dfrac{an^2}{V^2}$

$P = \dfrac{(0.500 \text{ mol})\left(0.082\ 06\ \dfrac{\text{L} \cdot \text{atm}}{\text{K} \cdot \text{mol}}\right)(300 \text{ K})}{[(0.600 \text{ L}) - (0.500 \text{ mol})(0.0387 \text{ L/mol})]} - \dfrac{\left(1.35\ \dfrac{\text{L}^2 \cdot \text{atm}}{\text{mol}^2}\right)(0.500 \text{ mol})^2}{(0.600 \text{ L})^2} = 20.3$ atm

10.114 UF_6, 352.0; 70 °C = 70 + 273 = 343 K

(a) $P = \dfrac{nRT}{V} = \dfrac{\left(512.9 \text{ g} \times \dfrac{1 \text{ mol}}{352.0 \text{ g}}\right)\left(0.082\ 06\ \dfrac{\text{L} \cdot \text{atm}}{\text{K} \cdot \text{mol}}\right)(343 \text{ K})}{(22.9 \text{ L})} = 1.79$ atm

(b) $P = \dfrac{nRT}{V - nb} - \dfrac{an^2}{V^2}$

$n = 512.9 \text{ g} \times \dfrac{1 \text{ mol } UF_6}{352.0 \text{ g } UF_6} = 1.457 \text{ mol } UF_6$

$$P = \frac{(1.457 \text{ mol})\left(0.082\ 06\ \dfrac{\text{L}\cdot\text{atm}}{\text{K}\cdot\text{mol}}\right)(343 \text{ K})}{[(22.9 \text{ L}) - (1.457 \text{ mol})(0.1128 \text{ L/mol})]} - \frac{\left(15.80\ \dfrac{\text{L}^2\cdot\text{atm}}{\text{mol}^2}\right)(1.457 \text{ mol})^2}{(22.9 \text{ L})^2} = 1.74 \text{ atm}$$

The Earth's Atmosphere, Greenhouse Gases, and Climate Change (Sections 10.9–10.11)

10.116 Troposphere, stratosphere, mesosphere, and thermosphere. Temperature changes are used to distinguish between different regions of the atmosphere.

10.118 The force of gravity is strongest at the earth's surface and becomes weaker at higher altitude. Because of this, the troposphere contains about 75% of the mass of the entire atmosphere.

10.120 (a) visible and UV   (b) UV   (c) infrared   (d) infrared

10.122 Because of nuclear fusion, the Sun emits all forms of electromagnetic radiation. The Earth absorbs visible radiation, warms up, and radiates infrared radiation back to space.

10.124 Nitrogen and oxygen are diatomic molecules that do not have a dipole moment. As the bond stretches in a vibration, the molecule still does not have a dipole moment. Since no change in dipole moment occurs, IR radiation will not be absorbed.

10.126 The symmetric stretch does not absorb IR radiation; the asymmetric stretch absorbs IR radiation.

10.128 Although N_2O has a greater potential for warming on a per mass basis, the atmosphere has a much higher concentration of CO_2.

10.130 CO_2, N_2O, and CH_4

10.132 The Earth's average temperature has risen about 1 °C since 1900.

Multiconcept Problems

10.134 $15.0 \text{ gal} \times \dfrac{3.7854 \text{ L}}{1 \text{ gal}} = 56.8 \text{ L}; \quad 25 \text{ °C} = 25 + 273 = 298 \text{ K}$

$$n = \frac{PV}{RT} = \frac{\left(743 \text{ mm Hg} \times \dfrac{1.00 \text{ atm}}{760 \text{ mm Hg}}\right)(56.8 \text{ L})}{\left(0.082\ 06\ \dfrac{\text{L}\cdot\text{atm}}{\text{K}\cdot\text{mol}}\right)(298 \text{ K})} = 2.27 \text{ mol gasoline}$$

$2.27 \text{ mol gasoline} \times \dfrac{105 \text{ g gasoline}}{1 \text{ mol gasoline}} = 238 \text{ g gasoline}$

$238 \text{ g} \times \dfrac{1 \text{ mL}}{0.75 \text{ g}} \times \dfrac{1 \text{ L}}{1000 \text{ mL}} \times \dfrac{1 \text{ gal}}{3.7854 \text{ L}} \times \dfrac{20.0 \text{ mi}}{1 \text{ gal}} = 1.68 \text{ mi}$

10.136 $n = \dfrac{PV}{RT} = \dfrac{(2.15 \text{ atm})(7.35 \text{ L})}{\left(0.082\ 06\ \dfrac{\text{L}\cdot\text{atm}}{\text{K}\cdot\text{mol}}\right)(293 \text{ K})} = 0.657 \text{ mol Ar}$

$0.657 \text{ mol Ar} \times \dfrac{39.948 \text{ g Ar}}{1 \text{ mol Ar}} = 26.2 \text{ g Ar}$

$m_{total} = 478.1 \text{ g} + 26.2 \text{ g} = 504.3 \text{ g}$

10.138 (a) $n_{total} = \dfrac{PV}{RT} = \dfrac{\left(258 \text{ mm Hg} \times \dfrac{1.00 \text{ atm}}{760 \text{ mm Hg}}\right)(0.500 \text{ L})}{\left(0.082\ 06\ \dfrac{\text{L}\cdot\text{atm}}{\text{K}\cdot\text{mol}}\right)(293 \text{ K})} = 0.007\ 06 \text{ mol}$

(b) $n_B = \dfrac{PV}{RT} = \dfrac{\left(344 \text{ mm Hg} \times \dfrac{1 \text{ atm}}{760 \text{ mm Hg}}\right)(0.250 \text{ L})}{\left(0.082\ 06\ \dfrac{\text{L}\cdot\text{atm}}{\text{K}\cdot\text{mol}}\right)(293 \text{ K})} = 0.004\ 71 \text{ moles}$

(c) $d = \dfrac{0.218 \text{ g}}{0.250 \text{ L}} = 0.872 \text{ g/L}$

(d) molar mass $= \dfrac{0.218 \text{ g}}{0.004\ 71 \text{ mol}} = 46.3 \text{ g/mol}, NO_2;$ molecular weight $= 46.3$

(e) $Hg_2CO_3(s) + 6 \, HNO_3(aq) \rightarrow 2 \, Hg(NO_3)_2(aq) + 3 \, H_2O(l) + CO_2(g) + 2 \, NO_2(g)$

10.140 $PV = nRT$

$n_{total(initial)} = \dfrac{PV}{RT} = \dfrac{(3.00 \text{ atm})(10.0 \text{ L})}{\left(0.082\ 06\ \dfrac{\text{L}\cdot\text{atm}}{\text{K}\cdot\text{mol}}\right)(373.1 \text{ K})} = 0.980 \text{ mol}$

$n_{total(final)} = \dfrac{PV}{RT} = \dfrac{(2.40 \text{ atm})(10.0 \text{ L})}{\left(0.082\ 06\ \dfrac{\text{L}\cdot\text{atm}}{\text{K}\cdot\text{mol}}\right)(373.1 \text{ K})} = 0.784 \text{ mol}$

	$CS_2(g)$	+	$3 \, O_2(g)$	$\rightarrow$	$CO_2(g)$	+	$2 \, SO_2(g)$
before reaction (mol)	y		0.980 − y		0		0
change (mol)	−x		−3x		+x		+2x
after reaction (mol)	y − x = 0		0.980 − y − 3x		x		2x

$n_{total(final)} = (y - x) + (0.980 - y - 3x) + x + 2x = 0.784 \text{ mol}$

$0.980 \text{ mol} - 4x + 3x = 0.784 \text{ mol}$

$x = 0.980 \text{ mol} - 0.784 \text{ mol} = 0.196 \text{ mol}$

mol $CO_2 = x = 0.196 \text{ mol}$

$P_{CO_2} = \dfrac{nRT}{V} = \dfrac{(0.196 \text{ mol})\left(0.082\ 06\ \dfrac{\text{L}\cdot\text{atm}}{\text{K}\cdot\text{mol}}\right)(373.1 \text{ K})}{(10.0 \text{ L})} = 0.600 \text{ atm}$

mol $SO_2 = 2x = 2(0.196 \text{ mol}) = 0.392 \text{ mol}$

$$P_{SO_2} = \frac{nRT}{V} = \frac{(0.392 \text{ mol})\left(0.082\ 06\ \dfrac{L \cdot atm}{K \cdot mol}\right)(373.1 \text{ K})}{(10.0 \text{ L})} = 1.20 \text{ atm}$$

mol O_2 = 0.980 mol – y – 3x = 0.980 mol – x – 3x = 0.980 – 4(0.196 mol) = 0.196 mol

$P_{O_2} = P_{CO_2} = 0.600$ atm

10.142 PCl_3, 137.3; O_2, 32.00; $POCl_3$, 153.3

$2\ PCl_3(g) + O_2(g) \rightarrow 2\ POCl_3(g)$

$$\text{mol } PCl_3 = 25.0 \text{ g} \times \frac{1 \text{ mol } PCl_3}{137.3 \text{ g } PCl_3} = 0.182 \text{ mol } PCl_3$$

$$\text{mol } O_2 = 3.00 \text{ g} \times \frac{1 \text{ mol } O_2}{32.00 \text{ g } O_2} = 0.0937 \text{ mol } O_2$$

Check for limiting reactant.

$$\text{mol } O_2 \text{ needed } = 0.182 \text{ mol } PCl_3 \times \frac{1 \text{ mol } O_2}{2 \text{ mol } PCl_3} = 0.0910 \text{ mol } O_2 \text{ needed}$$

There is a slight excess of O_2. PCl_3 is the limiting reactant.

$$\text{mol } POCl_3 = 0.182 \text{ mol } PCl_3 \times \frac{2 \text{ mol } POCl_3}{2 \text{ mol } PCl_3} = 0.182 \text{ mol } POCl_3$$

mol O_2 left over = 0.0937 mol – 0.0910 mol = 0.0027 mol O_2 left over

T = 200.0 °C = 200.0 + 273.15 = 473.1 K; PV = nRT

$$P = \frac{nRT}{V} = \frac{(0.182 \text{ mol} + 0.0027 \text{ mol})\left(0.082\ 06\ \dfrac{L \cdot atm}{K \cdot mol}\right)(473.1 \text{ K})}{(5.00 \text{ L})} = 1.43 \text{ atm}$$

10.144 NO_2, 46.01

Calculate the NO_2 pressure before any NO_2 dimerization.

$$P = \frac{nRT}{V} = \frac{\left(9.66 \text{ g} \times \dfrac{1 \text{ mol}}{46.01 \text{ g}}\right)\left(0.082\ 06\ \dfrac{L \cdot atm}{K \cdot mol}\right)(298 \text{ K})}{(6.51 \text{ L})} = 0.789 \text{ atm}$$

	$2\ NO_2(g)$	$\rightarrow$	$N_2O_4(g)$
before reaction (atm)	0.789		0
change (atm)	–2x		+x
after reaction (atm)	0.789 – 2x		x

0.487 = 0.789 – 2x + x

0.487 = 0.789 – x

x = 0.789 – 0.487 = 0.302 atm

$P_{NO_2} = 0.789 – 2x = 0.789 – 2(0.302) = 0.185$ atm; $\qquad P_{N_2O_4} = x = = 0.302$ atm

$$X_{NO_2} = \frac{0.185 \text{ atm}}{0.487 \text{ atm}} = 0.380 \text{ and } X_{N_2O_4} = \frac{0.302 \text{ atm}}{0.487 \text{ atm}} = 0.620$$

10.146 $X + 3 O_2 \rightarrow 2 CO_2 + 3 H_2O$

(a) $X = C_2H_6O$

$C_2H_6O + 3 O_2 \rightarrow 2 CO_2 + 3 H_2O$

(b) It is an empirical formula because it is the smallest whole number ratio of atoms. It is also a molecular formula because any higher multiple such as $C_4H_{12}O_2$ does not correspond to a stable electron-dot structure.

(c)

(d) C_2H_6O, 46.07

$$mol\ C_2H_6O = 5.000\ g\ C_2H_6O \times \frac{1\ mol\ C_2H_6O}{46.07\ g\ C_2H_6O} = 0.1085\ mol\ C_2H_6O$$

$$\Delta H_{combustion} = \frac{-144.2\ kJ}{0.1085\ mol} = -1328.6\ kJ/mol$$

$$\Delta H_{combustion} = [2\ \Delta H°_f(CO_2) + 3\ \Delta H°_f(H_2O)] - \Delta H°_f(C_2H_6O)$$

$$\Delta H°_f(C_2H_6O) = [2\ \Delta H°_f(CO_2) + 3\ \Delta H°_f(H_2O)] - \Delta H_{combustion}$$
$$= [(2\ mol)(-393.5\ kJ/mol) + (3\ mol)(-241.8\ kJ/mol)] - (-1328.6\ kJ)$$
$$= -183.8\ kJ/mol$$

10.148 (a) Freezing point of H_2O on the Rankine scale is $(9/5)(273.15) = 492\ °R$.

(b) $R = \dfrac{PV}{nT} = \dfrac{(1.00\ atm)(22.414\ L)}{(1.00\ mol)(492\ °R)} = 0.0456\ \dfrac{L \cdot atm}{°R \cdot mol}$

(c) $P = \dfrac{(2.50\ mol)\left(0.0456\ \dfrac{L \cdot atm}{°R \cdot mol}\right)(525\ °R)}{[(0.4000\ L) - (2.50\ mol)(0.04278\ L/mol)]} - \dfrac{\left(2.253\ \dfrac{L^2 \cdot atm}{mol^2}\right)(2.50\ mol)^2}{(0.4000\ L)^2}$

$P = 204.2\ atm - 88.0\ atm = 116\ atm$

10.150 CO_2, 44.01; H_2O, 18.02

(a) $mol\ C = 0.3744\ g\ CO_2 \times \dfrac{1\ mol\ CO_2}{44.01\ g\ CO_2} \times \dfrac{1\ mol\ C}{1\ mol\ CO_2} = 0.008\ 507\ mol\ C$

$mass\ C = 0.008\ 507\ mol\ C \times \dfrac{12.011\ g\ C}{1\ mol\ C} = 0.1022\ g\ C$

$mol\ H = 0.1838\ g\ H_2O \times \dfrac{1\ mol\ H_2O}{18.02\ g\ H_2O} \times \dfrac{2\ mol\ H}{1\ mol\ H_2O} = 0.020\ 400\ mol\ H$

$mass\ H = 0.020\ 400\ mol\ H \times \dfrac{1.008\ g\ H}{1\ mol\ H} = 0.02056\ g\ H$

$mass\ O = 0.1500\ g - 0.1022\ g - 0.02056\ g = 0.0272\ g\ O$

$$\text{mol O} = 0.0272 \text{ g O} \times \frac{1 \text{ mol O}}{16.00 \text{ g O}} = 0.001\ 70 \text{ mol O}$$

$C_{0.008\ 507}H_{0.020\ 400}O_{0.001\ 70}$; divide each subscript by the smallest, 0.001 70.

$C_{0.008\ 507\ /\ 0.001\ 70}H_{0.020\ 400\ /\ 0.001\ 70}O_{0.001\ 70\ /\ 0.001\ 70}$

The empirical formula is $C_5H_{12}O$.

The empirical formula mass is 88 g/mol.

(b) 1 atm = 101,325 Pa; T = 54.8 °C = 54.8 + 273.15 = 327.9 K

PV = nRT

$$n = \frac{PV}{RT} = \frac{\left(100.0 \text{ kPa} \times \dfrac{1.00 \text{ atm}}{101.325 \text{ kPa}}\right)(1.00 \text{ L})}{\left(0.082\ 06 \dfrac{\text{L} \cdot \text{atm}}{\text{K} \cdot \text{mol}}\right)(327.9 \text{ K})} = 0.0367 \text{ mol methyl } \textit{tert}\text{-butyl ether}$$

methyl *tert*-butyl ether molar mass $= \dfrac{3.233 \text{ g}}{0.0367 \text{ mol}} = 88.1$ g/mol

The empirical formula weight and the molar mass are the same, so the molecular formula and empirical formula are the same. $C_5H_{12}O$ is the molecular formula and 88.15 is the molecular weight for methyl *tert*-butyl ether.

(c) $C_5H_{12}O(l) + 15/2\ O_2(g) \rightarrow 5\ CO_2(g) + 6\ H_2O(l)$

(d) $\Delta H°_{combustion} = [5\ \Delta H°_f\,(CO_2) + 6\ \Delta H°_f\,(H_2O(l))] - \Delta H°_f\,(C_5H_{12}O) = -3368.7$ kJ

-3368.7 kJ $= [(5 \text{ mol})(-393.5 \text{ kJ/mol}) + (6 \text{ mol})(-285.8 \text{ kJ/mol})] - (1 \text{ mol})\Delta H°_f\,(C_5H_{12}O)$

$(1 \text{ mol})\Delta H°_f\,(C_5H_{12}O) = [(5 \text{ mol})(-393.5 \text{ kJ/mol}) + (6 \text{ mol})(-285.8 \text{ kJ/mol})] + 3368.7$ kJ

$\Delta H°_f\,(C_5H_{12}O) = -313.6$ kJ/mol

11 Liquids and Phase Changes

11.1 $\Delta H_{vap} = \dfrac{(\ln P_2 - \ln P_1)(R)}{\left(\dfrac{1}{T_1} - \dfrac{1}{T_2}\right)}$

$P_1 = 400$ mm Hg; $T_1 = 41.0\ ^{\circ}C = 314.2$ K
$P_2 = 760$ mm Hg; $T_2 = 331.9$ K

$\Delta H_{vap} = \dfrac{[\ln(760) - \ln(400)]\left(8.3145\ \dfrac{J}{K \cdot mol}\right)}{\left(\dfrac{1}{314.2\ K} - \dfrac{1}{331.9\ K}\right)} = 31{,}442$ J/mol $= 31.4$ kJ/mol

11.2 $\ln\left(\dfrac{P_2}{P_1}\right) = \ln P_2 - \ln P_1 = \dfrac{\Delta H_{vap}}{R}\left(\dfrac{1}{T_1} - \dfrac{1}{T_2}\right)$

$(\ln P_2 - \ln P_1)\left(\dfrac{R}{\Delta H_{vap}}\right) = \dfrac{1}{T_1} - \dfrac{1}{T_2}$

$\dfrac{1}{T_1} - (\ln P_2 - \ln P_1)\left(\dfrac{R}{\Delta H_{vap}}\right) = \dfrac{1}{T_2}$

$P_1 = 760$ mm Hg; $T_1 = 100.0\ ^{\circ}C = 373.1$ K
$P_2 = 407$ mm Hg;
Solve for T_2, the boiling point for H_2O on top of Pikes Peak

$\dfrac{1}{373.1\ K} - [\ln(407) - \ln(760)]\left(\dfrac{8.3145 \times 10^{-3}\ \dfrac{kJ}{K \cdot mol}}{40.7\ kJ/mol}\right) = \dfrac{1}{T_2}$

$\dfrac{1}{T_2} = 0.002\ 808;$ $T_2 = 356.1$ K $= 83.0\ ^{\circ}C$

11.3 boiling point $= 78.4\ ^{\circ}C = 351.6$ K
$\Delta G = \Delta H_{vap} - T\Delta S_{vap}$; At the boiling point (phase change), $\Delta G = 0$
$\Delta H_{vap} = T\Delta S_{vap}$

$\Delta S_{vap} = \dfrac{\Delta H_{vap}}{T} = \dfrac{38.56\ kJ/mol}{351.6\ K} = 0.1097$ kJ/(K $\cdot$ mol) $= 109.7$ J/(K $\cdot$ mol)

11.4 $\Delta G = \Delta H - T\Delta S$; at the boiling point (phase change), $\Delta G = 0$.

$\Delta H = T\Delta S$; $T = \dfrac{\Delta H_{vap}}{\Delta S_{vap}} = \dfrac{29.2\ kJ/mol}{87.5 \times 10^{-3}\ kJ/(K \cdot mol)} = 334$ K

11.5 C_6H_6, 78.1; $15.0 \text{ g } C_6H_6 \times \dfrac{1 \text{ mol } C_6H_6}{78.1 \text{ g } C_6H_6} = 0.192 \text{ mol } C_6H_6$

$q_1 = (0.192 \text{ mol})[136.0 \times 10^{-3} \text{ kJ/(K} \cdot \text{mol)}](80.1 \,^{\circ}\text{C} - 50 \,^{\circ}\text{C}) = 0.786 \text{ kJ}$
$q_2 = (0.192 \text{ mol})(30.72 \text{ kJ/mol}) = 5.90 \text{ kJ}$
$q_3 = (0.192 \text{ mol})(82.4 \times 10^{-3} \text{ kJ/(K} \cdot \text{mol)}](100 \,^{\circ}\text{C} - 80.1 \,^{\circ}\text{C}) = 0.315 \text{ kJ}$
$q_{\text{total}} = q_1 + q_2 + q_3 = 7.00 \text{ kJ};$ 7.00 kJ of heat is required.

11.6 H_2O, 18.0; $10.0 \text{ g } H_2O \times \dfrac{1 \text{ mol } H_2O}{18.0 \text{ g } H_2O} = 0.556 \text{ mol } H_2O$

$q_1 = (0.556 \text{ mol})[75.4 \times 10^{-3} \text{ kJ/(K} \cdot \text{mol)}](0 \,^{\circ}\text{C} - 25 \,^{\circ}\text{C}) = -1.05 \text{ kJ}$
$q_2 = (0.556 \text{ mol})(-6.01 \text{ kJ/mol}) = -3.34 \text{ kJ}$
$q_3 = (0.556 \text{ mol})(36.6 \times 10^{-3} \text{ kJ/(K} \cdot \text{mol)}](-10 \,^{\circ}\text{C} - 0 \,^{\circ}\text{C}) = -0.203 \text{ kJ}$
$q_{\text{total}} = q_1 + q_2 + q_3 = -4.59 \text{ kJ};$ 4.59 kJ of heat is removed.

11.7 $H_2O(g) \;\rightarrow\; H_2O(l) \;\rightarrow\; H_2O(g)$

11.8   (a)

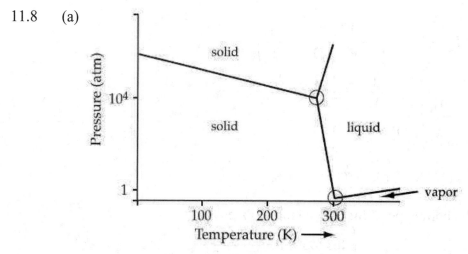

(b) Gallium has two triple points. The one below 1 atm is a solid, liquid, vapor triple point. The one at 10^4 atm is a solid(1), solid(2), liquid triple point.
(c) Increasing the pressure favors the liquid phase, giving the solid/liquid boundary a negative slope. At 1 atm pressure the liquid phase is more dense than the solid phase.

11.9   (a) Benzene, C_6H_6, was originally used to extract caffeine from coffee beans. Benzene is highly toxic and carcinogenic. Residual benzene in coffee poses a serious health threat. A safer, nontoxic solvent was needed to replace benzene.
(b) Caffeine is very soluble in supercritical CO_2 because caffeine is a nonpolar molecule and supercritical CO_2 is a nonpolar solvent. If the polarity of solute and solvent are matched, then solubility will be high.
(c) The supercritical phase of CO_2 can be reached at a relatively moderate temperature and pressure (31.1 $^{\circ}$C and 73.0 atm). The easily attainable critical point for CO_2 makes it the most widely used supercritical fluid.

11.10　75 °F = 24 °C;　At 70 atm and 24 °C, CO_2 is a liquid.

11.11　(a) $CO_2(s) \rightarrow CO_2(g)$
　　　(b) $CO_2(l) \rightarrow CO_2(g)$
　　　(c) $CO_2(g) \rightarrow CO_2(l) \rightarrow$ supercritical CO_2

11.12　(a) 5.11 atm　　　(b) −56.4 °C　　　(c) 31.1 °C

11.13　(a) For the phase transition, $CO_2(s) \rightarrow CO_2(g)$, the system is becoming more disordered, therefore ΔS > 0.
　　　(b) −78.5 °C
　　　(c) ΔG = ΔH − TΔS; ΔG = 0 at the sublimation temperature (−78.5 °C = 194.6 K). Set ΔG = 0 and solve for ΔS.
　　　ΔG = 0 = ΔH − TΔS

$$\Delta S = \frac{\Delta H}{T} = \frac{26.1 \text{ kJ/mol}}{194.6 \text{ K}} = 0.1341 \text{ kJ/(K·mol)} = 134.1 \text{ J/(K·mol)}$$

11.14　CO_2, 44.01

$$100.0 \text{ g} \times \frac{1 \text{ mol } CO_2}{44.01 \text{ g } CO_2} = 2.272 \text{ mol } CO_2$$

q_1 = (2.272 mol)(26.1 kJ/mol) = 59.3 kJ
q_2 = (2.272 mol)[35.0 × 10⁻³ kJ/(mol · °C)][33 °C − (−78.5 °C)] = 8.86 kJ
q_{total} = q_1 + q_2 = 68.2 kJ

Conceptual Problems

11.16　(a) normal boiling point ≈ 300 K;　normal melting point ≈ 180 K
　　　(b) (i) solid　　　(ii) gas　　　(iii) supercritical fluid

11.18　(a) The smectic and nematic liquid crystal phases are between the crystalline solid and the isotropic liquid phases.
　　　(b) The phase transition is from the smectic to the nematic phase. The molecules become more disordered as a result of the increase in temperature.
　　　(c) The phase transition is from the nematic to the smectic phase. The molecules become more ordered in a smaller volume as a result of the increase in pressure.

Section Problems
Properties of Liquids (Section 11.1)

11.20　It's water's surface tension that keeps the needle on top.

11.22　(a) CH_2Br_2 has the higher surface tension because it is polar while CCl_4 is nonpolar.
　　　(b) Ethylene glycol has the higher surface tension because it can hydrogen bond from both ends of the molecule while ethanol can only hydrogen bond from one end of the molecule.

　　　　　　　　　　　　　　　　　　　　163

11.24   Oleic acid has a higher viscosity than H_2O because it is a much larger molecule than H_2O with larger dispersion forces. It can also hydrogen bond.

Vapor Pressure and Boiling Point (Section 11.2)

11.26 $\Delta H_{vap} = \dfrac{(\ln P_2 - \ln P_1)(R)}{\left(\dfrac{1}{T_1} - \dfrac{1}{T_2}\right)}$

$T_1 = -5.1\ °C = 268.0\ K;\qquad P_1 = 100\ mm\ Hg$
$T_2 = 46.5\ °C = 319.6\ K;\qquad P_2 = 760\ mm\ Hg$

$\Delta H_{vap} = \dfrac{[\ln(760) - \ln(100)][8.3145 \times 10^{-3}\ kJ/(K \cdot mol)]}{\left(\dfrac{1}{268.0\ K} - \dfrac{1}{319.6\ K}\right)} = 28.0\ kJ/mol$

11.28 $\ln P_2 = \ln P_1 + \dfrac{\Delta H_{vap}}{R}\left(\dfrac{1}{T_1} - \dfrac{1}{T_2}\right)$

$\Delta H_{vap} = 28.0\ kJ/mol$
$P_1 = 100\ mm\ Hg;\qquad T_1 = -5.1\ °C = 268.0\ K;\qquad T_2 = 20.0\ °C = 293.2\ K$
Solve for P_2.

$\ln P_2 = \ln(100) + \dfrac{28.0\ kJ/mol}{[8.3145 \times 10^{-3}\ kJ/(K \cdot mol)]}\left(\dfrac{1}{268.0\ K} - \dfrac{1}{293.2\ K}\right)$

$\ln P_2 = 5.6852;\quad P_2 = e^{5.6852} = 294.5\ mm\ Hg = 294\ mm\ Hg$

11.30

T(K)	P_{vap}(mm Hg)	$\ln P_{vap}$	1/T
263	80.1	4.383	0.003 802
273	133.6	4.8949	0.003 663
283	213.3	5.3627	0.003 534
293	329.6	5.7979	0.003 413
303	495.4	6.2054	0.003 300
313	724.4	6.5853	0.003 195

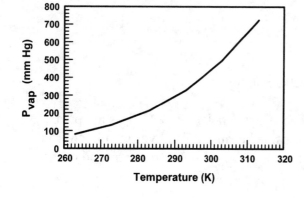

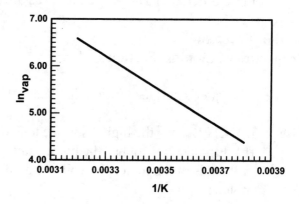

$$\ln P_{vap} = \left(-\frac{\Delta H_{vap}}{R} \right) \frac{1}{T} + C; \quad C = 18.2$$

$$\text{slope} = -3628 \text{ K} = -\frac{\Delta H_{vap}}{R}$$

$$\Delta H_{vap} = (3628 \text{ K})(R) = (3628 \text{ K})[8.3145 \times 10^{-3} \text{ kJ/(K}\cdot\text{mol)}] = 30.1 \text{ kJ/mol}$$

11.32 $\Delta H_{vap} = 30.1 \text{ kJ/mol}$

11.34 $\Delta H_{vap} = \dfrac{(\ln P_2 - \ln P_1)(R)}{\left(\dfrac{1}{T_1} - \dfrac{1}{T_2} \right)}$

$P_1 = 80.1 \text{ mm Hg};$ $\qquad$ $T_1 = 263 \text{ K}$
$P_2 = 724.4 \text{ mm Hg};$ $\qquad$ $T_2 = 313 \text{ K}$

$$\Delta H_{vap} = \frac{[\ln(724.4) - \ln(80.1)][8.3145 \times 10^{-3} \text{ kJ/(K}\cdot\text{mol)}]}{\left(\dfrac{1}{263 \text{ K}} - \dfrac{1}{313 \text{ K}} \right)} = 30.1 \text{ kJ/mol}$$

The calculated ΔH_{vap} and that obtained from the plot in Problem 11.30 are the same.

11.36 $\Delta H_{vap} = \dfrac{(\ln P_2 - \ln P_1)(R)}{\left(\dfrac{1}{T_1} - \dfrac{1}{T_2} \right)}$

$P_1 = 40.0 \text{ mm Hg};$ $\qquad$ $T_1 = -81.6 \text{ °C} = 191.6 \text{ K}$
$P_2 = 400 \text{ mm Hg};$ $\qquad$ $T_2 = -43.9 \text{ °C} = 229.2 \text{ K}$

$$\Delta H_{vap} = \frac{[\ln(400) - \ln(40.0)]\left(8.3145 \times 10^{-3} \dfrac{\text{kJ}}{\text{K}\cdot\text{mol}} \right)}{\left(\dfrac{1}{191.6 \text{ K}} - \dfrac{1}{229.2 \text{ K}} \right)} = 22.36 \text{ kJ/mol}$$

Using $\Delta H_{vap} = 22.36 \text{ kJ/mol}$

$$\ln P_2 = \ln P_1 + \frac{\Delta H_{vap}}{R} \left(\frac{1}{T_1} - \frac{1}{T_2} \right)$$

$$(\ln P_2 - \ln P_1)\left(\frac{R}{\Delta H_{vap}} \right) = \frac{1}{T_1} - \frac{1}{T_2}$$

$$\frac{1}{T_1} - (\ln P_2 - \ln P_1)\left(\frac{R}{\Delta H_{vap}} \right) = \frac{1}{T_2}$$

$P_1 = 40.0 \text{ mm Hg};$ $\quad$ $T_1 = 191.6 \text{ K}$
$P_2 = 760 \text{ mm Hg}$
Solve for T_2, the normal boiling point.

$$\frac{1}{191.6 \text{ K}} - [\ln(760) - \ln(40.0)]\left(\frac{8.3145 \times 10^{-3} \dfrac{\text{kJ}}{\text{K} \cdot \text{mol}}}{22.36 \text{ kJ/mol}}\right) = \frac{1}{T_2}$$

$$\frac{1}{T_2} = 0.004\ 124\ 33; \quad T_2 = 242.46 \text{ K} = -30.7 \text{ °C}$$

Phase Changes (Sections 11.3–11.4)

11.38 ΔH_{vap} is usually larger than ΔH_{fusion} because ΔH_{vap} is the heat required to overcome all intermolecular forces.

11.40   boiling point $= 218 \text{ °C} = 491 \text{ K}$

$\Delta G = \Delta H_{vap} - T\Delta S_{vap}$; At the boiling point (phase change), $\Delta G = 0$

$\Delta H_{vap} = T\Delta S_{vap}$; $\Delta S_{vap} = \dfrac{\Delta H_{vap}}{T} = \dfrac{43.3 \text{ kJ/mol}}{491 \text{ K}} = 0.0882 \text{ kJ/(K} \cdot \text{mol)} = 88.2 \text{ J/(K} \cdot \text{mol)}$

11.42

11.44   (a) $Hg(l) \rightarrow Hg(g)$
(b) no change of state, Hg remains a liquid
(c) $Hg(g) \rightarrow Hg(l) \rightarrow Hg(s)$

11.46 H_2O, 18.02; $5.00 \text{ g } H_2O \times \dfrac{1 \text{ mol } H_2O}{18.02 \text{ g } H_2O} = 0.2775 \text{ mol } H_2O$

$q_1 = (0.2775 \text{ mol})[36.6 \times 10^{-3} \text{ kJ/(K} \cdot \text{mol)}](273 \text{ K} - 263 \text{ K}) = 0.1016 \text{ kJ}$
$q_2 = (0.2775 \text{ mol})(6.01 \text{ kJ/mol}) = 1.668 \text{ kJ}$
$q_3 = (0.2775 \text{ mol})(75.3 \times 10^{-3} \text{ kJ/(K} \cdot \text{mol)}](303 \text{ K} - 273 \text{ K}) = 0.6269 \text{ kJ}$
$q_{total} = q_1 + q_2 + q_3 = 2.40 \text{ kJ};$ 2.40 kJ of heat is required.

11.48 H_2O, 18.02; $7.55 \text{ g } H_2O \times \dfrac{1 \text{ mol } H_2O}{18.02 \text{ g } H_2O} = 0.4190 \text{ mol } H_2O$

$q_1 = (0.4190 \text{ mol})[75.3 \times 10^{-3} \text{ kJ/(K} \cdot \text{mol)}](273.15 \text{ K} - 306.65 \text{ K}) = -1.057 \text{ kJ}$
$q_2 = -(0.4190 \text{ mol})(6.01 \text{ kJ/mol}) = -2.518 \text{ kJ}$

$q_3 = (0.4190 \text{ mol})[36.6 \times 10^{-3} \text{ kJ/(K} \cdot \text{mol)}](263.15 \text{ K} - 273.15 \text{ K}) = -0.1534 \text{ kJ}$

$q_{total} = q_1 + q_2 + q_3 = -3.73 \text{ kJ}$

3.73 kJ of heat is released.

11.50

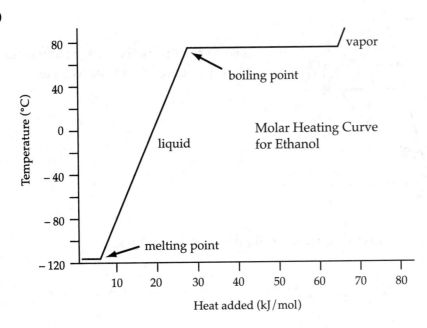

Phase Diagrams (Section 11.5)

11.52   As the pressure over the liquid H_2O is lowered, H_2O vapor is removed by the pump. As H_2O vapor is removed, more of the liquid H_2O is converted to H_2O vapor. This conversion is an endothermic process and the temperature decreases. The combination of both a decrease in pressure and temperature takes the system across the liquid/solid boundary in the phase diagram so the H_2O that remains turns to ice.

11.54   (a) gas   (b) liquid   (c) solid

11.56

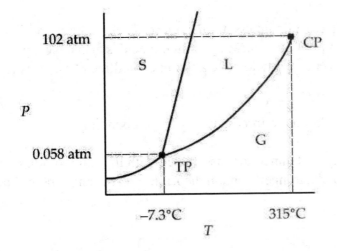

11.58   (a) $Br_2(s)$ (b) $Br_2(l)$

11.60   Solid O_2 does not melt when pressure is applied because the solid is denser than the liquid, and the solid/liquid boundary in the phase diagram slopes to the right.

11.62

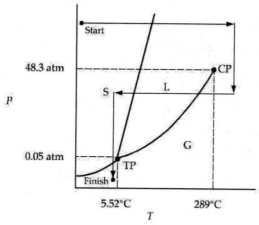

The starting phase is benzene as a solid, and the final phase is benzene as a gas.

11.64   solid → liquid → supercritical fluid → liquid → solid → gas

11.66

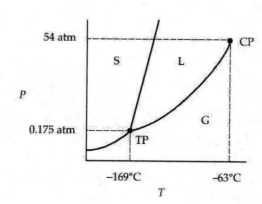

Kr cannot be liquefied at room temperature because room temperature is above T_c (−63 °C).

Liquid Crystals (Section 11.6)

11.68   Both (a) and (c) are likely to have a liquid crystalline phase. They both are rod-like and have 6-membered rings which promotes stacking. Both molecules also have a polar terminal group. Molecule (b) is not likely to have a liquid crystalline phase. It is rigid but not elongated and nonpolar.

11.70   London dispersion forces, dipole-dipole forces, and hydrogen bonding

11.72   Compound II would have a higher temperature associated with the solid to liquid crystal phase change because it has both stronger London dispersion forces and dipole-dipole forces.

Multiconcept Problems

11.74 Al_2O_3, ionic (greater lattice energy than NaCl because of higher ion charges); F_2, dispersion; H_2O, H–bonding, dipole-dipole; Br_2, dispersion (larger and more polarizable than F_2), ICl, dipole-dipole, NaCl, ionic
rank according to normal boiling points: $F_2 < Br_2 < ICl < H_2O < NaCl < Al_2O_3$

11.76 $C_2H_5OH(l) \rightarrow C_2H_5OH(g)$
Calculate ΔH and ΔS for this process and assume they do not change as a function of temperature.
$\Delta H° = \Delta H°_f(C_2H_5OH(g)) - \Delta H°_f(C_2H_5OH(l))$
$\Delta H° = [(1 \text{ mol})(-235.1 \text{ kJ/mol})] - [(1 \text{ mol})(-277.7 \text{ kJ/mol})] = 42.6 \text{ kJ}$

$\Delta S° = S°(C_2H_5OH(g)) - S°(C_2H_5OH(l))$
$\Delta S° = [(1 \text{ mol})(282.6 \text{ J/(K} \cdot \text{mol))}] - [(1 \text{ mol})(161 \text{ J/(K} \cdot \text{mol))}] = 122 \text{ J/K}$
$\Delta S° = 122 \times 10^{-3} \text{ kJ/K})$

$\Delta G° = \Delta H° - T\Delta S°$ and at the boiling point, $\Delta G = 0$
$0 = \Delta H° - T_{bp}\Delta S°$
$T_{bp}\Delta S° = \Delta H°$
$T_{bp} = \dfrac{\Delta H°}{\Delta S°} = \dfrac{42.6 \text{ kJ}}{122 \times 10^{-3} \text{ kJ/K}} = 349 \text{ K}$
$T_{bp} = 349 - 273 = 76 \text{ °C}$
$\ln P_2 = \ln P_1 + \dfrac{\Delta H_{vap}}{R}\left(\dfrac{1}{T_1} - \dfrac{1}{T_2}\right)$

$\Delta H_{vap} = 42.6 \text{ kJ/mol}$
At 1 atm, C_2H_5OH boils at 349 K; therefore set
$T_1 = 349$ K, and $P_1 = 1.00$ atm.
Let $T_2 = 25$ °C $= 298$ K, and solve for P_2.
P_2 is the vapor pressure of C_2H_5OH at 25 °C.
$\ln P_2 = \ln(1.00) + \dfrac{42.6 \text{ kJ/mol}}{[8.3145 \times 10^{-3} \text{ kJ/(K} \cdot \text{mol)}]}\left(\dfrac{1}{349 \text{ K}} - \dfrac{1}{298 \text{ K}}\right)$
$\ln P_2 = -2.512; \quad P_2 = e^{-2.512} = 0.0811 \text{ atm}$
$P_2 = 0.0811 \text{ atm} \times \dfrac{760 \text{ mm Hg}}{1.00 \text{ atm}} = 61.6 \text{ mm Hg}$

12 Solids and Solid-State Materials

12.1 The face-centered cube is the unit cell for cubic closest packing.

$$r = \sqrt{\frac{d^2}{8}}; \quad d = r \cdot 2 \cdot \sqrt{2} = 197 \text{ pm} \cdot 2 \cdot \sqrt{2} = 557 \text{ pm}$$

12.2 For a simple cube, $d = 2r$; $\quad r = \dfrac{d}{2} = \dfrac{334 \text{ pm}}{2} = 167 \text{ pm}$

12.3 For a simple cube, there is one atom per unit cell.

$$\text{mass of one Po atom} = 209 \text{ g/mol} \times \frac{1 \text{ mol}}{6.022 \times 10^{23} \text{ atoms}} = 3.4706 \times 10^{-22} \text{ g/atom}$$

unit cell edge = d = 334 pm = 334×10^{-12} m = 3.34×10^{-8} cm

unit cell volume = $d^3 = (3.34 \times 10^{-8} \text{ cm})^3 = 3.7260 \times 10^{-23} \text{ cm}^3$

$$\text{density} = \frac{\text{mass}}{\text{volume}} = \frac{3.4706 \times 10^{-22} \text{ g}}{3.7260 \times 10^{-23} \text{ cm}^3} = 9.31 \text{ g/cm}^3$$

12.4 The unit cell is a face-centered cube. There are four atoms in the unit cell.

$$\text{unit cell volume} = \left(383.3 \times 10^{-12} \text{ m} \times \frac{1 \text{ cm}}{1 \times 10^{-2} \text{ m}} \right)^3 = 5.631 \times 10^{-23} \text{ cm}^3$$

unit cell mass = $(5.631 \times 10^{-23} \text{ cm}^3)(22.67 \text{ g/cm}^3) = 1.277 \times 10^{-21}$ g

$$\text{atom mass} = \frac{1.227 \times 10^{-21} \text{ g}}{4 \text{ atoms}} = 3.192 \times 10^{-22} \text{ g/atom}$$

atomic mass = $(3.192 \times 10^{-22} \text{ g/atom})(6.022 \times 10^{23} \text{ atoms/mol}) = 192.2 \text{ g/mol}$

The metal is Ir.

12.5 (a) 1/8 S^{2-} at 8 corners = 1 S^{2-}; 1/2 S^{2-} at 6 faces = 3 S^{2-}; 4 Zn^{2+} inside
(b) The formula for zinc sulfide is ZnS.
(b) The oxidation state of Zn is 2+.
(c) The geometry around each zinc is tetrahedral.

12.6 (a) 1/8 Ca^{2+} at 8 corners = 1 Ca^{2+}; 1/2 O^{2-} at 6 faces = 3 O^{2-}; 1 Ti^{4+} inside
The formula for perovskite is $CaTiO_3$.
(b) The oxidation number of Ti is +4 to maintain charge neutrality in the unit cell.
(c) The geometry around titanium, oxygen, and calcium are all octahedral.

12.7 The electron configuration for Hg is [Xe] $4f^{14} 5d^{10} 6s^2$. Assuming the 5d and 6s bands overlap, the composite band can accommodate 12 valence electrons per metal atom. Weak bonding and a low melting point are expected for Hg because both the bonding and antibonding MOs are occupied.

12.8   (a)   The composite s-d band can accommodate 12 valence electrons per metal atom.
Hf [Xe] $6s^2 4f^{14} 5d^2$, 4 valence electrons (4 bonding, 0 antibonding)
The s-d band is 1/4 full, so Hf is picture (1).
Pt [Xe] $6s^2 4f^{14} 5d^8$, 10 valence electrons (6 bonding, 4 antibonding)
The s-d band is 5/6 full, so Pt is picture (2).
Re [Xe] $6s^2 4f^{14} 5d^5$, 7 valence electrons (6 bonding, 1 antibonding)
The s-d band is 7/12 full, so Re is picture (3).
(b) Re has an excess of 5 bonding electrons and it has the highest melting point and is the hardest of the three.
(c) Pt has an excess of only 2 bonding electrons and it has the lowest melting point and is the softest of the three.

12.9   Ge doped with As is an n-type semiconductor because As has an additional valence electron. The extra electrons are in the conduction band. The number of electrons in the conduction band of the doped Ge is much higher than for pure Ge, and the conductivity of the doped semiconductor is higher.

12.10   (a) (1), silicon; (2), white tin; (3), diamond; (4), silicon doped with aluminum
(b) (3) < (1) < (4) < (2)
Diamond (3) is an insulator with a large band gap. Silicon (1) is a semiconductor with a band gap smaller than diamond. The conduction band is partially occupied with a few electrons and the valence band is partially empty. Silicon doped with aluminum (4) is a p-type semiconductor that has fewer electrons than needed for bonding and has vacancies (positive holes) in the valence band. White tin (2) has a partially filled s-p composite band and is a metallic conductor.

12.11 $Si(OCH_3)_4 + 4 H_2O \rightarrow Si(OH)_4 + 4 HOCH_3$

12.12 $Ba[OCH(CH_3)_2]_2 + Ti[OCH(CH_3)_2]_4 + 6 H_2O \rightarrow BaTi(OH)_6(s) + 6 HOCH(CH_3)_2$

$$BaTi(OH)_6(s) \xrightarrow{\text{heat}} BaTiO_3(s) + 3 H_2O(g)$$

12.13   The color of the quantum dots depends on the wavelength of light they absorb, which is determined by band-gap energy. Different sizes of CdSe nanoparticles have different band-gap energies.

12.14   (a) 5.0 nm   (b) 2.2 nm   (c) 3.5 nm

12.15   CdSe diameter = 2 nm = 2×10^{-9} m and human hair diameter = 50 μm = 50×10^{-6} m
$\left(\dfrac{50 \times 10^{-6} \text{ m}}{2 \times 10^{-9} \text{ m}} \right)$ =25,000 CdSe nanoparticles can fit across a human hair.

12.16   (a) Size (a) absorbs red light so it appears green, size (b) absorbs orange light so it appears blue, size (c) absorbs yellow light so it appears violet, and size (d) absorbs green light so it appears red.
(b) Particle sizes from smallest to largest are: d < c < b < a

12.17 The smaller the particle, the larger the band gap and the greater the shift in the color of the emitted light from the red to the violet. The yellow quantum dot is larger because yellow is closer to the red than is the blue.

Conceptual Problems

12.18 (a) cubic closest-packed   (b) simple cubic
      (c) hexagonal closest-packed   (d) body-centered cubic

12.20 (a) $1/8$ Ti^{4+} at 8 corners $= 1$ Ti^{4+}; 1 Ti^{4+} inside; $1/2$ O^{2-} at 2 faces (2/face) $= 2$ O^{2-}; 2 O^{2-} inside.
      (b) TiO_2 (c) The oxidation state of titanium is $+4$.

12.22 (a) (1) and (4) are semiconductors; (2) is a metal; (3) is an insulator
      (b) (3) < (1) < (4) < (2). The conductivity increases with decreasing band gap.
      (c) (1) and (4) increases; (2) decreases; (3) not much change.

12.24

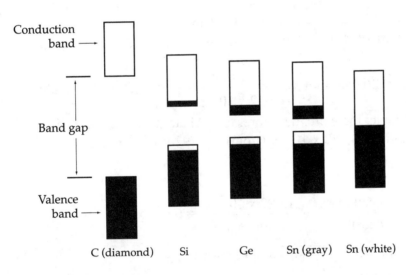

Section Problems
Types of Solids (Section 12.1)

12.26 molecular solid, CO_2, I_2;   metallic solid, any metallic element; covalent network solid, diamond;   ionic solid, NaCl

12.28 (a) rubber   (b) Na_3PO_4 (c) CBr_4 (d) quartz   (e) Au

12.30 Silicon carbide is a covalent network solid.

X-Ray Crystallography (Section 12.2)

12.32 $d = \dfrac{n\lambda}{2\sin\theta} = \dfrac{(1)(154.2 \text{ pm})}{2\sin(22.5°)} = 201 \text{ pm}$

The Packing of Spheres in Crystalline Solids: Unit Cells (Section 12.3)

12.34 From Table 12.2
Hexagonal and cubic closest packing are the most efficient because 74% of the available space is used. Simple cubic packing is the least efficient because only 52% of the available space is used.

12.36 Cu is face-centered cubic. d = 362 pm

$$r = \sqrt{\frac{d^2}{8}} = \sqrt{\frac{(362 \text{ pm})^2}{8}} = 128 \text{ pm}$$

362 pm = 362 x 10^{-12} m = 3.62 x 10^{-8} cm
unit cell volume = (3.62 x 10^{-8} cm)^3 = 4.74 x 10^{-23} cm^3

mass of one Cu atom = 63.55 g/mol x $\dfrac{1 \text{ mol}}{6.022 \times 10^{23} \text{ atom}}$ = 1.055 x 10^{-22} g/atom

Cu is face-centered cubic; there are, therefore, four Cu atoms in the unit cell.
unit cell mass = (4 atoms)(1.055 x 10^{-22} g/atom) = 4.22 x 10^{-22} g

$$\text{density} = \frac{\text{mass}}{\text{volume}} = \frac{4.22 \times 10^{-22} \text{g}}{4.74 \times 10^{-23} \text{ cm}^3} = 8.90 \text{ g/cm}^3$$

12.38 mass of one Al atom = 26.98 g/mol x $\dfrac{1 \text{ mol}}{6.022 \times 10^{23} \text{ atom}}$ = 4.480 x 10^{-23} g/atom

Al is face-centered cubic; there are, therefore, four Al atoms in the unit cell.
unit cell mass = (4 atoms)(4.480 x 10^{-23} g/atom) = 1.792 x 10^{-22} g

$$\text{density} = \frac{\text{mass}}{\text{volume}}$$

$$\text{unit cell volume} = \frac{\text{unit cell mass}}{\text{density}} = \frac{1.792 \times 10^{-22} \text{ g}}{2.699 \text{ g/cm}^3} = 6.640 \times 10^{-23} \text{ cm}^3$$

unit cell edge = d = $\sqrt[3]{6.640 \times 10^{-23} \text{ cm}^3}$ = 4.049 x 10^{-8} cm

d = 4.049 x 10^{-8} cm x $\dfrac{1 \text{m}}{100 \text{ cm}}$ = 4.049 x 10^{-10} m = 404.9 x 10^{-12} m = 404.9 pm

12.40 unit cell body diagonal = 4r = 549 pm

For W, r = $\dfrac{549 \text{ pm}}{4}$ = 137 pm

12.42 mass of one Ti atom = 47.88 g/mol x $\dfrac{1 \text{ mol}}{6.022 \times 10^{23} \text{ atoms}}$ = 7.951 x 10^{-23} g/atom

r = 144.8 pm = 144.8 x 10^{-12} m

r = 144.8 x 10^{-12} m x $\dfrac{100 \text{ cm}}{1 \text{ m}}$ = 1.448 x 10^{-8} cm

Calculate the volume and then the density for Ti assuming it is primitive cubic, body-centered cubic, and face-centered cubic. Compare the calculated density with the actual density to identify the unit cell.

For primitive cubic:
$$d = 2r; \text{ volume} = d^3 = [2(1.448 \times 10^{-8} \text{ cm})]^3 = 2.429 \times 10^{-23} \text{ cm}^3$$
$$\text{density} = \frac{\text{unit cell mass}}{\text{volume}} = \frac{7.951 \times 10^{-23} \text{ g}}{2.429 \times 10^{-23} \text{ cm}^3} = 3.273 \text{ g/cm}^3$$

For face-centered cubic:
$$d = 2\sqrt{2}r; \text{ volume} = d^3 = [2\sqrt{2}(1.448 \times 10^{-8} \text{ cm})]^3 = 6.870 \times 10^{-23} \text{ cm}^3$$
$$\text{density} = \frac{4(7.951 \times 10^{-23} \text{ g})}{6.870 \times 10^{-23} \text{ cm}^3} = 4.630 \text{ g/cm}^3$$

For body-centered cubic:
$$d = \frac{4r}{\sqrt{3}}; \text{ volume} = d^3 = \left[\frac{4(1.448 \times 10^{-8} \text{ cm})}{\sqrt{3}}\right]^3 = 3.739 \times 10^{-23} \text{ cm}^3$$
$$\text{density} = \frac{2(7.951 \times 10^{-23} \text{ g})}{3.739 \times 10^{-23} \text{ cm}^3} = 4.253 \text{ g/cm}^3$$

The calculated density for a face-centered cube (4.630 g/cm^3) is closest to the actual density of 4.54 g/cm^3. Ti crystallizes in the face-centered cubic unit cell.

12.44 mass of one Pb atom = 207.2 g/mol x $\dfrac{1 \text{ mol}}{6.022 \times 10^{23} \text{ atom}}$ = 3.441 x 10^{-22} g/atom

If the unit cell for Pb is the primitive cube it would contain one Pb atom and weigh 3.441 x 10^{-22} g.

If the unit cell for Pb is the face-centered cube it would contain four Pb atom and weigh (4)(3.441 x 10^{-22} g) = 1.376 x 10^{-21} g.

For a primitive cubic unit cell:
unit cell edge = d = 2r = 2(175 pm) = 3.50 x 10^{-8} cm
unit cell volume = d^3 = (3.50 x 10^{-8} cm)^3 = 4.29 x 10^{-23} cm^3
unit cell mass = (4.29 x 10^{-23} cm^3)(11.34 g/cm^3) = 4.86 x 10^{-22} g

For a face-centered cubic unit cell:
unit cell edge = d = 2$\sqrt{2}$ r = 2$\sqrt{2}$ (175 pm) = 4.95 x 10^{-8} cm
unit cell volume = d^3 = (4.95 x 10^{-8} cm)^3 = 1.21 x 10^{-22} cm^3
unit cell mass = (1.21 x 10^{-22} cm^3)(11.34 g/cm^3) = 1.38 x 10^{-21} g

The masses agree for the face-centered cube, therefore it is the unit cell for Pb.

12.46 Unit cell volume = $(7900 \times 10^{-12}\ \text{m})^2 (3800 \times 10^{-12}\ \text{m}) \left(\dfrac{100\ \text{cm}}{1\ \text{m}} \right)^3 = 2.37 \times 10^{-19}\ \text{cm}^3$

total lysozyme mass = $8 \times 1.44 \times 10^4\ \text{u} \times \dfrac{1.660\ 54 \times 10^{-27}\ \text{kg}}{1\ \text{u}} \times \dfrac{1000\ \text{g}}{1\ \text{kg}} = 1.91 \times 10^{-19}\ \text{g}$

total lysozyme volume = $1.91 \times 10^{-19}\ \text{g} \times \dfrac{1\ \text{cm}^3}{1.35\ \text{g}} = 1.42 \times 10^{-19}\ \text{cm3}$

% occupied unit cell volume = $\dfrac{1.42 \times 10^{-19}\ \text{cm}^3}{2.37 \times 10^{-19}\ \text{cm}^3} \times 100\% = 60\%$

12.48 unit cell edge = $d = 287\ \text{pm} = 287 \times 10^{-12}\ \text{m} = 2.87 \times 10^{-8}\ \text{cm}$
unit cell volume = $d^3 = (2.87 \times 10^{-8}\ \text{cm})^3 = 2.364 \times 10^{-23}\ \text{cm}^3$
unit cell mass = $(2.364 \times 10^{-23}\ \text{cm}^3)(7.86\ \text{g/cm}^3) = 1.858 \times 10^{-22}\ \text{g}$
Fe is body-centered cubic; therefore there are two Fe atoms per unit cell.

mass of one Fe atom = $\dfrac{1.858 \times 10^{-22}\ \text{g}}{2\ \text{Fe atoms}} = 9.290 \times 10^{-23}\ \text{g/atom}$

Avogadro's number = $55.85\ \text{g/mol} \times \dfrac{1\ \text{atom}}{9.290 \times 10^{-23}\ \text{g}} = 6.01 \times 10^{23}\ \text{atoms/mol}$

Structures of Ionic Solids (Section 12.4)

12.50 Six Na^+ ions touch each H^- ion and six H^- ions touch each Na^+ ion.

12.52 $Na^+ \quad H^- \quad Na^+$
$\leftarrow 488\ \text{pm} \rightarrow$ unit cell edge = $d = 488\ \text{pm}$; Na–H bond = $d/2 = 244\ \text{pm}$

Structures of Covalent Network Solids (Section 12.5)

12.54

12.56 Carbon nanotubes are tubular structures made of repeating six-membered carbon rings, as if a sheet of graphite were rolled up. Each carbon atom is sp^2 hybridized and is bonded with trigonal planar geometry to three other carbons. Carbon nanotubes are good conductors of electricity because the carbon atoms are sp^2 hybridized. Electrons can move in the extended system of overlapping unhybridized p orbitals to conduct an electric current.

Bonding in Metals (Section 12.6)

12.58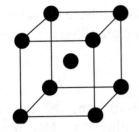

Each K has a single valence electron and has eight nearest neighbor K atoms. The valence electrons cannot be localized in an electron-pair bond between any particular pair of K atoms.

12.60 Malleability and ductility of metals follow from the fact that the delocalized bonding extends in all directions. When a metallic crystal is deformed, no localized bonds are broken. Instead, the electron sea simply adjusts to the new distribution of cations, and the energy of the deformed structure is similar to that of the original. Thus, the energy required to deform a metal is relatively small.

12.62 Ionic bonding is much stronger than metallic bonding.

12.64 The energy required to deform a transition metal like W is greater than that for Cs because W has more valence electrons and hence more electrostatic "glue".

12.66 The difference in energy between successive MOs in a metal decreases as the number of metal atoms increases so that the MOs merge into an almost continuous band of energy levels. Consequently, MO theory for metals is often called band theory.

12.68 The energy levels within a band occur in degenerate pairs; one set of energy levels applies to electrons moving to the right, and the other set applies to electrons moving to the left. In the absence of an electrical potential, the two sets of levels are equally populated. As a result there is no net electric current. In the presence of an electrical potential those electrons moving to the right are accelerated, those moving to the left are slowed down, and some change direction. Thus, the two sets of energy levels are now unequally populated. The number of electrons moving to the right is now greater than the number moving to the left, and so there is a net electric current.

12.70

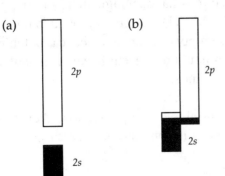

Diagram (b) shows the 2s and 2p bands overlapping in energy and the resulting composite band is only partially filled. Thus, Be is a good electrical conductor.

12.72   Transition metals have a d band that can overlap the s band to give a composite band consisting of six MOs per metal atom. Half of the MOs are bonding and half are antibonding, and thus one expects maximum bonding for metals that have six valence electrons per metal atom. Accordingly, the melting points of the transition metals go through a maximum at or near group 6B.

12.74   W   [Xe] $4f^{14} 5d^4 6s^2$ Au   [Xe] $4f^{14} 5d^{10} 6s^1$
These facts are better explained with the MO band model. Transition metals have a d band that can overlap the s band to give a composite band consisting of six MOs per metal atom. Half of the MOs are bonding and half are antibonding, and thus one expects maximum bonding for metals that have six valence electrons per metal atom. Accordingly, the hardness and melting points of the transition metals go through a maximum at or near group 6B.

Semiconductors (Sections 12.7–12.8)

12.76   A semiconductor is a material that has an electrical conductivity intermediate between that of a metal and that of an insulator. Si, Ge, and Sn (gray) are semiconductors.

12.78

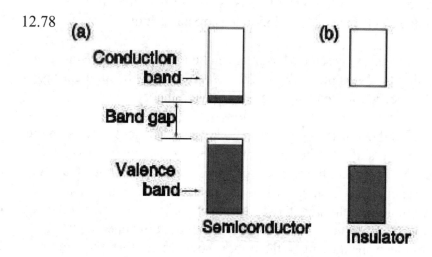

The MOs of a semiconductor are similar to those of an insulator, but the band gap in a semiconductor is smaller. As a result, a few electrons have enough energy to jump the gap and occupy the higher-energy, conduction band. The conduction band is thus partially filled, and the valence band is partially empty. When an electrical potential is applied to a semiconductor, it conducts a small amount of current because the potential can accelerate the electrons in the partially filled bands.

12.80   As the band gap increases, the number of electrons able to jump the gap and occupy the higher-energy conduction band decreases, and thus the conductivity decreases.

12.82 An n-type semiconductor is a semiconductor doped with a substance with more valence electrons than the semiconductor itself. Si doped with P is an example.

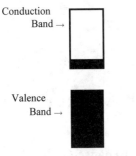

Conduction
Band →

Valence
Band →

n-Type semiconductor

12.84 In the MO picture, the extra electrons occupy the conduction band. The number of electrons in the conduction band of the doped Ge is much greater than for pure Ge, and the conductivity of the doped semiconductor is correspondingly higher.

12.86 (a) p-type (In is electron deficient with respect to Si)
       (b) n-type (Sb is electron rich with respect to Ge)
       (c) n-type (As is electron rich with respect to gray Sn)

12.88 $Cd(CH_3)_2(g) + H_2Se(g) \rightarrow CdSe(s) + 2 CH_4(g)$

12.90 $Al_2O_3 < Ge < Ge$ doped with $In < Fe < Cu$

12.92 In a diode, current flows only when the junction is under a forward bias (negative battery terminal on the n-type side). A p-n junction that is part of a circuit and subjected to an alternating potential acts as a rectifier, allowing current to flow in only one direction, thereby converting alternating current to direct current.

12.94 Both an LED and a photovoltaic cell contain p-n junctions, but the two devices involve opposite processes. An LED converts electrical energy to light; a photovoltaic, or solar, cell converts light to electricity.

12.96 $E = 193 \text{ kJ/mol} \times \dfrac{1000 \text{ J}}{1 \text{ kJ}} \times \dfrac{1 \text{ mol}}{6.02 \times 10^{23}} = 3.21 \times 10^{-19} \text{ J}$

$\nu = \dfrac{E}{h} = \dfrac{3.21 \times 10^{-19} \text{ J}}{6.626 \times 10^{-34} \text{ J·s}} = 4.84 \times 10^{14} \text{ s}^{-1}$

$\lambda = \dfrac{c}{\nu} = \dfrac{3.00 \times 10^8 \text{ m/s}}{4.84 \times 10^{14} \text{ s}^{-1}} = 6.20 \times 10^{-7} \text{ m} = 620 \times 10^{-9} \text{ m} = 620 \text{ nm}$

12.98 (a) InN has the smaller band-gap energy because In is larger than Ga.
       (b) GaN would emit ultraviolet light and InN would emit red light.

12.100 $GaP_{0.50}As_{0.50} < GaP_{0.80}As_{0.20} < GaP_{1.00}As_{0.00}$

12.102 (a) $E = 107 \text{ kJ/mol} \times \dfrac{1000 \text{ J}}{1 \text{ kJ}} \times \dfrac{1 \text{ mol}}{6.02 \times 10^{23}} = 1.78 \times 10^{-19} \text{ J}$

$\nu = \dfrac{E}{h} = \dfrac{1.78 \times 10^{-19} \text{ J}}{6.626 \times 10^{-34} \text{ J·s}} = 2.68 \times 10^{14} \text{ s}^{-1}$

$\lambda = \dfrac{c}{\nu} = \dfrac{3.00 \times 10^8 \text{ m/s}}{2.68 \times 10^{14} \text{ s}^{-1}} = 1.12 \times 10^{-6} \text{ m} = 1120 \times 10^{-9} \text{ m} = 1120 \text{ nm}$

(b) The wavelength is in the near IR and does not correspond to the highest intensity wavelength in the solar emission spectrum.

12.104 With a band gap of 130 kJ/mol, GaAs is a semiconductor. Because Ge lies between Ga and As in the periodic table, GaAs is isoelectronic with Ge.

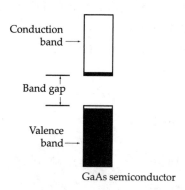

GaAs semiconductor

12.106 (a) In an n-type InP semiconductor the valence band is completely filled and the conduction band is partially full. Cd has only two 5s and no 5p electrons. Adding Cd to the n-type InP semiconductor would add positive holes that would combine with free electrons. This results in a decrease in the number of electrons in the conduction band and a decrease in the conductivity.

(b) In a p-type InP semiconductor the valence band is partially filled and the conduction band is empty. The charge carriers are positive holes. Se has 6 valence electrons. Adding Se to the p-type InP semiconductor would add electrons that would combine with positive holes. This results in a decrease in the number of positive holes and a decrease in the conductivity.

12.108 (a)

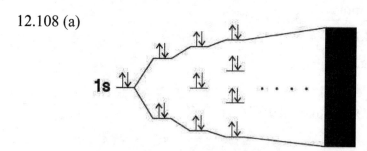

This material is an insulator because all MOs are filled, preventing the movement of electrons.

(b)

Neutral hydrogen atoms have only 1 valence electron, compared with 2 in H$^-$. Partially empty antibonding MOs will allow the movement of electrons, so the doped material will be a conductor.

(c) The missing electrons in the doped material create "holes" that are positive charge carriers. This type of doped material is a p-type semiconductor.

Superconductors (Section 12.9)

12.110 (1) A superconductor is able to levitate a magnet.

(2) In a superconductor, once an electric current is started, it flows indefinitely without loss of energy. A superconductor has no electrical resistance.

12.112 Some K$^+$ ions are surrounded octahedrally by six C_{60}^{3-} ions; others are surrounded tetrahedrally by four C_{60}^{3-} ions.

12.114 77 K is the boiling point of the readily available liquid N_2.

Ceramics and Composites (Section 12.10)

12.116 Ceramics are inorganic, nonmetallic, nonmolecular solids, including both crystalline and amorphous materials. Ceramics have higher melting points, and they are stiffer, harder, and more resistant to wear and corrosion than are metals.

12.118 Ceramics have higher melting points, and they are stiffer, harder, and more wear resistant than metals because they have stronger bonding. They maintain much of their strength at high temperatures, where metals either melt or corrode because of oxidation.

12.120 The brittleness of ceramics is due to strong chemical bonding. In silicon nitride each Si atom is bonded to four N atoms and each N atom is bonded to three Si atoms. The strong, highly directional covalent bonds prevent the planes of atoms from sliding over one another when the solid is subjected to a stress. As a result, the solid cannot deform to relieve the stress. It maintains its shape up to a point, but then the bonds give way suddenly and the material fails catastrophically when the stress exceeds a certain threshold value. By contrast, metals are able to deform under stress because their planes of metal cations can slide easily in the electron sea.

12.122 Ceramic processing is the series of steps that leads from raw material to the finished ceramic object.

12.124 $Zr[OCH(CH_3)_2]_4 + 4\,H_2O \rightarrow Zr(OH)_4 + 4\,HOCH(CH_3)_2$

12.126 $(HO)_3Si-O-H + H-O-Si(OH)_3 \rightarrow (HO)_3Si-O-Si(OH)_3 + H_2O$
Further reactions of this sort give a three-dimensional network of Si–O–Si bridges.
On heating, SiO_2 is obtained.

12.128 $2\,Ti(BH_4)_3(soln) \rightarrow 2\,TiB_2(s) + B_2H_6(g) + 9\,H_2(g)$

12.130 $3\,SiCl_4(g) + 4\,NH_3(g) \rightarrow Si_3N_4(s) + 12\,HCl(g)$

12.132   Graphite/epoxy composites are good materials for making tennis rackets and golf
clubs because of their high strength-to-weight ratios.

Multiconcept Problems

12.134   (a) Let the formula of magnetite be Fe_xO_y, then $Fe_xO_y + y\,CO \rightarrow x\,Fe + y\,CO_2$

$$n_{CO_2} = y = \frac{PV}{RT} = \frac{\left(751\ mm\ Hg \times \dfrac{1.00\ atm}{760\ mm\ Hg}\right)(1.136\ L)}{\left(0.082\ 06\ \dfrac{L \cdot atm}{K \cdot mol}\right)(298\ K)} = 0.04590\ mol\ CO_2$$

$0.04590\ mol\ CO_2 = mol\ of\ O\ in\ Fe_xO_y$

mass of O in $Fe_xO_y = 0.04590\ mol\ O \times \dfrac{16.0\ g\ O}{1\ mol\ O} = 0.7345\ g\ O$

mass of Fe in $Fe_xO_y = 2.660\ g - 0.7345\ g = 1.926\ g\ Fe$

(b) mol Fe in magnetite $= 1.926\ g\ Fe \times \dfrac{1\ mol\ Fe}{55.85\ g\ Fe} = 0.0345\ mol\ Fe$

formula of magnetite: $Fe_{0.0345}\ O_{0.0459}$ (divide each subscript by the smaller)
$Fe_{0.0345/0.0345}\ O_{0.0459/0.0345}$
$FeO_{1.33}$ (multiply both subscripts by 3)
$Fe_{(1 \times 3)}\ O_{(1.33 \times 3)};\ \ Fe_3O_4$

(c) unit cell edge $= d = 839\ pm = 839 \times 10^{-12}\ m$

$d = 839 \times 10^{-12}\ m \times \dfrac{100\ cm}{1\ m} = 8.39 \times 10^{-8}\ cm$

unit cell volume $= d^3 = (8.39 \times 10^{-8}\ cm)^3 = 5.91 \times 10^{-22}\ cm^3$
unit cell mass $= (5.91 \times 10^{-22}\ cm^3)(5.20\ g/cm^3) = 3.07 \times 10^{-21}\ g$

mass of Fe in unit cell $= \left(\dfrac{1.926\ g\ Fe}{2.660\ g}\right)(3.07 \times 10^{-21}\ g) = 2.22 \times 10^{-21}\ g\ Fe$

mass of O in unit cell $= \left(\dfrac{0.7345\ g\ O}{2.660\ g}\right)(3.07 \times 10^{-21}\ g) = 8.47 \times 10^{-22}\ g\ O$

Fe atoms in unit cell $= 2.22 \times 10^{-21}\ g \times \dfrac{6.022 \times 10^{23}\ atoms/mol}{55.847\ g/mol} = 24\ Fe\ atoms$

O atoms in unit cell $= 8.47 \times 10^{-22}\ g \times \dfrac{6.022 \times 10^{23}\ atoms/mol}{16.00\ g/mol} = 32\ O\ atoms$

12.136 (a) $n_{H_2} = \dfrac{PV}{RT} = \dfrac{\left(740 \text{ mm Hg} \times \dfrac{1.00 \text{ atm}}{760 \text{ mm Hg}}\right)(4.00 \text{ L})}{\left(0.08206 \dfrac{\text{L} \cdot \text{atm}}{\text{K} \cdot \text{mol}}\right)(296 \text{ K})} = 0.160 \text{ mol } H_2$

M = Group 3A metal; $\quad 2 \text{ M(s)} + 6 \text{ H}^+(aq) \rightarrow 2 \text{ M}^{3+}(aq) + 3 \text{ H}_2(g)$

$n_M = 0.160 \text{ mol } H_2 \times \dfrac{2 \text{ mol M}}{3 \text{ mol } H_2} = 0.107 \text{ mol M}$

mass M = 1.07 cm^3 x 2.70 g/cm^3 = 2.89 g M

molar mass M = $\dfrac{2.89 \text{ g M}}{0.107 \text{ mol M}}$ = 27.0 g/mol;   The Group 3A metal is Al

(b) mass of one Al atom = 26.98 g/mol x $\dfrac{1 \text{ mol}}{6.022 \times 10^{23} \text{ atoms}}$ = 4.48 x 10^{-23} g/atom

unit cell edge = d = 404 pm = 404 x 10^{-12} m

d = 404 x 10^{-12} m x $\dfrac{100 \text{ cm}}{1 \text{ m}}$ = 4.04 x 10^{-8} cm

unit cell volume = d^3 = (4.04 x 10^{-8} cm)^3 = 6.59 x 10^{-23} cm^3

Calculate the density of Al assuming it is primitive cubic, body-centered cubic, and face-centered cubic. Compare the calculated density with the actual density to identify the unit cell.

For primitive cubic:

density = $\dfrac{\text{unit cell mass}}{\text{unit cell volume}}$ = $\dfrac{(1 \text{ Al})(4.48 \times 10^{-23} \text{ g/Al atom})}{6.59 \times 10^{-23} \text{ cm}^3}$ = 0.680 g/cm^3

For body-centered cubic:

density = $\dfrac{\text{unit cell mass}}{\text{unit cell volume}}$ = $\dfrac{(2 \text{ Al})(4.48 \times 10^{-23} \text{ g/Al atom})}{6.59 \times 10^{-23} \text{ cm}^3}$ = 1.36 g/cm^3

For face-centered cubic:

density = $\dfrac{\text{unit cell mass}}{\text{unit cell volume}}$ = $\dfrac{(4 \text{ Al})(4.48 \times 10^{-23} \text{ g/Al atom})}{6.59 \times 10^{-23} \text{ cm}^3}$ = 2.72 g/cm^3

The calculated density for a face-centered cube (2.72 g/cm^3) is closest to the actual density of 2.70 g/cm^3. Al crystallizes in the face-centered cubic unit cell.

(c) r = $\sqrt{\dfrac{d^2}{8}}$ = $\sqrt{\dfrac{(404 \text{ pm})^2}{8}}$ = 143 pm

12.138 (a) $n_{X_2} = \dfrac{PV}{RT} = \dfrac{\left(755 \text{ mm Hg} \times \dfrac{1.00 \text{ atm}}{760 \text{ mm Hg}}\right)(0.500 \text{ L})}{\left(0.082 \, 06 \dfrac{\text{L} \cdot \text{atm}}{\text{K} \cdot \text{mol}}\right)(298 \text{ K})} = 0.0203 \text{ mol } X_2$

$$M(s) + 1/2 X_2(g) \rightarrow MX(s)$$

$$\text{mol M} = 0.0203 \text{ mol } X_2 \text{ x } \frac{1 \text{ mol M}}{1/2 \text{ mol } X_2} = 0.0406 \text{ mol M}$$

$$\text{molar mass M} = \frac{1.588 \text{ g M}}{0.0406 \text{ mol M}} = 39.1 \text{ g/mol; atomic mass} = 39.1 \text{ ; } M = K$$

(b) From Figure 6.2, the radius for K^+ is ~140 pm.
unit cell edge = 535 pm = $2r_{K^+} + 2r_{X^-}$

$$r_{X^-} = \frac{535 \text{ pm} - 2r_{K^+}}{2} = \frac{535 \text{ pm} - 2(140 \text{ pm})}{2} = 128 \text{ pm}$$

From Figure 6.3, $X^- = F^-$

(c) Because the cation and anion are of comparable size, the anions are not in contact with each other.

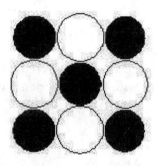

$\bigcirc = K^+$

$\bullet = F^-$

(d) unit cell contents: $1/8 \text{ F}^-$ at 8 corners and $1/2 \text{ F}^-$ at 6 faces = 4 F^-
                            $1/4 \text{ K}^+$ at 12 edges and 1 K^+ inside = 4 K^+

$$\text{mass of one } K^+ = \frac{39.098 \text{ g/mol}}{6.022 \text{ x } 10^{23} \text{ K}^+/\text{mol}} = 6.493 \text{ x } 10^{-23} \text{ g/K}^+$$

$$\text{mass of one } F^- = \frac{18.998 \text{ g/mol}}{6.022 \text{ x } 10^{23} \text{ F}^-/\text{mol}} = 3.155 \text{ x } 10^{-23} \text{ g/F}^-$$

unit cell mass = $(4 \text{ K}^+)(6.493 \text{ x } 10^{-23} \text{ g/K}^+) + (4 \text{ F}^-)(3.155 \text{ x } 10^{-23} \text{ g/F}^-) = 3.859 \text{ x } 10^{-22}$ g
unit cell volume = $[(535 \text{ x } 10^{-12} \text{ m})(100 \text{ cm/m})]^3 = 1.531 \text{ x } 10^{-22} \text{ cm}^3$

$$\text{density} = \frac{\text{mass of unit cell}}{\text{volume of unit cell}} = \frac{3.859 \text{ x } 10^{-22} \text{ g}}{1.531 \text{ x } 10^{-22} \text{ cm}^3} = 2.52 \text{ g/cm}^3$$

(e) $K(s) + 1/2 F_2(g) \rightarrow KF(s)$ is a formation reaction.

$$\Delta H^\circ_f(KF) = \frac{-22.83 \text{ kJ}}{0.0406 \text{ mol}} = -562 \text{ kJ/mol}$$

12.140   660 nm = 660 x 10^{-9} m and 3.0 mW = 3.0 x 10^{-3} W = 3.0 x 10^{-3} J/s

$$E = h\frac{c}{\lambda} = (6.626 \text{ x } 10^{-34} \text{ J·s})\left(\frac{3.00 \text{ x } 10^8 \text{ m/s}}{660 \text{ x } 10^{-9} \text{ m}}\right) = 3.0 \text{ x } 10^{-19} \text{ J/photon}$$

$$\text{\# of photons/s} = \frac{3.0 \times 10^{-3} \text{ J/s}}{3.0 \times 10^{-19} \text{ J/photon}} = 1.0 \times 10^{16} \text{ photons/s}$$

# of electrons/s = # of photons/s = 1.0×10^{16} electrons/s

$$\text{\# of moles of electrons/s} = 1.0 \times 10^{16} \text{ electrons/s} \times \frac{1 \text{ mol e}^-}{6.02 \times 10^{23} \text{ e}^-} = 1.7 \times 10^{-8} \text{ mol e}^-\text{/s}$$

$$A = 1.7 \times 10^{-8} \text{ mol e}^-\text{/s} \times \frac{96,500 \text{ C}}{1 \text{ mol e}^-} = 0.0016 \text{ C/s} = 0.0016 \text{ A} = 1.6 \times 10^{-3} \text{ A} = 1.6 \text{ mA}$$

12.142   (a)

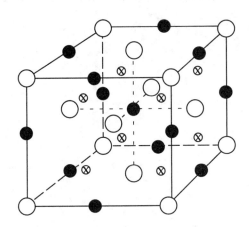

○ = C_{60}^{3-}

● = M^+ in an octahedral hole

⊗ = M^+ in a tetrahedral hole

(b)   There are 4 C_{60}^{3-} ions, 4 octahedral holes, and 8 tetrahedral holes per unit cell.

(c)   Octahedral holes: (1/2,1/2,1/2), (1/2,0,0), (0,1/2,0), (0,0,1/2)
Tetrahedral holes: (1/4,1/4,1/4), (3/4,1/4,1/4), (1/4,3/4,1/4), (3/4,3/4,1/4),
(1/4,1/4,3/4), (3/4,1/4,3/4), (1/4,3/4,3/4), (3/4,3/4,3/4)

(d)   Let the unit cell edge = a.
The face diagonal is equal to 4R = 4(500 pm) = 2000 pm

$a^2 + a^2 = (2000)^2$; $2a^2 = 4 \times 10^6$; $a^2 = 2 \times 10^6$; $a = \sqrt{2 \times 10^6} = 1414$ pm

$a = 2R(C_{60}^{3-}) + 2R(\text{octahedral hole}) = 1414$ pm

$$R(\text{octahedral hole}) = \frac{1414 \text{ pm} - 2R(C_{60}^{3-})}{2} = \frac{1414 \text{ pm} - 2(500 \text{ pm})}{2} = 207 \text{ pm}$$

The tetrahedron that defines the tetrahedral hole can be thought of as being found inside a cube with edge = a/2 = 707 pm. This cube is located in one corner of the unit cell. The face diagonal of this cube = $2R(C_{60}^{3-})$ = 1000 pm.

The body diagonal of this cube = $\sqrt{707^2 + 1000^2} = 1225$ pm

Body diagonal = $2R(C_{60}^{3-}) + 2R(\text{tetrahedral hole}) = 1225$ pm

$$R(\text{tetrahedral hole}) = \frac{1225 \text{ pm} - 2R(C_{60}^{3-})}{2} = \frac{1225 \text{ pm} - 2(500 \text{ pm})}{2} = 112 \text{ pm}$$

(e) Na^+ will fit into the octahedral and tetrahedral holes without expanding the C_{60}^{3-} framework. K^+ and Rb^+ will fit into the octahedral holes without expanding the C_{60}^{3-} framework but will fit into the tetrahedral holes only if the C_{60}^{3-} framework is expanded.

12.144   (a)

(b) $\Delta H^\circ = D_{C=C} - 2\,D_{C-C} = 611\ kJ - 2(350)\ kJ = -89\ kJ/unit$; exothermic

13

Solutions and Their Properties

13.1 KBr < 1,5 pentanediol < toluene

13.2 Vitamin E is more fat soluble.

13.3 NaCl, 58.44; 1.00 mol NaCl = 58.44 g
1.00 L H_2O = 1000 mL = 1000 g (assuming a density of 1.00 g/mL)

$$\text{mass \% NaCl} = \frac{58.44 \text{ g}}{1000 \text{ g} + 58.44 \text{ g}} \times 100\% = 5.52 \text{ mass \%}$$

13.4 $$7.50\% = \left(\frac{15.0 \text{ g}}{15.0 \text{ g} + x} \right)(100\%)$$

Let x equal the mass of water. Solve for x.
$(7.50\%)(15.0 \text{ g} + x) = (15.0 \text{ g})(100\%)$

$$x = \frac{(15.0 \text{ g})(100\%)}{7.50\%} - 15.0 \text{ g} = 185 \text{ g } H_2O$$

13.5 1.25 μg = 1.25 x 10^{-6} g;   (50.0 mL)(1.00 g/mL) = 50.0 g

$$\frac{1.25 \times 10^{-6} \text{ g}}{50.0 \text{ g}} \times 10^9 = 25.0 \text{ ppb}$$

This exceeds the 10.0 ppb limit for As.

13.6 $$\text{ppm} = \frac{\text{mass of } CO_2}{\text{total mass of solution}} \times 10^6 \text{ ppm}$$

total mass of solution = density x volume = (1.3 g/L)(1.0 L) = 1.3 g

$$35 \text{ ppm} = \frac{\text{mass of } CO_2}{1.3 \text{ g}} \times 10^6 \text{ ppm}$$

$$\text{mass of } CO_2 = \frac{(35 \text{ ppm})(1.3 \text{ g})}{10^6 \text{ ppm}} = 4.6 \times 10^{-5} \text{ g } CO_2$$

13.7 CH_3CO_2Na, 82.03

$$\text{kg } H_2O = (0.150 \text{ mol } CH_3CO_2Na)\left(\frac{1 \text{ kg } H_2O}{0.500 \text{ mol } CH_3CO_2Na} \right) = 0.300 \text{ kg } H_2O$$

$$\text{mass } CH_3CO_2Na = 0.150 \text{ mol } CH_3CO_2Na \times \frac{82.03 \text{ g } CH_3CO_2Na}{1 \text{ mol } CH_3CO_2Na} = 12.3 \text{ g } CH_3CO_2Na$$

mass of solution needed = 300 g + 12.3 g = 312 g

13.8 $C_{27}H_{46}O$, 386.7; $CHCl_3$, 119.4;   40.0 g x $\dfrac{1\ kg}{1000\ g}$ = 0.0400 kg

$$\text{molality} = \frac{\text{mol } C_{27}H_{46}O}{\text{kg } CHCl_3} = \frac{\left(0.385\ g\ x\ \dfrac{1\ mol}{386.7\ g}\right)}{0.0400\ kg} = 0.0249\ \text{mol/kg} = 0.0249\ m$$

$$X_{C_{27}H_{46}O} = \frac{\text{mol } C_{27}H_{46}O}{\text{mol } C_{27}H_{46}O\ +\ \text{mol } CHCl_3}$$

$$X_{C_{27}H_{46}O} = \frac{\left(0.385\ g\ x\ \dfrac{1\ mol}{386.7\ g}\right)}{\left[\left(0.385\ g\ x\ \dfrac{1\ mol}{386.7\ g}\right) + \left(40.0\ g\ x\ \dfrac{1\ mol}{119.4\ g}\right)\right]} = 2.96\ x\ 10^{-3}$$

13.9 Assume 1.00 L of solution.
mass of 1.00 L = (1.0042 g/mL)(1000 mL) = 1004.2 g of solution

0.500 mol CH_3CO_2H x $\dfrac{60.05\ g\ CH_3CO_2H}{1\ mol\ CH_3CO_2H}$ = 30.02 g CH_3CO_2H

mass of H_2O (solvent) = 1004.2 g – 30.02 g = 974.2 g = 0.9742 kg of H_2O

molality = $\dfrac{0.500\ mol}{0.9742\ kg}$ = 0.513 m

13.10 Assume you have a solution with 1.000 kg (1000 g) of H_2O. If this solution is 0.258 m, then it must also contain 0.258 mol glucose.

mass of glucose = 0.258 mol x $\dfrac{180.2\ g}{1\ mol}$ = 46.5 g glucose

mass of solution = 1000 g + 46.5 g = 1046.5 g
density = 1.0173 g/mL

volume of solution = 1046.5 g x $\dfrac{1\ mL}{1.0173\ g}$ = 1028.7 mL

volume = 1028.7 mL x $\dfrac{1\ L}{1000\ mL}$ = 1.029 L;   molarity = $\dfrac{0.258\ mol}{1.029\ L}$ = 0.251 M

13.11 M = k · P; k = $\dfrac{M}{P}$ = $\dfrac{3.2\ x\ 10^{-2}\ M}{1.0\ atm}$ = 3.2 x 10^{-2} mol/(L· atm)

M = k · P = [3.2 x 10^{-2} mol/(L· atm)](5.0 atm) = 0.16 M

13.12 (a) M = k · P = [3.2 x 10^{-2} mol/(L · atm)](2.5 atm) = 0.080 M
(b) P_{CO_2} = (0.0004)(1 atm) = 4.0 x 10^{-4} atm

M = k · P = [3.2 x 10^{-2} mol/(L · atm)](4.0 x 10^{-4} atm) = 1.3 x 10^{-5} M

13.13 H_2O, 18.02; $CaCl_2$, 110.0

$$100.0 \text{ g } H_2O \times \frac{1 \text{ mol } H_2O}{18.02 \text{ g } H_2O} = 5.549 \text{ mol } H_2O$$

$$10.00 \text{g } CaCl_2 \times \frac{1 \text{ mol } CaCl_2}{110.0 \text{ g } CaCl_2} = 0.090 \, 91 \text{ mol } CaCl_2$$

$$X_{H_2O} = \frac{5.549 \text{ mol}}{(2.7)(0.090 \, 91 \text{ mol}) + 5.549 \text{ mol}} = 0.9567$$

$$P_{soln} = P_{H_2O} \times X_{H_2O} = (233.7 \text{ mm Hg})(0.9576) = 223.8 \text{ mm Hg}$$

13.14 H_2O, 18.02; $MgCl_2$, 95.21

$$100.0 \text{ g } H_2O \times \frac{1 \text{ mol } H_2O}{18.02 \text{ g } H_2O} = 5.549 \text{ mol } H_2O$$

$$8.110 \text{g } MgCl_2 \times \frac{1 \text{ mol } MgCl_2}{95.21 \text{ g } MgCl_2} = 0.085 \, 18 \text{ mol } MgCl_2$$

$$X_{H_2O} = \frac{P_{soln}}{P_{H_2O}} = \frac{224.7 \text{ mm Hg}}{233.7 \text{ mm Hg}} = 0.9615$$

$$X_{H_2O} = 0.9615 = \frac{5.549 \text{ mol}}{(i)(0.085 \, 18 \text{ mol}) + 5.549 \text{ mol}} ; \text{ solve for the van't Hoff factor, i.}$$

$$(0.9615)[(i)(0.085 \, 18 \text{ mol}) + 5.549 \text{ mol})] = 5.549 \text{ mol}$$

$$i = \frac{\left(\dfrac{5.549 \text{ mol}}{0.9615} - 5.549 \text{ mol} \right)}{0.085 \, 18} = 2.6$$

13.15 $$100 \text{ g } C_2H_5OH \times \frac{1 \text{ mol } C_2H_5OH}{46.07 \text{ g } C_2H_5OH} = 2.171 \text{ mol } C_2H_6O$$

$$25.0 \text{ g } H_2O \times \frac{1 \text{ mol } H_2O}{18.02 \text{ g } H_2O} = 1.387 \text{ mol } H_2O$$

$$X_{C_2H_5OH} = \frac{2.171 \text{ mol}}{2.171 \text{ mol} + 1.387 \text{ mol}} = 0.6102$$

$$X_{H_2O} = \frac{1.387 \text{ mol}}{2.171 \text{ mol} + 1.387 \text{ mol}} = 0.3898$$

$$P_{soln} = X_{C_2H_5OH} P^o_{C_2H_5OH} + X_{H_2O} P^o_{H_2O}$$

$$P_{soln} = (0.6102)(61.2 \text{ mm Hg}) + (0.3898)(23.8 \text{ mm Hg}) = 46.6 \text{ mm Hg}$$

13.16 The red and blue curves are the pure liquids and the green curve is the mixture.

13.17 $C_2H_6O_2$, 62.07; $500.0 \text{ g} = 0.5000 \text{ kg}$

$$616.9 \text{ g } C_2H_6O_2 \times \frac{1 \text{ mol } C_2H_6O_2}{62.07 \text{ g } C_2H_6O_2} = 9.939 \text{ mol } C_2H_6O_2$$

$$\Delta T = K_b \cdot m = \left(0.51 \frac{°C \cdot kg}{mol}\right)\left(\frac{9.939 \text{ mol}}{0.5000 \text{ kg}}\right) = 10.1 °C$$

$$bp = 100.0 °C + 10.1 °C = 110.1 °C$$

13.18　The red curve represents the vapor pressure of pure chloroform.

(a)　The normal boiling point for a liquid is the temperature where the vapor pressure of the liquid equals 1 atm (760 mm Hg).　The approximate boiling point of pure chloroform is 62 °C.

(b)　The approximate boiling point of the solution is 69 °C.

$$\Delta T_b = 69 °C - 62 °C = 7 °C$$

$$\Delta T_b = K_b \cdot m$$

$$m = \frac{\Delta T_b}{K_b} = \frac{7 °C}{3.63 \frac{°C \cdot kg}{mol}} = 2 \text{ mol/kg} = 2 \ m$$

13.19　$C_6H_{12}O_6$, 180.2;　　37.0 °C = 310.1 K

$$\Pi = \left(\frac{50.0 \text{ g} \times \frac{1 \text{ mol}}{180.2 \text{ g}}}{1.00 \text{ L}}\right)\left(0.082\ 06 \frac{L \cdot atm}{K \cdot mol}\right)(310.1 \text{ K}) = 7.06 \text{ atm}$$

13.20　(a)　$\Pi = iMRT$;　　$M = \dfrac{\Pi}{iRT} = \dfrac{8.0 \text{ atm}}{(1.9)\left(0.082\ 06 \frac{L \cdot atm}{K \cdot mol}\right)(298 \text{ K})} = 0.17 \text{ M}$

(b)　There would be a net transfer of water from outside the cell to the inside of the cell and the cell would swell.

13.21　$\Pi = MRT$;　　$M = \dfrac{\Pi}{RT} = \dfrac{423.1 \text{ mm Hg} \times \frac{1 \text{ atm}}{760 \text{ mm Hg}}}{\left(0.082\ 06 \frac{L \cdot atm}{K \cdot mol}\right)(298 \text{ K})} = 0.0228 \text{ mol/L}$

200.0 mL = 0.2000 L

mol = (0.0228 mol/L)(0.2000 L) = 0.004 553 mol

$$molar\ mass = \frac{0.8220 \text{ g}}{0.004\ 553 \text{ mol}} = 180.5 \text{ g/mol}$$

13.22　$C_{12}H_{22}O_{11}$, 342.3

$$\Pi = MRT; \quad M = \frac{\Pi}{RT} = \frac{278 \text{ mm Hg} \times \frac{1 \text{ atm}}{760 \text{ mm Hg}}}{\left(0.082\ 06 \frac{L \cdot atm}{K \cdot mol}\right)(298 \text{ K})} = 0.014\ 96 \text{ mol/L}$$

100.0 mL = 0.1000 L

mol = (0.014 96 mol/L)(0.1000 L) = 0.001 496 mol

$$\text{molar mass} = \frac{0.512 \text{ g}}{0.001\ 496 \text{ mol}} = 342.3 \text{ g/mol}$$

The white powder is sucrose.

13.23   Both solvent molecules and small solute particles can pass through a semipermeable dialysis membrane. Only large colloidal particles such as proteins can't pass through. Only solvent molecules can pass through a semipermeable membrane used for osmosis.

13.24   Solvent–solvent is hydrogen bonding, solvent–solute is hydrogen bonding, and solute–solute is hydrogen bonding.

13.25   (a) Total conc $= (137 + 105 + 3.0 + 4.0 + 2.0 + 33 + 0.75 + 11.1)$ mmol/L $= 295.85$ mmol/L
295.85 mmol/L $= 295.85 \times 10^{-3}$ mol/L $= 0.295\ 85$ mol/L
25 °C $= 298$ K

$$\Pi = MRT = (0.295\ 85 \text{ mol/L})\left(0.082\ 06 \frac{\text{L} \cdot \text{atm}}{\text{K} \cdot \text{mol}}\right)(298\text{K}) = 7.23 \text{ atm}$$

(b) Solvent moves from the dialysis solution to blood.

13.26 $12.0 \text{ L} \times \dfrac{1 \text{ mL}}{1 \times 10^{-3} \text{ L}} \times \dfrac{1 \text{ g}}{\text{mL}} = 12{,}000 \text{ g}$

$2 \text{ ppm} = \dfrac{\text{mass of F}^-}{12{,}000 \text{ g}} \times 10^6$

$\text{mass of F}^- = \dfrac{(2)(12{,}000 \text{ g})}{10^6} = 0.024 \text{ g} = 24 \text{ mg}$

Conceptual Problems

13.28   (a) < (b) < (c)

13.30   The upper curve is pure ether.
(a) The normal boiling point for ether is the temperature where the upper curve intersects the 760 mm Hg line, ~ 37 °C.
(b) $\Delta T_b \approx 3$ °C
$\Delta T_b = K_b \cdot m$

$$m = \frac{\Delta T_b}{K_b} = \frac{3 \text{ °C}}{2.02 \dfrac{\text{°C} \cdot \text{kg}}{\text{mol}}} \approx 1.5 \text{ mol/kg} \approx 1.5 \ m$$

13.32   (a) The red curve represents the solution of a volatile solute and the green curve represents the solution of a nonvolatile solute.
(b) & (d)

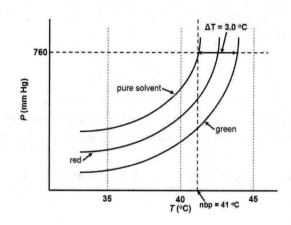

(c) $\Delta T = K_f \cdot m; \quad m = \dfrac{\Delta T}{K_f} = \dfrac{3.0 \,^\circ C}{2.0 \,^\circ C/m} = 1.5 \, m$

13.34 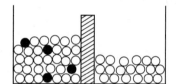 Assume that only the blue (open) spheres (solvent) can pass through the semipermeable membrane. There will be a net transfer of solvent from the right compartment (pure solvent) to the left compartment (solution) to achieve equilibrium.

Section Problems
Solutions, Energy Changes, and Solubility (Sections 13.1–13.3)

13.36 The surface area of a solid plays an important role in determining how rapidly a solid dissolves. The larger the surface area, the more solid–solvent interactions, and the more rapidly the solid will dissolve. Powdered NaCl has a much larger surface area than a large block of NaCl, and it will dissolve more rapidly.

13.38 (a) Na^+ has the larger hydration energy because of its smaller size and higher charge density.
(b) Ba^{2+} has the larger hydration energy because of its higher charge.

13.40 Solvent–solvent is hydrogen bonding, solvent–solute is dispersion, and solute–solute is dispersion. I_2 is not soluble in water.

13.42 $\Delta H_{soln} = \Delta H_{solute\text{-}solute} + \Delta H_{solvent\text{-}solvent} + \Delta H_{solute\text{-}solvent}$

$\Delta H_{soln} = 642 \text{ kJ/mol} + 31 \text{ kJ/mol} - 648 \text{ kJ/mol} = +25 \text{ kJ/mol}$

ΔH_{soln} is positive, the temperature of the solution will decrease.

13.44 Both Br_2 and CCl_4 are nonpolar, and intermolecular forces for both are dispersion forces. H_2O is a polar molecule with dipole-dipole forces and hydrogen bonding. Therefore, Br_2 is more soluble in CCl_4.

13.46 CH_3COOH and water are both polar molecules and they both can hydrogen bond. C_6H_6 is nonpolar. The solubility of CH_3COOH is greater in water.

13.48 Toluene is nonpolar and is insoluble in water.
Br_2 is nonpolar but because of its size, is polarizable and is soluble in water.
KBr is an ionic compound and is very soluble in water.
toluene < Br_2 < KBr   (solubility in H_2O)

13.50 There are three hydrogen bonds.

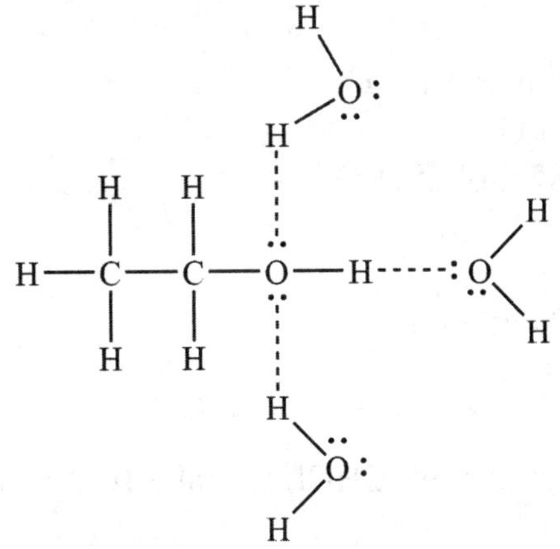

13.52 Ethyl alcohol and water are both polar with small dispersion forces. They both can hydrogen bond, and are miscible.
Pentyl alcohol is slightly polar and can hydrogen bond. It has, however, a relatively large solute-solute dispersion force because of its size, which limits its water solubility.

Units of Concentration (Section 13.4)

13.54 $CaCl_2$, 110.98
For a 1.00 *m* solution:
heat released = 81,300 J
mass of solution = 1000 g H_2O + 110.98 g $CaCl_2$ = 1110.98 g
$$\Delta T = \frac{q}{(\text{specific heat})(\text{mass of solution})} = \frac{81{,}300 \text{ J}}{[4.18 \text{ J/(K} \cdot \text{g)}](1110.98 \text{ g})} = 17.5 \text{ K} = 17.5 \,^{\circ}\text{C}$$
Final temperature = 25.0 °C + 17.5 °C = 42.5 °C

13.56 Assume 1.00 L of seawater.
mass of 1.00 L = (1000 mL)(1.025 g/mL) = 1025 g
$$\frac{\text{mass NaCl}}{1025 \text{ g}} \times 100\% = 3.50 \text{ mass \%};\qquad \text{mass NaCl} = \frac{1025 \text{ g} \times 3.50}{100} = 35.88 \text{ g}$$
There are 35.88 g NaCl per 1.00 L of solution.
$$M = \frac{\left(35.88 \text{ g NaCl} \times \dfrac{1 \text{ mol NaCl}}{58.44 \text{ g NaCl}}\right)}{1.00 \text{ L}} = 0.614 \text{ M}$$

13.58 $C_{16}H_{21}NO_2$, 259.3; 50 ng = 50 x 10^{-9} g
(a) mass of solution = (1.025 g/mL)(1000 mL) = 1025 g

$$ppb = \frac{50 \times 10^{-9} \text{ g}}{1025 \text{ g}} \times 10^9 = 0.049 \text{ ppb}$$

(b) 50 x 10^{-9} g x $\dfrac{1 \text{ mol } C_{16}H_{21}NO_2}{259.3 \text{ g } C_{16}H_{21}NO_2}$ = 1.9 x 10^{-10} mol

$$M = \frac{1.9 \times 10^{-10} \text{ mol}}{1.0 \text{ L}} = 1.9 \times 10^{-10} \text{ mol/L}$$

13.60   (a) Dissolve 0.150 mol of glucose in water; dilute to 1.00 L.
(b) Dissolve 1.135 mol of KBr in 1.00 kg of H_2O.
(c) Mix together 0.15 mol of CH_3OH with 0.85 mol of H_2O.

13.62 $C_7H_6O_2$, 122.12, 165 mL = 0.165 L
mol $C_7H_6O_2$ = (0.0268 mol/L)(0.165 L) = 0.004 42 mol

mass $C_7H_6O_2$ = 0.004 42 mol x $\dfrac{122.12 \text{ g}}{1 \text{ mol}}$ = 0.540 g

Dissolve 4.42 x 10^{-3} mol (0.540 g) of $C_7H_6O_2$ in enough $CHCl_3$ to make 165 mL of solution.

13.64   (a) KCl, 74.6
A 0.500 M KCl solution contains 37.3 g of KCl per 1.00 L of solution.
A 0.500 mass % KCl solution contains 5.00 g of KCl per 995 g of water.
The 0.500 M KCl solution is more concentrated (that is, it contains more solute per amount of solvent).
(b) Both solutions contain the same amount of solute. The 1.75 M solution contains less solvent than the 1.75 m solution. The 1.75 M solution is more concentrated.

13.66   (a) $C_6H_8O_7$, 192.12

0.655 mol $C_6H_8O_7$ x $\dfrac{192.12 \text{ g } C_6H_8O_7}{1 \text{ mol } C_6H_8O_7}$ = 126 g $C_6H_8O_7$

mass % $C_6H_8O_7$ = $\dfrac{126 \text{ g}}{126 \text{ g} + 1000 \text{ g}}$ x 100% = 11.2 mass %

(b) 0.135 mg = 0.135 x 10^{-3} g
(5.00 mL H_2O)(1.00 g/mL) = 5.00 g H_2O

mass % KBr = $\dfrac{0.135 \times 10^{-3} \text{ g}}{(0.135 \times 10^{-3} \text{ g}) + 5.00 \text{ g}}$ x 100% = 0.002 70 mass % KBr

(c) mass % aspirin = $\dfrac{5.50 \text{ g}}{5.50 \text{ g} + 145 \text{ g}}$ x 100% = 3.65 mass % aspirin

13.68 $P_{O_3} = P_{total} \cdot X_{O_3}$

$$X_{O_3} = \frac{P_{O_3}}{P_{total}} = \frac{1.6 \times 10^{-9} \text{ atm}}{1.3 \times 10^{-2} \text{ atm}} = 1.2 \times 10^{-7}$$

Assume one mole of air (29 g/mol)

mol $O_3 = n_{air} \cdot X_{O_3} = (1 \text{ mol})(1.2 \times 10^{-7}) = 1.2 \times 10^{-7}$ mol O_3

O_3, 48.00;   mass $O_3 = 1.2 \times 10^{-7}$ mol $\times \dfrac{48.0 \text{ g}}{1 \text{ mol}} = 5.8 \times 10^{-6}$ g O_3

ppm $O_3 = \dfrac{5.8 \times 10^{-6} \text{ g}}{29 \text{ g}} \times 10^6 = 0.20$ ppm

13.70 (a) H_2SO_4, 98.08;   molality $= \dfrac{\left(25.0 \text{ g} \times \dfrac{1 \text{ mol}}{98.08 \text{ g}}\right)}{1.30 \text{ kg}} = 0.196$ mol/kg $= 0.196$ m

(b) $C_{10}H_{14}N_2$, 162.23; CH_2Cl_2, 84.93

2.25 g $C_{10}H_{14}N_2$ $\times \dfrac{1 \text{ mol } C_{10}H_{14}N_2}{162.23 \text{ g } C_{10}H_{14}N_2} = 0.0139$ mol $C_{10}H_{14}N_2$

80.0 g CH_2Cl_2 $\times \dfrac{1 \text{ mol } CH_2Cl_2}{84.93 \text{ g } CH_2Cl_2} = 0.942$ mol CH_2Cl_2

$X_{C_{10}H_{14}N_2} = \dfrac{0.0139 \text{ mol}}{0.942 \text{ mol} + 0.0139 \text{ mol}} = 0.0145$; $X_{CH_2Cl_2} = \dfrac{0.942 \text{ mol}}{0.942 \text{ mol} + 0.0139 \text{ mol}} = 0.985$

13.72 16.0 mass % $= \dfrac{16.0 \text{ g } H_2SO_4}{16.0 \text{ g } H_2SO_4 + 84.0 \text{ g } H_2O}$

H_2SO_4, 98.08;   density $= 1.1094$ g/mL

volume of solution $= 100.0$ g $\times \dfrac{1 \text{ mL}}{1.1094 \text{ g}} = 90.14$ mL $= 0.090 \ 14$ L

molarity $= \dfrac{\left(16.0 \text{ g} \times \dfrac{1 \text{ mol}}{98.08 \text{ g}}\right)}{0.090 \ 14 \text{ L}} = 1.81$ M

13.74 molality $= \dfrac{\left(40.0 \text{ g} \times \dfrac{1 \text{ mol}}{62.07 \text{ g}}\right)}{0.0600 \text{ kg}} = 10.7$ mol/kg $= 10.7$ m

13.76 $C_{19}H_{21}NO_3$, 311.38; 1.5 mg $= 1.5 \times 10^{-3}$ g

1.3×10^{-3} mol/kg $= \dfrac{\left(1.5 \times 10^{-3} \text{ g} \times \dfrac{1 \text{ mol}}{311.38 \text{ g}}\right)}{\text{kg of solvent}}$;   solve for kg of solvent.

$$\text{kg of solvent} = \frac{\left(1.5 \times 10^{-3}\text{ g} \times \dfrac{1 \text{ mol}}{311.38 \text{ g}}\right)}{1.3 \times 10^{-3} \text{ mol/kg}} = 0.0037 \text{ kg}$$

Because the solution is very dilute, kg of solvent ≈ kg of solution.

$$\text{g of solution} = (0.0037 \text{ kg})\left(\frac{1000 \text{ g}}{1 \text{ kg}}\right) = 3.7 \text{ g}$$

13.78 $C_6H_{12}O_6$, 180.16; H_2O, 18.02;   Assume 1.00 L of solution.
mass of solution = (1000 mL)(1.0624 g/mL) = 1062.4 g

$$\text{mass of solute} = 0.944 \text{ mol} \times \frac{180.16 \text{ g}}{1 \text{ mol}} = 170.1 \text{ g } C_6H_{12}O_6$$

mass of H_2O = 1062.4 g – 170.1 g = 892.3 g H_2O

$$\text{mol } C_6H_{12}O_6 = 0.944 \text{ mol}; \quad \text{mol } H_2O = 892.3 \text{ g} \times \frac{1 \text{ mol}}{18.02 \text{ g}} = 49.5 \text{ mol}$$

(a) $X_{C_6H_{12}O_6} = \dfrac{\text{mol } C_6H_{12}O_6}{\text{mol } C_6H_{12}O_6 + \text{mol } H_2O} = \dfrac{0.944 \text{ mol}}{0.944 \text{ mol} + 49.5 \text{ mol}} = 0.0187$

(b) mass % $= \dfrac{\text{mass } C_6H_{12}O_6}{\text{total mass of solution}} \times 100\% = \dfrac{170.1 \text{ g}}{1062.4 \text{ g}} \times 100\% = 16.0\%$

(c) molality $= \dfrac{\text{mol } C_6H_{12}O_6}{\text{kg } H_2O} = \dfrac{0.944 \text{ mol}}{0.8923 \text{ kg}} = 1.06 \text{ mol/kg} = 1.06 \ m$

Solubility and Henry's Law (Section 13.5)

13.80 From Figure 13.7, first determine the solubility, in g/100 mL, for each compound.
(a) $CuSO_4$, 159.6, ~42 g/100 mL; NH_4Cl, 53.5, ~56 g/100 mL

$$M = \frac{\left(42 \text{ g} \times \dfrac{1 \text{ mol } CuSO_4}{159.6 \text{ g}}\right)}{0.100 \text{ L}} = 2.6 \text{ mol/L}$$

$$M = \frac{\left(56 \text{ g} \times \dfrac{1 \text{ mol } NH_4Cl}{53.5 \text{ g}}\right)}{0.100 \text{ L}} = 10.5 \text{ mol/L}$$

NH_4Cl has the higher molar solubility.
(b) CH_3CO_2Na, 82.0, ~48 g/100 mL; glucose ($C_6H_{12}O_6$), 180.2, ~90 g/100 mL

$$M = \frac{\left(48 \text{ g} \times \dfrac{1 \text{ mol } CH_3CO_2Na}{82.0 \text{ g}}\right)}{0.100 \text{ L}} = 5.9 \text{ mol/L}$$

$$M = \frac{\left(90 \text{ g} \times \dfrac{1 \text{ mol glucose}}{180.2 \text{ g}}\right)}{0.100 \text{ L}} = 5.0 \text{ mol/L}$$

CH_3CO_2Na has the higher molar solubility.

13.82 $M = k \cdot P = (0.091 \ \frac{mol}{L \cdot atm})(0.75 \ atm) = 0.068 \ M$

13.84 $M = k \cdot P$

$k = \frac{M}{P} = \frac{2.21 \times 10^{-3} \ mol/L}{1.00 \ atm} = 2.21 \times 10^{-3} \ \frac{mol}{L \cdot atm}$

Convert 4 mg/L to mol/L:

$4 \ mg = 4 \times 10^{-3} \ g$

$O_2 \ molarity = \frac{\left(4 \times 10^{-3} \ g \times \frac{1 \ mol}{32.00 \ g}\right)}{1.00 \ L} = 1.25 \times 10^{-4} \ M$

$P_{O_2} = \frac{M}{k} = \frac{1.25 \times 10^{-4} \ \frac{mol}{L}}{2.21 \times 10^{-3} \ \frac{mol}{L \cdot atm}} = 0.06 \ atm$

13.86 $k = 2.4 \times 10^{-4} \ mol/(L \cdot atm)$

$M = k \cdot P = [2.4 \times 10^{-4} \ mol/(L \cdot atm)](2.00 \ atm) = 4.8 \times 10^{-4} \ mol/L$

Colligative Properties (Sections 13.6–13.9)

13.88 The vapor pressure of toluene is lower than the vapor pressure of benzene at the same temperature. When 1 mL of toluene is added to 100 mL of benzene, the vapor pressure of the solution decreases, which means that the boiling point of the solution will increase. When 1 mL of benzene is added to 100 mL of toluene, the vapor pressure of the solution increases, which means that the boiling point of the solution will decrease.

13.90 $FeCl_3$ and $CaCl_2$ are a strong electrolytes with a van't Hoff factors of $i = 4$ and 3, respectively. Glucose is a nonelectrolyte. The effective molality of $FeCl_3$ is $(4)(0.10) = 0.40 \ m$. The effective molality of $CaCl_2$ is $(3)(0.15) = 0.45 \ m$. Glucose $= (1)(0.30) = 0.30 \ m$.
Freezing point ranking: $CaCl_2 \ < \ FeCl_3 \ < \ glucose$

13.92 $C_7H_6O_2$, 122.1; C_2H_6O, 46.07

$X_{solv} = \frac{mol \ C_2H_6O}{mol \ C_2H_6O + mol \ C_7H_6O_2} = \frac{\left(100 \ g \times \frac{1 \ mol}{46.07 \ g}\right)}{\left(100 \ g \times \frac{1 \ mol}{46.07 \ g}\right) + \left(5.00 \ g \times \frac{1 \ mol}{122.1 \ g}\right)} = 0.981$

$P_{soln} = P_{solv} \cdot X_{solv} = (100.5 \ mm \ Hg)(0.981) = 98.6 \ mm \ Hg$

13.94 $MgCl_2$, 95.21

$110 \ g \times \frac{1 \ kg}{1000 \ g} = 0.110 \ kg$

$$\Delta T_f = K_f \cdot m \cdot i = \left(1.86 \; \frac{^{\circ}C \cdot kg}{mol}\right)\left(\frac{\left(7.40 \; g \; x \; \dfrac{1 \; mol}{95.21 \; g}\right)}{0.110 \; kg}\right)(2.7) = 3.55 \; ^{\circ}C$$

Solution freezing point $= 0.00 \; ^{\circ}C - \Delta T_f = 0.00 \; ^{\circ}C - 3.55 \; ^{\circ}C = -3.55 \; ^{\circ}C$

13.96   HCl, 36.46; $\Delta T_f = K_f \cdot m \cdot i$

$$190 \; g \; x \; \frac{1 \; kg}{1000 \; g} = 0.190 \; kg$$

Solution freezing point $= -4.65 \; ^{\circ}C = 0.00 \; ^{\circ}C - \Delta T_f; \; \Delta T_f = 4.65 \; ^{\circ}C$

$$i = \frac{\Delta T_f}{K_f \cdot m} = \frac{4.65 \; ^{\circ}C}{\left(1.86 \; \dfrac{^{\circ}C \cdot kg}{mol}\right)\left(\dfrac{9.12 \; g \; x \; \dfrac{1 \; mol}{36.46 \; g}}{0.190 \; kg}\right)} = 1.9$$

13.98   NaCl is a nonvolatile solute. Methyl alcohol is a volatile solute. When NaCl is added to water, the vapor pressure of the solution is decreased, which means that the boiling point of the solution will increase. When methyl alcohol is added to water, the vapor pressure of the solution is increased, which means that the boiling point of the solution will decrease.

13.100

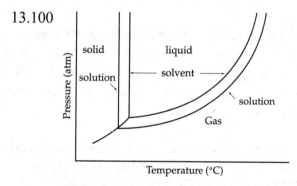

13.102 (a) CH_4N_2O, 60.06; H_2O, 18.02

$$10.0 \; g \; CH_4N_2O \; x \; \frac{1 \; mol \; CH_4N_2O}{60.06 \; g \; CH_4N_2O} = 0.167 \; mol \; CH_4N_2O$$

$$150.0 \; g \; H_2O \; x \; \frac{1 \; mol \; H_2O}{18.02 \; g \; H_2O} = 8.32 \; mol \; H_2O$$

$$X_{H_2O} = \frac{8.32 \; mol}{8.32 \; mol \; + \; 0.167 \; mol} = 0.980$$

$$P_{soln} = P^{o}_{H_2O} \cdot X_{H_2O} = (71.93 \; mm \; Hg)(0.980) = 70.5 \; mm \; Hg$$

(b) LiCl, 42.39; $10.0 \; g \; LiCl \; x \; \dfrac{1 \; mol \; LiCl}{42.39 \; g \; LiCl} = 0.236 \; mol \; LiCl$

LiCl dissociates into $Li^+(aq)$ and $Cl^-(aq)$ in H_2O.
mol Li^+ = mol Cl^- = mol LiCl = 0.236 mol

$$150.0 \text{ g H}_2\text{O} \times \frac{1 \text{ mol H}_2\text{O}}{18.02 \text{ g H}_2\text{O}} = 8.32 \text{ mol H}_2\text{O}$$

$$X_{\text{H}_2\text{O}} = \frac{8.32 \text{ mol}}{8.32 \text{ mol} + 0.236 \text{ mol} + 0.236 \text{ mol}} = 0.946$$

$$P_{\text{soln}} = P^{\text{o}}_{\text{H}_2\text{O}} \cdot X_{\text{H}_2\text{O}} = (71.93 \text{ mm Hg})(0.946) = 68.0 \text{ mm Hg}$$

13.104 For H_2O, $K_b = 0.51 \dfrac{^{\text{o}}\text{C} \cdot \text{kg}}{\text{mol}}$; 150.0 g = 0.1500 kg

(a) $\Delta T_b = K_b \cdot m = \left(0.51 \dfrac{^{\text{o}}\text{C} \cdot \text{kg}}{\text{mol}}\right)\left(\dfrac{0.167 \text{ mol}}{0.1500 \text{ kg}}\right) = 0.57 \,^{\text{o}}\text{C}$

Solution boiling point = $100.00 \,^{\text{o}}\text{C} + \Delta T_b = 100.00 \,^{\text{o}}\text{C} + 0.57 \,^{\text{o}}\text{C} = 100.57 \,^{\text{o}}\text{C}$

(b) $\Delta T_b = K_b \cdot m = \left(0.51 \dfrac{^{\text{o}}\text{C} \cdot \text{kg}}{\text{mol}}\right)\left(\dfrac{2(0.236 \text{ mol})}{0.1500 \text{ kg}}\right) = 1.6 \,^{\text{o}}\text{C}$

Solution boiling point = $100.00 \,^{\text{o}}\text{C} + \Delta T_b = 100.00 \,^{\text{o}}\text{C} + 1.6 \,^{\text{o}}\text{C} = 101.6 \,^{\text{o}}\text{C}$

13.106 Solution freezing point = $-4.3 \,^{\text{o}}\text{C} = 0.00 \,^{\text{o}}\text{C} - \Delta T_f$; $\Delta T_f = 4.3 \,^{\text{o}}\text{C}$

$$\Delta T_f = K_f \cdot m \cdot i; \qquad i = \frac{\Delta T_f}{K_f \cdot m} = \frac{4.3 \,^{\text{o}}\text{C}}{\left(1.86 \dfrac{^{\text{o}}\text{C} \cdot \text{kg}}{\text{mol}}\right)(1.0 \text{ mol/kg})} = 2.3$$

13.108 Let $X_{\text{heptane}} = x$ and $X_{\text{octane}} = 1 - x$

(428 mm Hg)x + (175 mm Hg)(1 − x) = 305 mm Hg

(428 mm Hg)x + 175 mm Hg − (175 mm Hg)x = 305 mm Hg

(428 mm Hg − 175 mm Hg)x = 305 mm Hg − 175 mm Hg

(253 mm Hg)x = 130 mm Hg

$$x = X_{\text{heptane}} = \frac{130 \text{ mm Hg}}{253 \text{ mm Hg}} = 0.514$$

13.110 Acetone, C_3H_6O, 58.08, $P^{\text{o}}_{C_3H_6O} = 285$ mm Hg

Ethyl acetate, $C_4H_8O_2$, 88.11, $P^{\text{o}}_{C_4H_8O_2} = 118$ mm Hg

$$25.0 \text{ g C}_3\text{H}_6\text{O} \times \frac{1 \text{ mol C}_3\text{H}_6\text{O}}{58.08 \text{ g C}_3\text{H}_6\text{O}} = 0.430 \text{ mol C}_3\text{H}_6\text{O}$$

$$25.0 \text{ g C}_4\text{H}_8\text{O}_2 \times \frac{1 \text{ mol C}_4\text{H}_8\text{O}_2}{88.11 \text{ g C}_4\text{H}_8\text{O}_2} = 0.284 \text{ mol C}_4\text{H}_8\text{O}_2$$

$$X_{C_3H_6O} = \frac{0.430 \text{ mol}}{0.430 \text{ mol} + 0.284 \text{ mol}} = 0.602; \quad X_{C_4H_8O_2} = \frac{0.284 \text{ mol}}{0.430 \text{ mol} + 0.284 \text{ mol}} = 0.398$$

$$P_{\text{soln}} = P^{\text{o}}_{C_3H_6O} \cdot X_{C_3H_6O} + P^{\text{o}}_{C_4H_8O_2} \cdot X_{C_4H_8O_2}$$

$$P_{\text{soln}} = (285 \text{ mm Hg})(0.602) + (118 \text{ mm Hg})(0.398) = 219 \text{ mm Hg}$$

13.112 In the liquid, $X_{acetone} = 0.602$ and $X_{ethyl\ acetate} = 0.398$

In the vapor, $P_{Total} = 219$ mm Hg

$$P_{acetone} = P^o_{acetone} \cdot X_{acetone} = (285\ mm\ Hg)(0.602) = 172\ mm\ Hg$$

$$P_{ethyl\ acetate} = P^o_{ethyl\ acetate} \cdot X_{ethyl\ acetate} = (118\ mm\ Hg)(0.398) = 47\ mm\ Hg$$

$$X_{acetone} = \frac{P_{acetone}}{P_{total}} = \frac{172\ mm\ Hg}{219\ mm\ Hg} = 0.785; \quad X_{ethyl\ acetate} = \frac{P_{ethyl\ acetate}}{P_{total}} = \frac{47\ mm\ Hg}{219\ mm\ Hg} = 0.215$$

13.114 $C_9H_8O_4$, 180.16; 215 g = 0.215 kg

$$\Delta T_b = K_b \cdot m = 0.47\ ^\circ C; \quad K_b = \frac{\Delta T_b}{m} = \frac{0.47\ ^\circ C}{\left(\dfrac{5.00\ g\ \times\ \dfrac{1\ mol}{180.16\ g}}{0.215\ kg}\right)} = 3.6\ \frac{^\circ C \cdot kg}{mol}$$

13.116 $\Delta T_b = K_b \cdot m = 1.76\ ^\circ C; \quad m = \dfrac{\Delta T_b}{K_b} = \dfrac{1.76\ ^\circ C}{3.07\ \dfrac{^\circ C \cdot kg}{mol}} = 0.573\ m$

13.118 $\Pi = MRT$

(a) NaCl 58.44; 350.0 mL = 0.3500 L

There are 2 moles of ions/mole of NaCl

$$\Pi = (2)\left(\frac{5.00\ g\ \times\ \dfrac{1\ mol}{58.44\ g}}{0.3500\ L}\right)\left(0.082\ 06\ \frac{L \cdot atm}{K \cdot mol}\right)(323\ K) = 13.0\ atm$$

(b) CH_3CO_2Na, 82.03; 55.0 mL = 0.0550 L

There are 2 moles of ions/mole of CH_3CO_2Na

$$\Pi = (2)\left(\frac{6.33\ g\ \times\ \dfrac{1\ mol}{82.03\ g}}{0.0550\ L}\right)\left(0.082\ 06\ \frac{L \cdot atm}{K \cdot mol}\right)(283K) = 65.2\ atm$$

13.120 $\Pi = MRT; \quad M = \dfrac{\Pi}{RT} = \dfrac{4.85\ atm}{\left(0.082\ 06\ \dfrac{L \cdot atm}{K \cdot mol}\right)(300\ K)} = 0.197\ M$

13.122 K_f for snow (H_2O) is 1.86 $\dfrac{^\circ C \cdot kg}{mol}$. Reasonable amounts of salt are capable of lowering

the freezing point (ΔT_f) of the snow below an air temperature of –2 °C. Reasonable amounts of salt, however, are not capable of causing a ΔT_f of more than 30 °C, which would be required if it is to melt snow when the air temperature is –30 °C.

13.124 $\Pi = 407.2$ mm Hg x $\dfrac{1 \text{ atm}}{760 \text{ mm Hg}} = 0.5358$ atm

$\Pi = MRT; \quad M = \dfrac{\Pi}{RT} = \dfrac{0.5358 \text{ atm}}{\left(0.082 \ 06 \ \dfrac{L \cdot atm}{K \cdot mol}\right)(298.15 \text{ K})} = 0.021 \ 90 \text{ M}$

200.0 mL x $\dfrac{1 \text{ L}}{1000 \text{ mL}} = 0.2000$ L

mol cellobiose $= (0.2000 \text{ L})(0.021 \ 90 \text{ mol/L}) = 4.380 \times 10^{-3}$ mol

molar mass of cellobiose $= \dfrac{1.500 \text{ g cellobiose}}{4.380 \times 10^{-3} \text{ mol cellobiose}} = 342.5$ g/mol

molecular weight $= 342.5$

13.126 HCl is a strong electrolyte in H_2O and completely dissociates into two solute particles per each HCl.

HF is a weak electrolyte in H_2O. Only a few percent of the HF molecules dissociates into ions.

13.128 Sucrose ($C_{12}H_{22}O_{11}$), 342.3; fructose ($C_6H_{12}O_6$), 180.2

$\Pi = MRT; \quad M = \dfrac{\Pi}{RT} = \dfrac{0.1843 \text{ atm}}{\left(0.082 \ 06 \ \dfrac{L \cdot atm}{K \cdot mol}\right)(298.0 \text{ K})} = 0.007 \ 537 \text{ M}$

$n_{total} = (0.007 \ 537 \text{ mol/L})(1.50 \text{ L}) = 0.0113$ mol

Let x = g of sucrose and 2.850 g – x = g of fructose

0.0113 mol = mol sucrose + mol fructose

$0.0113 \text{ mol} = (x)\left(\dfrac{1 \text{ mol sucrose}}{342.3 \text{ g}}\right) + (2.850 \text{ g} - x)\left(\dfrac{1 \text{ mol fructose}}{180.2 \text{ g}}\right)$

Solve for x.

$0.0113 \text{ mol} = (x)\left(\dfrac{1 \text{ mol}}{342.3 \text{ g}}\right) + (2.850 \text{ g} - x)\left(\dfrac{1 \text{ mol}}{180.2 \text{ g}}\right)$

$0.0113 \text{ mol} = (x)\left(\dfrac{1 \text{ mol}}{342.3 \text{ g}}\right) + 0.0158 \text{ mol} - (x)\left(\dfrac{1 \text{ mol}}{180.2 \text{ g}}\right)$

$0.0113 \text{ mol} - 0.0158 \text{ mol} = (x)\left(\dfrac{1 \text{ mol}}{342.3 \text{ g}}\right) - (x)\left(\dfrac{1 \text{ mol}}{180.2 \text{ g}}\right)$

$-0.0045 \text{ mol} = (x)\left[\left(\dfrac{1 \text{ mol}}{342.3 \text{ g}}\right) - \left(\dfrac{1 \text{ mol}}{180.2 \text{ g}}\right)\right]$

$x = \dfrac{-0.0045 \text{ mol}}{\left[\left(\dfrac{1 \text{ mol}}{342.3 \text{ g}}\right) - \left(\dfrac{1 \text{ mol}}{180.2 \text{ g}}\right)\right]} = 1.71$ g sucrose

1.71 g sucrose x $\dfrac{1 \text{ mol sucrose}}{342.3 \text{ g sucrose}} = 0.005 \ 00$ mol sucrose

$X_{sucrose} = \dfrac{n_{sucrose}}{n_{total}} = \dfrac{0.005 \ 00 \text{ mol}}{0.0113 \text{ mol}} = 0.442$

13.130 $C_{10}H_8$, 128.17; $\Delta T_f = 0.35\ ^\circ C$

$$\Delta T_f = K_f \cdot m;\quad m = \frac{\Delta T_f}{K_f} = \frac{0.35\ ^\circ C}{5.12\ \dfrac{^\circ C \cdot kg}{mol}} = 0.0684\ mol/kg = 0.0684\ m$$

$$150.0\ g \times \frac{1\ kg}{1000\ g} = 0.1500\ kg$$

mol $C_{10}H_8 = (0.1500\ kg)(0.0684\ mol/kg) = 0.0103$ mol $C_{10}H_8$

$$mass\ C_{10}H_8 = 0.0103\ mol\ C_{10}H_8 \times \frac{128.17\ g\ C_{10}H_8}{1\ mol\ C_{10}H_8} = 1.3\ g\ C_{10}H_8$$

13.132 NaCl, 58.44; there are 2 ions/NaCl

A 3.5 mass % aqueous solution of NaCl contains 3.5 g NaCl and 96.5 g H_2O.

$$molality = \frac{\left(3.5\ g\ \times\ \dfrac{1\ mol}{58.44\ g}\right)}{0.0965\ kg} = 0.62\ mol/kg = 0.62\ m$$

$$\Delta T_f = K_f \cdot 2 \cdot m = \left(1.86\ \frac{^\circ C \cdot kg}{mol}\right)(2)(0.62\ mol/kg) = 2.3\ ^\circ C$$

Solution freezing point $= 0.0\ ^\circ C - \Delta T_f = 0.0\ ^\circ C - 2.3\ ^\circ C = -2.3\ ^\circ C$

$$\Delta T_b = K_b \cdot 2 \cdot m = \left(0.51\ \frac{^\circ C \cdot kg}{mol}\right)(2)(0.62\ mol/kg) = 0.63\ ^\circ C$$

Solution boiling point $= 100.00\ ^\circ C + \Delta T_b = 100.00\ ^\circ C + 0.63\ ^\circ C = 100.63\ ^\circ C$

13.134 $\Delta T_f = K_f \cdot i \cdot m$; $CaCl_2$, 111.0

$$m = \frac{\Delta T_f}{K_f \cdot i} = \frac{1.14\ ^\circ C}{\left(1.86\ \dfrac{^\circ C \cdot kg}{mol}\right)(2.71)} = 0.226\ mol/kg$$

$$0.226\ mol\ CaCl_2 \times \frac{111.0\ g\ CaCl_2}{1\ mol\ CaCl_2} = 25.1\ g\ CaCl_2$$

$$mass\ \%\ CaCl_2 = \frac{mass\ CaCl_2}{mass\ CaCl_2\ +\ mass\ H_2O} \times 100\%$$

$$= \frac{25.1\ g}{25.1\ g + 1000.0\ g} \times 100\% = 2.45\%$$

13.136 H_2O, 18.02

A 0.62 m LiCl solution contains 0.62 mol of LiCl and 1.00 kg (= 1000 g) of H_2O.

(0.62 mol LiCl)(1.96) = 1.21 mol of solute particles

$$1000\ g\ H_2O \times \frac{1\ mol\ H_2O}{18.02\ g\ H_2O} = 55.5\ mol\ H_2O$$

$$P_{soln} = P^o_{H_2O} \cdot X_{H_2O} = 23.76 \text{ mm Hg x } \frac{55.5 \text{ mol H}_2O}{55.5 \text{ mol H}_2O + 1.21 \text{ mol solute}} = 23.25 \text{ mm Hg}$$

Vapor pressure depression = 23.76 mm Hg – 23.25 mm Hg = 0.51 mm Hg

13.138 KCl, 74.55; KNO$_3$, 101.10; Ba(NO$_3$)$_2$, 261.34

$$\Pi = MRT; \quad M = \frac{\Pi}{RT} = \frac{\left(744.7 \text{ mm Hg x } \frac{1.00 \text{ atm}}{760 \text{ mm Hg}}\right)}{\left(0.082 \ 06 \frac{L \cdot atm}{K \cdot mol}\right)(298 \text{ K})} = 0.040 \ 07 \text{ M}$$

$$0.040 \ 07 \text{ M} = \frac{\text{mol ions}}{0.500 \text{ L}}; \quad \text{mol ions} = (0.040 \ 07 \text{ mol/L})(0.500 \text{ L}) = 0.020 \ 035 \text{ mol ions}$$

mass Cl = 1.000 g x 0.2092 = 0.2092 g Cl

$$\text{mass KCl} = 0.2092 \text{ g Cl x } \frac{1 \text{ mol Cl}}{35.453 \text{ g Cl}} \text{ x } \frac{1 \text{ mol KCl}}{1 \text{ mol Cl}} \text{ x } \frac{74.55 \text{ g KCl}}{1 \text{ mol KCl}} = 0.440 \text{ g KCl}$$

$$\text{mol ions from KCl} = 0.440 \text{ g KCl x } \frac{1 \text{ mol KCl}}{74.55 \text{ g KCl}} \text{ x } \frac{2 \text{ mol ions}}{1 \text{ mol KCl}} = 0.0118 \text{ mol ions}$$

mol ions from KNO$_3$ and Ba(NO$_3$)$_2$ = 0.020 035 – 0.0118 = 0.008 235 mol ions
Let x = mass KNO$_3$ and y = mass Ba(NO$_3$)$_2$
x + y = 1.000 g – 0.440 g = 0.560 g

$$\left(\frac{x}{101.10}\right)2 + \left(\frac{y}{261.34}\right)3 = 0.008 \ 235 \text{ mol ions}$$

x = 0.560 – y
0.0198x + 0.0115y = 0.008 235
0.0198(0.560 – y) + 0.0115y = 0.008 235
0.011 09 – 0.0198y + 0.0115y = 0.008 235

$$0.002 \ 855 = 0.0083y; \quad y = \frac{0.002 \ 855}{0.0083} = 0.3440; \quad x = 0.560 - 0.3440 = 0.216$$

$$\text{mass \% KCl} = \frac{0.440 \text{ g}}{1.000 \text{ g}} \text{ x } 100\% = 44.0\%$$

$$\text{mass \% KNO}_3 = \frac{0.216 \text{ g}}{1.000 \text{ g}} \text{ x } 100\% = 21.6\%$$

$$\text{mass \% Ba(NO}_3)_2 = \frac{0.344 \text{ g}}{1.000 \text{ g}} \text{ x } 100\% = 34.4\%$$

13.140 Solution freezing point = –1.03 °C = 0.00 °C – ΔT_f; ΔT_f = 1.03 °C

$$\Delta T_f = K_f \cdot m; \quad m = \frac{\Delta T_f}{K_f} = \frac{1.03 \ ^oC}{1.86 \frac{^oC \cdot kg}{mol}} = 0.554 \text{ mol/kg} = 0.554 \ m$$

$$\Pi = MRT; \quad M = \frac{\Pi}{RT} = \frac{(12.16 \text{ atm})}{\left(0.082 \ 06 \frac{L \cdot atm}{K \cdot mol}\right)(298 \text{ K})} = 0.497 \frac{mol}{L}$$

Assume 1.000 L = 1000 mL of solution.

mass of solution = (1000 mL)(1.063 g/mL) = 1063 g

$$\text{mass of } H_2O \text{ in 1000 mL of solution} = \frac{1000 \text{ g } H_2O}{0.554 \text{ mol of solute}} \times 0.497 \text{ mol} = 897 \text{ g } H_2O$$

mass of solute = total mass – mass of H_2O = 1063 g – 897 g = 166 g solute

$$\text{molar mass} = \frac{166 \text{ g}}{0.497 \text{ mol}} = 334 \text{ g/mol}$$

13.142 (a) For NaCl, i = 2 and for $MgCl_2$, i = 3;   T = 25 °C = 25 + 273 = 298 K

$$\Pi = i \cdot MRT = [(2)(0.470 \text{ mol/L}) + (3)(0.068 \text{ mol/L})]\left(0.082\ 06 \frac{L \cdot atm}{K \cdot mol} \right)(298 \text{ K}) = 28.0 \text{ atm}$$

(b) Calculate the molarity for an osmotic pressure = 100.0 atm.

$$\Pi = MRT;\ \ M = \frac{\Pi}{RT} = \frac{(100.0 \text{ atm})}{\left(0.082\ 06 \dfrac{L \cdot atm}{K \cdot mol} \right)(298 \text{ K})} = 4.09 \text{ mol/L}$$

$$M_{conc} \times V_{conc} = M_{dil} \times V_{dil}$$

$$V_{conc} = \frac{M_{dil} \times V_{dil}}{M_{conc}} = \frac{[(2)(0.470 \text{ mol/L}) + (3)(0.068 \text{ mol/L})](1.00 \text{ L})}{4.09 \text{ mol/L}} = 0.28 \text{ L}$$

A volume of 1.00 L of seawater can be reduced to 0.28 L by an osmotic pressure of 100.0 atm. The volume of fresh water that can be obtained is (1.00 L – 0.28 L) = 0.72 L.

Multiconcept Problems

13.144 First, determine the empirical formula:
Assume 100.0 g of β-carotene.

10.51% H $10.51 \text{ g H} \times \dfrac{1 \text{ mol H}}{1.008 \text{ g H}} = 10.43 \text{ mol H}$

89.49% C $89.49 \text{ g C} \times \dfrac{1 \text{ mol C}}{12.01 \text{ g C}} = 7.45 \text{ mol C}$

$C_{7.45}H_{10.43}$; divide each subscript by the smaller, 7.45.
$C_{7.45/7.45}H_{10.43/7.45}$
$CH_{1.4}$
Multiply each subscript by 5 to obtain integers.
Empirical formula is C_5H_7, 67.1.

Second, calculate the molecular weight:

$$\Delta T_f = K_f \cdot m;\ \ \ m = \frac{\Delta T_f}{K_f} = \frac{1.17 \text{ °C}}{37.7 \dfrac{\text{°C} \cdot \text{kg}}{\text{mol}}} = 0.0310 \text{ mol/kg} = 0.0310\ m$$

$$1.50 \text{ g} \times \frac{1 \text{kg}}{1000 \text{ g}} = 1.50 \times 10^{-3} \text{ kg}$$

mol β-carotene = $(1.50 \times 10^{-3} \text{ kg})(0.0310 \text{ mol/kg}) = 4.65 \times 10^{-5} \text{ mol}$

$$\text{molar mass of }\beta\text{-carotene} = \frac{0.0250 \text{ g }\beta\text{-carotene}}{4.65 \text{ x } 10^{-5} \text{ mol }\beta\text{-carotene}} = 538 \text{ g/mol}$$

molecular weight = 538

Finally, determine the molecular formula:
Divide the molecular weight by the empirical formula weight.

$$\frac{538}{67.1} = 8$$

molecular formula is $C_{(8 \text{ x } 5)}H_{(8 \text{ x } 7)}$, or $C_{40}H_{56}$

13.146 First, determine the empirical formula.
3.47 mg = 3.47 x 10^{-3} g sample
10.10 mg = 10.10 x 10^{-3} g CO_2
2.76 mg = 2.76 x 10^{-3} g H_2O

$$\text{mass C} = 10.10 \text{ x } 10^{-3} \text{ g } CO_2 \text{ x } \frac{12.01 \text{ g C}}{44.01 \text{ g } CO_2} = 2.76 \text{ x } 10^{-3} \text{ g C}$$

$$\text{mass H} = 2.76 \text{ x } 10^{-3} \text{ g } H_2O \text{ x } \frac{2 \text{ x } 1.008 \text{ g H}}{18.02 \text{ g } H_2O} = 3.09 \text{ x } 10^{-4} \text{ g H}$$

mass O = 3.47 x 10^{-3} g – 2.76 x 10^{-3} g C – 3.09 x 10^{-4} g H = 4.01 x 10^{-4} g O

$$2.76 \text{ x } 10^{-3} \text{ g C x } \frac{1 \text{ mol C}}{12.01 \text{ g C}} = 2.30 \text{ x } 10^{-4} \text{ mol C}$$

$$3.09 \text{ x } 10^{-4} \text{ g H x } \frac{1 \text{ mol H}}{1.008 \text{ g H}} = 3.07 \text{ x } 10^{-4} \text{ mol H}$$

$$4.01 \text{ x } 10^{-4} \text{ g O x } \frac{1 \text{ mol O}}{16.00 \text{ g O}} = 2.51 \text{ x } 10^{-5} \text{ mol O} = 0.251 \text{ x } 10^{-4} \text{ mol O}$$

To simplify the empirical formula, divide each mol quantity by 10^{-4}.
$C_{2.30}H_{3.07}O_{0.251}$; divide all subscripts by the smallest, 0.251.
$C_{2.30/0.251}H_{3.07/0.251}O_{0.251/0.251}$
$C_{9.16}H_{12.23}O$
Empirical formula is $C_9H_{12}O$, 136.

Second, determine the molecular weight.

7.55 mg = 7.55 x 10^{-3} g estradiol; $0.500 \text{ g x } \frac{1 \text{ kg}}{1000 \text{ g}} = 5.00 \text{ x } 10^{-4} \text{ kg camphor}$

$$\Delta T_f = K_f \cdot m; \quad m = \frac{\Delta T_f}{K_f} = \frac{2.10 \text{ °C}}{37.7 \dfrac{\text{°C}\cdot\text{kg}}{\text{mol}}} = 0.0557 \text{ mol/kg} = 0.0557 \, m$$

$$m = \frac{\text{mol estradiol}}{\text{kg solvent}}$$

mol estradiol = m x (kg solvent) = (0.0557 mol/kg)(5.00 x 10^{-4} kg) = 2.79 x 10^{-5} mol

$$\text{molar mass} = \frac{7.55 \times 10^{-3} \text{ g estradiol}}{2.79 \times 10^{-5} \text{ mol estradiol}} = 271 \text{ g/mol; molecular weight} = 271$$

Finally, determine the molecular formula:
Divide the molecular weight by the empirical formula weight.
$$\frac{271}{136} = 2$$
molecular formula is $C_{(2 \times 9)}H_{(2 \times 12)}O_{(2 \times 1)}$, or $C_{18}H_{24}O_2$

13.148 (a) H_2SO_4, 98.08; $2.238 \text{ mol } H_2SO_4 \times \dfrac{98.08 \text{ g } H_2SO_4}{1 \text{ mol } H_2SO_4} = 219.50 \text{ g } H_2SO_4$

mass of 2.238 m solution = 219.50 g H_2SO_4 + 1000 g H_2O = 1219.50 g

volume of 2.238 m solution = $1219.50 \text{ g} \times \dfrac{1.0000 \text{ mL}}{1.1243 \text{ g}} = 1084.68 \text{ mL} = 1.0847 \text{ L}$

molarity of 2.238 m solution = $\dfrac{2.238 \text{ mol}}{1.0847 \text{ L}} = 2.063 \text{ M}$

The molarity of the H_2SO_4 solution is less than the molarity of the $BaCl_2$ solution. Because equal volumes of the two solutions are mixed, H_2SO_4 is the limiting reactant and the number of moles of H_2SO_4 determines the number of moles of $BaSO_4$ produced as the white precipitate.

$(0.05000 \text{ L}) \times (2.063 \text{ mol } H_2SO_4/\text{L}) \times \dfrac{1 \text{ mol } BaSO_4}{1 \text{ mol } H_2SO_4} \times \dfrac{233.39 \text{ g } BaSO_4}{1 \text{ mol } BaSO_4} = 24.07 \text{ g } BaSO_4$

(b) More precipitate will form because of the excess $BaCl_2$ in the solution.

13.150 (a) 20.00 mL = 0.02000 L
mol NaOH = (0.02000 L)(2.00 mol/L) = 0.0400 mol NaOH
$\text{mol } CO_2 = 0.0400 \text{ mol NaOH} \times \dfrac{1 \text{ mol } CO_2}{2 \text{ mol NaOH}} = 0.0200 \text{ mol } CO_2$

$\text{mol C} = 0.0200 \text{ mol } CO_2 \times \dfrac{1 \text{ mol C}}{1 \text{ mol } CO_2} = 0.0200 \text{ mol C}$

$\text{mass C} = 0.0200 \text{ mol C} \times \dfrac{12.011 \text{ g C}}{1 \text{ mol C}} = 0.240 \text{ g C}$

mass H = mass of compound – mass of C = 0.270 g – 0.240 g = 0.030 g H

$\text{mol H} = 0.030 \text{ g H} \times \dfrac{1 \text{ mol H}}{1.008 \text{ g H}} = 0.030 \text{ mol H}$

The mole ratio of C and H in the molecule is $C_{0.0200}H_{0.030}$.
$C_{0.0200}H_{0.030}$; divide both subscripts by the smaller of the two, 0.0200.
$C_{0.0200/0.0200}H_{0.030/0.0200}$
$C_1H_{1.5}$, multiply both subscripts by 2.
$C_{(2 \times 1)}H_{(2 \times 1.5)}$
C_2H_3 (27.05) is the empirical formula.

(b) $\Delta T_f = K_f \cdot m$; $m = \dfrac{\Delta T_f}{K_f} = \dfrac{(179.8\ ^\circ C - 177.9\ ^\circ C)}{37.7\ \dfrac{^\circ C \cdot kg}{mol}} = 0.050$ mol/kg $= 0.050\ m$

50.0 g x $\dfrac{1\ kg}{1000\ g} = 0.0500$ kg

mol solute $= (0.050\ \text{mol/kg})(0.0500\ \text{kg}) = 0.0025$ mol

molar mass $= \dfrac{0.270\ g}{0.0025\ mol} = 108$ g/mol; molecular weight $= 108$

(c) To find the molecular formula, first divide the molecular weight by the weight of the empirical formula unit.

$\dfrac{108}{27} = 4$

Multiply the subscripts in the empirical formula by the result of this division, 4.

$C_{(4\ x\ 2)}\ H_{(4\ x\ 3)}$

C_8H_{12} is the molecular formula of the compound.

13.152 AgCl, 143.32

Solution freezing point $= -4.42\ ^\circ C = 0.00\ ^\circ C - \Delta T_f$; $\Delta T_f = 0.00\ ^\circ C + 4.42\ ^\circ C = 4.42\ ^\circ C$

$\Delta T_f = K_f \cdot m$

total ion $m = \dfrac{\Delta T_f}{K_f} = \dfrac{4.42\ ^\circ C}{1.86\ \dfrac{^\circ C \cdot kg}{mol}} = 2.376$ mol/kg $= 2.376\ m$

150.0 g x $\dfrac{1\ kg}{1000\ g} = 0.1500$ kg

total mol of ions $= (2.376\ \text{mol/kg})(0.1500\ \text{kg}) = 0.3564$ mol of ions

An excess of $AgNO_3$ reacts with all Cl^- to produce 27.575 g AgCl.

total mol $Cl^- = 27.575$ g AgCl x $\dfrac{1\ mol\ AgCl}{143.32\ g\ AgCl}$ x $\dfrac{1\ mol\ Cl^-}{1\ mol\ AgCl} = 0.1924$ mol Cl^-

Let P = mol XCl and Q = mol YCl_2.

0.3564 mol ions = 2 x mol XCl + 3 x mol YCl_2 = (2 x P) + (3 x Q)

0.1924 mol Cl^- = mol XCl + 2 x mol YCl_2 = P + (2 x Q)

P = 0.1924 − (2 x Q)

0.3564 = 2 x [0.1924 − (2 x Q)] + (3 x Q) = 0.3848 − (4 x Q) + (3 x Q)

Q = 0.3848 − 0.3564 = 0.0284 mol YCl_2

P = 0.1924 − (2 x Q) = 0.1924 − (2 x 0.0284) = 0.1356 mol XCl

mass Cl in XCl = 0.1356 mol XCl x $\dfrac{1\ mol\ Cl}{1\ mol\ XCl}$ x $\dfrac{35.453\ g\ Cl}{1\ mol\ Cl} = 4.81$ g Cl

mass Cl in YCl_2 = 0.0284 mol YCl_2 x $\dfrac{2\ mol\ Cl}{1\ mol\ YCl_2}$ x $\dfrac{35.453\ g\ Cl}{1\ mol\ Cl} = 2.01$ g Cl

total mass of XCl and YCl_2 = 8.900 g

mass of X + Y = total mass − mass Cl = 8.900 g − 4.81 g − 2.01 g = 2.08 g

X is an alkali metal and there are 0.1356 mol of X in XCl.

If X = Li, then mass of X = (0.1356 mol)(6.941 g/mol) = 0.941 g

If X = Na, then mass of X = (0.1356 mol)(22.99 g/mol) = 3.12 g but this is not possible because 3.12 g is greater than the total mass of X + Y. Therefore, X is Li.

mass of Y = 2.08 – mass of X = 2.08 g – 0.941 g = 1.14 g

Y is an alkaline earth metal and there are 0.0284 mol of Y in YCl_2.

molar mass of Y = 1.14 g/0.0284 mol = 40.1 g/mol. Therefore, Y is Ca.

$$\text{mass LiCl} = 0.1356 \text{ mol LiCl} \times \frac{42.39 \text{ g LiCl}}{1 \text{ mol LiCl}} = 5.75 \text{ g LiCl}$$

$$\text{mass CaCl}_2 = 0.0284 \text{ mol CaCl}_2 \times \frac{110.98 \text{ g CaCl}_2}{1 \text{ mol CaCl}_2} = 3.15 \text{ g CaCl}_2$$

13.154 NaCl, 58.44; $C_{12}H_{22}O_{11}$, 342.3

Let X = mass NaCl and Y = mass $C_{12}H_{22}O_{11}$, then X + Y = 100.0 g.

$$500.0 \text{ g} \times \frac{1 \text{ kg}}{1000 \text{ g}} = 0.5000 \text{ kg}$$

Solution freezing point = –2.25 °C = 0.00 °C – ΔT_f; ΔT_f = 0.00 °C + 2.25 °C = 2.25 °C

$$\Delta T_f = K_f \cdot (m_{NaCl} \cdot i + m_{C_{12}H_{22}O_{11}})$$

$$\Delta T_b = \left(1.86 \frac{°C \cdot kg}{mol} \right) \left(\frac{(\text{mol NaCl} \cdot 2) + (\text{mol } C_{12}H_{22}O_{11})}{0.5000 \text{ kg}} \right) = 2.25 °C$$

$$\text{mol NaCl} = \text{X g NaCl} \times \frac{1 \text{ mol NaCl}}{58.44 \text{ g NaCl}} = \text{X}/58.44 \text{ mol}$$

$$\text{mol } C_{12}H_{22}O_{11} = \text{Y g } C_{12}H_{22}O_{11} \times \frac{1 \text{ mol } C_{12}H_{22}O_{11}}{342.3 \text{ g } C_{12}H_{22}O_{11}} = \text{Y}/342.3 \text{ mol}$$

$$\Delta T_b = \left(1.86 \frac{°C \cdot kg}{mol} \right) \left(\frac{((\text{X}/58.44) \cdot 2 \text{ mol}) + ((\text{Y}/342.3) \text{ mol})}{0.5000 \text{ kg}} \right) = 2.25 °C$$

X = 100 – Y

$$\left(1.86 \frac{°C \cdot kg}{mol} \right) \left(\frac{\{[(100 - Y)/58.44] \cdot 2 \text{ mol}]\} + [(Y/342.3) \text{ mol}]}{0.5000 \text{ kg}} \right) = 2.25 °C$$

$$\left(\frac{[(200/58.44) - (2Y/58.44) + (Y/342.3)] \text{ mol}}{0.5000 \text{ kg}} \right) = \frac{2.25 °C}{\left(1.86 \frac{°C \cdot kg}{mol} \right)} = 1.21 \text{ mol/kg}$$

$$\left(\frac{[(3.42) - (0.0313Y)] \text{ mol}}{0.5000 \text{ kg}} \right) = 1.21 \text{ mol/kg}$$

[(3.42) – (0.0313Y)] = (0.5000 kg)(1.21) = 0.605

– 0.0313 Y = 0.605 – 3.42 = – 2.81

Y = (– 2.81)/(– 0.0313) = 89.8 g of $C_{12}H_{22}O_{11}$

X = 100.0 g – Y = 100.0 g – 89.8 g = 10.2 g of NaCl

Chemical Kinetics

14.1 $3\,I^-(aq) + H_3AsO_4(aq) + 2\,H^+(aq) \rightarrow I_3^-(aq) + H_3AsO_3(aq) + H_2O(l)$

(a) $-\dfrac{\Delta[I^-]}{\Delta t} = 4.8 \times 10^{-4}$ M/s

$\dfrac{\Delta[I_3^-]}{\Delta t} = \dfrac{1}{3}\left(-\dfrac{\Delta[I^-]}{\Delta t}\right) = \left(\dfrac{1}{3}\right)(4.8 \times 10^{-4}\ \text{M/s}) = 1.6 \times 10^{-4}$ M/s

(b) $-\dfrac{\Delta[H^+]}{\Delta t} = 2\left(\dfrac{\Delta[I_3^-]}{\Delta t}\right) = (2)(1.6 \times 10^{-4}\ \text{M/s}) = 3.2 \times 10^{-4}$ M/s

14.2 (a) $\dfrac{-\Delta[A]}{\Delta t} = \dfrac{0.04\ \text{M} - 0.07\ \text{M}}{500\ \text{s}} = 6.0 \times 10^{-5}$ M/s

$\dfrac{-\Delta[B]}{\Delta t} = \dfrac{0.01\ \text{M} - 0.07\ \text{M}}{500\ \text{s}} = 1.2 \times 10^{-4}$ M/s

$\dfrac{\Delta[C]}{\Delta t} = \dfrac{0.03\ \text{M} - 0.0\ \text{M}}{500\ \text{s}} = 6.0 \times 10^{-5}$ M/s

(b) $A + 2B \rightarrow C$

14.3 The reaction is first order in BrO_3^-, first order in Br^-, second order in H^+, and fourth order overall.

14.4 For NO, when the [NO] doubles, the rate doubles. This indicates that the reaction order for NO is 1. For Cl_2, when the $[Cl_2]$ is halved, the rate decreases by a factor of 4. This indicates that the reaction order for Cl_2 is 2. The reaction is 3rd order overall.

14.5 (a) Rate = $k[NO_2]^m[CO]^n$

$$m = \frac{\ln\left(\dfrac{\text{Rate}_2}{\text{Rate}_1}\right)}{\ln\left(\dfrac{[NO_2]_2}{[NO_2]_1}\right)} = \frac{\ln\left(\dfrac{1.13 \times 10^{-2}}{5.00 \times 10^{-3}}\right)}{\ln\left(\dfrac{0.150}{0.100}\right)} = 2$$

$$n = \frac{\ln\left(\dfrac{\text{Rate}_3 \cdot [NO_2]_2}{\text{Rate}_2 \cdot [NO_2]_3}\right)}{\ln\left(\dfrac{[CO]_3}{[CO]_2}\right)} = \frac{\ln\left(\dfrac{(2.00 \times 10^{-2})(0.150)^2}{(1.13 \times 10^{-2})(0.200)^2}\right)}{\ln\left(\dfrac{0.200}{0.100}\right)} = 0$$

$$Rate = k[NO_2]^2$$

(b) From Experiment 1:

$$k = \frac{Rate}{[NO_2]^2} = \frac{5.00 \times 10^{-3} \text{ M/s}}{(0.100 \text{ M})^2} = 0.500 \text{ 1/(M} \cdot \text{s})$$

14.6   (a) $Rate = k[C_2H_4Br_2]^m[I^-]^n$

$$m = \frac{\ln\left(\dfrac{Rate_2}{Rate_1}\right)}{\ln\left(\dfrac{[C_2H_4Br_2]_2}{[C_2H_4Br_2]_1}\right)} = \frac{\ln\left(\dfrac{1.74 \times 10^{-4}}{6.45 \times 10^{-5}}\right)}{\ln\left(\dfrac{0.343}{0.127}\right)} = 1$$

$$n = \frac{\ln\left(\dfrac{Rate_3 \cdot [C_2H_4Br_2]_2}{Rate_2 \cdot [C_2H_4Br_2]_3}\right)}{\ln\left(\dfrac{[I^-]_3}{[I^-]_2}\right)} = \frac{\ln\left(\dfrac{(1.26 \times 10^{-4})(0.343)}{(1.74 \times 10^{-4})(0.203)}\right)}{\ln\left(\dfrac{0.125}{0.102}\right)} = 1$$

$$Rate = k[C_2H_4Br_2][I^-]$$

(b) From Experiment 1:

$$k = \frac{Rate}{[C_2H_4Br_2][I^-]} = \frac{6.45 \times 10^{-5} \text{ M/s}}{(0.127 \text{ M})(0.102 \text{ M})} = 4.98 \times 10^{-3}/(\text{M} \cdot \text{s})$$

(c) $Rate = \dfrac{\Delta[I_3^-]}{\Delta t} = k[C_2H_4Br_2][I^-] = [4.98 \times 10^{-3}/(\text{M} \cdot \text{s})](0.150 \text{ M})^2 = 1.12 \times 10^{-4} \text{ M/s}$

(d) $-\dfrac{1}{3}\dfrac{\Delta[I^-]}{\Delta t} = \dfrac{1}{1}\dfrac{\Delta[I_3^-]}{\Delta t} = 1.12 \times 10^{-4} \text{ M/s}$

$$-3\frac{\Delta[I_3^-]}{\Delta t} = (-3)(1.12 \times 10^{-4} \text{ M/s}) = -3.36 \times 10^{-4} \text{ M/s}$$

14.7   (a) The reactions in vessels (a) and (b) have the same rate, the same number of B molecules, but different numbers of A molecules. Therefore, the rate does not depend on A and its reaction order is zero. The same conclusion can be drawn from the reactions in vessels (c) and (d).
The rate for the reaction in vessel (c) is four times the rate for the reaction in vessel (a). Vessel (c) has twice as many B molecules than does vessel (a). Because the rate quadruples when the concentration of B doubles, the reaction order for B is two.
(b) $rate = k[B]^2$

14.8 The rate law is Rate = $k[A]^2[B]$. In vessel 1, the Rate = $k(2)^2(4) = 0.01$ M/s

$$k = \frac{0.01 \text{ M/s}}{(2)^2(4)} = 6.25 \times 10^{-4} \text{ M/s}$$

In vessel 2, Rate = $(6.25 \times 10^{-4}$ M/s$)(4)^2(2) = 0.02$ M/s
In vessel 3, Rate = $(6.25 \times 10^{-4}$ M/s$)(1)^2(8) = 0.005$ M/s

14.9 $k = 1.50 \times 10^{-6}$ M/s

$[A]_t = -kt + [A]_o$

$kt = [A]_o - [A]_t$

$$t = \frac{[A]_o - [A]_t}{k} = \frac{(5.00 \times 10^{-3} \text{ M}) - (1.00 \times 10^{-3} \text{ M})}{1.50 \times 10^{-6} \text{ M/s}} = 2667 \text{ s} = 2670 \text{ s}$$

$$t = 2667 \text{ s} \times \frac{1 \text{ min}}{60 \text{ s}} = 44.4 \text{ min}$$

14.10

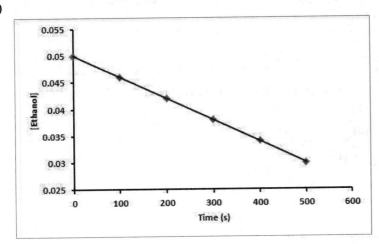

(a) A graph of $[C_2H_5OH]$ vs time is linear (slope = -4.0×10^{-5} M/s), which indicates that the reaction is zeroth order in $[C_2H_5OH]$.

(b) $k = -$slope = 4.0×10^{-5} M/s

(c) $t = 15.0$ min s $\times \dfrac{60 \text{ s}}{1 \text{ min}} = 900$ s

$[C_2H_5OH]_t = -kt + [C_2H_5OH]_o$

$[C_2H_5OH]_t = -(4.0 \times 10^{-5}$ M/s$)(900$ s$) + (5.0 \times 10^{-2}$ M$) = 0.014$ M

14.11 (a) $\ln \dfrac{[Co(NH_3)_5Br^{2+}]_t}{[Co(NH_3)_5Br^{2+}]_o} = -kt$

$k = 6.3 \times 10^{-6}$/s; $t = 10.0$ h $\times \dfrac{3600 \text{ s}}{1 \text{ h}} = 36,000$ s

$\ln[Co(NH_3)_5Br^{2+}]_t = -kt + \ln[Co(NH_3)_5Br^{2+}]_o$

$\ln[Co(NH_3)_5Br^{2+}]_t = -(6.3 \times 10^{-6}$/s$)(36,000$ s$) + \ln(0.100)$

$\ln[Co(NH_3)_5Br^{2+}]_t = -2.5294$;   After 10.0 h, $[Co(NH_3)_5Br^{2+}] = e^{-2.5294} = 0.080$ M

(b) $[Co(NH_3)_5Br^{2+}]_o = 0.100$ M

If 75% of the $Co(NH_3)_5Br^{2+}$ reacts then 25% remains.

$[Co(NH_3)_5Br^{2+}]_t = (0.25)(0.100$ M$) = 0.025$ M

$$\ln \frac{[Co(NH_3)_5Br^{2+}]_t}{[Co(NH_3)_5Br^{2+}]_o} = -kt; \quad t = \frac{\ln \dfrac{[Co(NH_3)_5Br^{2+}]_t}{[Co(NH_3)_5Br^{2+}]_o}}{-k}$$

$$t = \frac{\ln\left(\dfrac{0.025}{0.100}\right)}{-(6.3 \times 10^{-6}/s)} = 2.2 \times 10^5 \text{ s}; \quad t = 2.2 \times 10^5 \text{ s} \times \frac{1 \text{ h}}{3600 \text{ s}} = 61 \text{ h}$$

14.12 $\ln \dfrac{[A]_t}{[A]_o} = -kt$

(a) Let $[A]_o = 100$ and $[A]_t = 45$ (55% of 100 has reacted, 45 is left)

$$k = -\frac{\ln \dfrac{[A]_t}{[A]_o}}{t} = -\frac{\ln \dfrac{[45]_t}{[100]_o}}{14.2 \text{ h}} = 0.0562 \text{ h}^{-1}$$

(b) Let $[A]_o = 100$ and $[A]_t = 15$ (85% of 100 has reacted, 15 is left)

$$t = -\frac{\ln \dfrac{[A]_t}{[A]_o}}{k} = -\frac{\ln \dfrac{[15]_t}{[100]_o}}{0.0562 \text{ h}^{-1}} = 33.8 \text{ h}$$

14.13

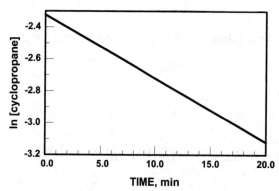

Slope $= -0.03989/\text{min} = -6.6 \times 10^{-4}/\text{s}$ and k = –slope

A plot of ln[cyclopropane] versus time is linear, indicating that the data fit the equation for a first-order reaction. $k = 6.6 \times 10^{-4}/\text{s}$ (0.040/min)

14.14 $\ln \dfrac{[N_2O_5]_t}{[N_2O_5]_0} = -kt$, k = – (slope) = – (– $9.8 \times 10^{-4}/\text{s}$) = $9.8 \times 10^{-4}/\text{s}$

$$t = 10 \text{ min x } \frac{60 \text{ s}}{1 \text{ min}} = 600 \text{ s}$$

$$\ln[N_2O_5]_t = -kt + \ln[N_2O_5]_0 = -(9.8 \times 10^{-4}/s)(600 \text{ s}) + \ln(0.100) = -2.891$$
$$[N_2O_5]_t = e^{-2.891} = 0.055 \text{ M}$$

	N_2O_5	→	$2 NO_2$	$+ 1/2 O_2$
initial (M)	0.100		0	0
change (M)	−x		+2x	+1/2x
	x = 0.100 − 0.055 = 0.045			
final (M)	0.055		0.090	0.023

After 10.0 min: $[N_2O_5] = 0.055$ M, $[NO_2] = 0.090$ M and $[O_2] = 0.023$ M

14.15 $k = 1.8 \times 10^{-5}/s$

$$t_{1/2} = \frac{0.693}{k} = \frac{0.693}{1.8 \times 10^{-5}/s} = 38{,}500 \text{ s}; \qquad t_{1/2} = 38{,}500 \text{ s x } \frac{1 \text{ h}}{3600 \text{ s}} = 11 \text{ h}$$

$$0.30 \text{ M} \xrightarrow{t_{1/2}} 0.15 \text{ M} \xrightarrow{t_{1/2}} 0.075 \text{ M} \xrightarrow{t_{1/2}} 0.0375 \text{ M} \xrightarrow{t_{1/2}} 0.019 \text{ M}$$

14.16 After one half-life, there would be four A molecules remaining. After two half-lives, there would be two A molecules remaining. This is represented by the drawing at t = 10 min. 10 min is equal to two half-lives, therefore, $t_{1/2} = 5$ min for this reaction. After 15 min (three half-lives) only one A molecule would remain.

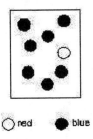

○ red ● blue

14.17

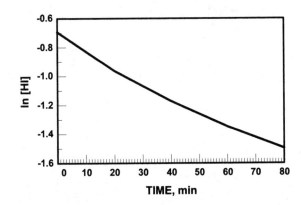

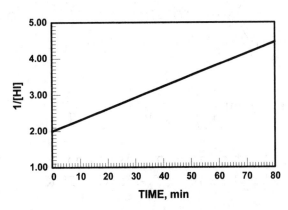

(a) A plot of 1/[HI] versus time is linear. The reaction is second-order.

k = slope = 0.0308/(M · min)

(b) $t = \dfrac{1}{k}\left[\dfrac{1}{[HI]_t} - \dfrac{1}{[HI]_o}\right] = \dfrac{1}{0.0308/(M \cdot min)}\left[\dfrac{1}{0.100\ M} - \dfrac{1}{0.500\ M}\right] = 260$ min

14.18

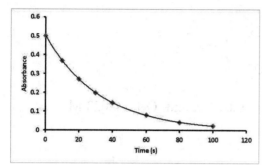

A graph of ln Abs vs time is linear (slope = –0.03071 /s), which indicates that the reaction is first-order in dye.

k = –slope = 0.03071 /s

$t_{1/2} = \dfrac{0.693}{k} = \dfrac{0.693}{0.03071/s} = 22.6$ s

14.19 (a) Because $\Delta E < 0$, the reaction is exothermic.

(b) E_a for the reverse reaction equals 132 kJ/mol + 226 kJ/mol = 358 kJ/mol

14.20 unsuccessful

successful

unsuccessful

unsuccessful

14.21 (a) $\ln\left(\dfrac{k_2}{k_1}\right) = \left(\dfrac{-E_a}{R}\right)\left(\dfrac{1}{T_2} - \dfrac{1}{T_1}\right)$

$k_1 = 3.7 \times 10^{-5}$/s, $T_1 = 25\ ^{\circ}C = 298$ K

$k_2 = 1.7 \times 10^{-3}$/s, $T_2 = 55\ ^{\circ}C = 328$ K

$$E_a = - \frac{[\ln k_2 - \ln k_1]R}{\left(\dfrac{1}{T_2} - \dfrac{1}{T_1}\right)}$$

$$E_a = - \frac{[\ln(1.7 \times 10^{-3}) - \ln(3.7 \times 10^{-5})][8.314 \times 10^{-3} \text{ kJ/(K·mol)}]}{\left(\dfrac{1}{328 \text{ K}} - \dfrac{1}{298 \text{ K}}\right)} = 104 \text{ kJ/mol}$$

(b) $k_1 = 3.7 \times 10^{-5}$/s, $T_1 = 25 \,^\circ$C $= 298$ K
  solve for k_2, $T_2 = 35 \,^\circ$C $= 308$ K

$$\ln k_2 = \left(\frac{-E_a}{R}\right)\left(\frac{1}{T_2} - \frac{1}{T_1}\right) + \ln k_1$$

$$\ln k_2 = \left(\frac{-104 \text{ kJ/mol}}{8.314 \times 10^{-3} \text{ kJ/(K·mol)}}\right)\left(\frac{1}{308 \text{ K}} - \frac{1}{298 \text{ K}}\right) + \ln (3.7 \times 10^{-5})$$

$\ln k_2 = -8.84$; $k_2 = e^{-8.84} = 1.4 \times 10^{-4}$/s

14.22 $\ln\left(\dfrac{k_2}{k_1}\right) = \left(\dfrac{-E_a}{R}\right)\left(\dfrac{1}{T_2} - \dfrac{1}{T_1}\right)$

$k_1 = 1$, $T_1 = 25 \,^\circ$C $= 298$ K
$k_2 = 3$, $T_2 = 36 \,^\circ$C $= 309$ K

$$E_a = - \frac{[\ln k_2 - \ln k_1]R}{\left(\dfrac{1}{T_2} - \dfrac{1}{T_1}\right)}$$

$$E_a = - \frac{[\ln(3) - \ln(1)][8.314 \times 10^{-3} \text{ kJ/(K·mol)}]}{\left(\dfrac{1}{309 \text{ K}} - \dfrac{1}{298 \text{ K}}\right)} = 76.5 \text{ kJ/mol}$$

14.23 (a) $NO_2(g) + F_2(g) \rightarrow NO_2F(g) + F(g)$
  $\underline{F(g) + NO_2(g) \rightarrow NO_2F(g)}$
Overall reaction $2\,NO_2(g) + F_2(g) \rightarrow 2\,NO_2F(g)$
Because F(g) is produced in the first reaction and consumed in the second, it is a reaction intermediate.
(b) In each reaction there are two reactants, so each elementary reaction is bimolecular.

14.24 (a) $H_2O_2(aq) \rightarrow 2\,OH(aq)$
  $H_2O_2(aq) + OH(aq) \rightarrow H_2O(l) + HO_2(aq)$
  $\underline{HO_2(aq) + OH(aq) \rightarrow H_2O(l) + O_2(g)}$
Overall reaction $2\,H_2O_2(aq) \rightarrow 2\,H_2O(l) + O_2(g)$

Because OH(aq) and HO_2(aq) are produced and then consumed, they are reaction intermediates.
(b) Step 1 is unimolecular because it has only one reactant. Steps 2 and 3 are bimolecular because in each reaction there are two reactants.

14.25 (a) Rate = $k[Br]^2[Ar]$ (b) Rate = $k[O_3][O]$ (c) Rate = $k[Co(CN)_5(H_2O)^{2-}]$

14.26 (a) ii)   (b) iii)   (c) i)

14.27

$$NO(g) + Cl_2(g) \rightarrow NOCl(g) + Cl(g) \quad \text{(slow)}$$
$$\underline{NO(g) + Cl(g) \rightarrow NOCl(g)} \quad \text{(fast)}$$

Overall reaction $2 NO(g) + Cl_2(g) \rightarrow 2 NOCl(g)$

The predicted rate law for the overall reaction is the rate law for the first (slow) elementary reaction: Rate = $k[NO][Cl_2]$
The predicted rate law is in accord with the observed rate law.

14.28

$$Co(CN)_5(H_2O)^{2-}(aq) \rightarrow Co(CN)_5^{2-}(aq) + H_2O(l) \quad \text{(slow)}$$
$$\underline{Co(CN)_5^{2-}(aq) + I^-(aq) \rightarrow Co(CN)_5I^{3-}(aq)} \quad \text{(fast)}$$

Overall reaction $Co(CN)_5(H_2O)^{2-}(aq) + I^-(aq) \rightarrow Co(CN)_5I^{3-}(aq) + H_2O(l)$

The predicted rate law for the overall reaction is the rate law for the first (slow) elementary reaction: Rate = $k[Co(CN)_5(H_2O)^{2-}]$
The predicted rate law is in accord with the observed rate law.

14.29

$$NO(g) + O_2(g) \underset{k_{-1}}{\overset{k_1}{\rightleftharpoons}} NO_3(g) \qquad \text{fast}$$

$$NO_3(g) + NO(g) \xrightarrow{k_2} 2NO_2(g) \quad \text{slow}$$

$Rate_{forward} = k_1[NO][O_2]$ and $Rate_{reverse} = k_{-1}[NO_3]$
Because of the equilibrium, $Rate_{forward} = Rate_{reverse}$, and $k_1[NO][O_2] = k_{-1}[NO_3]$.

$$[NO_3] = \frac{k_1}{k_{-1}}[NO][O_2]$$

The rate law for the rate determining step is Rate = $k_2[NO_3][NO]$. In this rate law substitute for $[NO_3]$.

Rate = $k_2 \dfrac{k_1}{k_{-1}}[NO]^2[O_2]$, which is consistent with the experimental rate law where

$$k = \frac{k_2 k_1}{k_{-1}}.$$

14.30

$$I_2(g) \underset{k_{-1}}{\overset{k_1}{\rightleftharpoons}} 2\, I(g) \qquad\qquad \text{fast}$$

$$H_2(g) + I(g) \underset{k_{-2}}{\overset{k_2}{\rightleftharpoons}} H_2I(g) \qquad \text{fast}$$

$$H_2I(g) + I(g) \xrightarrow{k_3} 2\, HI(g) \qquad \text{slow}$$

(a) $H_2(g) + I_2(g) \rightarrow 2\, HI(g)$

(b) From step 1, $\text{Rate}_{\text{forward}} = k_1[I_2]$ and $\text{Rate}_{\text{reverse}} = k_{-1}[I]^2$

Because of the equilibrium, $\text{Rate}_{\text{forward}} = \text{Rate}_{\text{reverse}}$, and $k_1[I_2] = k_{-1}[I]^2$.

$$[I]^2 = \frac{k_1}{k_{-1}}[I_2]$$

From step 2, $\text{Rate}_{\text{forward}} = k_2[H_2][I]$ and $\text{Rate}_{\text{reverse}} = k_{-2}[H_2I]$

Because of the equilibrium, $\text{Rate}_{\text{forward}} = \text{Rate}_{\text{reverse}}$, and $k_2[H_2][I] = k_{-2}[H_2I]$.

$$[H_2I] = \frac{k_2}{k_{-2}}[H_2][I]$$

The rate law for the rate determining step (step 3) is $\text{Rate} = k_3[H_2I][I]$.

Substitute for $[H_2I]$. $\text{Rate} = k_3\, \dfrac{k_2}{k_{-2}}[H_2][I][I] = k_3\, \dfrac{k_2}{k_{-2}}[H_2][I]^2$

Substitute for $[I]^2$. $\text{Rate} = k_3\, \dfrac{k_2}{k_{-2}}[H_2]\, \dfrac{k_1}{k_{-1}}[I_2] = k_3\, \dfrac{k_1}{k_{-1}}\, \dfrac{k_2}{k_{-2}}[H_2][I_2]$, which is consistent

with the experimental rate law.

(c) $k = \dfrac{k_1 k_2 k_3}{k_{-1} k_{-2}}$

14.31 Assume that concentration is proportional to the number of each molecule in a box.

(a) $\text{Rate} = k\,[B][C_2]$

(b) The mechanism agrees with the rate law. The rate law for the overall reaction is the rate law for the first (slow) elementary reaction: $\text{Rate} = k[B][C_2]$

(c) B doesn't appear in the overall reaction because it is consumed in the first step and regenerated in the third step. B is therefore a catalyst. C and CB are intermediates because they are formed in one step and then consumed in subsequent steps in the mechanism.

14.32 (a)

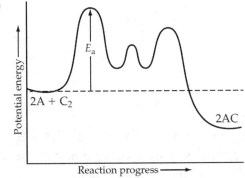

(b) $\text{B} + \text{C}_2 \rightarrow \text{BC}_2$ (slow)
        $\text{A} + \text{BC}_2 \rightarrow \text{AC} + \text{BC}$
        $\underline{\text{A} + \text{BC} \rightarrow \text{AC} + \text{B}}$
        $2\,\text{A} + \text{C}_2 \rightarrow 2\,\text{AC}$ (overall)

B doesn't appear in the overall reaction because it is consumed in the first step and regenerated in the third step. B is therefore a catalyst. BC_2 and BC are intermediates because they are formed in one step and then consumed in a subsequent step in the reaction.

14.33   Soluble enzymes are homogeneous catalysts and membrane-bound enzymes are heterogeneous.

14.34   (a) $\text{S} \rightarrow \text{P}$
        (b) E is a catalyst. ES is an intermediate.
        (c) $\text{Rate}_{forward} = k_1[\text{E}][\text{S}]$ and $\text{Rate}_{reverse} = k_{-1}[\text{ES}]$
        Because of the equilibrium, $\text{Rate}_{forward} = \text{Rate}_{reverse}$, and $k_1[\text{E}][\text{S}] = k_{-1}[\text{ES}]$.

$$\frac{k_1}{k_{-1}}[\text{E}][\text{S}] = [\text{ES}]$$

The rate law for the rate determining step is $\text{Rate} = k_2[\text{ES}]$. In this rate law substitute for [ES].

$$\text{Rate} = k_2 \frac{k_1}{k_{-1}}[\text{E}][\text{S}]$$

        (d) $k = \dfrac{k_2 k_1}{k_{-1}}$

14.35

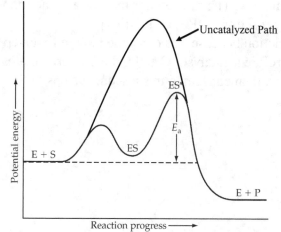

14.36   The substrate concentration is changed by a factor of $(3.4 \times 10^{-5}\ \text{M})/(1.4 \times 10^{-5}\ \text{M}) = 2.4$. Because the reaction is first-order at these (low) substrate concentrations, the enzyme catalyzed rate of product formation will increase by a factor of 2.4.

14.37 (a) Because the reaction is zeroth-order in substrate at high substrate concentration, there is no change in rate when the substrate concentration is changed from 2.8×10^{-3} M to 4.8×10^{-3} M.

(b) At high substrate concentrations, the enzyme becomes saturated (i.e., completely bound) with the substrate. At this point, the reaction rate reaches a maximum value and becomes independent of substrate concentration (zeroth order in substrate) because only substrate bound to the enzyme can react.

14.38 (a) $t_{1/2} = \dfrac{0.693}{k} = \dfrac{0.693}{1.7 \times 10^{-1}\,\text{s}^{-1}} = 4.1\ \text{s}$

(b) $t = \dfrac{\ln \dfrac{[\text{ester}]_t}{[\text{ester}]_o}}{-k} = \dfrac{\ln \dfrac{[1.5 \times 10^{-4}]}{[5.5 \times 10^{-3}]}}{-1.7 \times 10^{-1}\,\text{s}^{-1}} = 21\ \text{s}$

Conceptual Problems

14.40 (a) Because Rate $= k[\text{A}][\text{B}]$, the rate is proportional to the product of the number of A molecules and the number of B molecules. The relative rates of the reaction in vessels (a) – (d) are $2 : 1 : 4 : 2$.

(b) Because the same reaction takes place in each vessel, the k's are all the same.

14.42 (a) For the first-order reaction, half of the A molecules are converted to B molecules each minute.

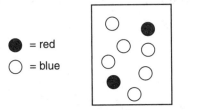
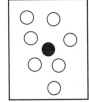

t = 2 min   t = 3 min

(b) Because half of the A molecules are converted to B molecules in 1 min, the half-life is 1 minute.

14.44 (a) Because the half-life is inversely proportional to the concentration of A molecules, the reaction is second-order in A.

(b) Rate $= k[\text{A}]^2$

(c) The second box represents the passing of one half-life, and the third box represents the passing of a second half-life for a second-order reaction. A relative value of k can be calculated.

$k = \dfrac{1}{t_{1/2}[\text{A}]} = \dfrac{1}{(1)(16)} = 0.0625$

$t_{1/2}$ in going from box 3 to box 4 is: $t_{1/2} = \dfrac{1}{k[\text{A}]} = \dfrac{1}{(0.0625)(4)} = 4\ \text{min}$

(For fourth box, $t = 7$ min)

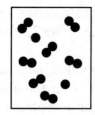

t = 3 min + 4 min = 7 min

14.46 Assume that concentration is proportional to the number of each molecule in a box.
(a) Comparing boxes (1) and (2), the concentration of B doubles, A remains the same and the rate does not change. This means the reaction is zeroth-order in B.
Comparing boxes (3) and (2), the concentration of A doubles, B remains the same and the rate quadruples. This means the reaction is second-order in A.
(b) Rate = k [A]².
(c) $\qquad$ 2 A $\rightarrow$ A₂ $\qquad$ (slow)
$\dfrac{A_2 + B \rightarrow AB + A}{A + B \rightarrow AB}$ $\qquad$ (overall)
(d) A₂ is an intermediate because it is formed in one step and then consumed in a subsequent step in the reaction.

14.48 (a) BC + D → B + CD
(b) 1. B–C + D (reactants), A (catalyst); 2. B---C---A (transition state), D (reactant);
3. A–C (intermediate), B (product), D (reactant); 4. A---C---D (transition state),
B (product); 5. A (catalyst), C–D + B (products)
(c) The first step is rate determining because the first maximum in the potential energy curve is greater than the second (relative) maximum; Rate = k[A][BC]
(d) Endothermic

Section Problems
Reaction Rates (Section 14.1)

14.50 2 N₂O₅(g) → 4 NO₂(g) + O₂(g)

time	[N₂O₅]	[O₂]
200 s	0.0142 M	0.0029 M
300 s	0.0120 M	0.0040 M

$$\text{Rate of decomposition of N}_2\text{O}_5 = -\frac{\Delta[\text{N}_2\text{O}_5]}{\Delta t} = -\frac{0.0120\ \text{M} - 0.0142\ \text{M}}{300\ \text{s} - 200\ \text{s}} = 2.2 \times 10^{-5}\ \text{M/s}$$

$$\text{Rate of formation of O}_2 = \frac{\Delta[\text{O}_2]}{\Delta t} = \frac{0.0040\ \text{M} - 0.0029\ \text{M}}{300\ \text{s} - 200\ \text{s}} = 1.1 \times 10^{-5}\ \text{M/s}$$

14.52 (a) Rate $= \dfrac{\Delta[\text{NO}_2]}{\Delta t} = \dfrac{(0.0278\ \text{M}) - (0\ \text{M})}{700\ \text{s} - 0\ \text{s}} = 4.0 \times 10^{-5}\ \text{M/s}$

(b) Rate $= \dfrac{\Delta[NO_2]}{\Delta t} = \dfrac{(0.0256\ M) - (0.0063\ M)}{600\ s - 100\ s} = 3.9 \times 10^{-5}\ M/s$

(c) Rate $= \dfrac{\Delta[NO_2]}{\Delta t} = \dfrac{(0.0229\ M) - (0.0115\ M)}{500\ s - 200\ s} = 3.8 \times 10^{-5}\ M/s$

(d) Rate $= \dfrac{\Delta[NO_2]}{\Delta t} = \dfrac{(0.0197\ M) - (0.0160\ M)}{400\ s - 300\ s} = 3.7 \times 10^{-5}\ M/s$

Rate (d) is the best estimate of the instantaneous rate because it is determined from measurements taken over the smallest time interval.

14.54 (a) The instantaneous rate of decomposition of N_2O_5 at t = 200 s is determined from the slope of the curve at t = 200 s.

Rate $= -\dfrac{\Delta[N_2O_5]}{\Delta t} = -\text{slope} = -\dfrac{(1.20 \times 10^{-2}\ M) - (1.69 \times 10^{-2}\ M)}{300\ s - 100\ s} = 2.4 \times 10^{-5}\ M/s$

(b) The initial rate of decomposition of N_2O_5 is determined from the slope of the curve at t = 0 s. This is equivalent to the slope of the curve from 0 s to 100 s because in this time interval the curve is almost linear.

Initial rate $= -\dfrac{\Delta[N_2O_5]}{\Delta t} = -\text{slope} = -\dfrac{(1.69 \times 10^{-2}\ M) - (2.00 \times 10^{-2}\ M)}{100\ s - 0\ s} = 3.1 \times 10^{-5}\ M/s$

14.56 (a) $-\dfrac{\Delta[H_2]}{\Delta t} = -3\dfrac{\Delta[N_2]}{\Delta t}$; The rate of consumption of H_2 is 3 times faster.

(b) $\dfrac{\Delta[NH_3]}{\Delta t} = -2\dfrac{\Delta[N_2]}{\Delta t}$; The rate of formation of NH_3 is 2 times faster.

14.58 (a) $-\dfrac{1}{2}\dfrac{\Delta[Br_2]}{\Delta t} = \dfrac{\Delta[ClO_2^-]}{\Delta t} = -2.4 \times 10^{-6}\ M/s$

(b) Rate $= -\dfrac{\Delta[Br^-]}{\Delta t} = -4\dfrac{\Delta[ClO_2^-]}{\Delta t} = -4(-2.4 \times 10^{-6}\ M/s) = 9.6 \times 10^{-6}\ M/s$

Rate Laws (Sections 14.2 and 14.3)

14.60 Rate $= k[H_2][ICl]$;   units for k are $\dfrac{L}{mol \cdot s}$ or $1/(M \cdot s)$

14.62 (a) Rate $= k[CH_3Br][OH^-]$
(b) Because the reaction is first-order in OH^-, if the $[OH^-]$ is decreased by a factor of 5, the rate will also decrease by a factor of 5.
(c) Because the reaction is first-order in each reactant, if both reactant concentrations are doubled, the rate will increase by a factor of $2 \times 2 = 4$.

14.64 (a) Rate = k[Cu($C_{10}H_8N_2)_2^+$]^2[O_2]
(b) The overall reaction order is 2 + 1 = 3.
(c) Because the reaction is second-order in Cu($C_{10}H_8N_2)_2^+$, if the [Cu($C_{10}H_8N_2)_2^+$] is decreased by a factor of four, the rate will decrease by a factor of 16 (1/4 x 1/4 = 1/16).

14.66 (a) Rate = k[NH_4^+]^m[NO_2^-]^n

$$m = \dfrac{\ln\left(\dfrac{Rate_2}{Rate_1}\right)}{\ln\left(\dfrac{[NH_4^+]_2}{[NH_4^+]_1}\right)} = \dfrac{\ln\left(\dfrac{3.6 \times 10^{-6}}{7.2 \times 10^{-6}}\right)}{\ln\left(\dfrac{0.12}{0.24}\right)} = 1; \quad n = \dfrac{\ln\left(\dfrac{Rate_3}{Rate_2}\right)}{\ln\left(\dfrac{[NO_2^-]_3}{[NO_2^-]_2}\right)} = \dfrac{\ln\left(\dfrac{5.4 \times 10^{-6}}{3.6 \times 10^{-6}}\right)}{\ln\left(\dfrac{0.15}{0.10}\right)} = 1$$

Rate = k[NH_4^+][NO_2^-]

(b) From Experiment 1: $k = \dfrac{Rate}{[NH_4^+][NO_2^-]} = \dfrac{7.2 \times 10^{-6}\ M/s}{(0.24\ M)(0.10\ M)} = 3.0 \times 10^{-4}/(M \cdot s)$

(c) Rate = k[NH_4^+][NO_2^-] = [3.0 x 10^{-4}/(M · s)](0.39 M)(0.052 M) = 6.1 x 10^{-6} M/s

14.68 $H_2O_2(aq) + 3\ I^-(aq) + 2\ H^+(aq) \rightarrow I_3^-(aq) + 2\ H_2O(l)$

Rate = k[H_2O_2]^m[I^-]^n

(a) $\dfrac{Rate_3}{Rate_1} = \dfrac{2.30 \times 10^{-4}\ M/s}{1.15 \times 10^{-4}\ M/s} = 2$ $\qquad \dfrac{[H_2O_2]_3}{[H_2O_2]_1} = \dfrac{0.200\ M}{0.100\ M} = 2$

Because both ratios are the same, m = 1.

$\dfrac{Rate_2}{Rate_1} = \dfrac{2.30 \times 10^{-4}\ M/s}{1.15 \times 10^{-4}\ M/s} = 2$ $\qquad \dfrac{[I^-]_2}{[I^-]_1} = \dfrac{0.200\ M}{0.100\ M} = 2$

Because both ratios are the same, n = 1.
The rate law is: Rate = k[H_2O_2][I^-]

(b) $k = \dfrac{Rate}{[H_2O_2][I^-]}$

Using data from Experiment 1: $k = \dfrac{1.15 \times 10^{-4}\ M/s}{(0.100\ M)(0.100\ M)} = 1.15 \times 10^{-2}/(M \cdot s)$

(c) Rate = k[H_2O_2][I^-] = [1.15 x 10^{-2}/(M · s)](0.300 M)(0.400 M) = 1.38 x 10^{-3} M/s

Integrated Rate Law; Half-Life (Sections 14.4–14.6)

14.70 $\ln\dfrac{[C_3H_6]_t}{[C_3H_6]_0} = -kt$, k = 6.7 x 10^{-4}/s

(a) t = 30 min x $\dfrac{60\ s}{1\ min}$ = 1800 s

$\ln[C_3H_6]_t = -kt + \ln[C_3H_6]_0 = -(6.7 \times 10^{-4}/s)(1800\ s) + \ln(0.0500) = -4.202$

$[C_3H_6]_t = e^{-4.202} = 0.015 \text{ M}$

(b) $\quad t = \dfrac{\ln \dfrac{[C_3H_6]_t}{[C_3H_6]_0}}{-k} = \dfrac{\ln\left(\dfrac{0.0100}{0.0500}\right)}{-(6.7 \times 10^{-4}/\text{s})} = 2402 \text{ s}; \quad t = 2402 \text{ s} \times \dfrac{1 \text{ min}}{60 \text{ s}} = 40 \text{ min}$

(c) $[C_3H_6]_0 = 0.0500 \text{ M};\qquad$ If 25% of the C_3H_6 reacts then 75% remains.

$[C_3H_6]_t = (0.75)(0.0500 \text{ M}) = 0.0375 \text{ M}$

$t = \dfrac{\ln \dfrac{[C_3H_6]_t}{[C_3H_6]_0}}{-k} = \dfrac{\ln\left(\dfrac{0.0375}{0.0500}\right)}{-(6.7 \times 10^{-4}/\text{s})} = 429 \text{ s}; \quad t = 429 \text{ s} \times \dfrac{1 \text{ min}}{60 \text{ s}} = 7.2 \text{ min}$

14.72 $\quad t_{1/2} = \dfrac{0.693}{k} = \dfrac{0.693}{6.7 \times 10^{-4}/\text{s}} = 1034 \text{ s} = 17 \text{ min}$

$t = \dfrac{\ln \dfrac{[C_3H_6]_t}{[C_3H_6]_0}}{-k} = \dfrac{\ln \dfrac{(0.0625)(0.0500)}{(0.0500)}}{-6.7 \times 10^{-4}/\text{s}} = 4140 \text{ s}$

$t = 4140 \text{ s} \times \dfrac{1 \text{ min}}{60 \text{ s}} = 69 \text{ min}$

This is also 4 half-lives. $\quad 100 \xrightarrow{t_{1/2}} 50 \xrightarrow{t_{1/2}} 25 \xrightarrow{t_{1/2}} 12.5 \xrightarrow{t_{1/2}} 6.25$

14.74 $\quad kt = \dfrac{1}{[C_4H_6]_t} - \dfrac{1}{[C_4H_6]_0}, \qquad k = 4.0 \times 10^{-2}/(\text{M} \cdot \text{s})$

(a) $t = 1.00 \text{ h} \times \dfrac{60 \text{ min}}{1 \text{ hr}} \times \dfrac{60 \text{ s}}{1 \text{ min}} = 3600 \text{ s}$

$\dfrac{1}{[C_4H_6]_t} = kt + \dfrac{1}{[C_4H_6]_0} = (4.0 \times 10^{-2}/(\text{M} \cdot \text{s}))(3600 \text{ s}) + \dfrac{1}{0.0200 \text{ M}}$

$\dfrac{1}{[C_4H_6]_t} = 194/\text{M} \text{ and } [C_4H_6] = 5.2 \times 10^{-3} \text{ M}$

(b) $\quad t = \dfrac{1}{k}\left[\dfrac{1}{[C_4H_6]_t} - \dfrac{1}{[C_4H_6]_0}\right]$

$t = \dfrac{1}{4.0 \times 10^{-2}/(\text{M} \cdot \text{s})}\left[\dfrac{1}{(0.0020 \text{ M})} - \dfrac{1}{(0.0200 \text{ M})}\right] = 11{,}250 \text{ s}$

$t = 11{,}250 \text{ s} \times \dfrac{1 \text{ min}}{60 \text{ s}} \times \dfrac{1 \text{ hr}}{60 \text{ min}} = 3.1 \text{ h}$

14.76 $t_{1/2} = \dfrac{1}{k[C_4H_6]_o} = \dfrac{1}{[4.0 \times 10^{-2}/(M \cdot s)](0.0200\ M)} = 1250\ s = 21\ min$

$t = \dfrac{1}{k}\left[\dfrac{1}{[C_4H_6]_t} - \dfrac{1}{[C_4H_6]_o}\right]$

$t = \dfrac{1}{4.0 \times 10^{-2}/(M \cdot s)}\left[\dfrac{1}{(0.0050\ M)} - \dfrac{1}{(0.0100\ M)}\right] = 2500\ s = 42\ min$

or, recognizing that the time in going from 0.0100 M to 0.0050 M is a $t_{1/2}$,

$t = t_{1/2} = \dfrac{1}{k[C_4H_6]_o} = \dfrac{1}{[4.0 \times 10^{-2}/(M \cdot s)](0.0100\ M)} = 2500\ s = 42\ min$

14.78 $k = \dfrac{0.693}{t_{1/2}} = \dfrac{0.693}{248\ s} = 2.79 \times 10^{-3}/s$

14.80 $2\ AB_2 \rightarrow A_2 + 2\ B_2$
(a) Measure the change in the concentration of AB_2 as a function of time.
(b) and (c) If a plot of $[AB_2]$ versus time is linear, the reaction is zeroth-order and
$k = -slope$. If a plot of $\ln[AB_2]$ versus time is linear, the reaction is first-order and
$k = -slope$. If a plot of $1/[AB_2]$ versus time is linear, the reaction is second-order and
$k = slope$.

14.82

time (s)	[NOBr]	ln[NOBr]	1/[NOBr]
0	0.0390	−3.244	25.6
10	0.0301	−3.503	33.2
40	0.0175	−4.046	57.1
120	0.00812	−4.813	123.2
320	0.00376	−5.583	266.0

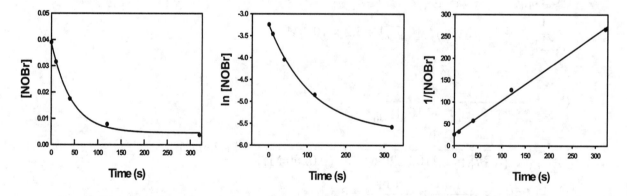

A plot of 1/[NOBr] versus time is linear. The reaction is second-order in NOBr.
$k = slope = 0.752/(M \cdot s)$

14.84 $kt = \dfrac{1}{[C_7H_{12}]_t} - \dfrac{1}{[C_7H_{12}]_0}$, $k = 0.030 \,/(M \cdot s)$

(a) $\dfrac{1}{[C_7H_{12}]_t} = kt + \dfrac{1}{[C_7H_{12}]_0} = (0.030/(M \cdot s))(1600 \, s) + \dfrac{1}{0.035 \, M}$

$\dfrac{1}{[C_7H_{12}]_t} = 79.6/M$ and $[C_7H_{12}]_t = 0.013 \, M$

(b) $t = \dfrac{1}{k}\left[\dfrac{1}{[C_7H_{12}]_t} - \dfrac{1}{[C_7H_{12}]_0}\right]$

$t = \dfrac{1}{0.030/(M \cdot s)}\left[\dfrac{1}{(0.035/20 \, M)} - \dfrac{1}{(0.035 \, M)}\right] = 1.8 \times 10^4 \, s$

(c) $t_{1/2} = \dfrac{1}{k[C_7H_{12}]_0} = \dfrac{1}{[0.030/(M \cdot s)](0.075 \, M)} = 4.4 \times 10^2 \, s$

The Arrhenius Equation (Sections 14.7 and 14.8)

14.86 Very few collisions involve a collision energy greater than or equal to the activation energy, and only a fraction of those have the proper orientation for reaction.

14.88 As the temperature of a gas is raised by 10 °C, even though the collision frequency increases by only ~2%, the reaction rate increases by 100% or more because there is an exponential increase in the fraction of the collisions that leads to products.

14.90

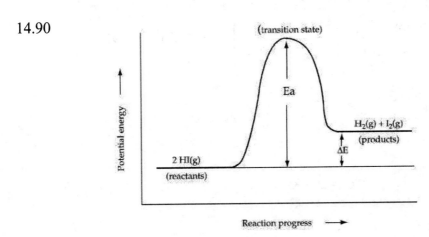

14.92 (a)

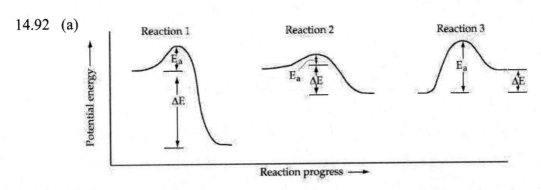

(b) Reaction 2 is the fastest (smallest E_a), and reaction 3 is the slowest (largest E_a).

(c) Reaction 3 is the most endothermic (positive ΔE), and reaction 1 is the most exothermic (largest negative ΔE).

14.94 (a) $\ln\left(\dfrac{k_2}{k_1}\right) = \left(\dfrac{-E_a}{R}\right)\left(\dfrac{1}{T_2} - \dfrac{1}{T_1}\right)$

$k_1 = 1.3/(M \cdot s),\ T_1 = 700\ K$
$k_2 = 23.0/(M \cdot s),\ T_2 = 800\ K$

$E_a = -\dfrac{[\ln k_2 - \ln k_1](R)}{\left(\dfrac{1}{T_2} - \dfrac{1}{T_1}\right)}$

$E_a = -\dfrac{[\ln(23.0) - \ln(1.3)][8.314 \times 10^{-3}\ kJ/(K \cdot mol)]}{\left(\dfrac{1}{800\ K} - \dfrac{1}{700\ K}\right)} = 134\ kJ/mol$

(b) $k_1 = 1.3/(M \cdot s),\ T_1 = 700\ K$
    solve for $k_2,\ T_2 = 750\ K$

$\ln k_2 = \left(\dfrac{-E_a}{R}\right)\left(\dfrac{1}{T_2} - \dfrac{1}{T_1}\right) + \ln k_1$

$\ln k_2 = \left(\dfrac{-133.8\ kJ/mol}{8.314 \times 10^{-3}\ kJ/(K \cdot mol)}\right)\left(\dfrac{1}{750\ K} - \dfrac{1}{700\ K}\right) + \ln(1.3) = 1.795$

$k_2 = e^{1.795} = 6.0/(M \cdot s)$

14.96 $\ln\left(\dfrac{k_2}{k_1}\right) = \left(\dfrac{-E_a}{R}\right)\left(\dfrac{1}{T_2} - \dfrac{1}{T_1}\right)$

assume $k_1 = 1.0/(M \cdot s)$ at $T_1 = 25\ ^\circ C = 298\ K$
assume $k_2 = 15/(M \cdot s)$ at $T_2 = 50\ ^\circ C = 323\ K$

$E_a = -\dfrac{[\ln k_2 - \ln k_1](R)}{\left(\dfrac{1}{T_2} - \dfrac{1}{T_1}\right)}$

$E_a = -\dfrac{[\ln(15) - \ln(1.0)][8.314 \times 10^{-3}\ kJ/(K \cdot mol)]}{\left(\dfrac{1}{323\ K} - \dfrac{1}{298\ K}\right)} = 87\ kJ/mol$

14.98 $\ln\left(\dfrac{k_2}{k_1}\right) = \left(\dfrac{-E_a}{R}\right)\left(\dfrac{1}{T_2} - \dfrac{1}{T_1}\right)$

$k_2 = 2.5k_1$
$k_1 = 1.0, \quad T_1 = 20\,°C = 293\ K$
$k_2 = 2.5, \quad T_2 = 30\,°C = 303\ K$

$E_a = -\dfrac{[\ln k_2 - \ln k_1](R)}{\left(\dfrac{1}{T_2} - \dfrac{1}{T_1}\right)}$

$E_a = -\dfrac{[\ln(2.5) - \ln(1.0)][8.314 \times 10^{-3}\ kJ/(K\cdot mol)]}{\left(\dfrac{1}{303\ K} - \dfrac{1}{293\ K}\right)} = 68\ kJ/mol$

$k_1 = 1.0, \quad T_1 = 120\,°C = 393\ K$
$k_2 = ?, \quad\ T_2 = 130\,°C = 403\ K$
Solve for k_2.

$\ln k_2 = \dfrac{-E_a}{R}\left(\dfrac{1}{T_2} - \dfrac{1}{T_1}\right) + \ln k_1$

$\ln k_2 = \dfrac{-68\ kJ/mol}{[8.314 \times 10^{-3}\ kJ/(K\cdot mol)]}\left(\dfrac{1}{403\ K} - \dfrac{1}{393\ K}\right) + \ln(1.0) = 0.516$

$k_2 = e^{0.516} = 1.7;$ \quad The rate increases by a factor of 1.7.

14.100 A → B + C
(a) Measure the change in the concentration of A as a function of time at several different temperatures.
(b) Plot ln [A] versus time, for each temperature. Straight line graphs will result and k at each temperature equals –slope. Graph ln k versus 1/K, where K is the kelvin temperature. Determine the slope of the line. $E_a = -R$(slope) where $R = 8.314 \times 10^{-3}\ kJ/(K \cdot mol)$.

14.102 Plot ln k versus 1/T to determine the activation energy, E_a.

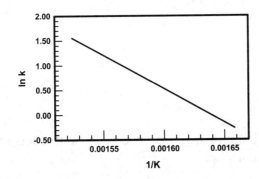

Slope $= -1.359 \times 10^4$ K

$E_a = -R(\text{slope}) = -[8.314 \times 10^{-3} \text{ kJ/(K} \cdot \text{mol)}](-1.359 \times 10^4 \text{ K}) = 113 \text{ kJ/mol}$

Reaction Mechanisms (Sections 14.9–14.11)

14.104 There is no relationship between the coefficients in a balanced chemical equation for an overall reaction and the exponents in the rate law unless the overall reaction occurs in a single elementary step, in which case the coefficients in the balanced equation are the exponents in the rate law.

14.106 (a)

$H_2(g) + ICl(g) \rightarrow HI(g) + HCl(g)$

$\underline{HI(g) + ICl(g) \rightarrow I_2(g) + HCl(g)}$

Overall reaction $\quad H_2(g) + 2 ICl(g) \rightarrow I_2(g) + 2 HCl(g)$

(b) Because HI(g) is produced in the first step and consumed in the second step, it is a reaction intermediate.

(c) In each reaction there are two reactant molecules, so each elementary reaction is bimolecular.

14.108 (a) bimolecular, Rate $= k[O_3][Cl]$ (b) unimolecular, Rate $= k[NO_2]$

       (c) bimolecular, Rate $= k[ClO][O]$ (d) termolecular, Rate $= k[Cl]^2[N_2]$

14.110 (a)

$NO_2Cl(g) \rightarrow NO_2(g) + Cl(g)$

$\underline{Cl(g) + NO_2Cl(g) \rightarrow NO_2(g) + Cl_2(g)}$

Overall reaction $\quad 2 NO_2Cl(g) \rightarrow 2 NO_2(g) + Cl_2(g)$

(b) 1. unimolecular; 2. bimolecular

(c) Rate $= k[NO_2Cl]$

14.112 $NO_2(g) + F_2(g) \rightarrow NO_2F(g) + F(g)$ (slow)

      $F(g) + NO_2(g) \rightarrow NO_2F(g)$ (fast)

14.114 (a) $2 NO(g) + O_2(g) \rightarrow 2 NO_2(g)$

(b) $\text{Rate}_{\text{forward}} = k_1[NO]^2$ and $\text{Rate}_{\text{reverse}} = k_{-1}[N_2O_2]$

Because of the equilibrium, $\text{Rate}_{\text{forward}} = \text{Rate}_{\text{reverse}}$, and $k_1[NO]^2 = k_{-1}[N_2O_2]$.

$[N_2O_2] = \dfrac{k_1}{k_{-1}} [NO]^2$

The rate law for the rate determining step is Rate $= 2k_2[N_2O_2][O_2]$ because two NO molecules are consumed in the overall reaction for every N_2O_2 that reacts in the second step. In this rate law substitute for $[N_2O_2]$.   Rate $= 2k_2 \dfrac{k_1}{k_{-1}} [NO]^2[O_2]$

(c) $k = \dfrac{2k_2 k_1}{k_{-1}}$

Catalysis (Sections 14.13–14.15)

14.116 A catalyst does participate in the reaction, but it is not consumed because it reacts in one step of the reaction and is regenerated in a subsequent step.

14.118 (a) $O_3(g) + O(g) \rightarrow 2 O_2(g)$
  (b) Cl acts as a catalyst.
  (c) ClO is a reaction intermediate.
  (d) A catalyst reacts in one step and is regenerated in a subsequent step. A reaction intermediate is produced in one step and consumed in another.

14.120 (a)

$$NH_2NO_2(aq) + OH^-(aq) \rightarrow NHNO_2^-(aq) + H_2O(l)$$
$$\underline{NHNO_2^-(aq) \rightarrow N_2O(g) + OH^-(aq)}$$

Overall reaction $\quad NH_2NO_2(aq) \rightarrow N_2O(g) + H_2O(l)$

  (b) OH^- acts as a catalyst because it is used in the first step and regenerated in the second. $NHNO_2^-$ is a reaction intermediate because it is produced in the first step and consumed in the second.
  (c) The rate will decrease because added acid decreases the concentration of OH^-, which appears in the rate law since it is a catalyst.

14.122 (a) Rate = $k[B_2][C]$
  (b) $B_2 + C \rightarrow CB + B \quad$ (slow)
    $CB + A \rightarrow AB + C \quad$ (fast)
  (c) C is a catalyst. C does not appear in the chemical equation because it is consumed in the first step and regenerated in the second step.

Multiconcept Problems

14.124 (a) The reaction rate will increase with an increase in temperature at constant volume.
  (b) The reaction rate will decrease with an increase in volume at constant temperature because reactant concentrations will decrease.
  (c) The reaction rate will increase with the addition of a catalyst.
  (d) Addition of an inert gas at constant volume will not affect the reaction rate.

14.126 (a) $I^-(aq) + OCl^-(aq) \rightarrow Cl^-(aq) + OI^-(aq)$
  (b) From the data in the table, initial rates $= -\dfrac{\Delta[I^-]}{\Delta t}$ have been calculated.

For example, from Experiment 1:

$$\text{Initial rate} = -\frac{\Delta[I^-]}{\Delta t} = -\frac{(2.17 \times 10^{-4}\,M - 2.40 \times 10^{-4}\,M)}{10\,s} = 2.30 \times 10^{-6}\,M/s$$

Initial concentrations and initial rate data have been collected in the table below.

EXPT	$[I^-]_0$ (M)	$[OCl^-]_0$ (M)	$[OH^-]_0$ (M)	Initial Rate (M/s)
1	2.40×10^{-4}	1.60×10^{-4}	1.00	2.30×10^{-6}
2	1.20×10^{-4}	1.60×10^{-4}	1.00	1.20×10^{-6}
3	2.40×10^{-4}	4.00×10^{-5}	1.00	6.00×10^{-7}
4	1.20×10^{-4}	1.60×10^{-4}	2.00	6.00×10^{-7}

Rate = $k[I^-]^m[OCl^-]^n[OH^-]^p$

From Expts 1 and 2, $[I^-]$ is cut in half and the initial rate is cut in half; therefore m = 1.
From Expts 1 and 3, $[OCl^-]$ is reduced by a factor of four and the initial rate is reduced by a factor of four; therefore n = 1.
From Expts 2 and 4, $[OH^-]$ is doubled and the initial rate is cut in half; therefore p = –1.

$$\text{Rate} = k \frac{[I^-][OCl^-]}{[OH^-]}$$

From data in Expt 1:

$$k = \frac{\text{Rate }[OH^-]}{[I^-][OCl^-]} = \frac{(2.30 \times 10^{-6}\text{ M/s})(1.00\text{ M})}{(2.40 \times 10^{-4}\text{ M})(1.60 \times 10^{-4}\text{ M})} = 60/s$$

(c) The reaction does not occur by a single-step mechanism because OH^- appears in the rate law but not in the overall reaction.

(d)
$$OCl^-(aq) + H_2O(l) \rightleftharpoons HOCl(aq) + OH^-(aq) \quad \text{(fast)}$$
$$HOCl(aq) + I^-(aq) \rightarrow HOI(aq) + Cl^-(aq) \quad \text{(slow)}$$
$$\underline{HOI(aq) + OH^-(aq) \rightarrow H_2O(l) + OI^-(aq) \quad \text{(fast)}}$$

Overall reaction $\quad I^-(aq) + OCl^-(aq) \rightarrow Cl^-(aq) + OI^-(aq)$

Because the forward and reverse rates in step 1 are equal, $k_1[OCl^-][H_2O] = k_{-1}[HOCl][OH^-]$. Solving for [HOCl] and substituting into the rate law for the second step gives

$$\text{Rate} = k_2[HOCl][I^-] = \frac{k_1 k_2}{k_{-1}} \frac{[OCl^-][H_2O][I^-]}{[OH^-]}$$

$[H_2O]$ is constant and can be combined into k.
Because the rate law for the overall reaction is equal to the rate law for the rate-determining step, the rate law for the overall reaction is

$$\text{Rate} = k \frac{[OCl^-][I^-]}{[OH^-]} \quad \text{where } k = \frac{k_1 k_2 [H_2O]}{k_{-1}}$$

14.128 $X \rightarrow$ products is a first-order reaction

$$t = 60 \text{ min} \times \frac{60\text{ s}}{1\text{ min}} = 3600 \text{ s}$$

$$\ln \frac{[X]_t}{[X]_o} = -kt; \quad k = \frac{\ln \dfrac{[X]_t}{[X]_o}}{-t}$$

At 25 °C, calculate k_1: $\quad k_1 = \dfrac{\ln\left(\dfrac{0.600\text{ M}}{1.000\text{ M}}\right)}{-3600\text{ s}} = 1.42 \times 10^{-4}\text{ s}^{-1}$

At 35 °C, calculate k_2: $\quad k_2 = \dfrac{\ln\left(\dfrac{0.200\text{ M}}{0.600\text{ M}}\right)}{-3600\text{ s}} = 3.05 \times 10^{-4}\text{ s}^{-1}$

At an unknown temperature calculate k_3: $k_3 = \dfrac{\ln\left(\dfrac{0.010\ M}{0.200\ M}\right)}{-3600\ s} = 8.32 \times 10^{-4}\ s^{-1}$

$T_1 = 25\ °C = 25 + 273 = 298\ K$
$T_2 = 35\ °C = 35 + 273 = 308\ K$

Calculate E_a using k_1 and k_2.

$$\ln\left(\frac{k_2}{k_1}\right) = \left(\frac{-E_a}{R}\right)\left(\frac{1}{T_2} - \frac{1}{T_1}\right)$$

$$E_a = -\frac{[\ln k_2 - \ln k_1](R)}{\left(\dfrac{1}{T_2} - \dfrac{1}{T_1}\right)}$$

$$E_a = -\frac{[\ln(3.05 \times 10^{-4}) - \ln(1.42 \times 10^{-4})][8.314 \times 10^{-3}\ kJ/(K\cdot mol)]}{\left(\dfrac{1}{308\ K} - \dfrac{1}{298\ K}\right)} = 58.3\ kJ/mol$$

Use E_a, k_1, and k_3 to calculate T_3.

$$\frac{1}{T_3} = \frac{\ln\left(\dfrac{k_3}{k_1}\right)}{\left(\dfrac{-E_a}{R}\right)} + \frac{1}{T_1} = \frac{\ln\left(\dfrac{8.32 \times 10^{-4}}{1.42 \times 10^{-4}}\right)}{\left(\dfrac{-58.3\ kJ/mol}{8.314 \times 10^{-3}\ kJ/(K\cdot mol)}\right)} + \frac{1}{298\ K} = 0.003104/K$$

$$T_3 = \frac{1}{0.003104/K} = 322\ K = 322 - 273 = 49\ °C$$

At 3:00 p.m. raise the temperature to 49 °C to finish the reaction by 4:00 p.m.

14.130   (a) When equal volumes of two solutions are mixed, both concentrations are cut in half.
$[H_3O^+]_o = [OH^-]_o = 1.0\ M$
When 99.999% of the acid is neutralized, $[H_3O^+] = [OH^-] = 1.0\ M - (1.0\ M \times 0.99999)$
$$= 1.0 \times 10^{-5}\ M$$

Using the 2nd order integrated rate law:

$$kt = \frac{1}{[H_3O^+]_t} - \frac{1}{[H_3O^+]_o}; \qquad t = \frac{1}{k}\left[\frac{1}{[H_3O^+]_t} - \frac{1}{[H_3O^+]_o}\right]$$

$$t = \frac{1}{(1.3 \times 10^{11}\ M^{-1}s^{-1})}\left[\frac{1}{(1.0 \times 10^{-5}\ M)} - \frac{1}{(1.0\ M)}\right] = 7.7 \times 10^{-7}\ s$$

(b) The rate of an acid-base neutralization reaction would be limited by the speed of mixing, which is much slower than the intrinsic rate of the reaction itself.

14.132   Looking at the two experiments at 600 K, when the NO_2 concentration is doubled, the rate increased by a factor of 4. Therefore, the reaction is 2nd order.

Rate = k $[NO_2]^2$
Calculate k_1 at 600 K: k_1 = Rate/$[NO_2]^2$ = 5.4 x 10^{-7} M s^{-1}/$(0.0010 M)^2$ = 0.54 $M^{-1} s^{-1}$
Calculate k_2 at 700 K: k_2 = Rate/$[NO_2]^2$ = 5.2 x 10^{-5} M s^{-1}/$(0.0020 M)^2$ = 13 $M^{-1} s^{-1}$
Calculate E_a using k_1 and k_2.

$$\ln\left(\frac{k_2}{k_1}\right) = \left(\frac{-E_a}{R}\right)\left(\frac{1}{T_2} - \frac{1}{T_1}\right)$$

$$E_a = -\frac{[\ln k_2 - \ln k_1](R)}{\left(\dfrac{1}{T_2} - \dfrac{1}{T_1}\right)}$$

$$E_a = -\frac{[\ln(13) - \ln(0.54)][8.314 \text{ x } 10^{-3} \text{ kJ/(K}\cdot\text{mol)}]}{\left(\dfrac{1}{700 \text{ K}} - \dfrac{1}{600 \text{ K}}\right)} = 111 \text{ kJ/mol}$$

Calculate k_3 at 650 K using E_a and k_1.
Solve for k_3.

$$\ln k_3 = \frac{-E_a}{R}\left(\frac{1}{T_3} - \frac{1}{T_1}\right) + \ln k_1$$

$$\ln k_3 = \frac{-111 \text{ kJ/mol}}{[8.314 \text{ x } 10^{-3} \text{ kJ/(K}\cdot\text{mol)}]}\left(\frac{1}{650 \text{ K}} - \frac{1}{600 \text{ K}}\right) + \ln(0.54) = 1.0955$$

$k_3 = e^{1.0955} = 3.0 \text{ M}^{-1} \text{ s}^{-1}$

$$k_3 t = \frac{1}{[NO_2]_t} - \frac{1}{[NO_2]_o}; \qquad t = \frac{1}{k_3}\left[\frac{1}{[NO_2]_t} - \frac{1}{[NO_2]_o}\right]$$

$$t = \frac{1}{(3.0 \text{ M}^{-1}\text{s}^{-1})}\left[\frac{1}{(0.0010 \text{ M})} - \frac{1}{(0.0050 \text{ M})}\right] = 2.7 \text{ x } 10^2 \text{ s}$$

14.134   Rate = k $[HI]^x$

$$\frac{\text{Rate}_2}{\text{Rate}_1} = \frac{k[0.30]^x}{k[0.10]^x}$$

$$x = \frac{\log\dfrac{\text{Rate}_2}{\text{Rate}_1}}{\log\dfrac{(0.30)}{(0.10)}} = \frac{\log\dfrac{1.6 \text{ x } 10^{-4}}{1.8 \text{ x } 10^{-5}}}{\log\dfrac{(0.30)}{(0.10)}} = \frac{0.949}{0.477} = 2$$

Rate = k $[HI]^2$

At 700 K, k_1 = $\dfrac{\text{Rate}}{[HI]^2}$ = $\dfrac{1.8 \text{ x } 10^{-5} \text{ M/s}}{(0.10 \text{ M})^2}$ = 1.8 x 10^{-3} M^{-1} s^{-1}

At 800 K, $k_2 = \dfrac{\text{Rate}}{[\text{HI}]^2} = \dfrac{3.9 \times 10^{-3}\ \text{M/s}}{(0.20\ \text{M})^2} = 9.7 \times 10^{-2}\ \text{M}^{-1}\ \text{s}^{-1}$

$$\ln\left(\frac{k_2}{k_1}\right) = \left(\frac{-E_a}{R}\right)\left(\frac{1}{T_2} - \frac{1}{T_1}\right)$$

$$E_a = -\frac{[\ln k_2 - \ln k_1](R)}{\left(\dfrac{1}{T_2} - \dfrac{1}{T_1}\right)}$$

$$E_a = -\frac{[\ln(9.7 \times 10^{-2}) - \ln(1.8 \times 10^{-3})][8.314 \times 10^{-3}\ \text{kJ/(K}\cdot\text{mol)}]}{\left(\dfrac{1}{800\ \text{K}} - \dfrac{1}{700\ \text{K}}\right)} = 186\ \text{kJ/mol}$$

Calculate k_4 at 650 K using E_a and k_1.
Solve for k_4.

$$\ln k_4 = \frac{-E_a}{R}\left(\frac{1}{T_4} - \frac{1}{T_1}\right) + \ln k_1$$

$$\ln k_4 = \frac{-186\ \text{kJ/mol}}{[8.314 \times 10^{-3}\ \text{kJ/(K}\cdot\text{mol)}]}\left(\frac{1}{650\ \text{K}} - \frac{1}{700\ \text{K}}\right) + \ln(1.8 \times 10^{-3}) = -8.778$$

$k_4 = e^{-8.778} = 1.5 \times 10^{-4}\ \text{M}^{-1}\ \text{s}^{-1}$

$$[\text{HI}] = \sqrt{\frac{\text{Rate}}{k_4}} = \sqrt{\frac{1.0 \times 10^{-5}\ \text{M/s}}{1.5 \times 10^{-4}\ \text{M}^{-1}\text{s}^{-1}}} = 0.26\ \text{M}$$

14.136   A $\rightarrow$ C   is a first-order reaction.
The reaction is complete at 200 s when the absorbance of C reaches 1.200.
Because there is a one to one stoichiometry between A and C, the concentration of A must be proportional to 1.200 – absorbance of C. Any two data points can be used to find k. Let $[A]_o \propto 1.200$ and at 100 s, $[A]_t \propto 1.200 - 1.188 = 0.012$

$$\ln\frac{[A]_t}{[A]_o} = -kt; \qquad k = \frac{\ln\dfrac{[A]_t}{[A]_o}}{-t}; \qquad k = \frac{\ln\left(\dfrac{0.012\ \text{M}}{1.200\ \text{M}}\right)}{-100\ \text{s}} = 0.0461\ \text{s}^{-1}$$

14.138   2 HI(g) $\rightarrow$ H$_2$(g) + I$_2$(g)

(a) mass HI $= 1.50\ \text{L} \times \dfrac{1000\ \text{mL}}{1\ \text{L}} \times \dfrac{0.0101\ \text{g}}{1\ \text{mL}} = 15.15\ \text{g HI}$

15.15 g HI $\times \dfrac{1\ \text{mol HI}}{127.91\ \text{g HI}} = 0.118\ \text{mol HI}$

$[\text{HI}] = \dfrac{0.118\ \text{mol}}{1.50\ \text{L}} = 0.0787\ \text{mol/L}$

$$-\frac{\Delta[HI]}{\Delta t} = k[HI]^2 = (0.031/(M \cdot min))(0.0787\ M)^2 = 1.92\ x\ 10^{-4}\ M/min$$

$$2\ HI(g) \ \rightarrow \ H_2(g) \ + \ I_2(g)$$

$$\frac{\Delta[I_2]}{\Delta t} = \frac{1}{2}\left(-\frac{\Delta[HI]}{\Delta t}\right) = \frac{1.92\ x\ 10^{-4}\ M/min}{2} = 9.60\ x\ 10^{-5}\ M/min$$

$$(9.60\ x\ 10^{-5}\ M/min)(1.50\ L)(6.022\ x\ 10^{23}\ molecules/mol) = 8.7\ x\ 10^{19}\ molecules/min$$

(b) Rate = $k[HI]^2$

$$\frac{1}{[HI]_t} = kt + \frac{1}{[HI]_o} = (0.031/(M \cdot min))\left(8.00\ h\ x\ \frac{60.0\ min}{1\ h}\right) + \frac{1}{0.0787\ M} = 27.59/M$$

$$[HI]_t = \frac{1}{27.59/M} = 0.0362\ M$$

From stoichiometry, $[H_2]_t = 1/2\ ([HI]_o - [HI]_t) = 1/2\ (0.0787\ M - 0.0362\ M) = 0.0212\ M$

410 °C = 683 K
PV = nRT

$$P_{H_2} = \left(\frac{n}{V}\right)RT = (0.0212\ mol/L)\left(0.082\ 06\ \frac{L \cdot atm}{K \cdot mol}\right)(683\ K) = 1.2\ atm$$

14.140   (a) N_2O_5, 108.01

$$[N_2O_5]_o = \frac{\left(2.70\ g\ N_2O_5\ x\ \dfrac{1\ mol\ N_2O_5}{108.01\ g\ N_2O_5}\right)}{2.00\ L} = 0.0125\ mol/L$$

$$\ln[N_2O_5]_t = -kt + \ln[N_2O_5]_o = -(1.7\ x\ 10^{-3}\ s^{-1})\left(13.0\ min\ x\ \frac{60.0\ s}{1\ min}\right) + \ln(0.0125) = -5.71$$

$[N_2O_5]_t = e^{-5.71} = 3.31\ x\ 10^{-3}\ mol/L$
After 13.0 min, mol $N_2O_5 = (3.31\ x\ 10^{-3}\ mol/L)(2.00\ L) = 6.62\ x\ 10^{-3}\ mol\ N_2O_5$

	$N_2O_5(g)$	$\rightarrow$	$2\ NO_2(g)$	+	$1/2\ O_2(g)$
before reaction (mol)	0.0250		0		0
change (mol)	−x		+2x		+1/2x
after reaction (mol)	0.0250 − x		2x		1/2x

After 13.0 min, mol $N_2O_5 = 6.62\ x\ 10^{-3} = 0.0250 - x$
x = 0.0184 mol

After 13.0 min, $n_{total} = n_{N_2O_5} + n_{NO_2} + n_{O_2} = (6.62\ x\ 10^{-3}) + 2(0.0184) + 1/2(0.0184)$

$n_{total} = 0.0526\ mol$
55 °C = 328 K

$$PV = nRT; \quad P_{total} = \frac{nRT}{V} = \frac{(0.0526 \text{ mol})\left(0.082\ 06 \dfrac{L \cdot atm}{K \cdot mol}\right)(328 \text{ K})}{2.00 \text{ L}} = 0.71 \text{ atm}$$

(b) $N_2O_5(g) \rightarrow 2 NO_2(g) + 1/2 O_2(g)$
$\Delta H°_{rxn} = 2 \Delta H°_f(NO_2) - \Delta H°_f(N_2O_5)$
$\Delta H°_{rxn} = (2 \text{ mol})(33.2 \text{ kJ/mol}) - (1 \text{ mol})(11 \text{ kJ/mol}) = 55.4 \text{ kJ} = 5.54 \times 10^4 \text{ J}$
initial rate $= k[N_2O_5]_o = (1.7 \times 10^{-3} \text{ s}^{-1})(0.0125 \text{ mol/L}) = 2.125 \times 10^{-5} \text{ mol/(L} \cdot \text{s})$
initial rate absorbing heat $= [2.125 \times 10^{-5} \text{ mol/(L} \cdot \text{s})](2.00 \text{ L})(5.54 \times 10^4 \text{ J/mol}) = 2.4 \text{ J/s}$

(c)

$$\ln [N_2O_5]_t = -kt + \ln [N_2O_5]_o = -(1.7 \times 10^{-3} \text{ s}^{-1})\left(10.0 \text{ min} \times \frac{60.0 \text{ s}}{1 \text{ min}}\right) + \ln (0.0125) = -5.40$$

$[N_2O_5]_t = e^{-5.40} = 4.52 \times 10^{-3} \text{ mol/L}$
After 10.0 min, mol $N_2O_5 = (4.52 \times 10^{-3} \text{ mol/L})(2.00 \text{ L}) = 9.03 \times 10^{-3} \text{ mol } N_2O_5$

	$N_2O_5(g)$	$\rightarrow$	$2 NO_2(g)$	$+$	$1/2 O_2(g)$
before reaction (mol)	0.0250		0		0
change (mol)	$-x$		$+2x$		$+1/2x$
after reaction (mol)	$0.0250 - x$		$2x$		$1/2x$

After 10.0 min, mol $N_2O_5 = 9.03 \times 10^{-3} = 0.0250 - x$
$x = 0.0160 \text{ mol}$
heat absorbed $= (0.0160 \text{ mol})(55.4 \text{ kJ/mol}) = 0.89 \text{ kJ}$

14.142 H_2O_2, 34.01
mass $H_2O_2 = (0.500 \text{ L})(1000 \text{ mL/1 L})(1.00 \text{ g/ 1 mL})(0.0300) = 15.0 \text{ g } H_2O_2$

$$\text{mol } H_2O_2 = 15.0 \text{ g } H_2O_2 \times \frac{1 \text{ mol } H_2O_2}{34.01 \text{ g } H_2O_2} = 0.441 \ H_2O_2$$

$$[H_2O_2]_o = \frac{0.441 \text{ mol}}{0.500 \text{ L}} = 0.882 \text{ mol /L}$$

$$k = \frac{0.693}{t_{1/2}} = \frac{0.693}{10.7 \text{ h}} = 6.48 \times 10^{-2}/\text{h}$$

$\ln [H_2O_2]_t = -kt + \ln [H_2O_2]_o$
$\ln [H_2O_2]_t = -(6.48 \times 10^{-2}/\text{h})(4.02 \text{ h}) + \ln (0.882)$
$\ln [H_2O_2]_t = -0.386; \quad [H_2O_2]_t = e^{-0.386} = 0.680 \text{ mol/L}$
mol $H_2O_2 = (0.680 \text{ mol/L})(0.500 \text{ L}) = 0.340 \text{ mol}$

	$2 H_2O_2(aq)$	$\rightarrow$	$2 H_2O(l)$	$+$	$O_2(g)$
before reaction (mol)	0.441		0		0
change (mol)	$- 2x$		$+2x$		$+x$
after reaction (mol)	$0.441 - 2x$		$2x$		x

After 4.02 h, mol $H_2O_2 = 0.340 \text{ mol} = 0.441 - 2x$; solve for x.

$2x = 0.101$

$x = 0.0505 \text{ mol} = \text{mol } O_2$

$P = 738 \text{ mm Hg} \times \dfrac{1.00 \text{ atm}}{760 \text{ mm Hg}} = 0.971 \text{ atm}$

$PV = nRT$

$V = \dfrac{nRT}{P} = \dfrac{(0.0505 \text{ mol})\left(0.082\ 06\ \dfrac{L \cdot atm}{K \cdot mol}\right)(293 \text{ K})}{0.971 \text{ atm}} = 1.25 \text{ L}$

$P\Delta V = (0.971 \text{ atm})(1.25 \text{ L}) = 1.21 \text{ L} \cdot \text{atm}$

$w = -P\Delta V = -1.21 \text{ L} \cdot \text{atm} = (-1.21 \text{ L} \cdot \text{atm})\left(101\ \dfrac{J}{L \cdot atm}\right) = -122 \text{ J}$

15 Chemical Equilibrium

15.1 $K_c = \dfrac{[SO_3]^2}{[SO_2]^2[O_2]}$; $\quad K_c' = \dfrac{[SO_2]^2[O_2]}{[SO_3]^2}$; $\quad K_c' = \dfrac{1}{K_c}$

15.2 (a) $K_c(\text{overall}) = \dfrac{[NO_2]^2}{[N_2][O_2]^2}$

(b) $K_c(\text{overall}) = K_{c1} \times K_{c2} = (4.3 \times 10^{-25})(6.4 \times 10^9) = 2.8 \times 10^{-15}$

15.3 $K_c = \dfrac{[SO_3]^2}{[SO_2]^2[O_2]} = \dfrac{(5.0 \times 10^{-2})^2}{(3.0 \times 10^{-3})^2(3.5 \times 10^{-3})} = 7.9 \times 10^4$

$K_c' = \dfrac{1}{K_c} = \dfrac{[SO_2]^2[O_2]}{[SO_3]^2} = \dfrac{(3.0 \times 10^{-3})^2(3.5 \times 10^{-3})}{(5.0 \times 10^{-2})^2} = 1.3 \times 10^{-5}$

15.4 (a) $K_c = \dfrac{[H^+][C_3H_5O_3^-]}{[C_3H_6O_3]} = \dfrac{(3.65 \times 10^{-3})(3.65 \times 10^{-3})}{(9.64 \times 10^{-2})} = 1.38 \times 10^{-4}$

(b) $[C3H6O3] = \dfrac{[H^+][C_3H_5O_3^-]}{K_c} = \dfrac{(1.17 \times 10^{-2})(1.17 \times 10^{-2})}{(1.38 \times 10^{-4})} = 0.992\ M$

15.5 From (1), $K_c = \dfrac{[AB][B]}{[A][B_2]} = \dfrac{(1)(2)}{(1)(2)} = 1$

For a mixture to be at equilibrium, $\dfrac{[AB][B]}{[A][B_2]}$ must be equal to 1.

For (2), $\dfrac{[AB][B]}{[A][B_2]} = \dfrac{(2)(1)}{(2)(1)} = 1$. This mixture is at equilibrium.

For (3), $\dfrac{[AB][B]}{[A][B_2]} = \dfrac{(1)(1)}{(4)(2)} = 0.125$. This mixture is not at equilibrium.

For (4), $\dfrac{[AB][B]}{[A][B_2]} = \dfrac{(2)(2)}{(4)(1)} = 1$. This mixture is at equilibrium.

15.6

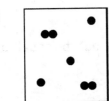

15.7 $K_p = \dfrac{(P_{CO_2})(P_{H_2})}{(P_{CO})(P_{H_2O})} = \dfrac{(6.12)(20.3)}{(1.31)(10.0)} = 9.48$

15.8 $K_p = 25 = \dfrac{(P_{HI})^2}{(P_{H_2})(P_{I_2})}$

$P_{HI} = \sqrt{(25)(P_{H_2})(P_{I_2})} = \sqrt{(25)(0.286)(0.286)} = 1.43$ atm

15.9 $2\,NO(g) + O_2(g) \rightleftharpoons 2\,NO_2(g);\qquad \Delta n = 2 - 3 = -1$
$K_p = K_c(RT)^{\Delta n},\qquad K_c = K_p(1/RT)^{\Delta n}$
at 500 K: $\quad K_p = (6.9 \times 10^5)[(0.082\,06)(500)]^{-1} = 1.7 \times 10^4$

15.10 $K_c' = \dfrac{1}{K_c} = \dfrac{1}{245}$

$2\,SO_3(g) \rightleftharpoons 2\,SO_2(g) + O_2(g);\quad \Delta n = 3 - 2 = 1$
$K_p = K_c'(RT)^{\Delta n}$
$K_p = (1/245)[(0.082\,06)(1,000)] = 0.335$

15.11 $K_p = \dfrac{(P_{HCl})^4}{(P_{SiCl_4})(P_{H_2})^2}$

15.12 $Mg(OH)_2(s) \rightleftharpoons Mg^{2+}(aq) + 2\,OH^-(aq)$
$K_c = [Mg^{2+}][OH^-]^2 = (1.65 \times 10^{-4})(3.30 \times 10^{-4})^2 = 1.80 \times 10^{-11}$

15.13 $K_c = 1.2 \times 10^{82}$ is very large. When equilibrium is reached, very little if any ethanol will remain because the reaction goes to completion.

15.14 $K_c = 1.2 \times 10^{-42}$. Because K_c is very small, the equilibrium mixture contains mostly H_2 molecules. H is in periodic group 1A. A very small value of K_c is consistent with strong bonding between 2 H atoms, each with one valence electron.

15.15 The container volume of 5.0 L must be included to calculate molar concentrations.

$Q_c = \dfrac{[NO_2]_t^2}{[NO]_t^2[O_2]_t} = \dfrac{(0.80\text{ mol}/5.0\text{ L})^2}{(0.060\text{ mol}/5.0\text{ L})^2(1.0\text{ mol}/5.0\text{ L})} = 890$

Because $Q_c < K_c$, the reaction is not at equilibrium. The reaction will proceed to the right to reach equilibrium.

15.16 $K_c = \dfrac{[AB]^2}{[A_2][B_2]} = 4$; For a mixture to be at equilibrium, $\dfrac{[AB]^2}{[A_2][B_2]}$ must be equal to 4.

For (1), $Q_c = \dfrac{[AB]^2}{[A_2][B_2]} = \dfrac{(6)^2}{(1)(1)} = 36$, $Q_c > K_c$

For (2), $Q_c = \dfrac{[AB]^2}{[A_2][B_2]} = \dfrac{(4)^2}{(2)(2)} = 4$, $Q_c = K_c$

For (3), $Q_c = \dfrac{[AB]^2}{[A_2][B_2]} = \dfrac{(2)^2}{(3)(3)} = 0.44$, $Q_c < K_c$

(a) (2)   (b) (1), reverse; (3), forward

15.17

	$CO(g)$	+	$H_2O(g)$	$\rightleftharpoons$	$CO_2(g)$	+	$H_2(g)$
initial (M)	0.150		0.150		0		0
change (M)	$-x$		$-x$		$+x$		$+x$
equil (M)	$0.150 - x$		$0.150 - x$		x		x

$K_c = 4.24 = \dfrac{[CO_2][H_2]}{[CO][H_2O]} = \dfrac{x^2}{(0.150 - x)^2}$

Take the square root of both sides and solve for x.

$\sqrt{4.24} = \sqrt{\dfrac{x^2}{(0.150 - x)^2}}$; $2.06 = \dfrac{x}{0.150 - x}$; $x = 0.101$

At equilibrium, $[CO_2] = [H_2] = x = 0.101$ M
$[CO] = [H_2O] = 0.150 - x = 0.150 - 0.101 = 0.049$ M

15.18

	$CO(g)$	+	$H_2O(g)$	$\rightleftharpoons$	$CO_2(g)$	+	$H_2(g)$
initial (M)	0.100		0.100		0.100		0.100
change (M)	$-x$		$-x$		$+x$		$+x$
equil (M)	$0.100 - x$		$0.100 - x$		$0.100 + x$		$0.100 + x$

$K_c = 4.24 = \dfrac{[CO_2][H_2]}{[CO][H_2O]} = \dfrac{(0.100 + x)^2}{(0.100 - x)^2}$

Take the square root of both sides and solve for x.

$\sqrt{4.24} = \sqrt{\dfrac{(0.100 + x)^2}{(0.100 - x)^2}}$; $2.06 = \dfrac{0.100 + x}{0.100 - x}$; $x = 0.035$

At equilibrium, $[CO_2] = [H_2] = 0.100 + x = 0.100 + 0.035 = 0.135$ M
$[CO] = [H_2O] = 0.100 - x = 0.100 - 0.035 = 0.065$ M

15.19

	$H_2(g)$	+	$I_2(g)$	$\rightleftharpoons$	$2\,HI(g)$
initial (M)	0.100		0.300		0
change (M)	$-x$		$-x$		$+2x$
equil (M)	$0.100 - x$		$0.300 - x$		$2x$

$K_c = \dfrac{[HI]^2}{[H_2][I_2]} = \dfrac{(2x)^2}{(0.100 - x)(0.300 - x)} = 57.0$

$53x^2 - 22.8x + 1.71 = 0$

Use the quadratic formula to solve for x.

$$x = \frac{-(-22.8) \pm \sqrt{(-22.8)^2 - 4(53)(1.71)}}{2(53)} = \frac{22.8 \pm 12.54}{106}$$

$x = 0.333$ and 0.0968

Discard 0.333 because it leads to negative concentrations and that is impossible.

$[H_2] = 0.100 - x = 0.100 - 0.0968 = 0.0032$ M

$[I_2] = 0.300 - x = 0.300 - 0.0968 = 0.2032$

$[HI] = 2x = (2)(0.0968) = 0.1936$ M

15.20

	$N_2O_4(g)$	$\rightleftarrows$	$2\ NO_2(g)$
initial (M)	0.500		0
change (M)	$-x$		$+2x$
equil (M)	$0.500 - x$		$2x$

$$K_c = \frac{[NO_2]^2}{[N_2O_4]} = \frac{(2x)^2}{(0.500 - x)} = 4.64 \times 10^{-3}$$

$4x^2 + (4.64 \times 10^{-3})x - (2.32 \times 10^{-3}) = 0$

Use the quadratic formula to solve for x.

$$x = \frac{-(4.64 \times 10^{-3}) \pm \sqrt{(4.64 \times 10^{-3})^2 - 4(4)(-2.32 \times 10^{-3})}}{2(4)} = \frac{-0.004\ 64 \pm 0.1927}{8}$$

$x = -0.0247$ and 0.0235

Discard the negative solution (-0.0247) because it leads to a negative concentration of NO_2 and that is impossible.

$[N_2O_4] = 0.500 - x = 0.500 - 0.0235 = 0.476$ M

$[NO_2] = 2x = 2(0.0235) = 0.0470$ M

15.21

	$CO_2(g)$	+	$H_2(g)$	$\rightleftarrows$	$CO(g)$	+	$H_2O(g)$
initial (M)	0.750		0.550		0		0
change (M)	$-x$		$-x$		$+x$		$+x$
equil (M)	$0.750 - x$		$0.550 - x$		x		x

Because K_c is very small, $-x$ can be neglected in the next equation.

$$K_c = \frac{[CO][H_2O]}{[CO_2][H_2]} = \frac{x^2}{(0.750 - x)(0.550 - x)} \cong \frac{x^2}{(0.750)(0.550)} = 9.71 \times 10^{-9}$$

$x^2 = (0.750)(0.550)(9.71 \times 10^{-9}) = 4.005 \times 10^{-9}$

$x = \sqrt{4.005 \times 10^{-9}} = 6.33 \times 10^{-5}$

$[CO_2] = 0.750$ M; $\quad [H_2] = 0.550$ M; $\quad [CO] = [H_2O] = x = 6.33 \times 10^{-5}$ M

15.22

	$H_2O(g)$	+	$Cl_2(g)$	$\rightleftarrows$	$2\ HCl(g)$	+	$O_2(g)$
initial (M)	0.130		0.250		0		0
change (M)	$-x$		$-x$		$+2x$		$+x$
equil (M)	$0.130 - x$		$0.250 - x$		$2x$		x

Because K_c is very small, $-x$ can be neglected in the next equation.

$$K_c = \frac{[HCl]^2[O_2]}{[H_2O][Cl_2]} = \frac{((2x)^2)x}{(0.130-x)(0.250-x)} \cong \frac{4x^3}{(0.130)(0.250)} = 8.96 \times 10^{-9}$$

$x^3 = (0.130)(0.250)(8.96 \times 10^{-9})/4 = 7.280 \times 10^{-11}$

$x = \sqrt[3]{7.280 \times 10^{-11}} = 4.18 \times 10^{-4}$

$[H_2O] = 0.130$ M; $[Cl_2] = 0.250$ M

$[HCl] = 2x = 2(4.18 \times 10^{-4}) = 8.36 \times 10^{-4}$ M

$[O_2] = x = 4.18 \times 10^{-4}$ M

15.23

	C(s)	+	H$_2$O(g)	⇌	CO(g)	+	H$_2$(g)
initial (atm)			0		1.00		1.00
change (atm)			+x		−x		−x
equil (atm)			x		1.00 − x		1.00 − x

$$K_p = \frac{(P_{CO})(P_{H_2})}{(P_{H_2O})} = 2.44 = \frac{(1.00-x)(1.00-x)}{x}$$

$x^2 - 4.44x + 1.00 = 0$

Use the quadratic formula to solve for x.

$$x = \frac{-(-4.44) \pm \sqrt{(-4.44)^2 - 4(1)(1.00)}}{2(1)} = \frac{4.44 \pm 3.96}{2}$$

x = 4.20 and 0.238

Discard 4.20 because it leads to negative partial pressures and that is impossible.

$P_{H_2O} = x = 0.238$ atm

$P_{CO} = P_{H_2} = 1.00 - x = 1.00 - 0.238 = 0.76$ atm

15.24 $K_p = \dfrac{(P_{CO})(P_{H_2})}{(P_{H_2O})} = 2.44$, $Q_p = \dfrac{(1.00)(1.40)}{(1.20)} = 1.17$, $Q_p < K_p$ and the reaction goes to

the right to reach equilibrium.

	C(s)	+	H$_2$O(g)	⇌	CO(g)	+	H$_2$(g)
initial (atm)			1.20		1.00		1.40
change (atm)			−x		+x		+x
equil (atm)			1.20 − x		1.00 + x		1.40 + x

$$K_p = \frac{(P_{CO})(P_{H_2})}{(P_{H_2O})} = 2.44 = \frac{(1.00+x)(1.40+x)}{(1.20-x)}$$

$x^2 + 4.84x - 1.53 = 0$

Use the quadratic formula to solve for x.

$$x = \frac{-(4.84) \pm \sqrt{(4.84)^2 - 4(1)(-1.53)}}{2(1)} = \frac{-4.84 \pm 5.44}{2}$$

x = −5.14 and 0.300

Discard the negative solution (−5.14) because it leads to negative partial pressures and that is impossible.

$$P_{H_2O} = 1.20 - x = 1.20 - 0.300 = 0.90 \text{ atm}$$

$$P_{CO} = 1.00 + x = 1.00 + 0.300 = 1.30 \text{ atm}$$

$$P_{H_2} = 1.40 + x = 1.40 + 0.300 = 1.70 \text{ atm}$$

15.25 (a) CO (reactant) added, H_2 concentration increases.
(b) CO_2 (product) added, H_2 concentration decreases.
(c) H_2O (reactant) removed, H_2 concentration decreases.
(d) CO_2 (product) removed, H_2 concentration increases.
(e) CO (reactant) added and CO_2 (product) removed, H_2 concentration increases.

(a), (d) and (e) will increase the H_2 concentration.

15.26 (a) The concentration of both Ca^{2+} and $C_2O_4^{2-}$ will increase as fluid is lost and the reaction will proceed toward products.
(b) Increasing the concentration of Ca^{2+} will cause the reaction to proceed toward products.
(c) Decreasing the concentration of $C_2O_4^{2-}$ will cause the reaction to proceed toward reactants and the kidney stone will dissolve.
(d) The concentration of both Ca^{2+} and $C_2O_4^{2-}$ will decrease causing the reaction proceed toward reactants and the kidney stone will dissolve.

15.27 Because there are 2 mol of gas on the left side and 1 mol of gas on the right side of the balanced equation, the stress of an increase in pressure is relieved by a shift in the reaction to the side with fewer moles of gas (in this case, to products). The number of moles of reaction products increases.

15.28 (a)

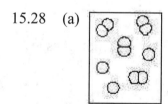

(b) If there is no change in temperature, K_c remains the same.

15.29 Le Châtelier's principle predicts that a stress of added heat will be relieved by net reaction in the direction that absorbs the heat. Since the reaction is endothermic, the equilibrium will shift from left to right (K_c will increase) with an increase in temperature. Therefore, the equilibrium mixture will contain more of the offending NO, the higher the temperature.

15.30 The reaction is exothermic. As the temperature is increased the reaction shifts from right to left. The amount of ethyl acetate decreases.

$$K_c = \frac{[CH_3CO_2C_2H_5][H_2O]}{[CH_3CO_2H][C_2H_5OH]}$$

As the temperature is decreased, the reaction shifts from left to right. The product concentrations increase, and the reactant concentrations decrease. This corresponds to an increase in K_c.

15.31   There are more AB(g) molecules at the higher temperature. The equilibrium shifted to the right at the higher temperature, which means the reaction is endothermic.

15.32   Heat + $BaCO_3(s)$ ⇌ $BaO(s)$ + $CO_2(g)$

(a) 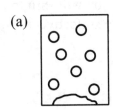   (b)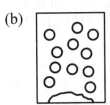

15.33   (a) Because K_c is so large, k_f is larger than k_r.

(b) $K_c = \dfrac{k_f}{k_r}$; $k_r = \dfrac{k_f}{K_c} = \dfrac{8.5 \times 10^6 \, M^{-1} s^{-1}}{3.4 \times 10^{34}} = 2.5 \times 10^{-28} \, M^{-1} \, s^{-1}$

15.34   (a) If A is similar for forward and reverse reactions, then k_r is larger because the activation energy for the reverse reaction is lower.

(b) Because $K_c = \dfrac{k_f}{k_r}$ and $k_r > k_f$, then $K_c < 1$.

(c) The rates of forward and reverse reactions both decrease. However, the rate of the forward reaction has a larger decrease than the rate of reverse reaction. K_c decreases when temperature is lowered in a endothermic process.

15.35   (a) 75%   (b) 20%
(c) An efficient unloading of oxygen occurs in working muscle tissue when the partial pressure of oxygen drops.

15.36   An increase in temperature will shift the position of the equilibrium toward reactants, releasing oxygen from hemoglobin and making it available for use by the muscles.

15.37   (a) Equilibrium constants for each sequential step in binding oxygen become larger.
(b) Each step becomes more product favored, meaning that when oxygen binds it increases the affinity of the remaining heme groups for other oxygen molecules.
(c) The oxygen carrying capacity is greatly increased.

15.38   (a) $K_c = K_{c1} \times K_{c2} \times K_{c3} \times K_{c4}$
$K_c = (1.5 \times 10^4)(3.5 \times 10^4)(5.9 \times 10^4)(1.7 \times 10^6) = 5.3 \times 10^{19}$
(b) Hb + O_2 ⇌ $Hb(O_2)_4$
$[O_2] = (1.61 \, \mu M/mm \, Hg)(95 \, mm \, Hg) = 153 \, \mu M = 153 \times 10^{-6} \, M$

$$K_c = \frac{[Hb(O_2)_4]}{[Hb][O_2]^4} = 5.3 \times 10^{19}$$

$$\frac{[Hb(O_2)_4]}{[Hb]} = K_c[O_2]^4 = (5.3 \times 10^{19})(153 \times 10^{-6})^4 = 2.9 \times 10^4$$

15.39 $Hb + O_2 \rightleftharpoons Hb(O_2)$

If CO binds to Hb, Hb is removed from the reaction and the reaction will shift to the left, resulting in O_2 being released from $Hb(O_2)$. This will decrease the effectiveness of Hb for carrying O_2.

15.40 $Hb(O_2)(aq) + CO(g) \rightleftharpoons Hb(CO)(aq) + O_2(g)$ $K = 207$

(a) $P_{O_2} = 0.20$ atm and $P_{CO} = 0.0015$ atm

$$K = \frac{[Hb(CO)](P_{O_2})}{[Hb(O_2)](P_{CO})} = 207$$

$$\frac{[Hb(CO)]}{[Hb(O_2)]} = \frac{(P_{CO})(207)}{(P_{O_2})} = \frac{(0.0015)(207)}{(0.20)} = 1.6$$

(b) An increase in the concentration of oxygen in the blood will shift the equilibrium toward reactants increasing the concentration of oxygen bound to hemoglobin.

Conceptual Problems

15.42   (a) $A_2 + C_2 \rightleftharpoons 2 AC$ (most product molecules)

        (b) $A_2 + B_2 \rightleftharpoons 2 AB$ (fewest product molecules)

15.44   (a) $A_2 + 2 B \rightleftharpoons 2 AB$

        (b) The number of AB molecules will increase, because as the volume is decreased at constant temperature, the pressure will increase and the reaction will shift to the side of fewer molecules to reduce the pressure.

15.46   As the temperature is raised, the reaction proceeds in the reverse direction. This is consistent with an exothermic reaction where "heat" can be considered as a product.

15.48   (a)   (b)   (c)

15.50   (a) $A \rightarrow 2 B$

        (b) (1) The reaction is exothermic. As the temperature is increased, the reaction shifts to the left. A increases, B decreases, and K_c decreases.

(2) When the volume decreases, the reaction shifts to the side with fewer gas molecules, which is towards the reactant. The amount of A increases.

(3) If there is no volume change, there is no change in the equilibrium composition, and the amount of A remains the same.

(4) A catalyst does not change the equilibrium composition, and the amount of A remains the same.

Section Problems
The Equilibrium State (Section 15.1)

15.52   (d) The rate of the forward reaction is equal to the rate of the reverse reaction.

Equilibrium Constant Expressions and Equilibrium Constants (Sections 15.2–15.4)

15.54 $K_c = \dfrac{[C_2H_5OC_2H_5][H_2O]}{[C_2H_5OH]^2}$

15.56   (a) $K_c = \dfrac{[CO][H_2]^3}{[CH_4][H_2O]}$ (b) $K_c = \dfrac{[ClF_3]^2}{[F_2]^3[Cl_2]}$ (c) $K_c = \dfrac{[HF]^2}{[H_2][F_2]}$

15.58   (a) $K_p = \dfrac{(P_{CO})(P_{H_2})^3}{(P_{CH_4})(P_{H_2O})}$, $\Delta n = 2$ and $K_p = K_c(RT)^2$

(b) $K_p = \dfrac{(P_{ClF_3})^2}{(P_{F_2})^3(P_{Cl_2})}$, $\Delta n = -2$ and $K_p = K_c(RT)^{-2}$

(c) $K_p = \dfrac{(P_{HF})^2}{(P_{H_2})(P_{F_2})}$, $\Delta n = 0$ and $K_p = K_c$

15.60   (a) This reaction is the reverse of the original reaction.

$K_c' = \dfrac{1}{K_c} = \dfrac{1}{7.5 \times 10^{-9}} = 1.3 \times 10^8$

(b) This reaction is the reverse of the original reaction multiplied by ½.

$K_c' = \sqrt{\dfrac{1}{K_c}} = \sqrt{\dfrac{1}{7.5 \times 10^{-9}}} = 1.2 \times 10^4$

(c) This reaction is the original reaction multiplied by 2.

$K_c' = (K_c)^2 = (7.5 \times 10^{-9})^2 = 5.6 \times 10^{-17}$

15.62 $K_p(1) = 7.2 \times 10^7$

(a) $K_p(2) = \dfrac{1}{K_p(1)} = \dfrac{1}{7.2 \times 10^7} = 1.4 \times 10^{-8}$

(b) $K_p(3) = (K_p(1))^2 = (7.2 \times 10^7)^2 = 5.2 \times 10^{15}$

(c) $K_p(4) = \dfrac{1}{(K_p(1))^3} = \dfrac{1}{(7.2 \times 10^7)^3} = 2.7 \times 10^{-24}$

15.64 $2\,Na(l) + O_2(g) \;\rightleftharpoons\; Na_2O_2(s)$ $K_c' = 1/K_c = 1/(5 \times 10^{-29})$

         $\underline{Na_2O(s) \;\rightleftharpoons\; 2\,Na(l) + \tfrac{1}{2}\,O_2(g)}$ $K_c = 2 \times 10^{-25}$

   Overall $Na_2O(s) + \tfrac{1}{2}\,O_2(g) \;\rightleftharpoons\; Na_2O_2(s)$

$$K_c(\text{overall}) = K_c' \times K_c = \frac{2 \times 10^{-25}}{5 \times 10^{-29}} = 4 \times 10^3$$

15.66 $K_c = \dfrac{[PCl_3][Cl_2]}{[PCl_5]} = \dfrac{(1.5 \times 10^{-2})(3.2 \times 10^{-2})}{(8.3 \times 10^{-3})} = 0.058$

15.68 $\Delta n = (1) - (1 + 1) = -1$

        $K_p = K_c(RT)^{\Delta n} = (2.2 \times 10^5)[(0.08206)(298)]^{-1} = 9.0 \times 10^3$

15.70 $K_p = P_{H_2O} = 0.0313$ atm; $\Delta n = 1$

        $K_c = K_p\left(\dfrac{1}{RT}\right)^{\Delta n} = (0.0313)\left(\dfrac{1}{(0.082\,06)(298)}\right) = 1.28 \times 10^{-3}$

15.72 $2\,ClO(g) \;\rightleftharpoons\; Cl_2O_2(g)$ $K_c = 4.96 \times 10^{11}$

        $K_c = \dfrac{[Cl_2O_2]}{[ClO]^2} = \dfrac{(6.00 \times 10^{-6})}{[ClO]^2} = 4.96 \times 10^{11}$

        $[ClO]^2 = \dfrac{(6.00 \times 10^{-6})}{(4.96 \times 10^{11})}$; $[ClO] = \sqrt{\dfrac{(6.00 \times 10^{-6})}{(4.96 \times 10^{11})}} = 3.48 \times 10^{-9}$ M

15.74   (a) $K_c = \dfrac{[CO_2]^3}{[CO]^3}$, $K_p = \dfrac{(P_{CO_2})^3}{(P_{CO})^3}$ (b) $K_c = \dfrac{1}{[O_2]^3}$, $K_p = \dfrac{1}{(P_{O_2})^3}$

       (c) $K_c = [SO_3]$, $K_p = P_{SO_3}$ (d) $K_c = [Ba^{2+}][SO_4^{2-}]$

Using the Equilibrium Constant (Section 15.5)

15.76   (a) Because K_c is very large, the equilibrium mixture contains mostly product.

       (b) Because K_c is very small, the equilibrium mixture contains mostly reactants.

15.78 K_p is large. The equilibrium lies far to the right and much SO_2 will be present at equilibrium.

15.80 The container volume of 10.0 L must be included to calculate molar concentrations.

$$Q_c = \frac{[CS_2]_t[H_2]_t^4}{[CH_4]_t[H_2S]_t^2} = \frac{(3.0 \text{ mol}/10.0 \text{ L})(3.0 \text{ mol}/10.0 \text{ L})^4}{(2.0 \text{ mol}/10.0 \text{ L})(4.0 \text{ mol}/10.0 \text{ L})^2} = 7.6 \times 10^{-2}; \quad K_c = 2.5 \times 10^{-3}$$

The reaction is not at equilibrium because $Q_c > K_c$. The reaction will proceed in the reverse direction to attain equilibrium.

15.82 (a) $Q_p = \dfrac{(P_{P_2})(P_{H_2})^3}{(P_{PH_3})^2} = \dfrac{(0.871)(0.517)^3}{(0.0260)^2} = 178; \quad K_p = 398$

Because $Q_p < K_p$, the reaction will proceed in the forward direction to attain equilibrium.

(b) $K_p = \dfrac{(P_{P_2})(P_{H_2})^3}{(P_{PH_3})^2} = 398$

$$(P_{PH_3})^2 = \frac{(P_{P_2})(P_{H_2})^3}{398}$$

$$P_{PH_3} = \sqrt{\frac{(P_{P_2})(P_{H_2})^3}{398}} = \sqrt{\frac{(0.412)(0.822)^3}{398}} = 0.0240 \text{ atm}$$

15.84

	$N_2O_4(g)$	$\rightleftharpoons$	$2 NO_2(g)$
initial (M)	0.0500		0
change (M)	−x		+2x
equil (M)	0.0500 − x		2x

$$K_c = 4.64 \times 10^{-3} = \frac{[NO_2]^2}{[N_2O_4]} = \frac{(2x)^2}{(0.0500 - x)}$$

$4x^2 + (4.64 \times 10^{-3})x - (2.32 \times 10^{-4}) = 0$
Use the quadratic formula to solve for x.

$$x = \frac{-(4.64 \times 10^{-3}) \pm \sqrt{(4.64 \times 10^{-3})^2 - 4(4)(-2.32 \times 10^{-4})}}{2(4)} = \frac{-0.00464 \pm 0.06110}{8}$$

x = −0.008 22 and 0.007 06
Discard the negative solution (−0.008 22) because it leads to a negative concentration of NO_2 and that is impossible.
$[N_2O_4] = 0.0500 - x = 0.0500 - 0.007 \ 06 = 0.0429$ M
$[NO_2] = 2x = 2(0.007 \ 06) = 0.0141$ M

15.86 The container volume of 2.00 L must be included to calculate molar concentrations.
Initial [HI] = 9.30×10^{-3} mol/2.00 L = 4.65×10^{-3} M = 0.004 65 M

	$H_2(g)$	+	$I_2(g)$	$\rightleftharpoons$	$2 HI(g)$
initial (M)	0		0		0.004 65
change (M)	+x		+x		−2x
equil (M)	x		x		0.004 65 − 2x

$x = [H_2] = [I_2] = 6.29 \times 10^{-4} \, M = 0.000\ 629 \, M$

$[HI] = 0.004\ 65 - 2x = 0.004\ 65 - 2(0.000\ 629) = 0.003\ 39 \, M$

$K_c = \dfrac{[HI]^2}{[H_2][I_2]} = \dfrac{(0.003\ 39)^2}{(0.000\ 629)^2} = 29.0$

15.88 $CH_3CO_2C_2H_5(soln) + H_2O(soln) \rightleftharpoons CH_3CO_2H(soln) + C_2H_5OH(soln)$

$K_c(\text{hydrolysis}) = \dfrac{1}{K_c(\text{forward})} = \dfrac{1}{3.4} = 0.29$

15.90 (a) $Q_c = \dfrac{[Br_2][Cl_2]}{[BrCl]^2} = \dfrac{(0.035)(0.035)}{(0.050)^2} = 0.49; \quad K_c = 0.145$

Because $Q_c > K_c$, the reaction will proceed in the reverse direction to attain equilibrium.

(b)

	2 BrCl(*soln*)	$\rightleftharpoons$	Br$_2$(*soln*)	+	Cl$_2$(*soln*)
initial (M)	0.050		0.035		0.035
change (M)	+2x		−x		−x
equil (M)	0.050 + 2x		0.035 − x		0.035 − x

$K_c = \dfrac{[Br_2][Cl_2]}{[BrCl]^2} = 0.145 = \dfrac{(0.035 - x)(0.035 - x)}{(0.050 + 2x)^2}$

Take the square root of both sides and solve for x.

$\sqrt{0.145} = \sqrt{\dfrac{(0.035 - x)^2}{(0.050 + 2x)^2}}$

$0.381 = \dfrac{(0.035 - x)}{(0.050 + 2x)}; \quad x = 0.009$

$[BrCl] = 0.050 + 2x = 0.050 + 2(0.009) = 0.068 \, M$

$[Br_2] = [Cl_2] = 0.035 - x = 0.035 - 0.009 = 0.026 \, M$

15.92 $K_c = 2.7 \times 10^2 = \dfrac{[SO_3]^2}{[SO_2]^2[O_2]}; \quad$ Because $[SO_3] = [SO_2]$, then $2.7 \times 10^2 = \dfrac{1}{[O_2]}$

$[O_2] = 3.7 \times 10^{-3} \, M$

15.94

	N$_2$(g)	+	O$_2$(g)	$\rightleftharpoons$	2 NO(g)
initial (M)	2.24		0.56		0
change (M)	−x		−x		+2x
equil (M)	2.24 − x		0.56 − x		2x

$K_c = \dfrac{[NO]^2}{[N_2][O_2]} = 1.7 \times 10^{-3} = \dfrac{(2x)^2}{(2.24 - x)(0.56 - x)}$

$4x^2 + (4.8 \times 10^{-3})x - (2.1 \times 10^{-3}) = 0$

Use the quadratic formula to solve for x.

$x = \dfrac{-(4.8 \times 10^{-3}) \pm \sqrt{(4.8 \times 10^{-3})^2 - 4(4)(-2.1 \times 10^{-3})}}{2(4)} = \dfrac{-0.0048 \pm 0.1834}{8}$

x = –0.0235 and 0.0223

Discard the negative solution (–0.0235) because it gives a negative NO concentration and that is impossible.

$[N_2] = 2.24 - x = 2.24 - 0.0223 = 2.22$ M; $[O_2] = 0.56 - x = 0.56 - 0.0223 = 0.54$ M

$[NO] = 2x = 2(0.0223) = 0.045$ M

15.96 (a) $K_c = \dfrac{[CH_3CO_2C_2H_5][H_2O]}{[CH_3CO_2H][C_2H_5OH]} = 3.4 = \dfrac{(x)(12.0)}{(4.0)(6.0)}$;   x = 6.8 moles $CH_3CO_2C_2H_5$

Note that the volume cancels because the same number of molecules appear on both sides of the chemical equation.

(b) $CH_3CO_2H(soln) + C_2H_5OH(soln) \rightleftarrows CH_3CO_2C_2H_5(soln) + H_2O(soln)$

initial (mol)   1.00   10.00   0   0

change (mol)   –x   –x   +x   +x

equil (mol)   1.00 – x   10.00 – x   x   x

$K_c = 3.4 = \dfrac{x^2}{(1.00 - x)(10.00 - x)}$

$2.4x^2 - 37.4x + 34 = 0$

Use the quadratic formula to solve for x.

$x = \dfrac{-(-37.4) \pm \sqrt{(-37.4)^2 - 4(2.4)(34)}}{2(2.4)} = \dfrac{37.4 \pm 32.75}{4.8}$

x = 0.969 and 14.6

Discard the larger solution (14.6) because it leads to negative concentrations and that is impossible.

mol CH_3CO_2H = 1.00 – x = 1.00 – 0.969 = 0.03 mol

mol C_2H_5OH = 10.00 – x = 10.00 – 0.969 = 9.03 mol

mol $CH_3CO_2C_2H_5$ = mol H_2O = x = 0.97 mol

15.98 $ClF_3(g) \rightleftarrows ClF(g) + F_2(g)$

initial (atm)   1.47   0   0

change (atm)   –x   +x   +x

equil (atm)   1.47 – x   x   x

$K_p = \dfrac{(P_{ClF})(P_{F_2})}{(P_{ClF_3})} = 0.140 = \dfrac{(x)(x)}{1.47 - x}$; solve for x.

$x^2 + 0.140x - 0.2058 = 0$

Use the quadratic formula to solve for x.

$x = \dfrac{-(0.140) \pm \sqrt{(0.140)^2 - (4)(1)(-0.2058)}}{2(1)}$

$x = \dfrac{-0.140 \pm 0.918}{2}$

x = 0.389 and –0.529

Discard the negative solution (–0.529) because it gives negative partial pressures and that is impossible.

$P_{ClF} = P_{F_2} = x = 0.389$ atm

$P_{ClF_3} = 1.47 - x = 1.47 - 0.389 = 1.08$ atm

15.100 $2 HI(g) \rightleftarrows H_2(g) + I_2(g)$

Calculate K_c. $K_c = \dfrac{[H_2][I_2]}{[HI]^2} = \dfrac{(0.13)(0.70)}{(2.1)^2} = 0.0206;$ $[HI] = \dfrac{0.20 \text{ mol}}{0.5000 \text{ L}} = 0.40$ M

	$2 HI(g)$	$\rightleftarrows$	$H_2(g)$	$+$	$I_2(g)$
initial (M)	0.40		0		0
change (M)	$-2x$		$+x$		$+x$
equil (M)	$0.40 - 2x$		x		x

$K_c = 0.0206 = \dfrac{[H_2][I_2]}{[HI]^2} = \dfrac{x^2}{(0.40 - 2x)^2}$

Take the square root of both sides, and solve for x.

$\sqrt{0.0206} = \sqrt{\dfrac{x^2}{(0.40 - 2x)^2}};$ $0.144 = \dfrac{x}{0.40 - 2x};$ $x = 0.045$

At equilibrium, $[H_2] = [I_2] = x = 0.045$ M; $[HI] = 0.40 - 2x = 0.40 - 2(0.045) = 0.31$ M

15.102 $[H_2O] = \dfrac{6.00 \text{ mol}}{5.00 \text{ L}} = 1.20$ M

	$C(s)$	$+$	$H_2O(g)$	$\rightleftarrows$	$CO(g)$	$+$	$H_2(g)$
initial (M)			1.20		0		0
change (M)			$-x$		$+x$		$+x$
equil (M)			$1.20 - x$		x		x

$K_c = \dfrac{[CO][H_2]}{[H_2O]} = 3.0 \times 10^{-2} = \dfrac{x^2}{1.20 - x}$

$x^2 + (3.0 \times 10^{-2})x - 0.036 = 0$

Use the quadratic formula to solve for x.

$x = \dfrac{-(0.030) \pm \sqrt{(0.030)^2 - 4(-0.036)}}{2(1)} = \dfrac{-0.030 \pm 0.381}{2}$

$x = 0.176$ and -0.206

Discard the negative solution (-0.206) because it leads to negative concentrations and that is impossible.

$[CO] = [H_2] = x = 0.18$ M

$[H_2O] = 1.20 - x = 1.20 - 0.18 = 1.02$ M

15.104

	$C(s)$	$+$	$CO_2(g)$	$\rightleftarrows$	$2 CO(g)$
initial (M)	excess		1.50 mol/20.0 L		0
change (M)			$-x$		$+2x$
equil (M)			$0.0750 - x$		$2x$

$[CO] = 2x = 7.00 \times 10^{-2}$ M; $x = 0.0350$ M

(a) $[CO_2] = 0.0750 - x = 0.0750 - 0.0350 = 0.0400$ M

(b) $K_c = \dfrac{[CO]^2}{[CO_2]} = \dfrac{(7.00 \times 10^{-2})^2}{(0.0400)} = 0.122$

15.106 NO_2, 46.01

mol $NO_2 = 4.60$ g $NO_2 \times \dfrac{1 \text{ mol } NO_2}{46.01 \text{ g } NO_2} = 0.100$ mol NO_2

$[NO_2] = 0.100$ mol/10.0 L $= 0.0100$ M

$$2\,NO_2(g) \quad \rightleftarrows \quad N_2O_4(g)$$

initial (M)	0.0100	0
change (M)	–2x	+x
equil (M)	0.0100 – 2x	x

$K_c = \dfrac{[N_2O_4]}{[NO_2]^2} = 4.72 = \dfrac{x}{(0.0100 - 2x)^2}$

$18.88x^2 - 1.189x + 0.000\,472 = 0$

Use the quadratic formula to solve for x.

$x = \dfrac{-(-1.189) \pm \sqrt{(-1.189)^2 - 4(18.88)(0.000\,472)}}{2(18.88)} = \dfrac{1.189 \pm 1.1739}{37.76}$

$x = 0.0626$ and 4.00×10^{-4}

Discard the larger solution (0.0626) because it leads to a negative concentration of NO_2 and that is impossible.

$n_{total} = (0.0100 - 2x + x)(10.0 \text{ L}) = (0.0100 - x)(10.0 \text{ L})$
    $= [0.0100 \text{ mol/L} - (4.00 \times 10^{-4} \text{ mol/L})](10.0 \text{ L}) = 0.0960$ mol

$100\ ^\circ C = 100 + 273 = 373$ K

$P_{total} = \dfrac{n_{total}RT}{V} = \dfrac{(0.0960 \text{ mol})\left(0.082\,06\ \dfrac{L \cdot atm}{K \cdot mol}\right)(373 \text{ K})}{10.0 \text{ L}} = 0.294$ atm

15.108 $500\ ^\circ C = 500 + 273 = 773$ K and $840\ ^\circ C = 840 + 273 = 1113$ K

Calculate the undissociated pressure of F_2 at 1113 K.

$\dfrac{P_2}{T_2} = \dfrac{P_1}{T_1}$; $P_2 = \dfrac{P_1 T_2}{T_1} = \dfrac{(0.600 \text{ atm})(1113 \text{ K})}{773 \text{ K}} = 0.864$ atm

$$F_2(g) \quad \rightleftarrows \quad 2\,F(g)$$

initial (atm)	0.864	0
change (atm)	–x	+2x
equil (atm)	0.864 – x	2x

$P_{total} = (0.864 \text{ atm} - x) + 2x = 0.864 \text{ atm} + x = 0.984$ atm

$x = 0.984 \text{ atm} - 0.864 \text{ atm} = 0.120$ atm

$P_{F_2} = (0.864 \text{ atm} - x) = (0.864 \text{ atm} - 0.120 \text{ atm}) = 0.744$ atm

$P_F = 2x = 2(0.120 \text{ atm}) = 0.240$ atm

$$K_p = \frac{(P_F)^2}{P_{F_2}} = \frac{(0.240)^2}{0.744} = 0.0774$$

15.110 $[COCl] = 1.00 \text{ mol}/50.0 \text{ L} = 0.0200 \text{ M}$

	$COCl_2(g)$	$\rightleftharpoons$	$CO(g)$	$+$	$Cl_2(g)$
initial (M)	0.0200		0		0
change (M)	$-x$		$+x$		$+x$
equil (M)	$0.0200 - x$		x		x

$$K_c = \frac{[CO][Cl_2]}{[COCl_2]} = 8.4 \times 10^{-4} = \frac{x^2}{0.0200 - x}$$

$x^2 + 8.4 \times 10^{-4}x - 1.68 \times 10^{-5} = 0$

Use the quadratic formula to solve for x.

$$x = \frac{-(8.4 \times 10^{-4}) \pm \sqrt{(8.4 \times 10^{-4})^2 - 4(1)(-1.68 \times 10^{-5})}}{2(1)} = \frac{(-8.4 \times 10^{-4}) \pm (8.2 \times 10^{-3})}{2}$$

$x = 0.0037$ and -0.0045

Discard the negative solution (-0.0045) because it leads to a negative concentrations of CO and Cl_2 and that is impossible.

$[COCl_2] = 0.0200 - x = 0.0200 - 0.0037 = 0.0163 \text{ M}$

$[CO] = [Cl_2] = x = 0.0037 \text{ M}$

mol $COCl_2 = (0.0163 \text{ mol/L})(50.0 \text{ L}) = 0.815 \text{ mol}$

mol CO = mol $Cl_2 = (0.0037 \text{ mol/L})(50.0 \text{ L}) = 0.185 \text{ mol}$

$360\ ^\circ C = 360 + 273 = 633 \text{ K}$

$$P = \frac{nRT}{V} = \frac{(0.815 \text{ mol} + 0.185 \text{ mol} + 0.185 \text{ mol})\left(0.082\ 06\ \frac{L \cdot atm}{K \cdot mol}\right)(633 \text{ K})}{50.0 \text{ L}} = 1.23 \text{ atm}$$

15.112 N_2O_4, 92.01

$$\text{mol } N_2O_4 = 46.0 \text{ g} \times \frac{1 \text{ mol } N_2O_4}{92.01 \text{ g}} = 0.500 \text{ mol } N_2O_4$$

$45\ ^\circ C = 318 \text{ K}$

$K_p = K_c(RT)^{\Delta n}$

$\Delta n = 2 - 1 = 1$

$K_p = K_c(RT) = (0.619)(0.08\ 206)(318) = 16.2$

$$P_{N_2O_4} = \frac{nRT}{V} = \frac{(0.500 \text{ mol})\left(0.082\ 06\ \frac{L \cdot atm}{K \cdot mol}\right)(318 \text{ K})}{2.00 \text{ L}} = 6.52 \text{ atm}$$

	$N_2O_4(g)$	$\rightleftharpoons$	$2 NO_2(g)$
initial (atm)	6.52		0
change (atm)	$-x$		$+2x$
equil (atm)	$6.52 - x$		$2x$

$$K_p = \frac{(P_{NO_2})^2}{P_{N_2O_4}} = 16.2 = \frac{(2x)^2}{6.52 - x}$$

$(16.2)(6.52 - x) = 4x^2$

$105.6 - 16.2x = 4x^2$

$4x^2 + 16.2x - 105.6 = 0$

Use the quadratic formula to solve for x.

$$x = \frac{-(16.2) \pm \sqrt{(16.2)^2 - 4(4)(-105.6)}}{2(4)} = \frac{-16.2 \pm 44.2}{8}$$

$x = -7.55$ and 3.50

Discard the negative solution (–7.55) because it leads to a negative partial pressure and that is impossible.

$P_{N_2O_4} = 6.52 - x = 6.52 - 3.50 = 3.02$ atm; $P_{NO_2} = 2x = 2(3.50) = 7.00$ atm

15.114

	L-Aspartate	$\rightleftharpoons$	Fumarate	+	NH_4^+
initial (M)	0.008 32		0		0
change (M)	–x		+x		+x
equil (M)	0.008 32 – x		x		x

$$K_c = \frac{[Fumarate][NH_4^+]}{[L\text{-}Aspartate]} = 6.95 \times 10^{-3} = \frac{x^2}{0.008\ 32 - x}$$

$x^2 + 0.006\ 95x - 5.78 \times 10^{-5} = 0$

Use the quadratic formula to solve for x.

$$x = \frac{-(0.006\ 95) \pm \sqrt{(0.006\ 95)^2 - 4(1)(-5.78 \times 10^{-5})}}{2(1)} = \frac{-0.006\ 95 \pm 0.0167}{2}$$

$x = 0.004\ 87$ and -0.0118

Discard the negative solution (–0.0118) because it leads to a negative concentrations and that is impossible.

[L-Aspartate] = 0.008 32 – x = 0.008 32 – 0.004 87 = 0.003 45 M

[Fumarate] = [NH_4^+] = x = 0.004 87 M

15.116

	$SO_2(g)$	+	$Cl_2(g)$	$\rightleftharpoons$	$SO_2Cl_2(g)$
initial (M)	1.50		0.85		0
change (M)	–x		–x		+x
equil (M)	1.50 – x		0.85 – x		x

Because K_c is small, –x can be neglected in the next equation.

$$K_c = \frac{[SO_2Cl_2]}{[SO_2][Cl_2]} = \frac{x}{(1.50 - x)(0.85 - x)} \cong \frac{x}{(1.50)(0.85)} = 8.40 \times 10^{-3}$$

$x = (1.50)(0.85)(8.40 \times 10^{-3}) = 0.0107$

[SO_2] = 1.50 – x = 1.50 – 0.0107 = 1.49 M

[Cl_2] = 0.85 – x = 0.85 – 0.0107 = 0.84 M

[SO_2Cl_2] = 0.0107 M

Le Châtelier's Principle (Sections 15.6–15.9)

15.118 (a) Cl^- (reactant) added, $AgCl(s)$ increases
(b) Ag^+ (reactant) added, $AgCl(s)$ increases
(c) Ag^+ (reactant) removed, $AgCl(s)$ decreases
(d) Cl^- (reactant) removed, $AgCl(s)$ decreases

Disturbing the equilibrium by decreasing $[Cl^-]$ increases Q_c $\left(Q_c = \dfrac{1}{[Ag^+]_t[Cl^-]_t} \right)$ to a

value greater than K_c. To reach a new state of equilibrium, Q_c must decrease, which means that the denominator must increase; that is, the reaction must go from right to left, thus decreasing the amount of solid $AgCl$.

15.120 (a) Because there are 2 mol of gas on the left side and 3 mol of gas on the right side of the balanced equation, the stress of an increase in pressure is relieved by a shift in the reaction to the side with fewer moles of gas (in this case, to reactants). The number of moles of reaction products decreases.
(b) Because there are 2 mol of gas on both sides of the balanced equation, the composition of the equilibrium mixture is unaffected by a change in pressure. The number of moles of reaction product remains the same.
(c) Because there are 2 mol of gas on the left side and 1 mol of gas on the right side of the balanced equation, the stress of an increase in pressure is relieved by a shift in the reaction to the side with fewer moles of gas (in this case, to products). The number of moles of reaction products increases.

15.122 $CO(g) + H_2O(g) \rightleftharpoons CO_2(g) + H_2(g)$ $\qquad \Delta H° = -41.2 \text{ kJ}$
The reaction is exothermic. $[H_2]$ decreases when the temperature is increased.
As the temperature is decreased, the reaction shifts to the right. $[CO_2]$ and $[H_2]$ increase, $[CO]$ and $[H_2O]$ decrease, and K_c increases.

15.124 (a) HCl is a source of Cl^- (product), the reaction shifts left, the equilibrium $[CoCl_4^{2-}]$ increases.
(b) $Co(NO_3)_2$ is a source of $Co(H_2O)_6^{2+}$ (product), the reaction shifts left, the equilibrium $[CoCl_4^{2-}]$ increases.
(c) All concentrations will initially decrease and the reaction will shift to the right; the equilibrium $[CoCl_4^{2-}]$ decreases.
(d) For an exothermic reaction, the reaction shifts to the left when the temperature is increased; the equilibrium $[CoCl_4^{2-}]$ increases.

15.126 (a) The reaction is exothermic. The amount of CH_3OH (product) decreases as the temperature increases.
(b) When the volume decreases, the reaction shifts to the side with fewer gas molecules. The amount of CH_3OH increases.
(c) Addition of an inert gas (He) does not affect the equilibrium composition. There is no change.

(d) Addition of CO (reactant) shifts the reaction toward product. The amount of CH_3OH increases.
(e) Addition or removal of a catalyst does not affect the equilibrium composition. There is no change.

15.128 (a) add Au (a solid); no shift
(b) OH^- (product) increased; shift toward reactants.
(c) O_2 (reactant) partial pressure increased; shift toward products.
(d) Fe^{3+} decreases CN^- (reactant) by forming $Fe(CN)_6^{3-}$; shift toward reactants.

15.130 (a) Because K_p is larger at the higher temperature, the reaction has shifted toward products at the higher temperature, which means the reaction is endothermic.
(b) (i) Increasing the volume causes the reaction to shift toward the side with more mol of gas (product side). The equilibrium amounts of PCl_3 and Cl_2 increase while that of PCl_5 decreases.
(ii) If there is no volume change, there is no change in equilibrium concentrations.
(iii) Addition of a catalyst does not affect the equilibrium concentrations.

Chemical Equilibrium and Chemical Kinetics (Section 15.10)

15.132 (a) A catalyst does not affect the equilibrium composition. The amount of CO remains the same.
(b) The reaction is exothermic. An increase in temperature shifts the reaction toward reactants. The amount of CO increases.
(c) Because there are 3 mol of gas on the left side and 2 mol of gas on the right side of the balanced equation, the stress of an increase in pressure is relieved by a shift in the reaction to the side with fewer moles of gas (in this case, to products). The amount of CO decreases.
(d) An increase in pressure as a result of the addition of an inert gas (with no volume change) does not affect the equilibrium composition. The amount of CO remains the same.
(e) Adding O_2 increases the O_2 concentration and shifts the reaction toward products. The amount of CO decreases.

15.134 Because the rate of the forward reaction is greater than the rate of the reverse reaction, the reaction will proceed in the forward direction to attain equilibrium.

15.136 $A + B \rightleftharpoons C$
$rate_f = k_f[A][B]$ and $rate_r = k_r[C]$; at equilibrium, $rate_f = rate_r$
$k_f[A][B] = k_r[C]$; $\dfrac{k_f}{k_r} = \dfrac{[C]}{[A][B]} = K_c$

15.138 $K_c = \dfrac{k_f}{k_r} = \dfrac{0.13}{6.2 \times 10^{-4}} = 210$

15.140 k_r increases more than k_f, this means that E_a (reverse) is greater than E_a (forward). The reaction is exothermic when E_a (reverse) > E_a (forward).

15.142 (a) The Arrhenius equation gives for the forward and reverse reactions:
$$k_f = A_f e^{-E_{a,f}/RT} \quad \text{and} \quad k_r = A_r e^{-E_{a,r}/RT}$$
Addition of a catalyst decreases the activation energies by ΔE_a, so
$$k_f = A_f e^{-(E_{a,f} - \Delta E_a)/RT} = A_f e^{-E_{a,f}/RT} \times e^{\Delta E_a/RT}$$
and
$$k_r = A_r e^{-(E_{a,r} - \Delta E_a)/RT} = A_r e^{-E_{a,r}/RT} \times e^{\Delta E_a/RT}$$
Therefore, the rate constants for both the forward and reverse reactions increase by the same factor, $e^{\Delta E_a/RT}$.

(b) The equilibrium constant is given by $K_c = \dfrac{k_f}{k_r} = \dfrac{A_f e^{-E_{a,f}/RT}}{A_r e^{-E_{a,r}/RT}} = \dfrac{A_f}{A_r} e^{-\Delta E/RT}$

where $\Delta E = E_{a,f} - E_{a,r}$. Addition of a catalyst decreases the activation energies by ΔE_a.

So, $K_c = \dfrac{k_f}{k_r} = \dfrac{A_f e^{-E_{a,f}/RT} \times e^{\Delta E_a/RT}}{A_r e^{-E_{a,r}/RT} \times e^{\Delta E_a/RT}} = \dfrac{A_f}{A_r} e^{-\Delta E/RT}$

K_c is unchanged because of cancellation of $e^{\Delta E_a/RT}$, the factor by which the two rate constants increase.

Multiconcept Problems

15.144 (a) $K_c = \dfrac{[H^+][CH_3CO_2^-]}{[CH_3CO_2H]}$

(b)

	$CH_3CO_2H(aq)$	$\rightleftharpoons$	$H^+(aq)$	+	$CH_3CO_2^-(aq)$
initial (M)	1.0		0		0
change (M)	−0.0042		+0.0042		+0.0042
equil (M)	1.0 − 0.0042		0.0042		0.0042

$$K_c = \frac{(0.0042)(0.0042)}{(1.0 - 0.0042)} = 1.8 \times 10^{-5}$$

15.146 (a) $K_c = \dfrac{[C_2H_6][C_2H_4]}{[C_4H_{10}]}$ $\qquad K_p = \dfrac{(P_{C_2H_6})(P_{C_2H_4})}{P_{C_4H_{10}}}$

(b) $K_p = 12$; $\Delta n = 1$; $K_c = K_p\left(\dfrac{1}{RT}\right) = (12)\left(\dfrac{1}{(0.082\ 06)(773)}\right) = 0.19$

(c)

	$C_4H_{10}(g)$	$\rightleftharpoons$	$C_2H_6(g)$	+	$C_2H_4(g)$
initial (atm)	50		0		0
change (atm)	−x		+x		+x
equil (atm)	50 − x		x		x

$K_p = 12 = \dfrac{x^2}{50 - x}$; $\qquad x^2 + 12x - 600 = 0$

Use the quadratic formula to solve for x.

$$x = \frac{(-12) \pm \sqrt{(12)^2 - 4(1)(-600)}}{2(1)} = \frac{-12 \pm 50.44}{2}$$

x = –31.22 and 19.22

Discard the negative solution (–31.22) because it leads to negative concentrations and that is impossible.

% C_4H_{10} converted = $\dfrac{19.22}{50}$ x 100% = 38%

$P_{total} = P_{C_4H_{10}} + P_{C_2H_6} + P_{C_2H_4} = (50 - x) + x + x = (50 - 19) + 19 + 19 = 69$ atm

(d) A decrease in volume would decrease the % conversion of C_4H_{10}.

15.148 (a) PV = nRT, $n_{total} = \dfrac{PV}{RT} = \dfrac{(0.588 \text{ atm})(1.00 \text{ L})}{\left(0.082\ 06\ \dfrac{\text{L} \cdot \text{atm}}{\text{K} \cdot \text{mol}}\right)(300 \text{ K})} = 0.0239$ mol

	2 NOBr(g)	⇌	2 NO(g)	+	Br_2(g)
initial (mol)	0.0200		0		0
change (mol)	–2x		+2x		+x
equil (mol)	0.0200 – 2x		2x		x

n_{total} = 0.0239 mol = (0.0200 – 2x) + 2x + x = 0.0200 + x

x = 0.0239 – 0.0200 = 0.0039 mol

Because the volume is 1.00 L, the molarity equals the number of moles.

[NOBr] = 0.0200 – 2x = 0.0200 – 2(0.0039) = 0.0122 M

[NO] = 2x = 2(0.0039) = 0.0078 M

[Br_2] = x = 0.0039 M

$$K_c = \frac{[NO]^2[Br_2]}{[NOBr]^2} = \frac{(0.0078)^2(0.0039)}{(0.0122)^2} = 1.6 \text{ x } 10^{-3}$$

(b) Δn = (3) – (2) = 1, $K_p = K_c(RT) = (1.6 \text{ x } 10^{-3})(0.082\ 06)(300) = 0.039$

15.150 (a)

	$(NH_4)(NH_2CO_2)$(s)	⇌	2 NH_3(g)	+	CO_2(g)
initial (atm)			0		0
change (atm)			+2x		+x
equil (atm)			2x		x

P_{total} = 2x + x = 3x = 0.116 atm

x = 0.116 atm/3 = 0.0387 atm

P_{NH_3} = 2x = 2(0.0387 atm) = 0.0774 atm

P_{CO_2} = x = 0.0387 atm

$K_p = (P_{NH_3})^2(P_{CO_2}) = (0.0774)^2(0.0387) = 2.32 \text{ x } 10^{-4}$

(b) (i) The total quantity of NH_3 would decrease. When product, CO_2, is added, the equilibrium will shift to the left.

(ii) The total quantity of NH_3 would remain unchanged. Adding a pure solid, $(NH_4)(NH_2CO_2)$, to a heterogeneous equilibrium will not affect the position of the equilibrium.

(iii) The total quantity of NH_3 would increase. When product, CO_2, is removed, the equilibrium will shift to the right.

(iv) The total quantity of NH_3 would increase. When the total volume is increased, the reaction will shift to the side with more total moles of gas, which in this case is the product side.

(v) The total quantity of NH_3 would remain unchanged. Neon is an inert gas that will have no effect on the reaction or on the position of equilibrium.

(vi) The total quantity of NH_3 would increase. Because the reaction is endothermic, an increase in temperature will shift the equilibrium to the right.

15.152 (a) $[N_2O_4] = \dfrac{0.500 \text{ mol}}{4.00 \text{ L}} = 0.125$ M

	$N_2O_4(g)$	$\rightleftharpoons$	$2 NO_2(g)$
initial (M)	0.125		0
change (M)	$-(0.793)(0.125)$		$+(2)(0.793)(0.125)$
equil (M)	$0.125 - (0.793)(0.125)$		$(2)(0.793)(0.125)$

At equilibrium, $[N_2O_4] = 0.125 - (0.793)(0.125) = 0.0259$ M
$[NO_2] = (2)(0.793)(0.125) = 0.198$ M

$K_c = \dfrac{[NO_2]^2}{[N_2O_4]} = \dfrac{(0.198)^2}{(0.0259)} = 1.51$

$\Delta n = 2 - 1 = 1$ and $K_p = K_c(RT)^{\Delta n}$; $K_p = K_c(RT) = (1.51)(0.082\ 06)(400) = 49.6$

(b)

15.154 2 monomer $\rightleftharpoons$ dimer
(a) In benzene, $K_c = 1.51 \times 10^2$

	2 monomer	$\rightleftharpoons$	dimer
initial (M)	0.100		0
change (M)	$-2x$		$+x$
equil (M)	$0.100 - 2x$		x

$K_c = \dfrac{[\text{dimer}]}{[\text{monomer}]^2} = 1.51 \times 10^2 = \dfrac{x}{(0.100 - 2x)^2}$

$604x^2 - 61.4x + 1.51 = 0$

Use the quadratic formula to solve for x.

$x = \dfrac{-(-61.4) \pm \sqrt{(-61.4)^2 - (4)(604)(1.51)}}{2(604)} = \dfrac{61.4 \pm 11.04}{1208}$

$x = 0.0600$ and 0.0417

Discard the larger solution (0.0600) because it gives a negative concentration of the monomer and that is impossible.

$[\text{monomer}] = 0.100 - 2x = 0.100 - 2(0.0417) = 0.017$ M; $\quad$ $[\text{dimer}] = x = 0.0417$ M

$$\frac{[\text{dimer}]}{[\text{monomer}]} = \frac{0.0417 \text{ M}}{0.017 \text{ M}} = 2.5$$

(b) In H_2O, $K_c = 3.7 \times 10^{-2}$

	2 monomer	$\rightleftharpoons$	dimer
initial (M)	0.100		0
change (M)	$-2x$		$+x$
equil (M)	$0.100 - 2x$		x

$$K_c = \frac{[\text{dimer}]}{[\text{monomer}]^2} = 3.7 \times 10^{-2} = \frac{x}{(0.100 - 2x)^2}$$

$0.148x^2 - 1.0148x + 0.000\,37 = 0$

Use the quadratic formula to solve for x.

$$x = \frac{-(-1.0148) \pm \sqrt{(-1.0148)^2 - (4)(0.148)(0.00037)}}{2(0.148)} = \frac{1.0148 \pm 1.0147}{0.296}$$

$x = 6.86$ and 3.4×10^{-4}

Discard the larger solution (6.86) because it gives a negative concentration of the monomer and that is impossible.

$[\text{monomer}] = 0.100 - 2x = 0.100 - 2(3.4 \times 10^{-4}) = 0.099 \text{ M}$; $[\text{dimer}] = x = 3.4 \times 10^{-4} \text{ M}$

$$\frac{[\text{dimer}]}{[\text{monomer}]} = \frac{3.4 \times 10^{-4} \text{ M}}{0.099 \text{ M}} = 0.0034$$

(c) K_c for the water solution is so much smaller than K_c for the benzene solution because H_2O can hydrogen bond with acetic acid, thus preventing acetic acid dimer formation. Benzene cannot hydrogen bond with acetic acid.

15.156 (a) CO_2, 44.01; CO, 28.01

$$79.2 \text{ g } CO_2 \times \frac{1 \text{ mol } CO_2}{44.01 \text{ g } CO_2} = 1.80 \text{ mol } CO_2$$

	$CO_2(g)$	+	C(s)	$\rightleftharpoons$	2 CO(g)
initial (mol)	1.80				0
change (mol)	$-x$				$+2x$
equil (mol)	$1.80 - x$				$2x$

total mass of gas in flask $= (16.3 \text{ g/L})(5.00 \text{ L}) = 81.5 \text{ g}$

$81.5 = (1.80 - x)(44.01) + (2x)(28.01)$

$81.5 = 79.22 - 44.01x + 56.02x$; $2.28 = 12.01x$; $x = 2.28/12.01 = 0.19$

$n_{CO_2} = 1.80 - x = 1.80 - 0.19 = 1.61 \text{ mol } CO_2$; $n_{CO} = 2x = 2(0.19) = 0.38 \text{ mol CO}$

$$P_{CO_2} = \frac{nRT}{V} = \frac{(1.61 \text{ mol})\left(0.082\,06 \dfrac{\text{L} \cdot \text{atm}}{\text{K} \cdot \text{mol}}\right)(1000 \text{ K})}{5.0 \text{ L}} = 26.4 \text{ atm}$$

$$P_{CO} = \frac{nRT}{V} = \frac{(0.38 \text{ mol})\left(0.082\,06 \dfrac{\text{L} \cdot \text{atm}}{\text{K} \cdot \text{mol}}\right)(1000 \text{ K})}{5.0 \text{ L}} = 6.24 \text{ atm}$$

$$K_p = \frac{(P_{CO})^2}{(P_{CO_2})} = \frac{(6.24)^2}{(26.4)} = 1.47$$

(b) At 1100K, the total mass of gas in flask = (16.9 g/L)(5.00 L) = 84.5 g

84.5 = (1.80 − x)(44.01) + (2x)(28.01)

84.5 = 79.22 − 44.01x + 56.02x; 5.28 = 12.01x; x = 5.28/12.01 = 0.44

n_{CO_2} = 1.80 − x = 1.80 − 0.44 = 1.36 mol CO_2; n_{CO} = 2x = 2(0.44) = 0.88 mol CO

$$P_{CO_2} = \frac{nRT}{V} = \frac{(1.36\ mol)\left(0.082\ 06\ \frac{L \cdot atm}{K \cdot mol}\right)(1100\ K)}{5.0\ L} = 24.6\ atm$$

$$P_{CO} = \frac{nRT}{V} = \frac{(0.88\ mol)\left(0.082\ 06\ \frac{L \cdot atm}{K \cdot mol}\right)(1100\ K)}{5.0\ L} = 15.9\ atm$$

$$K_p = \frac{(P_{CO})^2}{(P_{CO_2})} = \frac{(15.9)^2}{(24.6)} = 10.3$$

(c) In agreement with Le Châtelier's principle, the reaction is endothermic because K_p increases with increasing temperature.

15.158 (a) N_2O_4, 92.01

$$14.58\ g\ N_2O_4 \times \frac{1\ mol\ N_2O_4}{92.01\ g\ N_2O_4} = 0.1585\ mol\ N_2O_4$$

$$PV = nRT \qquad P_{N_2O_4} = \frac{nRT}{V} = \frac{(0.1585\ mol)\left(0.082\ 06\ \frac{L \cdot atm}{K \cdot mol}\right)(400\ K)}{1.000\ L} = 5.20\ atm$$

$$\begin{array}{lcc} & N_2O_4(g) & \rightleftarrows \quad 2\ NO_2(g) \\ \text{initial (atm)} & 5.20 & 0 \\ \text{change (atm)} & -x & +2x \\ \text{equil (atm)} & 5.20 - x & 2x \end{array}$$

$P_{total} = P_{N_2O_4} + P_{NO_2} = (5.20 - x) + (2x) = 9.15\ atm$

5.20 + x = 9.15 atm

x = 3.95 atm

$P_{N_2O_4}$ = 5.20 − x = 5.20 − 3.95 = 1.25 atm

P_{NO_2} = 2x = 2(3.95) = 7.90 atm

$$K_p = \frac{(P_{NO_2})^2}{(P_{N_2O_4})} = \frac{(7.90)^2}{(1.25)} = 49.9$$

$$\Delta n = 1\ and\ K_c = K_p\left(\frac{1}{RT}\right) = \frac{(49.9)}{(0.082\ 06)(400)} = 1.52$$

(b) $\Delta H°_{rxn} = [2\ \Delta H°_f(NO_2)] - \Delta H°_f(N_2O_4)$

$\Delta H°_{rxn}$ = [(2 mol)(33.2 kJ/mol)] − [(1mol)(11.1 kJ/mol)] = 55.3 kJ

$$\text{moles } N_2O_4 \text{ reacted} = n = \frac{PV}{RT} = \frac{(3.95 \text{ atm})(1.000 \text{ L})}{\left(0.082\ 06 \dfrac{L \cdot atm}{K \cdot mol}\right)(400 \text{ K})} = 0.1203 \text{ mol } N_2O_4$$

$$q = (55.3 \text{ kJ/mol } N_2O_4)(0.1203 \text{ mol } N_2O_4) = 6.65 \text{ kJ}$$

15.160 The atmosphere is 21% (0.21) O_2; $P_{O_2} = (0.21)\left(720 \text{ mm Hg x } \dfrac{1 \text{ atm}}{760 \text{ mm Hg}}\right) = 0.199 \text{ atm}$

$$2 \text{ O}_3(g) \ \rightleftharpoons \ 3 \text{ O}_2(g)$$

$$K_p = \frac{(P_{O_2})^3}{(P_{O_3})^2}; \quad P_{O_3} = \sqrt{\frac{(P_{O_2})^3}{K_p}} = \sqrt{\frac{(0.199)^3}{1.3 \times 10^{57}}} = 2.46 \times 10^{-30} \text{ atm}$$

$$\text{vol} = 10 \times 10^6 \text{ m}^3 \times \left(\frac{100 \text{ cm}}{1 \text{ m}}\right)^3 \times \frac{1 \text{ L}}{1000 \text{ cm}^3} = 1.0 \times 10^{10} \text{ L}$$

$$n_{O_3} = \frac{PV}{RT} = \frac{(2.46 \times 10^{-30} \text{ atm})(1.0 \times 10^{10} \text{ L})}{\left(0.082\ 06 \dfrac{L \cdot atm}{K \cdot mol}\right)(298 \text{ K})} = 1.0 \times 10^{-21} \text{ mol } O_3$$

$$\text{O}_3 \text{ molecules} = 1.0 \times 10^{-21} \text{ mol } O_3 \times \frac{6.022 \times 10^{23} \text{ O}_3 \text{ molecules}}{1 \text{ mol } O_3} = 6.0 \times 10^2 \text{ O}_3 \text{ molecules}$$

15.162 $PCl_5(g) \ \rightleftharpoons \ PCl_3(g) \ + \ Cl_2(g)$
$\Delta n = (2) - (1) = 1$ and at 700 K, $K_p = K_c(RT) = (46.9)(0.082\ 06)(700) = 2694$
(a) Because K_p is larger at the higher temperature, the reaction has shifted toward products at the higher temperature, which means the reaction is endothermic. Because the reaction involves breaking two P–Cl bonds and forming just one Cl–Cl bond, it should be endothermic.
(b) PCl_5, 208.24

$$\text{mol } PCl_5 = 1.25 \text{ g } PCl_5 \times \frac{1 \text{ mol } PCl_5}{208.24 \text{ g } PCl_5} = 6.00 \times 10^{-3} \text{ mol}$$

$$PV = nRT, \quad P_{PCl_5} = \frac{nRT}{V} = \frac{(6.00 \times 10^{-3} \text{ mol})\left(0.082\ 06 \dfrac{L \cdot atm}{K \cdot mol}\right)(700 \text{ K})}{0.500 \text{ L}} = 0.689 \text{ atm}$$

Because K_p is so large, first assume the reaction goes to completion and then allow for a small back reaction.

	$PCl_5(g)$	$\rightleftharpoons$	$PCl_3(g)$	+	$Cl_2(g)$
before rxn (atm)	0.689		0		0
100% rxn (atm)	−0.689		+0.689		+0.689
after rxn (atm)	0		0.689		0.689
back rxn (atm)	+x		−x		−x
equil (atm)	x		0.689 − x		0.689 − x

$$K_p = \frac{(P_{PCl_3})(P_{Cl_2})}{P_{PCl_5}} = 2694 = \frac{(0.689 - x)^2}{x} \approx \frac{(0.689)^2}{x}$$

$$x = P_{PCl_5} = \frac{(0.689)^2}{2694} = 1.76 \times 10^{-4} \text{ atm}$$

$$P_{total} = P_{PCl_5} + P_{PCl_3} + P_{Cl_2}$$

$$P_{total} = x + (0.689 - x) + (0.689 - x) = 0.689 + 0.689 - 1.76 \times 10^{-4} = 1.38 \text{ atm}$$

$$\% \text{ dissociation} = \frac{(P_{PCl_5})_o - (P_{PCl_5})}{(P_{PCl_5})_o} \times 100\% = \frac{0.689 - (1.76 \times 10^{-4})}{0.689} \times 100\% = 99.97\%$$

(c)

The molecular geometry is trigonal bipyramidal. There is no dipole moment because of a symmetrical distribution of Cl's around the central P.

The molecular geometry is trigonal pyramidal. There is a dipole moment because of the lone pair of electrons on the P and an unsymmetrical distribution of Cl's around the central P.

16 | Aqueous Equilibria: Acids and Bases

16.1 $HSO_4^-(aq) + H_2O(l) \rightleftharpoons H_3O^+(aq) + SO_4^{2-}(aq)$
$SO_4^{2-}(aq)$ is the conjugate base of $HSO_4^-(aq)$.
$H_2O(l)$ is the conjugate base of $H_3O^+(aq)$.

16.2 $CO_3^{2-}(aq) + H_2O(l) \rightleftharpoons HCO_3^-(aq) + OH^-(aq)$

16.3 $HCl(aq) + NH_3(aq) \rightleftharpoons NH_4^+(aq) + Cl^-(aq)$
acid   base   acid   base
conjugate acid-base pairs

16.4 (a) $CH_3NH_2(aq) + H_2O(l) \rightleftharpoons OH^-(aq) + CH_3NH_3^+(aq)$
base   acid   base   acid
conjugate acid-base pairs

(b) $HNO_3(aq) + H_2O(l) \rightleftharpoons H_3O^+(aq) + NO_3^-(aq)$
acid   base   acid   base
conjugate acid-base pairs

16.5 $HF(aq) + NO_3^-(aq) \rightleftharpoons HNO_3(aq) + F^-(aq)$
HNO_3 is a stronger acid than HF, and F^- is a stronger base than NO_3^- (see Table 16.1). Because proton transfer occurs from the stronger acid to the stronger base, the reaction proceeds from right to left.

16.6 (a) Both HX and HY have the same initial concentration. HY is more dissociated than HX. Therefore, HY is the stronger acid.
(b) The conjugate base (X^-) of the weaker acid (HX) is the stronger base.
(c) $HX + Y^- \rightleftharpoons HY + X^-$; Proton transfer occurs from the stronger acid to the stronger base. The reaction proceeds to the left.

16.7 (a) H_2Se is a stronger acid than H_2S because Se is below S in the 6A group and the H–Se bond is weaker than the H–S bond.
(b) HNO_3 is a stronger acid than HNO_2 because acid strength increases with increasing oxidation number of N. The oxidation number for N is +5 in HNO_3 and +3 in HNO_2.
(c) HI is a stronger acid than H_2Te because I is to the right of Te in the same row of the periodic table, I is more electronegative than Te, and the H–I bond is more polar.

(d) H_2SO_3 is a stronger acid than H_2SeO_3 because acid strength increases with increasing electronegativity of the central atom. S is more electronegative than Se.

Only (d) has the stronger acid listed first.

16.8 H_3PO_4 and H_3AsO_4 are oxoacids that contain the same number of O atoms and acid strength is determined by the electronegativity of the central atom. Because P is more electronegative than As it can pull more electron density from the O–H bond resulting in a weaker and more polar bond in H_3PO_4. Therefore, H_3PO_4 is the stronger acid.

16.9 $[OH^-] = \dfrac{K_w}{[H_3O^+]} = \dfrac{1.0 \times 10^{-14}}{1.4 \times 10^{-4}} = 7.1 \times 10^{-11}$ M

Because $[H_3O^+] > [OH^-]$, the solution is acidic.

16.10 $K_w = [H_3O^+][OH^-]$; In a neutral solution, $[H_3O^+] = [OH^-]$

At 50 °C, $[H_3O^+] = [OH^-] = \sqrt{K_w} = \sqrt{5.5 \times 10^{-14}} = 2.3 \times 10^{-7}$ M

16.11 $[H_3O^+] = \dfrac{K_w}{[OH^-]} = \dfrac{1.0 \times 10^{-14}}{1.58 \times 10^{-6}} = 6.3 \times 10^{-9}$ M

$pH = -\log[H_3O^+] = -\log(6.3 \times 10^{-9}) = 8.20$

16.12 $pH = -\log[H_3O^+] = -\log(6.3) = -0.80$

16.13 $[H_3O^+] = 10^{-pH} = 10^{-2.8} = 2 \times 10^{-3}$ M

$[OH^-] = \dfrac{K_w}{[H_3O^+]} = \dfrac{1.0 \times 10^{-14}}{2 \times 10^{-3}} = 5 \times 10^{-12}$ M

16.14 milk: $[H_3O^+] = 10^{-pH} = 10^{-6.6} = 2.5 \times 10^{-7}$ M

black coffee: $[H_3O^+] = 10^{-pH} = 10^{-5.0} = 1.0 \times 10^{-5}$ M

$\dfrac{[H_3O^+]_{black\ coffee}}{[H_3O^+]_{milk}} = \dfrac{1.0 \times 10^{-5}}{2.5 \times 10^{-7}} = 40$

The $[H_3O^+]$ in coffee is 40 times greater.

16.15 (a) Because $HClO_4$ is a strong acid, $[H_3O^+] = 0.050$ M.

$pH = -\log[H_3O^+] = -\log(0.050) = 1.30$

(b) Because $Ba(OH)_2$ is a strong base, $[OH^-] = 2(0.010\ M) = 0.020$ M.

$[H_3O^+] = \dfrac{K_w}{[OH^-]} = \dfrac{1.0 \times 10^{-14}}{0.020} = 5.0 \times 10^{-13}$ M

$pH = -\log[H_3O^+] = -\log(5.0 \times 10^{-13}) = 12.30$

16.16 $CaO(s) + H_2O(l) \rightarrow Ca(OH)_2(aq)$; CaO, 56.08

$$0.25 \text{ g CaO x } \frac{1 \text{ mol CaO}}{56.08 \text{ g CaO}} \text{ x } \frac{1 \text{ mol Ca(OH)}_2}{1 \text{ mol CaO}} \text{ x } \frac{2 \text{ mol OH}^-}{1 \text{ mol Ca(OH)}_2} = 8.92 \text{ x } 10^{-3} \text{ mol OH}^-$$

$$[OH^-] = \frac{8.92 \text{ x } 10^{-3} \text{ mol OH}^-}{1.50 \text{ L}} = 5.94 \text{ x } 10^{-3} \text{ M}$$

$$[H_3O^+] = \frac{K_w}{[OH^-]} = \frac{1.0 \text{ x } 10^{-14}}{5.94 \text{ x } 10^{-3}} = 1.68 \text{ x } 10^{-12} \text{ M}$$

$pH = -\log[H_3O^+] = -\log(1.68 \text{ x } 10^{-12}) = 11.77$

16.17

$$HC_3H_5O_3(aq) + H_2O(l) \rightleftharpoons H_3O^+(aq) + C_3H_5O_3^-(aq)$$

initial (M)	0.10	~0	0
change (M)	−x	+x	+x
equil (M)	0.10 − x	x	x

$x = [H_3O^+] = 10^{-pH} = 10^{-2.42} = 3.8 \text{ x } 10^{-3} \text{ M}$

$[C_3H_5O_3^-] = x = 3.8 \text{ x } 10^{-3} \text{ M}$; $[HC_3H_5O_3] = 0.10 − x = (0.10 − 3.8 \text{ x } 10^{-3}) \text{ M}$

$$K_a = \frac{[H_3O^+][C_3H_5O_3^-]}{[HC_3H_5O_3]} = \frac{(3.8 \text{ x } 10^{-3})(3.8 \text{ x } 10^{-3})}{(0.10 − 3.8 \text{ x } 10^{-3})} = 1.50 \text{ x } 10^{-4}$$

$pK_a = -\log K_a = -\log (1.50 \text{ x } 10^{-4}) = 3.824$

16.18 (a) HZ is completely dissociated. HX and HY are at the same concentration and HX is more dissociated than HY. The strongest acid is HZ, the weakest is HY.
K_a (HY) < K_a (HX) < K_a (HZ)
(b) HZ
(c) HY has the highest pH; HX has the lowest pH (highest $[H_3O^+]$).

16.19

$$CH_3CO_2H(aq) + H_2O(l) \rightleftharpoons H_3O^+(aq) + CH_3CO_2^-(aq)$$

initial (M)	0.100	~0	0
change (M)	−x	+x	+x
equil (M)	0.100 − x	x	x

$$K_a = \frac{[H_3O^+][CH_3CO_2^-]}{[CH_3CO_2H]} = 1.8 \text{ x } 10^{-5} = \frac{x^2}{0.100 − x} \approx \frac{x^2}{0.100}$$

Solve for x. $x = [H_3O^+] = 1.34 \text{ x } 10^{-3} \text{ M}$

$pH = -\log[H_3O^+] = -\log(1.34 \text{ x } 10^{-3}) = 2.87$

$[CH_3CO_2^-] = x = 1.34 \text{ x } 10^{-3} \text{ M}$; $[CH_3CO_2H] = 0.100 − x = 0.099 \text{ M}$

$$[OH^-] = \frac{K_w}{[H_3O^+]} = \frac{1.00 \text{ x } 10^{-14}}{1.34 \text{ x } 10^{-3}} = 7.46 \text{ x } 10^{-12} \text{ M}$$

16.20 $[H_3O^+] = 10^{-pH} = 10^{-2.00} = 0.010 \text{ M}$

$$HCO_2H(aq) + H_2O(l) \rightleftharpoons H_3O^+(aq) + HCO_2^-(aq) \qquad Ka = 1.8 \text{ x } 10^{-4}$$

(M)	x − 0.010	0.010	0.010

$$K_a = \frac{[H_3O^+][HCO_2^-]}{[HCO_2H]} = 1.8 \times 10^{-4} = \frac{(0.010)^2}{(x-0.010)}$$

$1.8 \times 10^{-4} \, x - 1.8 \times 10^{-6} = 1.0 \times 10^{-4}$

$x = (1.0 \times 10^{-4} + 1.8 \times 10^{-6})/(1.8 \times 10^{-4}) = 0.57$ M

16.21

$$H_2C_6H_6O_6(aq) + H_2O(l) \rightleftarrows H_3O^+(aq) + HC_6H_6O_6^-(aq)$$

initial (M)	0.10	~0	0
change (M)	–x	+x	+x
equil (M)	0.10 – x	x	x

$$K_{a1} = \frac{[H_3O^+][HC_6H_6O_6^-]}{[H_2C_6H_6O_6]} = 8.0 \times 10^{-5} = \frac{x^2}{0.10-x} \approx \frac{x^2}{0.10}$$

Solve for x. $x = 2.8 \times 10^{-3}$

$[H_3O^+] = [HC_6H_6O_6^-] = x = 2.8 \times 10^{-3}$ M; $[H_2C_6H_6O_6] = 0.10 - x = 0.10$ M

The second dissociation of $H_2C_6H_6O_6$ produces a negligible amount of H_3O^+ compared with that from the first dissociation.

$$HC_6H_6O_6^-(aq) + H_2O(l) \rightleftarrows H_3O^+(aq) + C_6H_6O_6^{2-}(aq)$$

$$K_{a2} = \frac{[H_3O^+][C_6H_6O_6^{2-}]}{[HC_6H_6O_6^-]} = 1.6 \times 10^{-12} = \frac{(2.8 \times 10^{-3})[C_6H_6O_6^{2-}]}{(2.8 \times 10^{-3})}$$

$[C_6H_6O_6^{2-}] = K_{a2} = 1.6 \times 10^{-12}$ M

$$[OH^-] = \frac{K_w}{[H_3O^+]} = \frac{1.0 \times 10^{-14}}{2.8 \times 10^{-3}} = 3.6 \times 10^{-12} \text{ M}$$

$pH = -\log[H_3O^+] = -\log(2.8 \times 10^{-3}) = 2.55$

16.22

$$H_2CO_3(aq) + H_2O(l) \rightleftarrows H_3O^+(aq) + HCO_3^-(aq)$$

initial (M)	0.16	~0	0
change (M)	–x	+x	+x
equil (M)	0.16 – x	x	x

$$K_{a1} = \frac{[H_3O^+][HCO_3^-]}{[H_2CO_3]} = 4.3 \times 10^{-7} = \frac{x^2}{0.16-x} \approx \frac{x^2}{0.16}$$

Solve for x. $x = 2.6 \times 10^{-4}$

$[H_3O^+] = [HCO_3^-] = x = 2.6 \times 10^{-4}$ M; $[H_2CO_3] = 0.16 - x = 0.16$ M

The second dissociation of H_2CO_3 produces a negligible amount of H_3O^+ compared with that from the first dissociation.

$$HCO_3^-(aq) + H_2O(l) \rightleftarrows H_3O^+(aq) + CO_3^{2-}(aq)$$

$$K_{a2} = \frac{[H_3O^+][CO_3^{2-}]}{[HCO_3^-]} = 5.6 \times 10^{-11} = \frac{(2.6 \times 10^{-4})[CO_3^{2-}]}{(2.6 \times 10^{-4})}$$

$[CO_3^{2-}] = K_{a2} = 5.6 \times 10^{-11}$ M

$$[OH^-] = \frac{K_w}{[H_3O^+]} = \frac{1.0 \times 10^{-14}}{2.6 \times 10^{-4}} = 3.8 \times 10^{-11} \text{ M}$$

$$pH = -\log[H_3O^+] = -\log(2.6 \times 10^{-4}) = 3.59$$

16.23

	$NH_3(aq)$	$+$	$H_2O(l)$	$\rightleftharpoons$	$NH_4^+(aq)$	$+$	$OH^-(aq)$
initial (M)	0.40				0		~0
change (M)	$-x$				$+x$		$+x$
equil (M)	$0.40 - x$				x		x

$$K_b = \frac{[NH_4^+][OH^-]}{[NH_3]} = 1.8 \times 10^{-5} = \frac{x^2}{0.40 - x} \approx \frac{x^2}{0.40}$$

Solve for x. $x = [OH^-] = 2.7 \times 10^{-3}$ M

$[NH_4^+] = x = 2.7 \times 10^{-3}$ M; $[NH_3] = 0.40 - x = 0.40$ M

$$[H_3O^+] = \frac{K_w}{[OH^-]} = \frac{1.0 \times 10^{-14}}{2.7 \times 10^{-3}} = 3.7 \times 10^{-12} \text{ M}$$

$$pH = -\log[H_3O^+] = -\log(3.7 \times 10^{-12}) = 11.43$$

16.24 $[H_3O^+] = 10^{-pH} = 10^{-8.15} = 7.1 \times 10^{-9}$ M

$$[OH^-] = \frac{K_w}{[H_3O^+]} = \frac{1.0 \times 10^{-14}}{7.1 \times 10^{-9}} = 1.4 \times 10^{-6} \text{ M}$$

	$C_3H_5O_3^-(aq)$	$+$	$H_2O(l)$	$\rightleftharpoons$	$C_3H_5O_3H(aq)$	$+$	$OH^-(aq)$
(M)	$0.028 - 1.4 \times 10^{-6}$				1.4×10^{-6}		1.4×10^{-6}

$$K_b = \frac{[C_3H_5O_3H][OH^-]}{[C_3H_5O_3^-]} = \frac{(1.4 \times 10^{-6})^2}{(0.028 - 1.4 \times 10^{-6})} \approx \frac{(1.4 \times 10^{-6})^2}{0.028} = 7.0 \times 10^{-11}$$

16.25 $$K_a = \frac{K_w}{K_b \text{ for } C_5H_{11}N} = \frac{1.0 \times 10^{-14}}{1.3 \times 10^{-3}} = 7.7 \times 10^{-12}$$

16.26   (a) The strongest acid has the largest K_a. HX is the strongest acid because it is the most dissociated. Therefore, HX has the largest K_a.
(b) The weakest acid has the strongest conjugate base. HY is the weakest acid because it is the least dissociated. Therefore, Y^- has the largest K_b.
(c) The strongest acid has the weakest conjugate base. HX is the strongest acid because it is the most dissociated. Therefore, X^- is the weakest base.

16.27   0.20 M $NaNO_2$

For NO_2^-, $$K_b = \frac{K_w}{K_a \text{ for } HNO_2} = \frac{1.0 \times 10^{-14}}{4.6 \times 10^{-4}} = 2.2 \times 10^{-11}$$

	$NO_2^-(aq)$	$+$	$H_2O(l)$	$\rightleftharpoons$	$HNO_2(aq)$	$+$	$OH^-(aq)$
initial (M)	0.20				0		~0
change (M)	$-x$				$+x$		$+x$
equil (M)	$0.20 - x$				x		x

$$K_b = \frac{[HNO_2][OH^-]}{[NO_2^-]} = 2.2 \times 10^{-11} = \frac{x^2}{0.20-x} \approx \frac{x^2}{0.20}$$

Solve for x. $x = [OH^-] = 2.1 \times 10^{-6}$ M

$$[H_3O^+] = \frac{K_w}{[OH^-]} = \frac{1.0 \times 10^{-14}}{2.1 \times 10^{-6}} = 4.8 \times 10^{-9} \text{ M}$$

$$pH = -\log[H_3O^+] = -\log(4.8 \times 10^{-9}) = 8.32$$

16.28   (a) Zn^{2+} is an acidic cation. Cl^- is a neutral anion. The salt solution is acidic.

$$Zn(H_2O)_6{}^{2+}(aq) + H_2O(l) \rightleftharpoons H_3O^+(aq) + Zn(H_2O)_5(OH)^+(aq)$$

initial (M)	0.40	~0	0
change (M)	–x	+x	+x
equil(M)	0.40 – x	x	x

$$K_a = \frac{[H_3O^+][Zn(H_2O)_5(OH)^+]}{[Zn(H_2O)_6{}^{2+}]} = 2.5 \times 10^{-10} = \frac{x^2}{0.40-x} \approx \frac{x^2}{0.40}$$

Solve for x. $x = [H_3O^+] = 1.0 \times 10^{-5}$ M

$$pH = -\log[H_3O^+] = -\log(1.0 \times 10^{-5}) = 5.00$$

$$\% \text{ dissociation} = \frac{[Zn(H_2O)_6{}^{2+}]_{diss}}{[Zn(H_2O)_6{}^{2+}]_{initial}} \times 100\% = \frac{1.0 \times 10^{-5} \text{ M}}{0.40 \text{ M}} \times 100\% = 2.5 \times 10^{-3} \%$$

(b) $Fe(H_2O)_6{}^{3+}$ is a stronger acid than $Zn(H_2O)_6{}^{2+}$ because of the higher 3+ charge. The stronger acid, $Fe(H_2O)_6{}^{3+}$, has the higher percent dissociation.

16.29   Lewis acid, $AlCl_3$; Lewis base, Cl^-

16.30

16.31 H_2SO_4, sulfuric acid and HNO_3, nitric acid

16.32   1985, pH = 4.1 to 4.7;   2016, pH = 5.1 to 5.7

16.33 $CaO(s) + H_2SO_4(aq) \rightarrow CaSO_4(aq) + H_2O(l)$

16.34   (a) $\Delta pH = 5.6 - 3.6 = 2.0$
A decrease of 2 pH units represents an increase in $[H_3O^+]$ by a factor of 100.
(b) $[H_3O^+] = 10^{-pH} = 10^{-5.10} = 7.9 \times 10^{-6}$ M

$$[OH^-] = \frac{K_w}{[H_3O^+]} = \frac{1.0 \times 10^{-14}}{7.9 \times 10^{-6}} = 1.3 \times 10^{-9} \text{ M}$$

16.35 NO_2, 46.0

$$5.50 \text{ mg} \times \frac{1 \times 10^{-3} \text{ g}}{1 \text{ mg}} \times \frac{1 \text{ mol NO}_2}{46.0 \text{ g NO}_2} = 1.20 \times 10^{-4} \text{ mol NO}_2$$

$4 NO_2(g) + 2 H_2O(l) + O_2(g) \rightarrow 4 HNO_3(aq)$

mol HNO_3 = mol NO_2 = 1.20×10^{-4} mol
$[HNO_3] = 1.20 \times 10^{-4}$ mol/1.00 L = 1.20×10^{-4} M
Because HNO_3 is a strong acid, $[H_3O^+] = [HNO_3] = 1.20 \times 10^{-4}$ M
pH = $-\log [H_3O^+] = -\log (1.20 \times 10^{-4}) = 3.921$

16.36   Lewis acids include not only H^+ but also other cations and neutral molecules having vacant valence orbitals that can accept a share in a pair of electrons donated by a Lewis base. The O^{2-} from CaO is the Lewis base and SO_2 is the Lewis acid.

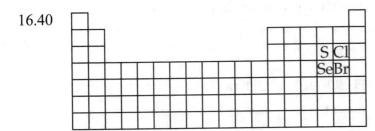

16.37   For NH_4^+, $K_a = \dfrac{K_w}{K_b \text{ for NH}_3} = \dfrac{1.0 \times 10^{-14}}{1.8 \times 10^{-5}} = 5.6 \times 10^{-10}$

For SO_4^{2-}, $K_b = \dfrac{K_w}{K_a \text{ for HSO}_4^-} = \dfrac{1.0 \times 10^{-14}}{1.2 \times 10^{-2}} = 8.3 \times 10^{-13}$

NH_4^+ is acidic, NO_3^- is neutral. NH_4NO_3 is an acidic salt.
NH_4^+ is acidic, SO_4^{2-} is basic. $K_a > K_b$, $(NH_4)_2SO_4$ is an acidic salt.

Conceptual Problems

16.38   (a) acids, HCO_3^- and H_3O^+; bases, H_2O and CO_3^{2-}
        (b) acids, HF and H_2CO_3; bases HCO_3^- and F^-

16.40

(a) H_2S, weakest; HBr, strongest. Acid strength for H_nX increases with increasing polarity of the H–X bond and with increasing size of X.
(b) H_2SeO_3, weakest; $HClO_3$, strongest. Acid strength for H_nYO_3 increases with increasing electronegativity of Y.

16.42   For H_2SO_4, there is complete dissociation of only the first H^+. At equilibrium, there should be H_3O^+, HSO_4^-, and a small amount of SO_4^{2-}. This is best represented by (b).

16.44　(a) $M(H_2O)_6^{n+}(aq) + H_2O(l) \rightleftharpoons H_3O^+(aq) + M(H_2O)_5(OH)^{(n-1)+}(aq)$

$$K_a = \frac{[H_3O^+][M(H_2O)_5(OH)^{(n-1)+}]}{[M(H_2O)_6^{n+}]}$$

(b) As the charge of the metal cation increases, the equilibrium constant increases because the OH bonds in coordinated water molecules are more polar, facilitating the dissociation of a H^+.

(c) $M(H_2O)_6^{3+}$ is the strongest acid. $M(H_2O)_6^+$, the weakest acid, has the strongest conjugate base.

16.46　(a) $H_3BO_3(aq) + H_2O(l) \rightleftharpoons H_3O^+(aq) + H_2BO_3^-(aq)$
(b) $H_3BO_3(aq) + 2\,H_2O(l) \rightleftharpoons H_3O^+(aq) + B(OH)_4^-(aq)$

Section Problems
Acid–Base Concepts (Sections 16.1)

16.48　A Brønsted-Lowry base is a proton acceptor. An Arrhenius base dissociates producing OH^-.
(a) NH_3 and (c) HCO_3^-

16.50　(a) SO_4^{2-}　(b) HSO_3^-　(c) HPO_4^{2-}　(d) NH_3　(e) OH^-　(f) NH_2^-

16.52　(a) $CH_3CO_2H(aq) + NH_3(aq) \rightleftharpoons NH_4^+(aq) + CH_3CO_2^-(aq)$
　　　　　acid　　　　　base ——— acid　　　　　base

(b) $CO_3^{2-}(aq) + H_3O^+(aq) \rightleftharpoons H_2O(l) + HCO_3^-(aq)$
　　　base　　　　acid ——— base　　　acid

(c) $HSO_3^-(aq) + H_2O(l) \rightleftharpoons H_3O^+(aq) + SO_3^{2-}(aq)$
　　　acid　　　　　base ——— acid　　　　base

(d) $HSO_3^-(aq) + H_2O(l) \rightleftharpoons H_2SO_3(aq) + OH^-(aq)$
　　　base　　　　acid　　　acid　　　base

16.54　In aqueous solution:
H_2S acts as an acid only.
HS^- can act as both an acid and a base.
S^{2-} can act as a base only.
H_2O can act as both an acid and a base.
H_3O^+ acts as an acid only.
OH^- acts as a base only.

Acid and Base Strength (Section 16.2)

16.56   The direction of the reaction for $K_c > 1$, is proton transfer from the stronger acid to the stronger base to give the weaker base and the weaker acid:
$$HSO_4^- + NO_2^- \rightarrow SO_4^{2-} + HNO_2$$

16.58   The weaker acid of the pair has the stronger conjugate base.
(a) HCl is a strong acid. HF is a weak acid. HF has the stronger conjugate base.
(b) NH_4^+ is a weaker acid than HF. NH_4^+ has the stronger conjugate base.
(c) HCN is a weaker acid than HSO_4^-. HCN has the stronger conjugate base.

Factors That Affect Acid Strength (Section 16.3)

16.60   (a) $PH_3 < H_2S < HCl$; electronegativity increases from P to Cl
(b) $NH_3 < PH_3 < AsH_3$; X–H bond strength decreases from N to As (down a group)
(c) $HBrO < HBrO_2 < HBrO_3$; acid strength increases with the number of O atoms

16.62   (a) HCl; The strength of a binary acid H_nA increases as A moves from left to right and from top to bottom in the periodic table.
(b) $HClO_3$; The strength of an oxoacid increases with increasing electronegativity and increasing oxidation state of the central atom.
(c) HBr; The strength of a binary acid H_nA increases as A moves from left to right and from top to bottom in the periodic table.

16.64   (a) H_2Te, weaker X–H bond
(b) H_3PO_4, P has higher electronegativity
(c) $H_2PO_4^-$, lower negative charge
(d) NH_4^+, higher positive charge and N is more electronegative than C

16.66   (b)

```
        O
        ||
H ——— C ——— O ——— H
```

The HCO group is more electronegative than the CH_3 group, (b) is more acidic.

Dissociation of Water; pH (Sections 16.4–16.6)

16.68 $[H_3O^+] = \dfrac{K_w}{[OH^-]} = \dfrac{1.0 \times 10^{-14}}{2.0 \times 10^{-6}} = 5.0 \times 10^{-9} M$

Because $[OH^-] > [H_3O^+]$, the solution is basic.

16.70   If $[H_3O^+] > 1.0 \times 10^{-7} M$, solution is acidic.
If $[H_3O^+] < 1.0 \times 10^{-7} M$, solution is basic.
If $[H_3O^+] = [OH^-] = 1.0 \times 10^{-7} M$, solution is neutral.
If $[OH^-] > 1.0 \times 10^{-7} M$, solution is basic
If $[OH^-] < 1.0 \times 10^{-7} M$, solution is acidic.

(a) $\quad [OH^-] = \dfrac{K_w}{[H_3O^+]} = \dfrac{1.0 \times 10^{-14}}{3.4 \times 10^{-9}} = 2.9 \times 10^{-6}$ M, basic

(b) $\quad [H_3O^+] = \dfrac{K_w}{[OH^-]} = \dfrac{1.0 \times 10^{-14}}{0.010} = 1.0 \times 10^{-12}$ M, basic

(c) $\quad [H_3O^+] = \dfrac{K_w}{[OH^-]} = \dfrac{1.0 \times 10^{-14}}{1.0 \times 10^{-10}} = 1.0 \times 10^{-4}$ M, acidic

(d) $\quad [OH^-] = \dfrac{K_w}{[H_3O^+]} = \dfrac{1.0 \times 10^{-14}}{1.0 \times 10^{-7}} = 1.0 \times 10^{-7}$ M, neutral

(e) $\quad [OH^-] = \dfrac{K_w}{[H_3O^+]} = \dfrac{1.0 \times 10^{-14}}{8.6 \times 10^{-5}} = 1.2 \times 10^{-10}$ M, acidic

16.72   At 200 °C and 750 atm, $K_w = [H_3O^+][OH^-] = 1.5 \times 10^{-11}$ and in pure water $[H_3O^+] = [OH^-]$.
$[H_3O^+]^2 = K_w$

$[H_3O^+] = [OH^-] = \sqrt{K_w} = \sqrt{1.5 \times 10^{-11}} = 3.9 \times 10^{-6}$ M

Because $[H_3O^+] = [OH^-]$, the solution is neutral.

16.74   (a) pH $= -\log[H_3O^+] = -\log(2.0 \times 10^{-5}) = 4.70$

(b) $[H_3O^+] = \dfrac{K_w}{[OH^-]} = \dfrac{1.0 \times 10^{-14}}{4 \times 10^{-3}} = 2.5 \times 10^{-12}$ M

pH $= -\log[H_3O^+] = -\log(2.5 \times 10^{-12}) = 11.6$
(c) pH $= -\log[H_3O^+] = -\log(3.56 \times 10^{-9}) = 8.449$
(d) pH $= -\log[H_3O^+] = -\log(10^{-3}) = 3$

(e) $[H_3O^+] = \dfrac{K_w}{[OH^-]} = \dfrac{1.0 \times 10^{-14}}{12} = 8.3 \times 10^{-16}$ M

pH $= -\log[H_3O^+] = -\log(8.3 \times 10^{-16}) = 15.08$

16.76 $[H_3O^+] = 10^{-pH}$;   (a) 8×10^{-5} M   (b) 1.5×10^{-11} M   (c) 1.0 M
                (d) 5.6×10^{-15} M   (e) 10 M   (f) 5.78×10^{-6} M

16.78   (a) chlorphenol red   (b) thymol blue   (c) methyl orange

Strong Acids and Strong Bases (Section 16.7)

16.80   From data in Table 16.1:   Strong acids: HNO_3;   Strong bases: none

16.82   (a) Because $Sr(OH)_2$ is a strong base, $[OH^-] = 2(1.0 \times 10^{-3}$ M$) = 2.0 \times 10^{-3}$ M.

$[H_3O^+] = \dfrac{K_w}{[OH^-]} = \dfrac{1.0 \times 10^{-14}}{2.0 \times 10^{-3}} = 5.0 \times 10^{-12}$ M

pH $= -\log[H_3O^+] = -\log(5.0 \times 10^{-12}) = 11.30$
(b)   Because HNO_3 is a strong acid, $[H_3O^+] = 0.015$ M.

$pH = -\log[H_3O^+] = -\log(0.015) = 1.82$

(c) Because NaOH is a strong base, $[OH^-] = 0.035$ M.

$$[H_3O^+] = \frac{K_w}{[OH^-]} = \frac{1.0 \times 10^{-14}}{0.035} = 2.9 \times 10^{-13} \text{ M}$$

$pH = -\log[H_3O^+] = -\log(2.9 \times 10^{-13}) = 12.54$

16.84 (a) LiOH, 23.95;   250 mL = 0.250 L

$$\text{molarity of LiOH(aq)} = \frac{\left(4.8 \text{ g} \times \dfrac{1 \text{ mol}}{23.95 \text{ g}}\right)}{0.250 \text{ L}} = 0.80 \text{ M}$$

LiOH is a strong base; therefore $[OH^-] = 0.80$ M

$$[H_3O^+] = \frac{K_w}{[OH^-]} = \frac{1.0 \times 10^{-14}}{0.80} = 1.25 \times 10^{-14} \text{ M}$$

$pH = -\log[H_3O^+] = -\log(1.25 \times 10^{-14}) = 13.90$

(b) HCl, 36.46

$$\text{molarity of HCl(aq)} = \frac{\left(0.93 \text{ g} \times \dfrac{1 \text{ mol}}{36.46 \text{ g}}\right)}{0.40 \text{ L}} = 0.064 \text{ M}$$

HCl is a strong acid; therefore $[H_3O^+] = 0.064$ M

$pH = -\log[H_3O^+] = -\log(0.064) = 1.19$

(c) $M_f \cdot V_f = M_i \cdot V_i$

$$M_f = \frac{M_i \cdot V_i}{V_f} = \frac{(0.10 \text{ M})(50 \text{ mL})}{(1000 \text{ mL})} = 5.0 \times 10^{-3} \text{ M}$$

$pH = -\log[H_3O^+] = -\log(5.0 \times 10^{-3}) = 2.30$

(d) For HCl, $M_f = \dfrac{M_i \cdot V_i}{V_f} = \dfrac{(2.0 \times 10^{-3} \text{ M})(100 \text{ mL})}{(500 \text{ mL})} = 4.0 \times 10^{-4} \text{ M}$

For HClO$_4$, $M_f = \dfrac{M_i \cdot V_i}{V_f} = \dfrac{(1.0 \times 10^{-3} \text{ M})(400 \text{ mL})}{(500 \text{ mL})} = 8.0 \times 10^{-4} \text{ M}$

$[H_3O^+] = (4.0 \times 10^{-4} \text{ M}) + (8.0 \times 10^{-4} \text{ M}) = 1.2 \times 10^{-3} \text{ M}$

$pH = -\log[H_3O^+] = -\log(1.2 \times 10^{-3}) = 2.92$

16.86 CaO, 56.08

$[H_3O^+] = 10^{-pH} = 10^{-(10.50)} = 3.16 \times 10^{-11}$ M

$$[OH^-] = \frac{K_w}{[H_3O^+]} = \frac{1.0 \times 10^{-14}}{3.16 \times 10^{-11}} = 3.16 \times 10^{-4} \text{ M} = 3.16 \times 10^{-4} \text{ mol/L}$$

$CaO(s) + H_2O(l) \rightarrow Ca^{2+}(aq) + 2 \, OH^-(aq)$

3.16×10^{-4} mol OH$^-$ $\times \dfrac{1 \text{ mol CaO}}{2 \text{ mol OH}^-} \times \dfrac{56.08 \text{ g CaO}}{1 \text{ mol CaO}} = 0.0089$ g CaO

Weak Acids (Sections 16.8–16.10)

16.88   (a) The larger the K_a, the stronger the acid.
$$C_6H_5OH \; < \; HOCl \; < \; CH_3CO_2H \; < \; HNO_3$$
(b) The larger the K_a, the larger the percent dissociation for the same concentration.
$$HNO_3 \; > \; CH_3CO_2H \; > \; HOCl \; > \; C_6H_5OH$$
1 M HNO_3, $[H_3O^+] = 1$ M

1 M CH_3CO_2H, $[H_3O^+] = \sqrt{[HA] \times K_a} = \sqrt{(1\ M)(1.8 \times 10^{-5})} = 4 \times 10^{-3}$ M

1 M $HOCl$, $[H_3O^+] = \sqrt{[HA] \times K_a} = \sqrt{(1\ M)(3.5 \times 10^{-8})} = 2 \times 10^{-4}$ M

1 M C_6H_5OH, $[H_3O^+] = \sqrt{[HA] \times K_a} = \sqrt{(1\ M)(1.3 \times 10^{-10})} = 1 \times 10^{-5}$ M

16.90

	$HOBr(aq)$	+	$H_2O(l)$	⇌	$H_3O^+(aq)$	+	$OBr^-(aq)$
initial (M)	0.040				~0		0
change (M)	$-x$				$+x$		$+x$
equil (M)	$0.040 - x$				x		x

$x = [H_3O^+] = 10^{-pH} = 10^{-5.05} = 8.9 \times 10^{-6}$ M

$$K_a = \frac{[H_3O^+][OBr^-]}{[HOBr]} = \frac{x^2}{0.040 - x} = \frac{(8.9 \times 10^{-6})^2}{0.040 - (8.9 \times 10^{-6})} = 2.0 \times 10^{-9}$$

16.92

	$HA(aq)$	+	$H_2O(l)$	⇌	$H_3O^+(aq)$	+	$A^-(aq)$
initial (M)	0.050				~0		0
change (M)	$-x$				$+x$		$+x$
equil (M)	$0.050 - x$				x		x

$x = [H_3O^+] = 10^{-pH} = 10^{-2.86} = 1.38 \times 10^{-3}$ M

$$K_a = \frac{[H_3O^+][A^-]}{[HA]} = \frac{x^2}{0.050 - x} = \frac{(1.38 \times 10^{-3})^2}{0.050 - (1.38 \times 10^{-3})} = 3.92 \times 10^{-5} = 3.9 \times 10^{-5}$$

$pK_a = -\log K_a = -\log(3.92 \times 10^{-5}) = 4.41$

16.94 $C_6H_8O_6$, 176.13;   250 mg = 0.250 g;   250 mL = 0.250 L

$$[C_6H_8O_6] = \frac{\left(0.250\ g \times \dfrac{1\ mol}{176.13\ g} \right)}{0.250\ L} = 5.68 \times 10^{-3}\ M$$

	$C_6H_8O_6(aq)$	+	$H_2O(l)$	⇌	$H_3O^+(aq)$	+	$C_6H_7O_6^-(aq)$
initial (M)	5.68×10^{-3}				~0		0
change (M)	$-x$				$+x$		$+x$
equil (M)	$(5.68 \times 10^{-3}) - x$				x		x

$$K_a = \frac{[H_3O^+][C_6H_7O_6^-]}{[C_6H_8O_6]} = 8.0 \times 10^{-5} = \frac{x^2}{(5.68 \times 10^{-3}) - x}$$

$x^2 + (8.0 \times 10^{-5})x - (4.54 \times 10^{-7}) = 0$

Use the quadratic formula to solve for x.

$$x = \frac{-(8.0 \times 10^{-5}) \pm \sqrt{(8.0 \times 10^{-5})^2 - (4)(-4.54 \times 10^{-7})}}{2(1)} = \frac{(-8.0 \times 10^{-5}) \pm 0.001\ 35}{2}$$

$x = 6.35 \times 10^{-4}$ and -7.15×10^{-4}

Of the two solutions for x, only the positive value of x has physical meaning because x is the $[H_3O^+]$.

$x = [H_3O^+] = 6.35 \times 10^{-4}$ M

$pH = -\log[H_3O^+] = -\log(6.35 \times 10^{-4}) = 3.20$

16.96 $K_a = 10^{-pK_a} = 10^{-4.25} = 5.6 \times 10^{-5}$

$$HC_3H_3O_2(aq) + H_2O(l) \rightleftharpoons H_3O^+(aq) + C_3H_3O_2^-(aq)$$

initial (M)	0.150	~0	0
change (M)	−x	+x	+x
equil (M)	0.150 − x	x	x

$$K_a = \frac{[H_3O^+][C_3H_3O_2^-]}{[HC_3H_3O_2]} = 5.6 \times 10^{-5} = \frac{x^2}{0.150 - x} \approx \frac{x^2}{0.150}$$

Solve for x. $x = 0.0029$ M $= [H_3O^+] = [C_3H_3O_2^-]$

$[HC_3H_3O_2] = 0.150 - x = 0.150 - 0.0029 = 0.147$ M

$pH = -\log[H_3O^+] = -\log(0.0029) = 2.54$

$$[OH^-] = \frac{K_w}{[H_3O^+]} = \frac{1.0 \times 10^{-14}}{0.0029} = 3.4 \times 10^{-12}\ M$$

(b) $$HC_3H_3O_2(aq) + H_2O(l) \rightleftharpoons H_3O^+(aq) + C_3H_3O_2^-(aq)$$

initial (M)	0.0500	~0	0
change (M)	−x	+x	+x
equil (M)	0.0500 − x	x	x

$$K_a = \frac{[H_3O^+][C_3H_3O_2^-]}{[HC_3H_3O_2]} = 5.6 \times 10^{-5} = \frac{x^2}{0.0500 - x} \approx \frac{x^2}{0.0500}$$

Solve for x. $x = 0.001\ 67$ M $= [H_3O^+] = [HC_3H_3O_2]_{diss}$

$$\% \text{ dissociation} = \frac{[HC_3H_3O_2]_{diss}}{[HC_3H_3O_2]_{initial}} \times 100\% = \frac{0.001\ 67\ M}{0.0500\ M} \times 100\% = 3.3\%$$

16.98 $$HNO_2(aq) + H_2O(l) \rightleftharpoons H_3O^+(aq) + NO_2^-(aq)$$

initial (M)	1.5	~0	0
change (M)	−x	+x	+x
equil (M)	1.5 − x	x	x

$$K_a = \frac{[H_3O^+][NO_2^-]}{[HNO_2]} = 4.5 \times 10^{-4} = \frac{x^2}{1.5 - x} \approx \frac{x^2}{1.5}$$

Solve for x. $x = 0.026$ M $= [H_3O^+];$ $pH = -\log[H_3O^+] = -\log(0.026) = 1.59$

$$\% \text{ dissociation} = \frac{[HNO_2]_{diss}}{[HNO_2]_{initial}} \times 100\% = \frac{0.026\ M}{1.5\ M} \times 100\% = 1.7\%$$

16.100 (a) From Worked Example 16.10 in the text:

$[H_3O^+] = [HF]_{diss} = 4.0 \times 10^{-3}$ M

% dissociation $= \dfrac{[HF]_{diss}}{[HF]_{initial}} \times 100\% = \dfrac{4.0 \times 10^{-3} \text{ M}}{0.050 \text{ M}} \times 100\% = 8.0\%$ dissociation

(b)

	HF(aq)	$+$	H$_2$O(l)	$\rightleftharpoons$	H$_3$O$^+$(aq)	$+$	F$^-$(aq)
initial (M)	0.50				~0		0
change (M)	$-x$				$+x$		$+x$
equil (M)	$0.50 - x$				x		x

$K_a = \dfrac{[H_3O^+][F^-]}{[HF]} = 3.5 \times 10^{-4} = \dfrac{x^2}{0.50 - x}$

$x^2 + (3.5 \times 10^{-4})x - (1.75 \times 10^{-4}) = 0$

Use the quadratic formula to solve for x.

$x = \dfrac{-(3.5 \times 10^{-4}) \pm \sqrt{(3.5 \times 10^{-4})^2 - 4(1)(-1.75 \times 10^{-4})}}{2(1)} = \dfrac{(-3.5 \times 10^{-4}) \pm 0.0265}{2}$

$x = 0.0131$ and -0.0134

Of the two solutions for x, only the positive value of x has physical meaning, because x is the $[H_3O^+]$.　　$[H_3O^+] = [HF]_{diss} = 0.013$ M

% dissociation $= \dfrac{[HF]_{diss}}{[HF]_{initial}} \times 100\% = \dfrac{0.013 \text{ M}}{0.50 \text{ M}} \times 100\% = 2.6\%$ dissociation

Polyprotic Acids (Section 16.11)

16.102　$H_2SeO_4(aq) + H_2O(l) \rightleftharpoons H_3O^+(aq) + HSeO_4^-(aq);\ \ K_{a1} = \dfrac{[H_3O^+][HSeO_4^-]}{[H_2SeO_4]}$

$HSeO_4^-(aq) + H_2O(l) \rightleftharpoons H_3O^+(aq) + SeO_4^{2-}(aq);\ \ K_{a2} = \dfrac{[H_3O^+][SeO_4^{2-}]}{[HSeO_4^-]}$

16.104

	H$_2$CO$_3$(aq)	$+$	H$_2$O(l)	$\rightleftharpoons$	H$_3$O$^+$(aq)	$+$	HCO$_3^-$(aq)
initial (M)	0.010				~0		0
change (M)	$-x$				$+x$		$+x$
equil (M)	$0.010 - x$				x		x

$K_{a1} = \dfrac{[H_3O^+][HCO_3^-]}{[H_2CO_3]} = 4.3 \times 10^{-7} = \dfrac{x^2}{0.010 - x} \approx \dfrac{x^2}{0.010}$

Solve for x.　$x = 6.6 \times 10^{-5}$

$[H_3O^+] = [HCO_3^-] = x = 6.6 \times 10^{-5}$ M;　　　$[H_2CO_3] = 0.010 - x = 0.010$ M

The second dissociation of H_2CO_3 produces a negligible amount of H_3O^+ compared with that from the first dissociation.

$HCO_3^-(aq) + H_2O(l) \rightleftharpoons H_3O^+(aq) + CO_3^{2-}(aq)$

$K_{a2} = \dfrac{[H_3O^+][CO_3^{2-}]}{[HCO_3^-]} = 5.6 \times 10^{-11} = \dfrac{(6.6 \times 10^{-5})[CO_3^{2-}]}{(6.6 \times 10^{-5})}$

$[CO_3^{2-}] = K_{a2} = 5.6 \times 10^{-11}$ M

$[OH^-] = \dfrac{K_w}{[H_3O^+]} = \dfrac{1.0 \times 10^{-14}}{6.6 \times 10^{-5}} = 1.5 \times 10^{-10}$ M

$pH = -\log[H_3O^+] = -\log(6.6 \times 10^{-5}) = 4.18$

16.106 For the dissociation of the first proton, the following equilibrium must be considered:

$$H_2C_2O_4(aq) \ + \ H_2O(l) \ \rightleftarrows \ H_3O^+(aq) \ + \ HC_2O_4^-(aq)$$

initial (M)	0.20	~0	0
change (M)	−x	+x	+x
equil (M)	0.20 − x	x	x

$K_{a1} = \dfrac{[H_3O^+][HC_2O_4^-]}{[H_2C_2O_4]} = 5.9 \times 10^{-2} = \dfrac{x^2}{0.20 - x}$

$x^2 + 0.059x - 0.0118 = 0$

Use the quadratic formula to solve for x.

$x = \dfrac{-(0.059) \pm \sqrt{(0.059)^2 - 4(1)(-0.0118)}}{2(1)} = \dfrac{-0.059 \pm 0.225}{2}$

$x = 0.083$ and -0.142

Of the two solutions for x, only the positive value of x has physical meaning, because x is the $[H_3O^+]$.

$[H_3O^+] = [HC_2O_4^-] = 0.083$ M

For the dissociation of the second proton, the following equilibrium must be considered:

$$HC_2O_4^-(aq) \ + \ H_2O(l) \ \rightleftarrows \ H_3O^+(aq) \ + \ C_2O_4^{2-}(aq)$$

initial (M)	0.083	0.083	0
change (M)	−x	+x	+x
equil (M)	0.083 − x	0.083 + x	x

$K_{a2} = \dfrac{[H_3O^+][C_2O_4^{2-}]}{[HC_2O_4^-]} = 6.4 \times 10^{-5} = \dfrac{(0.083 + x)(x)}{0.083 - x} \approx \dfrac{(0.083)(x)}{0.083} = x$

$[H_3O^+] = 0.083 + x = 0.083$ M

$pH = -\log[H_3O^+] = -\log(0.083) = 1.08$

$[C_2O_4^{2-}] = x = 6.4 \times 10^{-5}$ M

16.108 From the complete dissociation of the first proton, $[H_3O^+] = [HSeO_4^-] = 0.50$ M.

For the dissociation of the second proton, the following equilibrium must be considered:

$$HSeO_4^-(aq) \ + \ H_2O(l) \ \rightleftarrows \ H_3O^+(aq) \ + \ SeO_4^{2-}(aq)$$

initial (M)	0.50	0.50	0
change (M)	−x	+x	+x
equil (M)	0.50 − x	0.50 + x	x

$K_{a2} = \dfrac{[H_3O^+][SeO_4^{2-}]}{[HSeO_4^-]} = 1.2 \times 10^{-2} = \dfrac{(0.50 + x)(x)}{0.50 - x}$

$x^2 + 0.512x - 0.0060 = 0$

Use the quadratic formula to solve for x.

$$x = \frac{-(0.512) \pm \sqrt{(0.512)^2 - 4(-0.0060)}}{2(1)} = \frac{-0.512 \pm 0.535}{2}$$

$x = 0.011$ and -0.524

Of the two solutions for x, only the positive value of x has physical meaning, since x is the $[SeO_4^{2-}]$.

$[H_2SeO_4] = 0\ M$; $[HSeO_4^-] = 0.50 - x = 0.49\ M$; $[SeO_4^{2-}] = x = 0.011\ M$

$[H_3O^+] = 0.50 + x = 0.51\ M$

$pH = -\log[H_3O^+] = -\log(0.51) = 0.29$

$$[OH^-] = \frac{K_w}{[H_3O^+]} = \frac{1.0 \times 10^{-14}}{0.51} = 2.0 \times 10^{-14}\ M$$

16.110 $K_{a1} = 10^{-pK_{a1}} = 10^{-2.43} = 3.7 \times 10^{-3}$; $K_{a2} = 10^{-pK_{a2}} = 10^{-4.78} = 1.7 \times 10^{-5}$

	$H_2C_7H_3NO_4(aq) + H_2O(l)$	$\rightleftarrows$	$H_3O^+(aq)$	$+$	$HC_7H_3NO_4^-(aq)$
initial (M)	0.050		~0		0
change (M)	$-x$		$+x$		$+x$
equil (M)	$0.050 - x$		x		x

$$K_{a1} = \frac{[H_3O^+][HC_7H_3NO_4^-]}{[H_2C_7H_3NO_4]} = 3.7 \times 10^{-3} = \frac{x^2}{0.050 - x}$$

$x^2 + (3.7 \times 10^{-3})x - (1.85 \times 10^{-4}) = 0$

Use the quadratic formula to solve for x.

$$x = \frac{-(3.7 \times 10^{-3}) \pm \sqrt{(3.7 \times 10^{-3})^2 - (4)(1)(-1.85 \times 10^{-4})}}{2(1)} = \frac{-(3.7 \times 10^{-3}) \pm 0.0275}{2}$$

$x = 0.012$ and -0.016

Of the two solutions for x, only the positive value of x has physical meaning because x is the $[H_3O^+]$.

$x = 0.012\ M = [H_3O^+] = [HC_7H_3NO_4^-]$

$[H_2C_7H_3NO_4] = 0.050 - x = 0.050 - 0.012 = 0.038\ M$

The second dissociation of $H_2C_7H_3NO_4$ produces a negligible amount of H_3O^+ compared with that from the first dissociation.

$HC_7H_3NO_4^-(aq) + H_2O(l) \rightleftarrows H_3O^+(aq) + C_7H_3NO_4^{2-}(aq)$

$$K_{a2} = \frac{[H_3O^+][C_7H_3NO_4^{2-}]}{[HC_7H_3NO_4^-]} = 1.7 \times 10^{-5} = \frac{(0.012)[C_7H_3NO_4^{2-}]}{(0.012)}$$

$[C_7H_3NO_4^{2-}] = K_{a2} = 1.7 \times 10^{-5}\ M$

$$[OH^-] = \frac{K_w}{[H_3O^+]} = \frac{1.0 \times 10^{-14}}{0.012} = 8.3 \times 10^{-13}\ M$$

$pH = -\log[H_3O^+] = -\log(0.012) = 1.92$

Weak Bases (Section 16.12)

16.112 (a) $(CH_3)_2NH(aq) + H_2O(l) \rightleftarrows (CH_3)_2NH_2^+(aq) + OH^-(aq);$ $\quad K_b = \dfrac{[(CH_3)_2NH_2^+][OH^-]}{[(CH_3)_2NH]}$

(b) $C_6H_5NH_2(aq) + H_2O(l) \rightleftarrows C_6H_5NH_3^+(aq) + OH^-(aq);$ $\quad K_b = \dfrac{[C_6H_5NH_3^+][OH^-]}{[C_6H_5NH_2]}$

(c) $CN^-(aq) + H_2O(l) \rightleftarrows HCN(aq) + OH^-(aq);$ $\quad K_b = \dfrac{[HCN][OH^-]}{[CN^-]}$

16.114 $C_{21}H_{22}N_2O_2$, 334.42; $\quad$ 16 mg = 0.016 g

$$\text{molarity} = \frac{\left(0.016 \text{ g} \times \dfrac{1 \text{ mol}}{334.42 \text{ g}}\right)}{0.100 \text{ L}} = 4.8 \times 10^{-4} \text{ M}$$

$$C_{21}H_{22}N_2O_2(aq) \;+\; H_2O(l) \;\rightleftarrows\; C_{21}H_{23}N_2O_2^+(aq) \;+\; OH^-(aq)$$

initial (M)	4.8×10^{-4}	0	~0
change (M)	$-x$	$+x$	$+x$
equil (M)	$(4.8 \times 10^{-4}) - x$	x	x

$$K_b = \frac{[C_{21}H_{23}N_2O_2^+][OH^-]}{[C_{21}H_{22}N_2O_2]} = 1.8 \times 10^{-6} = \frac{x^2}{(4.8 \times 10^{-4}) - x}$$

$x^2 + (1.8 \times 10^{-6})x - (8.6 \times 10^{-10}) = 0$

Use the quadratic formula to solve for x.

$$x = \frac{-(1.8 \times 10^{-6}) \pm \sqrt{(1.8 \times 10^{-6})^2 - (4)(-8.6 \times 10^{-10})}}{2(1)} = \frac{(-1.8 \times 10^{-6}) \pm (5.87 \times 10^{-5})}{2}$$

$x = 2.84 \times 10^{-5}$ and -3.02×10^{-5}

Of the two solutions for x, only the positive value of x has physical meaning, because x is the $[OH^-]$.

$[OH^-] = 2.84 \times 10^{-5}$ M

$$[H_3O^+] = \frac{K_w}{[OH^-]} = \frac{1.0 \times 10^{-14}}{2.84 \times 10^{-5}} = 3.52 \times 10^{-10} \text{ M}$$

$pH = -\log[H_3O^+] = -\log(3.52 \times 10^{-10}) = 9.45$

16.116 $[H_3O^+] = 10^{-pH} = 10^{-9.5} = 3.16 \times 10^{-10}$ M

$$[OH^-] = \frac{K_w}{[H_3O^+]} = \frac{1.0 \times 10^{-14}}{3.16 \times 10^{-10}} = 3.16 \times 10^{-5} \text{ M}$$

$$C_{17}H_{19}NO_3(aq) \;+\; H_2O(l) \;\rightleftarrows\; C_{17}H_{20}NO_3^+(aq) \;+\; OH^-(aq)$$

initial (M)	7.0×10^{-4}	0	~0
change (M)	$-x$	$+x$	$+x$
equil (M)	$(7.0 \times 10^{-4}) - x$	x	x

$x = [OH^-] = 3.16 \times 10^{-5}$ M

$$K_b = \frac{[C_{17}H_{20}NO_3^+][OH^-]}{[C_{17}H_{19}NO_3]} = \frac{x^2}{(7.0 \times 10^{-4}) - x} = \frac{(3.16 \times 10^{-5})^2}{(7.0 \times 10^{-4}) - (3.16 \times 10^{-5})} = 1.49 \times 10^{-6}$$

$K_b = 1 \times 10^{-6}$

$pK_b = -\log K_b = -\log(1.49 \times 10^{-6}) = 5.827 = 5.8$

16.118 $K_b = 10^{-pK_b} = 10^{-5.47} = 3.4 \times 10^{-6}$

$$C_{18}H_{21}NO_4(aq) + H_2O(l) \rightleftarrows HC_{18}H_{21}NO_4^+(aq) + OH^-(aq)$$

initial (M)	2.50×10^{-3}	0	~0
change (M)	$-x$	$+x$	$+x$
equil (M)	$(2.50 \times 10^{-3}) - x$	x	x

$$K_b = \frac{[HC_{18}H_{21}NO_4^+][OH^-]}{[C_{18}H_{21}NO_4]} = 3.4 \times 10^{-6} = \frac{x^2}{(2.50 \times 10^{-3}) - x}$$

$x^2 + (3.4 \times 10^{-6})x - (8.5 \times 10^{-9}) = 0$

Use the quadratic formula to solve for x.

$$x = \frac{-(3.4 \times 10^{-6}) \pm \sqrt{(3.4 \times 10^{-6})^2 - 4(1)(-8.5 \times 10^{-9})}}{2(1)}$$

$$x = = \frac{-(3.4 \times 10^{-6}) \pm (1.84 \times 10^{-4})}{2}$$

$x = 9.0 \times 10^{-5}$ and -1.9×10^{-4}

Of the two solutions for x, only the positive value of x has physical meaning, because x is the $[OH^-]$.

$x = 9.0 \times 10^{-5}$ M $= [OH^-] = [HC_{18}H_{21}NO_4^+]$

$[C_{18}H_{21}NO_4] = (2.50 \times 10^{-3}) - x = (2.50 \times 10^{-3}) - (9.0 \times 10^{-5}) = 0.0024$ M

$$[H_3O^+] = \frac{K_w}{[OH^-]} = \frac{1.0 \times 10^{-14}}{9.0 \times 10^{-5}} = 1.1 \times 10^{-10} \text{ M}$$

$pH = -\log[H_3O^+] = -\log(1.1 \times 10^{-10}) = 9.96$

Relation Between K_a and K_b (Section 16.13)

16.120 (a) $K_a = \dfrac{K_w}{K_b \text{ for } C_3H_7NH_2} = \dfrac{1.0 \times 10^{-14}}{5.1 \times 10^{-4}} = 2.0 \times 10^{-11}$

(b) $K_a = \dfrac{K_w}{K_b \text{ for } NH_2OH} = \dfrac{1.0 \times 10^{-14}}{9.1 \times 10^{-9}} = 1.1 \times 10^{-6}$

(c) $K_a = \dfrac{K_w}{K_b \text{ for } C_6H_5NH_2} = \dfrac{1.0 \times 10^{-14}}{4.3 \times 10^{-10}} = 2.3 \times 10^{-5}$

(d) $K_a = \dfrac{K_w}{K_b \text{ for } C_5H_5N} = \dfrac{1.0 \times 10^{-14}}{1.8 \times 10^{-9}} = 5.6 \times 10^{-6}$

16.122 For $C_{10}H_{14}N_2H^+$, $K_{a1} = \dfrac{K_w}{K_{b1} \text{ for } C_{10}H_{14}N_2} = \dfrac{1.0 \times 10^{-14}}{1.0 \times 10^{-6}} = 1.0 \times 10^{-8}$

For $C_{10}H_{14}N_2H_2^{2+}$, $K_{a2} = \dfrac{K_w}{K_{b2} \text{ for } C_{10}H_{14}N_2H^+} = \dfrac{1.0 \times 10^{-14}}{1.3 \times 10^{-11}} = 7.7 \times 10^{-4}$

Acid–Base Properties of Salts (Section 16.14)

16.124 (a) $CH_3NH_3{}^+(aq) + H_2O(l) \rightleftharpoons H_3O^+(aq) + CH_3NH_2(aq)$

        acid   base ——— acid   base

(b) $Cr(H_2O)_6{}^{3+}(aq) + H_2O(l) \rightleftharpoons H_3O^+(aq) + Cr(H_2O)_5(OH)^{2+}(aq)$

        acid   base ——— acid   base

(c) $CH_3CO_2{}^-(aq) + H_2O(l) \rightleftharpoons CH_3CO_2H(aq) + OH^-(aq)$

        base   acid   acid   base

(d) $PO_4{}^{3-}(aq) + H_2O(l) \rightleftharpoons HPO_4{}^{2-}(aq) + OH^-(aq)$

        base   acid   acid   base

16.126 (a) F^- (conjugate base of a weak acid), basic solution
     (b) Br^- (anion of a strong acid), neutral solution
     (c) $NH_4{}^+$ (conjugate acid of a weak base), acidic solution
     (d) $K(H_2O)_6{}^+$ (neutral cation), neutral solution
     (e) $SO_3{}^{2-}$ (conjugate base of a weak acid), basic solution
     (f) $Cr(H_2O)_6{}^{3+}$ (acidic cation), acidic solution

16.128 (a) $(C_2H_5NH_3)NO_3$: $C_2H_5NH_3{}^+$, acidic cation; $NO_3{}^-$, neutral anion
     $C_2H_5NH_2$, $K_b = 6.4 \times 10^{-4}$

$C_2H_5NH_3{}^+$, $K_a = \dfrac{K_w}{K_b \text{ for } C_2H_5NH_2} = \dfrac{1.0 \times 10^{-14}}{6.4 \times 10^{-4}} = 1.56 \times 10^{-11}$

	$C_2H_5NH_3{}^+(aq)$ + $H_2O(l)$ $\rightleftharpoons$	$H_3O^+(aq)$ +	$C_2H_5NH_2(aq)$
initial (M)	0.10	~0	0
change (M)	−x	+x	+x
equil (M)	0.10 − x	x	x

$K_a = \dfrac{[H_3O^+][C_2H_5NH_2]}{[C_2H_5NH_3{}^+]} = 1.56 \times 10^{-11} = \dfrac{x^2}{0.10 - x} \approx \dfrac{x^2}{0.10}$

Solve for x. $x = 1.25 \times 10^{-6} \text{ M} = 1.2 \times 10^{-6} \text{ M} = [H_3O^+] = [C_2H_5NH_2]$
$pH = -\log[H_3O^+] = -\log(1.25 \times 10^{-6}) = 5.90$
$[C_2H_5NH_3{}^+] = 0.10 - x = 0.10 \text{ M};$ $[NO_3{}^-] = 0.10 \text{ M}$

$$[OH^-] = \frac{K_w}{[H_3O^+]} = \frac{1.0 \times 10^{-14}}{1.25 \times 10^{-6} \text{ M}} = 8.0 \times 10^{-9}$$

(b) $Na(CH_3CO_2)$: Na^+, neutral cation; $CH_3CO_2^-$, basic anion
CH_3CO_2H, $K_a = 1.8 \times 10^{-5}$

$$CH_3CO_2^-, K_b = \frac{K_w}{K_a \text{ for } CH_3CO_2H} = \frac{1.0 \times 10^{-14}}{1.8 \times 10^{-5}} = 5.6 \times 10^{-10}$$

	$CH_3CO_2^-(aq)$	+	$H_2O(aq)$	⇌	$CH_3CO_2H(aq)$	+	$OH^-(aq)$
initial (M)	0.10				0		~0
change (M)	−x				+x		+x
equil (M)	0.10 − x				x		x

$$K_b = \frac{[CH_3CO_2H][OH^-]}{[CH_3CO_2^-]} = 5.6 \times 10^{-10} = \frac{x^2}{0.10-x} \approx \frac{x^2}{0.10}$$

Solve for x. $x = 7.5 \times 10^{-6} \text{ M} = [CH_3CO_2H] = [OH^-]$
$[CH_3CO_2^-] = 0.10 - x = 0.10 \text{ M};$ $[Na^+] = 0.10 \text{ M}$

$$[H_3O^+] = \frac{K_w}{[OH^-]} = \frac{1.0 \times 10^{-14}}{7.5 \times 10^{-6}} = 1.3 \times 10^{-9} \text{ M}$$

$pH = -\log[H_3O^+] = -\log(1.3 \times 10^{-9}) = 8.89$
(c) $NaNO_3$: Na^+, neutral cation; NO_3^-, neutral anion
$[Na^+] = [NO_3^-] = 0.10 \text{ M}$
$[H_3O^+] = [OH^-] = 1.0 \times 10^{-7} \text{ M};$ $pH = 7.00$

16.130 For NH_4^+, $K_a = \dfrac{K_w}{K_b \text{ for } NH_3} = \dfrac{1.0 \times 10^{-14}}{1.8 \times 10^{-5}} = 5.6 \times 10^{-10}$

For CN^-, $K_b = \dfrac{K_w}{K_a \text{ for } HCN} = \dfrac{1.0 \times 10^{-14}}{4.9 \times 10^{-10}} = 2.0 \times 10^{-5}$

Because $K_b > K_a$, the solution is basic.

16.132 (a) $A^-(aq) + H_2O(l) ⇌ HA(aq) + OH^-(aq)$; basic
(b) $M(H_2O)_6^{3+}(aq) + H_2O(l) ⇌ H_3O^+(aq) + M(H_2O)_5(OH)^{2+}(aq)$; acidic
(c) $2 H_2O(l) ⇌ H_3O^+(aq) + OH^-(aq)$; neutral
(d) $M(H_2O)_6^{3+}(aq) + A^-(aq) ⇌ HA(aq) + M(H_2O)_5(OH)^{2+}(aq)$;
acidic because K_a for $M(H_2O)_6^{3+}$ (10^{-4}) is greater than K_b for A^- (10^{-9})

16.134

$NH_4^+(aq) + H_2O(l) ⇌ H_3O^+(aq) + NH_3(aq)$	$K_1 = K_a$
$F^-(aq) + H_2O(l) ⇌ HF(aq) + OH^-(aq)$	$K_2 = K_b$
$H_3O^+(aq) + OH^-(aq) ⇌ 2 H_2O(l)$	$K_3 = 1/K_w$
Overall reaction $NH_4^+(aq) + F^-(aq) ⇌ HF(aq) + NH_3(aq)$	$K = K_1K_2K_3$

$$K = K_1K_2K_3 = \frac{K_aK_b}{K_w}$$

The equilibrium constant for the transfer of a proton from the cation of a salt to the anion of the salt is equal to $(K_aK_b)/K_w$.

(a)

	$NH_4^+(aq)$	$+$	$F^-(aq)$	$\rightleftharpoons$	$HF(aq)$	$+$	$NH_3(aq)$
initial (M)	0.25		0.25		0		0
change (M)	$-x$		$-x$		$+x$		$+x$
equil (M)	$0.25 - x$		$0.25 - x$		x		x

$$K = \frac{K_aK_b}{K_w} = \frac{(5.56 \times 10^{-10})(2.86 \times 10^{-11})}{(1.0 \times 10^{-14})} = \frac{[HF][NH_3]}{[NH_4^+][F^-]} = \frac{x^2}{(0.25 - x)^2}$$

where K_a is K_a for NH_4^+ and K_b is K_b for F^-.
Take the square root of both sides and solve for x.
$x = 3.15 \times 10^{-4}$
$[NH_4^+] = [F^-] = 0.25 - x = 0.25$ M; $[NH_3] = [HF] = x = 3.15 \times 10^{-4}$ M

$$K_a(NH_4^+) = \frac{[H_3O^+][NH_3]}{[NH_4^+]} = \frac{[H_3O^+](3.15 \times 10^{-4})}{0.25} = 5.56 \times 10^{-10}$$

$[H_3O^+] = 4.4 \times 10^{-7}$ M

$$[OH^-] = \frac{K_w}{[H_3O^+]} = \frac{1.0 \times 10^{-14}}{4.4 \times 10^{-7}} = 2.3 \times 10^{-8}$ M$$

$pH = -\log[H_3O^+] = -\log(4.4 \times 10^{-7}) = 6.36$

(b)

	$NH_4^+(aq)$	$+$	$SO_3^{2-}(aq)$	$\rightleftharpoons$	$HSO_3^-(aq)$	$+$	$NH_3(aq)$
initial (M)	0.50		0.25		0		0
change (M)	$-x$		$-x$		$+x$		$+x$
equil (M)	$0.50 - x$		$0.25 - x$		x		x

$$K = \frac{K_aK_b}{K_w} = \frac{(5.56 \times 10^{-10})(1.59 \times 10^{-7})}{(1.0 \times 10^{-14})} = \frac{[HSO_3^-][NH_3]}{[NH_4^+][SO_3^{2-}]} = \frac{x^2}{(0.50 - x)(0.25 - x)}$$

where K_a is K_a for NH_4^+ and K_b is K_b for SO_3^{2-}.
$0.9912x^2 + 0.006\,63x - 0.001\,11 = 0$
Use the quadratic formula to solve for x.

$$x = \frac{-(0.006\,63) \pm \sqrt{(0.006\,63)^2 - 4(0.9912)(-0.001\,11)}}{2(0.9912)} = \frac{-0.006\,63 \pm 0.066\,67}{2(0.9912)}$$

$x = 0.0303$ and -0.0370
Of the two solutions for x, only the positive value of x has physical meaning, because x is the $[NH_3]$ and $[HSO_3^-]$.
$x = 0.030$
$[NH_4^+] = 0.50 - x = 0.47$ M; $[SO_3^{2-}] = 0.25 - x = 0.22$ M
$[NH_3] = [HSO_3^-] = x = 0.030$ M

$$K_b(SO_3^{2-}) = \frac{[HSO_3^-][OH^-]}{[SO_3^{2-}]} = \frac{(0.030)[OH^-]}{0.22} = 1.59 \times 10^{-7}$$

$[OH^-] = 1.166 \times 10^{-6}\ M = 1.2 \times 10^{-6}\ M$

$[H_3O^+] = \dfrac{K_w}{[OH^-]} = \dfrac{1.0 \times 10^{-14}}{1.166 \times 10^{-6}} = 8.6 \times 10^{-9}\ M$

$pH = -\log[H_3O^+] = -\log(8.6 \times 10^{-9}) = 8.07$

16.136 $HCO_3^-(aq) + Al(H_2O)_6^{3+}(aq) \rightarrow H_2O(l) + CO_2(g) + Al(H_2O)_5(OH)^{2+}(aq)$

Lewis Acids and Bases (Section 16.15)

16.138 (a) Lewis acid, SiF_4; Lewis base, F^- (b) Lewis acid, Zn^{2+}; Lewis base, NH_3
(c) Lewis acid, $HgCl_2$; Lewis base, Cl^- (d) Lewis acid, CO_2; Lewis base, H_2O

16.140 (a) $2\ :\!\overset{..}{\underset{..}{F}}\!:^- \ \ + \ SiF_4 \longrightarrow SiF_6^{\,2-}$

(b) $4\ \overset{..}{N}H_3 \ \ + \ Zn^{2+} \longrightarrow Zn(NH_3)_4^{\,2+}$

(c) $2\ :\!\overset{..}{\underset{..}{Cl}}\!:^- \ \ + \ HgCl_2 \longrightarrow HgCl_4^{\,2-}$

(d) $H_2\overset{..}{\underset{..}{O}}: \ + \ CO_2 \longrightarrow H_2CO_3$

16.142 (a) CN^-, Lewis base   (b) H^+, Lewis acid   (c) H_2O, Lewis base
(d) Fe^{3+}, Lewis acid   (e) OH^-, Lewis base   (f) CO_2, Lewis acid
(g) $P(CH_3)_3$, Lewis base   (h) $B(CH_3)_3$, Lewis acid

Multiconcept Problems

16.144 H_2O, 18.02

at $0\ °C$, $[H_2O] = \dfrac{\left(0.9998\ g \times \dfrac{1\ mol}{18.02\ g}\right)}{0.001\ L} = 55.48\ M$

$K_w = [H_3O^+][OH^-]$, for a neutral solution $[H_3O^+] = [OH^-]$

$[H_3O^+] = \sqrt{K_w} = \sqrt{1.14 \times 10^{-15}} = 3.376 \times 10^{-8}\ M$

$pH = -\log[H_3O^+] = -\log(3.376 \times 10^{-8}) = 7.472$

$\text{fraction dissociated} = \dfrac{[H_2O]_{diss}}{[H_2O]_{initial}} = \dfrac{3.376 \times 10^{-8}\ M}{55.48\ M} = 6.09 \times 10^{-10}$

$\%\ \text{dissociation} = \dfrac{[H_2O]_{diss}}{[H_2O]_{initial}} \times 100\% = \dfrac{3.376 \times 10^{-8}\ M}{55.48\ M} \times 100\% = 6.09 \times 10^{-8}\%$

16.146 HCl is a strong acid. $[H_3O^+]_{initial} = 0.10\ M$
The dissociation of HF produces a negligible amount of H_3O^+ compared with that produced from HCl.

$HF(aq) + H_2O(l) \rightleftharpoons H_3O^+(aq) + F^-(aq)$

$$K_a = \frac{[H_3O^+][F^-]}{[HF]} = 3.5 \times 10^{-4} = \frac{(0.10)[F^-]}{(0.10)}$$

$[F^-] = K_a = 3.5 \times 10^{-4}$ M; $\quad$ [HF] = 0.10 M; $\quad$ $[H_3O^+] = 0.10$ M

$$[OH^-] = \frac{K_w}{[H_3O^+]} = \frac{1.0 \times 10^{-14}}{0.10} = 1.0 \times 10^{-13} \text{ M}$$

$pH = -\log[H_3O^+] = -\log(0.10) = 1.00$

16.148 (a) $\quad P_{CO_2} = \left(\dfrac{750}{10^6} \cdot 1.0 \text{ atm} \right) = 7.5 \times 10^{-4}$ atm

Solubility $= k \cdot P = [3.2 \times 10^{-2} \text{ mol/(L} \cdot \text{atm)}](7.5 \times 10^{-4} \text{ atm}) = 2.4 \times 10^{-5}$ M

(b) $\qquad\qquad\quad$ $H_2CO_3(aq) + H_2O(l) \rightleftharpoons H_3O^+(aq) + HCO_3^-(aq)$

initial (M)	2.4×10^{-5}	~0	0
change (M)	$-x$	$+x$	$+x$
equil (M)	$2.4 \times 10^{-5} - x$	x	x

$$K_{a1} = \frac{[H_3O^+][HCO_3^-]}{[H_2CO_3]} = 4.3 \times 10^{-7} = \frac{x^2}{2.4 \times 10^{-5} - x}$$

$x^2 + 4.3 \times 10^{-7}x - 1.0 \times 10^{-11} = 0$

Use the quadratic formula to solve for x.

$$x = \frac{-(4.3 \times 10^{-7}) \pm \sqrt{(4.3 \times 10^{-7})^2 - (4)(1)(-1.0 \times 10^{-11})}}{2(1)} = \frac{(-4.3 \times 10^{-7}) \pm (6.3 \times 10^{-6})}{2}$$

$x = -3.4 \times 10^{-6}$ and 3.0×10^{-6}

Of the two solutions for x, only the positive value of x has physical meaning because x is the $[H_3O^+]$.

$x = [H_3O^+] = 3.0 \times 10^{-6}$ M

The second dissociation of H_2CO_3 produces a negligible amount of H_3O^+ compared with that from the first dissociation.

$pH = -\log[H_3O^+] = -\log(3.0 \times 10^{-6}) = 5.52$

(c) The acidity of rain will increase but only slightly.

16.150 Both reactions occur together.

Let $x = [H_3O^+]$ from CH_3CO_2H and $y = [H_3O^+]$ from $C_6H_5CO_2H$

The following two equilibria must be considered:

$\qquad\qquad\qquad$ $CH_3CO_2H(aq) + H_2O(l) \rightleftharpoons H_3O^+(aq) + CH_3CO_2^-(aq)$

initial (M)	0.10	y	0
change (M)	$-x$	$+x$	$+x$
equil (M)	$0.10 - x$	$x + y$	x

$\qquad\qquad\qquad$ $C_6H_5CO_2H(aq) + H_2O(l) \rightleftharpoons H_3O^+(aq) + C_6H_5CO_2^-(aq)$

initial (M)	0.10	x	0
change (M)	$-y$	$+y$	$+y$
equil (M)	$0.10 - y$	$x + y$	y

$$K_a(\text{for } CH_3CO_2H) = \frac{[H_3O^+][CH_3CO_2^-]}{[CH_3CO_2H]} = 1.8 \times 10^{-5} = \frac{(x+y)(x)}{0.10-x} \approx \frac{(x+y)(x)}{0.10}$$

$$1.8 \times 10^{-6} = (x+y)(x)$$

$$K_a(\text{for } C_6H_5CO_2H) = \frac{[H_3O^+][C_6H_5CO_2^-]}{[C_6H_5CO_2H]} = 6.5 \times 10^{-5} = \frac{(x+y)(y)}{0.10-y} \approx \frac{(x+y)(y)}{0.10}$$

$$6.5 \times 10^{-6} = (x+y)(y)$$

$$1.8 \times 10^{-6} = (x+y)(x)$$

$$6.5 \times 10^{-6} = (x+y)(y)$$

These two equations must be solved simultaneously for x and y. Divide the first equation by the second.

$$\frac{x}{y} = \frac{1.8 \times 10^{-6}}{6.5 \times 10^{-6}}; \quad x = 0.277y$$

$6.5 \times 10^{-6} = (x+y)(y)$; substitute $x = 0.277y$ into this equation and solve for y.

$$6.5 \times 10^{-6} = (0.277y + y)(y) = 1.277y^2$$

$$y = 0.002\ 256$$

$$x = 0.277y = (0.277)(0.002\ 256) = 0.000\ 624\ 9$$

$$[H_3O^+] = (x+y) = (0.000\ 624\ 9 + 0.002\ 256) = 0.002\ 881\ M$$

$$pH = -\log[H_3O^+] = -\log(0.002\ 881) = 2.54$$

16.152 H_3PO_4, 98.00

Assume 1.000 L of solution.

$$\text{Mass of solution} = 1.000\ L \times \frac{1000\ mL}{1\ L} \times \frac{1.0353\ g}{1\ mL} = 1035.3\ g$$

$$\text{Mass } H_3PO_4 = (0.070)(1035.3\ g) = 72.47\ g\ H_3PO_4$$

$$\text{mol } H_3PO_4 = 72.47\ g\ H_3PO_4 \times \frac{1\ mol\ H_3PO_4}{98.00\ g\ H_3PO_4} = 0.740\ mol\ H_3PO_4$$

$$[H_3PO_4] = \frac{0.740\ mol\ H_3PO_4}{1.000\ L} = 0.740\ M$$

For the dissociation of the first proton, the following equilibrium must be considered:

$$H_3PO_4(aq) + H_2O(l) \rightleftharpoons H_3O^+(aq) + H_2PO_4^-(aq)$$

initial (M)	0.740	~0	0
change (M)	−x	+x	+x
equil (M)	0.740 − x	x	x

$$K_{a1} = \frac{[H_3O^+][H_2PO_4^-]}{[H_3PO_4]} = 7.5 \times 10^{-3} = \frac{x^2}{0.740-x}$$

$$x^2 + (7.5 \times 10^{-3})x - (5.55 \times 10^{-3}) = 0$$

Solve for x using the quadratic formula.

$$x = \frac{-(7.5 \times 10^{-3}) \pm \sqrt{(7.5 \times 10^{-3})^2 - (4)(-5.55 \times 10^{-3})}}{2(1)} = \frac{(-7.5 \times 10^{-3}) \pm 0.149}{2}$$

x = 0.0708 and −0.0783

Of the two solutions for x, only the positive value of x has physical meaning, because x is the $[H_3O^+]$.

x = 0.0708 M = $[H_2PO_4^-]$ = $[H_3O^+]$

For the dissociation of the second proton, the following equilibrium must be considered:

$$H_2PO_4^-(aq) + H_2O(l) \rightleftharpoons H_3O^+(aq) + HPO_4^{2-}(aq)$$

initial (M)	0.0708	0.0708	0
change (M)	−y	+y	+y
equil (M)	0.0708 − y	0.0708 + y	y

$$K_{a2} = \frac{[H_3O^+][HPO_4^{2-}]}{[H_2PO_4^-]} = 6.2 \times 10^{-8} = \frac{(0.0708 + y)(y)}{0.0708 - y} \approx \frac{(0.0708)(y)}{0.0708} = y$$

y = 6.2×10^{-8} M = $[HPO_4^{2-}]$

For the dissociation of the third proton, the following equilibrium must be considered:

$$HPO_4^{2-}(aq) + H_2O(l) \rightleftharpoons H_3O^+(aq) + PO_4^{3-}(aq)$$

initial (M)	6.2×10^{-8}	0.0708	0
change (M)	−z	+z	+z
equil (M)	$(6.2 \times 10^{-8}) - z$	0.0708 + z	z

$$K_{a3} = \frac{[H_3O^+][PO_4^{3-}]}{[HPO_4^{2-}]} = 4.8 \times 10^{-13} = \frac{(0.0708 + z)(z)}{(6.2 \times 10^{-8}) - z} \approx \frac{(0.0708)(z)}{6.2 \times 10^{-8}}$$

z = 4.2×10^{-19} M = $[PO_4^{3-}]$

$[H_3PO_4]$ = 0.740 − x = 0.740 − 0.0708 = 0.67 M

$[H_2PO_4^-]$ = $[H_3O^+]$ = 0.0708 M = 0.071 M

$[HPO_4^{2-}]$ = 6.2×10^{-8} M

$[PO_4^{3-}]$ = 4.2×10^{-19} M

$$[OH^-] = \frac{K_w}{[H_3O^+]} = \frac{1.0 \times 10^{-14}}{0.0708} = 1.4 \times 10^{-13} \text{ M}$$

pH = −log$[H_3O^+]$ = −log(0.0708) = 1.15

16.154 $[H_3O^+]$ = 10^{-pH} = $10^{-9.07}$ = 8.51×10^{-10} M

$[H_3O^+][OH^-] = K_w$

$$[OH^-] = \frac{K_w}{[H_3O^+]} = \frac{1.0 \times 10^{-14}}{8.51 \times 10^{-10}} = 1.18 \times 10^{-5} \text{ M} = \text{x below.}$$

$$K_a = 1.8 \times 10^{-5} \text{ for CH}_3\text{CO}_2\text{H and } K_b = \frac{K_w}{K_a} = \frac{1.0 \times 10^{-14}}{1.8 \times 10^{-5}} = 5.56 \times 10^{-10}$$

Use the equilibrium associated with a weak base to solve for $[CH_3CO_2^-]$ = y below.

$$CH_3CO_2^-(aq) + H_2O(l) \rightleftharpoons CH_3CO_2H(aq) + OH^-(aq)$$

initial (M)	y	~0	0
change (M)	−x	+x	+x
equil (M)	y − x	x	x

$$K_b = \frac{[CH_3CO_2H][OH^-]}{[CH_3CO_2^-]} = 5.56 \times 10^{-10} = \frac{(1.18 \times 10^{-5})^2}{[y - (1.18 \times 10^{-5})]}$$

Solve for y.
$(5.56 \times 10^{-10})[y - (1.18 \times 10^{-5})] = (1.18 \times 10^{-5})^2$
$(5.56 \times 10^{-10})y - 6.56 \times 10^{-15} = 1.39 \times 10^{-10}$

$(5.56 \times 10^{-10})y = 1.39 \times 10^{-10}$
$y = (1.39 \times 10^{-10})/(5.56 \times 10^{-10}) = [CH_3CO_2^-] = 0.25$ M

In 1.00 L of solution, the mass of CH_3CO_2Na solute $= (0.25 \text{ mol/L})\left(\dfrac{82.035 \text{ g } CH_3CO_2Na}{1 \text{ mol } CH_3CO_2Na}\right) = 20.5$ g

$$\text{mass of solution} = (1000 \text{ mL})\left(\frac{1.0085 \text{ g}}{1 \text{ mL}}\right) = 1008.5 \text{ g}$$

mass of solvent $= 1008.5$ g $- 20.5$ g $= 988$ g $= 0.988$ kg

$$m = \frac{0.25 \text{ mol } CH_3CO_2Na}{0.988 \text{ kg}} = 0.25 \ m$$

Because CH_3CO_2Na is a strong electrolyte, the ionic compound is completely dissociated and $[CH_3CO_2^-] = [Na^+]$. The contribution of CH_3CO_2H and OH^- to the total molality of the solution is negligible.
$\Delta T_f = K_f \cdot (2 \cdot m) = (1.86 \text{ °C/m})(2)(0.25 \ m) = 0.93$ °C
Solution freezing point $= 0.00$ °C $- \Delta T_f = 0.00$ °C $- 0.93$ °C $= -0.93$ °C

16.156 Na_3PO_4, 163.94

$$3.28 \text{ g } Na_3PO_4 \times \frac{1 \text{ mol } Na_3PO_4}{163.94 \text{ g } Na_3PO_4} = 0.0200 \text{ mol} = 20.0 \text{ mmol } Na_3PO_4$$

300.0 mL x 0.180 mmol/mL = 54.0 mmol HCl

	$H_3O^+(aq)$ +	$PO_4^{3-}(aq)$	$\rightleftarrows$	$HPO_4^{2-}(aq)$ +	$H_2O(l)$
before (mmol)	54.0	20.0		0	
change (mmol)	−20.0	−20.0		+20.0	
after (mmol)	34.0	0		20.0	

	$H_3O^+(aq)$ +	$HPO_4^{2-}(aq)$	$\rightleftarrows$	$H_2PO_4^-(aq)$ +	$H_2O(l)$
before (mmol)	34.0	20.0		0	
change (mmol)	−20.0	−20.0		+20.0	
after (mmol)	14.0	0		20.0	

	$H_3O^+(aq)$ +	$H_2PO_4^-(aq)$	$\rightleftarrows$	$H_3PO_4(aq)$ +	$H_2O(l)$
before (mmol)	14.0	20.0		0	
change (mmol)	−14.0	−14.0		+14.0	
after (mmol)	0	6.0		14.0	

$$[H_3PO_4] = \frac{14.0 \text{ mmol}}{300.0 \text{ mL}} = 0.047 \text{ M}; \quad [H_2PO_4^-] = \frac{6.0 \text{ mmol}}{300.0 \text{ mL}} = 0.020 \text{ M}$$

$$H_3PO_4(aq) + H_2O(l) \rightleftharpoons H_3O^+(aq) + H_2PO_4^-(aq)$$

initial (M)	0.047	~0	0.020
change (M)	−x	+x	+x
equil (M)	0.047 − x	x	0.020 + x

$$K_a = \frac{[H_3O^+][H_2PO_4^{2-}]}{[H_3PO_4]} = 7.5 \times 10^{-3} = \frac{x(0.020 + x)}{(0.047 - x)}$$

$x^2 + 0.0275x - (3.525 \times 10^{-4}) = 0$

Solve for x using the quadratic formula.

$$x = \frac{-(0.0275) \pm \sqrt{(0.0275)^2 - (4)(-3.525 \times 10^{-4})}}{2(1)} = \frac{-0.0275 \pm 0.0465}{2}$$

x = 0.009 52 and −0.0370

Of the two solutions for x, only the positive value of x has physical meaning, because x is the $[H_3O^+]$.

pH = −log$[H_3O^+]$ = −log(0.009 52) = 2.02

16.158 (a) PV = nRT; $\quad n = \dfrac{PV}{RT} = \dfrac{(0.601 \text{ atm})(1.000 \text{ L})}{\left(0.082\ 06\ \dfrac{\text{L} \cdot \text{atm}}{\text{K} \cdot \text{mol}}\right)(293.1 \text{ K})} = 0.0250 \text{ mol HF}$

$50.0 \text{ mL} \times \dfrac{1.00 \text{ L}}{1000 \text{ mL}} = 0.0500 \text{ L}$

$[HF] = \dfrac{0.0250 \text{ mol HF}}{0.0500 \text{ L}} = 0.500 \text{ M}$

$$HF(aq) + H_2O(l) \rightleftharpoons H_3O^+(aq) + F^-(aq)$$

initial (M)	0.500	~0	0
change (M)	−x	+x	+x
equil (M)	0.500 − x	x	x

$$K_a = \frac{[H_3O^+][F^-]}{[HF]} = 3.5 \times 10^{-4} = \frac{x^2}{0.500 - x}$$

$x^2 + (3.5 \times 10^{-4})x - (1.75 \times 10^{-4}) = 0$

Solve for x using the quadratic formula.

$$x = \frac{-(3.5 \times 10^{-4}) \pm \sqrt{(3.5 \times 10^{-4})^2 - (4)(-1.75 \times 10^{-4})}}{2(1)} = \frac{(-3.5 \times 10^{-4}) \pm 0.0265}{2}$$

x = −0.0134 and 0.0131

Of the two solutions for x, only the positive value of x has physical meaning, because x is the $[H_3O^+]$.

pH = −log$[H_3O^+]$ = −log(0.0131) = 1.883 = 1.88

(b) % dissociation = $\dfrac{0.0131 \text{ M}}{0.500} \times 100\% = 2.62\% = 2.6\%$

New % dissociation = (3)(2.62%) = 7.86%

Let X equal the concentration of HF dissociated and Y the new volume (in liters) that would triple the % dissociation.

$$K_a = \frac{X^2}{(0.0250/Y) - X} = 3.5 \times 10^{-4}$$

$$\% \text{ dissociation} = \frac{X}{(0.0250/Y)} \times 100\% = 7.86\% \quad \text{and} \quad \frac{X}{(0.0250/Y)} = 0.0786$$

$X = 1.965 \times 10^{-3}/Y$

Substitute X into the K_a equation.

$$\frac{(1.965 \times 10^{-3}/Y)^2}{(0.0250/Y) - (1.965 \times 10^{-3}/Y)} = 3.5 \times 10^{-4}$$

$$\frac{3.861 \times 10^{-6}/Y^2}{0.0230/Y} = 3.5 \times 10^{-4}$$

$$\frac{3.861 \times 10^{-6}/Y}{0.0230} = 3.5 \times 10^{-4}$$

$$\frac{3.861 \times 10^{-6}}{8.05 \times 10^{-6}} = Y = 0.48 \text{ L}$$

16.160 (a) Rate $= k[OCl^-]^x[NH_3]^y[OH^-]^z$

From experiments 1 & 2, the $[OCl^-]$ doubles and the rate doubles, therefore $x = 1$.
From experiments 2 & 3, the $[NH_3]$ triples and the rate triples, therefore $y = 1$.
From experiments 3 & 4, the $[OH^-]$ goes up by a factor of 10 and the rate goes down by a factor of 10, therefore $z = -1$.

$$\text{Rate} = k \frac{[OCl^-][NH_3]}{[OH^-]}$$

$[H_3O^+] = 10^{-pH} = 10^{-12} = 1 \times 10^{-12} \text{ M}; \qquad [OH^-] = \dfrac{K_w}{[H_3O^+]} = \dfrac{1.0 \times 10^{-14}}{1 \times 10^{-12}} = 0.01 \text{ M}$

Using experiment 1: $k = \dfrac{(\text{Rate})[OH^-]}{[OCl^-][NH_3]} = \dfrac{(0.017 \text{ M/s})(0.01 \text{ M})}{(0.001 \text{ M})(0.01 \text{ M})} = 17 \text{ s}^{-1}$

(b) $K_1 = \dfrac{[HOCl][OH^-]}{[OCl^-]} = K_b(OCl^-) = \dfrac{K_w}{K_a(HOCl)} = \dfrac{1.0 \times 10^{-14}}{3.5 \times 10^{-8}} = 2.9 \times 10^{-7}$

For second step, Rate $= k_2 [HOCl][NH_3]$

Multiply the Rate by $\dfrac{K_1}{K_1} = \dfrac{K_1}{\left(\dfrac{[HOCl][OH^-]}{[OCl^-]}\right)}$

$$\text{Rate} = K_1 k_2 \frac{[HOCl][NH_3]}{\left(\dfrac{[HOCl][OH^-]}{[OCl^-]}\right)} = K_1 k_2 \frac{[HOCl][NH_3][OCl^-]}{[HOCl][OH^-]} = K_1 k_2 \frac{[OCl^-][NH_3]}{[OH^-]}$$

$K_1 k_2 = k = 17 \text{ s}^{-1}$

$k_2 = \dfrac{17 \text{ s}^{-1}}{2.9 \times 10^{-7} \text{ M}} = 5.9 \times 10^7 \text{ M}^{-1}\text{s}^{-1}$

17

Applications of Aqueous Equilibria

17.1 $HNO_2(aq) + OH^-(aq) \rightleftharpoons NO_2^-(aq) + H_2O(l);$ NO_2^- (basic anion), pH > 7.00

17.2   (a) $HF(aq) + OH^-(aq) \rightleftharpoons H_2O(l) + F^-(aq)$

$$K_n = \frac{K_a}{K_w} = \frac{3.5 \times 10^{-4}}{1.0 \times 10^{-14}} = 3.5 \times 10^{10}$$

(b) $H_3O^+(aq) + OH^-(aq) \rightleftharpoons 2\,H_2O(l)$

$$K_n = \frac{1}{K_w} = \frac{1}{1.0 \times 10^{-14}} = 1.0 \times 10^{14}$$

(c) $HF(aq) + NH_3(aq) \rightleftharpoons NH_4^+(aq) + F^-(aq)$

$$K_n = \frac{K_a K_b}{K_w} = \frac{(3.5 \times 10^{-4})(1.8 \times 10^{-5})}{1.0 \times 10^{-14}} = 6.3 \times 10^5$$

The tendency to proceed to completion is determined by the magnitude of K_n. The larger the value of K_n, the further does the reaction proceed to completion.

The tendency to proceed to completion is: reaction (c) < reaction (a) < reaction (b)

17.3 $HCN(aq) + H_2O(l) \rightleftharpoons H_3O^+(aq) + CN^-(aq)$

initial (M)	0.025	~0	0.010
change (M)	–x	+x	+x
equil (M)	0.025 – x	x	0.010 + x

$$K_a = \frac{[H_3O^+][CN^-]}{[HCN]} = 4.9 \times 10^{-10} = \frac{x(0.010 + x)}{0.025 - x} \approx \frac{x(0.010)}{0.025}$$

Solve for x. $x = 1.23 \times 10^{-9}\ M = 1.2 \times 10^{-9}\ M = [H_3O^+]$

$pH = -\log[H_3O^+] = -\log(1.23 \times 10^{-9}) = 8.91$

$$\%\ \text{dissociation} = \frac{[HCN]_{diss}}{[HCN]_{initial}} \times 100\% = \frac{1.23 \times 10^{-9}\ M}{0.025\ M} \times 100\% = 4.9 \times 10^{-6}\ \%$$

17.4   On mixing equal volumes of two solutions, both concentrations are cut in half.

$[CH_3NH_2] = 0.10\ M;$ $[CH_3NH_3Cl] = 0.30\ M$

                $CH_3NH_2(aq) + H_2O(l) \rightleftharpoons CH_3NH_3^+(aq) + OH^-(aq)$

initial (M)	0.10	0.30	~0
change (M)	–x	+x	+x
equil (M)	0.10 – x	0.30 + x	x

$$K_b = \frac{[CH_3NH_3^+][OH^-]}{[CH_3NH_2]} = 3.7 \times 10^{-4} = \frac{(0.30 + x)x}{0.10 - x} \approx \frac{(0.30)x}{0.10}$$

Solve for x. $x = [OH^-] = 1.2 \times 10^{-4}$ M

$$[H_3O^+] = \frac{K_w}{[OH^-]} = \frac{1.0 \times 10^{-14}}{1.2 \times 10^{-4}} = 8.3 \times 10^{-11} \text{ M}$$

$$pH = -\log[H_3O^+] = -\log(8.3 \times 10^{-11}) = 10.08$$

17.5   Both solutions contain the same number of HF molecules but solution 2 also contains five F^- ions. The dissociation equilibrium lies farther to the left for solution 2, and therefore solution 2 has the lower $[H_3O^+]$ and the higher pH. For solution 1, no common ion is present to suppress the dissociation of HF, and therefore solution 1 has the larger percent dissociation.

17.6   Each solution contains the same number of B molecules. The presence of BH^+ from BHCl lowers the percent dissociation of B. Solution (2) contains no BH^+, therefore it has the largest percent dissociation. BH^+ is the conjugate acid of B. Solution (1) has the largest amount of BH^+, and it would be the most acidic solution and have the lowest pH.

17.7   For the buffer, pH = 3.76

mol HF = 0.025 mol; mol F^- = 0.050 mol; vol = 0.100 L

		100%		
	HF(aq)	+ OH$^-$(aq)	→ F$^-$(aq)	+ H$_2$O(l)
before (mol)	0.025	0.004	0.050	
change (mol)	−0.004	−0.004	+0.004	
after (mol)	0.021	0	0.054	

$$[H_3O^+] = K_a \frac{[HF]}{[F^-]} = (3.5 \times 10^{-4})\left(\frac{0.21}{0.54}\right) = 1.36 \times 10^{-4} \text{ M}$$

$$pH = -\log[H_3O^+] = -\log(1.36 \times 10^{-4}) = 3.87$$

17.8   (a)

	HF(aq) + H$_2$O(l)	⇌ H$_3$O$^+$(aq)	+ F$^-$(aq)
initial (M)	0.050	~0	0.100
change (M)	−x	+x	+x
equil (M)	0.050 − x	x	0.100 + x

$$K_a = \frac{[H_3O^+][F^-]}{[HF]} = 3.5 \times 10^{-4} = \frac{x(0.100 + x)}{0.050 - x} \approx \frac{x(0.100)}{0.050}$$

Solve for x. $x = [H_3O^+] = 1.75 \times 10^{-4}$ M

$pH = -\log[H_3O^+] = -\log(1.75 \times 10^{-4}) = 3.76$

mol HF = 0.050 mol/L x 0.100 L = 0.0050 mol HF

mol F^- = 0.100 mol/L x 0.100 L = 0.0100 mol F^-

mol HNO_3 = mol H_3O^+ = 0.002 mol

		100%		
Neutralization reaction:	F$^-$(aq)	+ H$_3$O$^+$(aq)	→ HF(aq)	+ H$_2$O(l)
before reaction (mol)	0.0100	0.002	0.0050	
change (mol)	−0.002	−0.002	+0.002	
after reaction (mol)	0.008	0	0.007	

$$[HF] = \frac{0.007 \text{ mol}}{0.100 \text{ L}} = 0.07 \text{ M}; \qquad [F^-] = \frac{0.008 \text{ mol}}{0.100 \text{ L}} = 0.08 \text{ M}$$

$$[H_3O^+] = K_a \frac{[HF]}{[F^-]} = (3.5 \times 10^{-4})\frac{(0.07)}{(0.08)} = 3 \times 10^{-4} \text{ M}$$

$$pH = -\log[H_3O^+] = -\log(3 \times 10^{-4}) = 3.5$$

(b) When the solution is diluted, the acid to conjugate base ratio remains the same and therefore the pH does not change.

17.9   When equal volumes of two solutions are mixed together, the concentration of each solution is cut in half.

$$pH = pK_a + \log\frac{[\text{base}]}{[\text{acid}]} = pK_a + \log\frac{[CO_3^{2-}]}{[HCO_3^-]}$$

For HCO_3^-, $K_a = 5.6 \times 10^{-11}$, $pK_a = -\log K_a = -\log(5.6 \times 10^{-11}) = 10.25$

$$pH = 10.25 + \log\left(\frac{0.050}{0.10}\right) = 10.25 - 0.30 = 9.95$$

17.10   (a) $pH = pK_a + \log\frac{[\text{base}]}{[\text{acid}]} = 9.15 + \log\left(\frac{1}{50}\right) = 7.45$

(b) $\dfrac{[\text{base}]}{[\text{acid}]} = \dfrac{1}{50}$; % dissociation $= \dfrac{1}{(50+1)} \times 100\% = 1.96\%$

17.11 $pH = pK_a + \log\dfrac{[\text{base}]}{[\text{acid}]} = pK_a + \log\dfrac{[CO_3^{2-}]}{[HCO_3^-]}$

For HCO_3^-, $K_a = 5.6 \times 10^{-11}$, $pK_a = -\log K_a = -\log(5.6 \times 10^{-11}) = 10.25$

$$10.40 = 10.25 + \log\frac{[CO_3^{2-}]}{[HCO_3^-]}$$

$$\log\frac{[CO_3^{2-}]}{[HCO_3^-]} = 10.40 - 10.25 = 0.15$$

$$\frac{[CO_3^{2-}]}{[HCO_3^-]} = 10^{0.15} = 1.4$$

To obtain a buffer solution with pH 10.40, make the Na_2CO_3 concentration 1.4 times the concentration of $NaHCO_3$.

17.12   (a) HOCl, $K_a = 3.5 \times 10^{-8}$, $pK_a = 7.46$;   HOBr, $K_a = 2.0 \times 10^{-9}$, $pK_a = 8.70$
Choose an acid with a pK_a within 1 pH unit of the desired pH. The desired pH is 7.00, so HOCl–NaOCl is the buffer of choice.

(b) $pH = 7.00 = pK_a + \log\dfrac{[\text{base}]}{[\text{acid}]} = 7.46 + \log\dfrac{[OCl^-]}{[HOCl]}$

$$7.00 = 7.46 + \log \frac{[OCl^-]}{[HOCl]}$$

$$\log \frac{[OCl^-]}{[HOCl]} = 7.00 - 7.46 = -0.46$$

$$\frac{[OCl^-]}{[HOCl]} = 10^{-0.46} = 0.35$$

17.13   mol HCl = mol H_3O^+ = 0.100 mol/L x 0.0400 L = 0.004 00 mol
mol NaOH = mol OH^- = 0.100 mol/L x 0.0450 L = 0.004 50 mol

Neutralization reaction:	H_3O^+(aq) +	OH^-(aq)	→ 2 H_2O(l)
before reaction (mol)	0.004 00	0.004 50	
change (mol)	−0.004 00	−0.004 00	
after reaction (mol)	0	0.000 50	

$$[OH^-] = \frac{0.000\ 50\ mol}{(0.0400\ L + 0.0450\ L)} = 5.9 \times 10^{-3}\ M$$

$$[H_3O^+] = \frac{K_w}{[OH^-]} = \frac{1.0 \times 10^{-14}}{5.9 \times 10^{-3}} = 1.7 \times 10^{-12}\ M$$

$$pH = -\log[H_3O^+] = -\log(1.7 \times 10^{-12}) = 11.77$$

17.14   (a) mol NaOH = mol OH^- = 0.100 mol/L x 0.0400 L = 0.004 00 mol
mol HCl = mol H_3O^+ = 0.0500 mol/L x 0.0600 L = 0.003 00 mol

Neutralization reaction:	H_3O^+(aq) +	OH^-(aq)	→ 2 H_2O(l)
before reaction (mol)	0.003 00	0.004 00	
change (mol)	−0.003 00	−0.003 00	
after reaction (mol)	0	0.001 00	

$$[OH^-] = \frac{0.001\ 00\ mol}{(0.0400\ L + 0.0600\ L)} = 1.0 \times 10^{-2}\ M$$

$$[H_3O^+] = \frac{K_w}{[OH^-]} = \frac{1.0 \times 10^{-14}}{1.0 \times 10^{-2}} = 1.0 \times 10^{-12}\ M$$

$$pH = -\log[H_3O^+] = -\log(1.0 \times 10^{-12}) = 12.00$$

(b) mol NaOH = mol OH^- = 0.100 mol/L x 0.0400 L = 0.004 00 mol
mol HCl = mol H_3O^+ = 0.0500 mol/L x 0.0802 L = 0.004 01 mol

Neutralization reaction:	H_3O^+(aq) +	OH^-(aq)	→ 2 H_2O(l)
before reaction (mol)	0.004 01	0.004 00	
change (mol)	−0.004 00	−0.004 00	
after reaction (mol)	0.000 01	0	

$$[H_3O^+] = \frac{0.000\ 01\ mol}{(0.0400\ L + 0.0802\ L)} = 8.3 \times 10^{-5}\ M$$

$$pH = -\log[H_3O^+] = -\log(8.3 \times 10^{-5}) = 4.08$$

(c) mol NaOH = mol OH^- = 0.100 mol/L x 0.0400 L = 0.004 00 mol
mol HCl = mol H_3O^+ = 0.0500 mol/L x 0.1000 L = 0.005 00 mol

Neutralization reaction: $H_3O^+(aq) + OH^-(aq) \rightarrow 2 H_2O(l)$

before reaction (mol)	0.005 00	0.004 00
change (mol)	–0.004 00	–0.004 00
after reaction (mol)	0.001 00	0

$$[H_3O^+] = \frac{0.001\ 00\ mol}{(0.0400\ L\ +\ 0.1000\ L)} = 7.1 \times 10^{-3}\ M$$

$$pH = -\log[H_3O^+] = -\log(7.1 \times 10^{-3}) = 2.15$$

17.15 $mol\ NaOH\ required = \left(\dfrac{0.0500\ mol\ HOCl}{L} \right)(0.100\ L)\left(\dfrac{1\ mol\ NaOH}{1\ mol\ HOCl} \right) = 0.00500\ mol$

$vol\ NaOH\ required = (0.00500\ mol)\left(\dfrac{1\ L}{0.100\ mol} \right) = 0.0500\ L = 50.0\ mL$

50.0 mL of 0.100 M NaOH are required to reach the equivalence point.

At the equivalence point the solution contains the salt, NaOCl.
mol NaOCl = initial mol HOCl = 0.00500 mol = 5.00 mmol

$$[OCl^-] = \frac{5.00\ mmol}{(100.0\ mL\ +\ 50.0\ mL)} = 3.33 \times 10^{-2}\ M$$

$$For\ OCl^-,\ K_b = \frac{K_w}{K_a\ for\ HOCl} = \frac{1.0 \times 10^{-14}}{3.5 \times 10^{-8}} = 2.9 \times 10^{-7}$$

	$OCl^-(aq)$	$+ H_2O(l)$	$\rightleftharpoons$	$HOCl(aq)$	$+ OH^-(aq)$
initial (M)	0.0333			0	~0
change (M)	–x			+x	+x
equil (M)	0.0333 – x			x	x

$$K_b = \frac{[HOCl][OH^-]}{[OCl^-]} = 2.9 \times 10^{-7} = \frac{x^2}{0.0333 - x} \approx \frac{x^2}{0.0333}$$

Solve for x. $x = [OH^-] = 9.83 \times 10^{-5}\ M$

$$[H_3O^+] = \frac{K_w}{[OH^-]} = \frac{1.0 \times 10^{-14}}{9.83 \times 10^{-5}} = 1.02 \times 10^{-10}\ M$$

$$pH = -\log[H_3O^+] = -\log(1.02 \times 10^{-10}) = 9.99$$

17.16   (a) (3), only HA present   (b) (1), HA and A^- present
(c) (4), only A^- present   (d) (2), A^- and OH^- present

17.17   (a) mol NaOH required to reach first equivalence point

$$= \left(\frac{0.0800\ mol\ H_2SO_3}{L} \right)(0.0400\ L)\left(\frac{1\ mol\ NaOH}{1\ mol\ H_2SO_3} \right) = 0.003\ 20\ mol$$

vol NaOH required to reach first equivalence point

$$= (0.003\ 20\ mol)\left(\frac{1\ L}{0.160\ mol} \right) = 0.020\ L = 20.0\ mL$$

20.0 mL is enough NaOH solution to reach the first equivalence point for the titration of the diprotic acid, H_2SO_3.

For H_2SO_3,

$K_{a1} = 1.5 \times 10^{-2}$, $pK_{a1} = -\log K_{a1} = -\log(1.5 \times 10^{-2}) = 1.82$

$K_{a2} = 6.3 \times 10^{-8}$, $pK_{a2} = -\log K_{a2} = -\log(6.3 \times 10^{-8}) = 7.20$

At the first equivalence point, $pH = \dfrac{pK_{a1} + pK_{a2}}{2} = \dfrac{1.82 + 7.20}{2} = 4.51$

(b) mol NaOH required to reach second equivalence point

$$= \left(\frac{0.0800 \text{ mol } H_2SO_3}{L} \right)(0.0400 \text{ L})\left(\frac{2 \text{ mol NaOH}}{1 \text{ mol } H_2SO_3} \right) = 0.006\ 40 \text{ mol}$$

vol NaOH required to reach second equivalence point

$$= (0.006\ 40 \text{ mol})\left(\frac{1 \text{ L}}{0.160 \text{ mol}} \right) = 0.040 \text{ L} = 40.0 \text{ mL}$$

30.0 mL is enough NaOH solution to reach halfway to the second equivalent point. Halfway to the second equivalence point

$pH = pK_{a2} = -\log K_{a2} = -\log(6.3 \times 10^{-8}) = 7.20$

17.18   Let H_2A^+ = valine cation

(a) mol NaOH required to reach first equivalence point

$$= \left(\frac{0.0250 \text{ mol } H_2A^+}{L} \right)(0.0400 \text{ L})\left(\frac{1 \text{ mol NaOH}}{1 \text{ mol } H_2A^+} \right) = 0.001\ 00 \text{ mol}$$

vol NaOH required to reach first equivalence point

$$= (0.001\ 00 \text{ mol})\left(\frac{1 \text{ L}}{0.100 \text{ mol}} \right) = 0.0100 \text{ L} = 10.0 \text{ mL}$$

10.0 mL is enough NaOH solution to reach the first equivalence point for the titration of the diprotic acid, H_2A^+.

For H_2A^+,

$K_{a1} = 4.8 \times 10^{-3}$, $pK_{a1} = -\log K_{a1} = -\log(4.8 \times 10^{-3}) = 2.32$

$K_{a2} = 2.4 \times 10^{-10}$, $pK_{a2} = -\log K_{a2} = -\log(2.4 \times 10^{-10}) = 9.62$

At the first equivalence point, $pH = \dfrac{pK_{a1} + pK_{a2}}{2} = \dfrac{2.32 + 9.62}{2} = 5.97$

(b) mol NaOH required to reach second equivalence point

$$= \left(\frac{0.0250 \text{ mol } H_2A^+}{L} \right)(0.0400 \text{ L})\left(\frac{2 \text{ mol NaOH}}{1 \text{ mol } H_2A^+} \right) = 0.002\ 00 \text{ mol}$$

vol NaOH required to reach second equivalence point

$$= (0.002\ 00 \text{ mol})\left(\frac{1 \text{ L}}{0.100 \text{ mol}} \right) = 0.0200 \text{ L} = 20.0 \text{ mL}$$

15.0 mL is enough NaOH solution to reach halfway to the second equivalent point. Halfway to the second equivalence point

$$pH = pK_{a2} = -\log K_{a2} = -\log(2.4 \times 10^{-10}) = 9.62$$

(c) 20.0 mL is enough NaOH to reach the second equivalence point.
At the second equivalence point

mmol $A^- = (0.0250$ mmol/mL$)(40.0$ mL$) = 1.00$ mmol A^-

solution volume = 40.0 mL + 20.0 mL = 60.0 mL

$$[A^-] = \frac{1.00 \text{ mmol}}{60.0 \text{ mL}} = 0.0167 \text{ M}$$

	$A^-(aq)$	$+$	$H_2O(l)$	$\rightleftarrows$	$HA(aq)$	$+$	$OH^-(aq)$
initial (M)	0.0167				0		~0
change (M)	$-x$				$+x$		$+x$
equil (M)	$0.0167 - x$				x		x

$$K_b = \frac{K_w}{K_a \text{ for HA}} = \frac{K_w}{K_{a2}} = \frac{1.0 \times 10^{-14}}{2.4 \times 10^{-10}} = 4.17 \times 10^{-5}$$

$$K_b = \frac{[HA][OH^-]}{[A^-]} = 4.17 \times 10^{-5} = \frac{x^2}{0.0167 - x}$$

$$x^2 + (4.17 \times 10^{-5})x - (6.964 \times 10^{-7}) = 0$$

Use the quadratic formula to solve for x.

$$x = \frac{-(4.17 \times 10^{-5}) \pm \sqrt{(4.17 \times 10^{-5})^2 - (4)(1)(-6.964 \times 10^{-7})}}{2(1)} = \frac{(-4.17 \times 10^{-5}) \pm (1.67 \times 10^{-3})}{2}$$

$x = 8.14 \times 10^{-4}$ and -8.56×10^{-4}

Of the two solutions for x, only the positive value has physical meaning because x is the $[OH^-]$.

$x = [OH^-] = 8.14 \times 10^{-4}$ M

$$[H_3O^+] = \frac{K_w}{[OH^-]} = \frac{1.0 \times 10^{-14}}{8.14 \times 10^{-4}} = 1.23 \times 10^{-11} \text{ M}$$

$$pH = -\log[H_3O^+] = -\log(1.23 \times 10^{-11}) = 10.91$$

17.19 $K_{sp} = [Ca^{2+}]^3[PO_4^{3-}]^2$

17.20   Let the number of ions be proportional to its concentration.
For AgX, $K_{sp} = [Ag^+][X^-] \propto (4)(4) = 16$
For AgY, $K_{sp} = [Ag^+][Y^-] \propto (1)(9) = 9$
For AgZ, $K_{sp} = [Ag^+][Z^-] \propto (3)(6) = 18$
(a) AgZ   (b) AgY

17.21 $K_{sp} = [Ca^{2+}]^3[PO_4^{3-}]^2 = (2.01 \times 10^{-8})^3(1.6 \times 10^{-5})^2 = 2.1 \times 10^{-33}$

17.22 CaC_2O_4, $K_{sp} = 2.3 \times 10^{-9}$; $[Ca^{2+}] = 3.0 \times 10^{-8}$ M
$K_{sp} = [Ca^{2+}][C_2O_4^{2-}] = (3.0 \times 10^{-8})[C_2O_4^{2-}] = 2.3 \times 10^{-9}$

$$[C_2O_4^{2-}] = \frac{2.3 \times 10^{-9}}{3.0 \times 10^{-8}} = 0.077 \text{ M}$$

minimum $[Na_2C_2O_4] = 0.077$ M

17.23
$$Ag_2CrO_4(s) \rightleftharpoons 2\,Ag^+(aq) + CrO_4^{2-}(aq)$$

equil (M)   2x   x

$K_{sp} = [Ag^+]^2[CrO_4^{2-}] = 1.1 \times 10^{-12} = (2x)^2(x) = 4x^3$

$$\text{molar solubility} = x = \sqrt[3]{\frac{1.1 \times 10^{-12}}{4}} = 6.5 \times 10^{-5}\ \text{mol/L}$$

17.24 $[Ba^{2+}] = [SO_4^{2-}] = 1.05 \times 10^{-5}\ M$; $K_{sp} = [Ba^{2+}][SO_4^{2-}] = (1.05 \times 10^{-5})^2 = 1.10 \times 10^{-10}$

17.25 $[Mg^{2+}]_0$ is from 0.10 M $MgCl_2$.
$$MgF_2(s) \rightleftharpoons Mg^{2+}(aq) + 2\,F^-(aq)$$

	Mg^{2+}	F^-
initial (M)	0.10	0
change (M)	+x	+2x
equil (M)	0.10 + x	2x

$K_{sp} = 7.4 \times 10^{-11} = [Mg^{2+}][F^-]^2 = (0.10 + x)(2x)^2 \approx (0.10)(4x^2)$

$x = 1.4 \times 10^{-5}$,   molar solubility = $x = 1.4 \times 10^{-5}\ M$

17.26   pH = 11; $[H_3O^+] = 10^{-pH} = 10^{-11} = 1.0 \times 10^{-11}\ M$

$$[OH^-] = \frac{K_w}{[H_3O^+]} = \frac{1.0 \times 10^{-14}}{1.0 \times 10^{-11}} = 0.0010\ M$$

$$Zn(OH)2(s) \rightleftharpoons Zn^{2+}(aq) + 2\,OH^-(aq)$$

equil (M)   x   0.0010

$K_{sp} = 4.1 \times 10^{-17} = [Zn^{2+}][OH^-]^2 = (x)(0.0010)^2$

$x = 4.1 \times 10^{-11}$,   molar solubility = $x = 4.1 \times 10^{-11}\ M$

17.27 $[Cu^{2+}] = (5.0 \times 10^{-3}\ \text{mol})/(0.500\ \text{L}) = 0.010\ M$

$$Cu^{2+}(aq) + 4\,NH_3(aq) \rightleftharpoons Cu(NH_3)_4^{2+}(aq)$$

	Cu^{2+}	NH_3	$Cu(NH_3)_4^{2+}$
before reaction (M)	0.010	0.40	0
assume 100% reaction (M)	−0.010	−4(0.010)	+0.010
after reaction (M)	0	0.36	0.010
assume small back reaction (M)	+x	+4x	−x
equil (M)	x	0.36 + 4x	0.010 − x

$$K_f = \frac{[Cu(NH_3)_4^{2+}]}{[Cu^{2+}][NH_3]^4} = 5.6 \times 10^{11} = \frac{(0.010 - x)}{(x)(0.36 + 4x)^4} \approx \frac{0.010}{x(0.36)^4}$$

Solve for x. $x = [Cu^{2+}] = 1.1 \times 10^{-12}\ M$

17.28   Total solution volume = 25.0 mL + 35.0 mL = 60.0 mL

$$[Au^{3+}] = \frac{(25.0\ \text{mL})(3.0 \times 10^{-2}\ M)}{(60.0\ \text{mL})} = 0.0125\ M$$

$$[CN^-] = \frac{(35.0\ \text{mL})(1.0\ M)}{(60.0\ \text{mL})} = 0.583\ M$$

$$Au^{3+}(aq) \; + \; 2\,CN^-(aq) \; \rightleftharpoons \; Au(CN)_2^-(aq)$$

	Au^{3+}	$2\,CN^-$	$Au(CN)_2^-$
before reaction (M)	0.0125	0.583	0
assume 100% reaction (M)	−0.0125	−2(0.0125)	+0.0125
after reaction (M)	0	0.558	0.0125
assume small back reaction (M)	+x	+2x	−x
equil (M)	x	0.558 + 2x	0.0125 − x

$$K_f = \frac{[Au(CN)_2^-]}{[Au^{3+}][CN^-]^2} = 2 \times 10^{38} = \frac{(0.0125 - x)}{(x)(0.558 + 2x)^2} \approx \frac{0.0125}{x(0.558)^2}$$

Solve for x. $x = [Au^{3+}] = 2 \times 10^{-40}$ M

17.29

$$AgBr(s) \; \rightleftharpoons \; Ag^+(aq) \; + \; Br^-(aq) \qquad\qquad K_{sp} = 5.4 \times 10^{-13}$$
$$\underline{Ag^+(aq) \; + \; 2\,S_2O_3^{2-} \; \rightarrow \; Ag(S_2O_3)_2^{3-}(aq)} \qquad K_f = 4.7 \times 10^{13}$$

dissolution $AgBr(s) \; + \; 2\,S_2O_3^{2-}(aq) \; \rightleftharpoons \; Ag(S_2O_3)_2^{3-}(aq) \; + \; Br^-(aq)$
reaction

$K = (K_{sp})(K_f) = (5.4 \times 10^{-13})(4.7 \times 10^{13}) = 25.4$

$$AgBr(s) + 2\,S_2O_3^{2-}(aq) \rightleftharpoons Ag(S_2O_3)_2^{3-}(aq) + Br^-(aq)$$

	$2\,S_2O_3^{2-}$	$Ag(S_2O_3)_2^{3-}$	Br^-
initial (M)	0.10	0	0
change (M)	−2x	+x	+x
equil (M)	0.10 − 2x	x	x

$$K = \frac{[Ag(S_2O_3)_2^{3-}][Br^-]}{[S_2O_3^{2-}]^2} = 25.4 = \frac{x^2}{(0.10 - 2x)^2}$$

Take the square root of both sides and solve for x.

$$\sqrt{25.4} = \sqrt{\frac{x^2}{(0.10 - 2x)^2}}; \qquad 5.04 = \frac{x}{0.10 - 2x}; \qquad x = \text{molar solubility} = 0.045 \text{ mol/L}$$

17.30 Step 1: The precipitate is $Cu(OH)_2(s)$. NH_3 is a base and OH^- ions are present in aqueous solution to react with the Cu^{2+} ions.

$$Cu^{2+}(aq) \; + \; 2\,OH^-(aq) \; \rightleftharpoons \; Cu(OH)_2(s)$$

Step 2: In the presence of additional NH_3, the complex ion $Cu(NH_3)_4^{2+}$ forms.

$$Cu(OH)_2(s) \; + \; 4\,NH_3(aq) \; \rightleftharpoons \; Cu(NH_3)_4^{2+}(aq) \; + \; 2\,OH^-(aq)$$

17.31 On mixing equal volumes of two solutions, the concentrations of both solutions are cut in half.
For $BaCO_3$, $K_{sp} = 2.6 \times 10^{-9}$
$IP = [Ba^{2+}][CO_3^{2-}] = (5.0 \times 10^{-6})(2.0 \times 10^{-5}) = 1.0 \times 10^{-10}$
$IP < K_{sp}$; no precipitate will form.

17.32 $pH = pK_a + \log \dfrac{[\text{base}]}{[\text{acid}]} = pK_a + \log \dfrac{[NH_3]}{[NH_4^+]}$

For NH_4^+, $K_a = 5.6 \times 10^{-10}$, $pK_a = -\log K_a = -\log(5.6 \times 10^{-10}) = 9.25$

$$pH = 9.25 + \log \frac{(0.20)}{(0.20)} = 9.25; \quad [H_3O^+] = 10^{-pH} = 10^{-9.25} = 5.6 \times 10^{-10} \text{ M}$$

$$[OH^-] = \frac{K_w}{[H_3O^+]} = \frac{1.0 \times 10^{-14}}{5.6 \times 10^{-10}} = 1.8 \times 10^{-5} \text{ M}$$

$$[Fe^{2+}] = [Mn^{2+}] = \frac{(25 \text{ mL})(1.0 \times 10^{-3} \text{ M})}{250 \text{ mL}} = 1.0 \times 10^{-4} \text{ M}$$

For $Mn(OH)_2$, $K_{sp} = 2.1 \times 10^{-13}$
   $IP = [Mn^{2+}][OH^-]^2 = (1.0 \times 10^{-4})(1.8 \times 10^{-5})^2 = 3.2 \times 10^{-14}$
   $IP < K_{sp}$; no precipitate will form.
For $Fe(OH)_2$, $K_{sp} = 4.9 \times 10^{-17}$
   $IP = [Fe^{2+}][OH^-]^2 = (1.0 \times 10^{-4})(1.8 \times 10^{-5})^2 = 3.2 \times 10^{-14}$
   $IP > K_{sp}$; a precipitate of $Fe(OH)_2$ will form.

17.33   (a) H_2CO_3

(b) $H_2CO_3(aq) + H_2O(l) \rightleftarrows H_3O^+(aq) + HCO_3^-(aq)$
   $HCO_3^-(aq) + H_2O(l) \rightleftarrows H_3O^+(aq) + CO_3^{2-}(aq)$

17.34   (a) pH = 8.10 (1988) to pH = 8.06 (2017)
(b)   1880, $[H_3O^+] = 10^{-pH} = 10^{-8.2} = 6.3 \times 10^{-9}$ M

$$\text{Percent change} = \frac{[H_3O^+]_{final} - [H_3O^+]_{initial}}{[H_3O^+]_{initial}} \times 100\%$$

$$[H_3O^+]_{final} = \frac{\text{Percent change} \times [H_3O^+]_{initial}}{100\%} + [H_3O^+]_{initial}$$

$$[H_3O^+]_{final} = \frac{150\% \times 6.3 \times 10^{-9} \text{M}}{100\%} + 6.3 \times 10^{-9} \text{M} = 1.6 \times 10^{-8} \text{ M}$$

2100, pH = $-\log[H_3O^+] = -\log(1.6 \times 10^{-8}) = 7.8$

17.35 HCO_3^-, $K_a = 5.6 \times 10^{-11}$, $pK_a = 10.25$

$$pH = pK_a + \log \frac{[\text{base}]}{[\text{acid}]} = 10.25 - \log \frac{[HCO_3^-]}{[CO_3^{2-}]}$$

$$pH = 8.2 = pK_a + \log \frac{[\text{base}]}{[\text{acid}]} = 8.2 = 10.25 - \log \frac{[HCO_3^-]}{[CO_3^{2-}]}$$

$$\log \frac{[HCO_3^-]}{[CO_3^{2-}]} = 10.25 - 8.2 = 2.05; \quad \frac{[HCO_3^-]}{[CO_3^{2-}]} = 10^{2.05} = 112$$

$$pH = 7.9 = pK_a + \log \frac{[\text{base}]}{[\text{acid}]} = 7.9 = 10.25 - \log \frac{[HCO_3^-]}{[CO_3^{2-}]}$$

$$\log \frac{[HCO_3^-]}{[CO_3^{2-}]} = 10.25 - 7.9 = 2.35$$

$$\frac{[HCO_3^-]}{[CO_3^{2-}]} = 10^{2.35} = 224$$

17.36
$$CaCO_3(s) \rightleftharpoons Ca^{2+}(aq) + CO_3^{2-}(aq)$$
equil (M)   x   x
$$K_{sp} = [Ca^{2+}][CO_3^{2-}] = 5.0 \times 10^{-9} = (x)(x)$$
$$\text{molar solubility} = x = \sqrt{K_{sp}} = 7.1 \times 10^{-5} \text{ mol/L}$$

17.37   (a) $CaCO_3(s) \rightleftharpoons Ca^{2+}(aq) + CO_3^{2-}(aq)$ $\qquad K_{sp} = 5.0 \times 10^{-9}$
$H_2CO_3(aq) + H_2O(l) \rightleftharpoons H_3O^+(aq) + HCO_3^-(aq)$ $\qquad K_{a1} = 4.3 \times 10^{-7}$
$\underline{CO_3^{2-}(aq) + H_3O^+(aq) \rightleftharpoons HCO_3^-(aq) + H_2O(l)} \qquad 1/K_{a2} = 1/5.6 \times 10^{-11}$
$CaCO_3(s) + H_2CO_3(aq) \rightleftharpoons Ca^{2+}(aq) + 2 HCO_3^-(aq) \qquad K = K_{sp} \cdot K_{a1} \cdot (1/K_{a2})$
$K = K_{sp} \cdot K_{a1} \cdot (1/K_{a2}) = (5.0 \times 10^{-9})(4.3 \times 10^{-7})(1/5.6 \times 10^{-11}) = 3.8 \times 10^{-5}$
(b) As atmospheric CO_2 levels rise, the concentration of H_2CO_3 will increase, shifting the equilibrium of the overall reaction to the products. The molar solubility of $CaCO_3$ will increase.

Conceptual Problems

17.38   (4); only A^- and water should be present

17.40   A buffer solution contains a conjugate acid-base pair in about equal concentrations.
(a) (1), (3), and (4)
(b) (4) because it has the highest buffer concentration.

17.42   (a) (i) (1), only B present $\qquad$ (ii) (4), equal amounts of B and BH^+ present
(iii) (3), only BH^+ present $\qquad$ (iv) (2), BH^+ and H_3O^+ present
(b) The pH is less than 7 because BH^+ is an acidic cation.

17.44   (a) (1) corresponds to (iii); (2) to (i); (3) to (ii); and (4) to (iv)
(b)

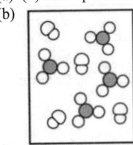

17.46   Let the number of ions be proportional to its concentration.
For Ag_2CrO_4, $K_{sp} = [Ag^+]^2[CrO_4^{2-}] \propto (4)^2(2) = 32$

For (2), IP $= [Ag^+]^2[CrO_4^{2-}] \propto (2)^2(4) = 16$
For (3), IP $= [Ag^+]^2[CrO_4^{2-}] \propto (6)^2(2) = 72$
For (4), IP $= [Ag^+]^2[CrO_4^{2-}] \propto (2)^2(6) = 24$
A precipitate will form when IP $> K_{sp}$. A precipitate will form only in (3).

Section Problems
Neutralization Reactions (Section 17.1)

17.48 (a) $C_6H_5CO_2H(aq) + NaOH(aq) \rightarrow H_2O(l) + C_6H_5CO_2Na(aq)$
net ionic equation: $C_6H_5CO_2H(aq) + OH^-(aq) \rightarrow H_2O(l) + C_6H_5CO_2^-(aq)$
The solution at neutralization contains a basic anion $(C_6H_5CO_2^-)$; pH > 7.00
(b) $HClO_4(aq) + NH_3(aq) \rightarrow NH_4ClO_4(aq)$
net ionic equation: $H_3O^+(aq) + NH_3(aq) \rightarrow H_2O(l) + NH_4^+(aq)$
The solution at neutralization contains an acidic cation (NH_4^+); pH < 7.00
(c) $HI(aq) + KOH(aq) \rightarrow H_2O(l) + KI(aq)$
net ionic equation: $H_3O^+(aq) + OH^-(aq) \rightarrow 2 H_2O(l)$
The solution at neutralization contains a neutral salt (KI); pH $= 7.00$
(d) $HOBr(aq) + (CH_3)_3N(aq) \rightarrow (CH_3)_3NHOBr(aq)$
net ionic equation: $HOBr(aq) + (CH_3)_3N(aq) \rightarrow (CH_3)_3NH^+(aq) + OBr^-(aq)$
The solution at neutralization contains the salt $(CH_3)_3NHOBr(aq)$.
$K_a((CH_3)_3NH^+) = 1.5 \times 10^{-10}$ and $K_b(OBr^-) = 5.0 \times 10^{-5}$
$K_b(OBr^-) > K_a((CH_3)_3NH^+)$; pH > 7.00

17.50 (a) After mixing, the solution contains the neutral salt, KI; pH $= 7.00$
(b) After mixing, the solution contains the acidic salt, NH_4I; pH < 7.00
Solution (b) has the lower pH.

17.52 Weak acid - weak base reaction $K_n = \dfrac{K_aK_b}{K_w} = \dfrac{(8.0 \times 10^{-5})(4.3 \times 10^{-10})}{1.0 \times 10^{-14}} = 3.4$

Because K_n is close to 1, there will be an appreciable amount of aniline present at equilibrium.

17.54 $K_n = \dfrac{K_aK_b}{K_w} = 24$; $K_b = \dfrac{K_nK_w}{K_a} = \dfrac{(24)(1.0 \times 10^{-14})}{(5.8 \times 10^{-10})} = 4.1 \times 10^{-4}$

The Common-Ion Effect (Section 17.2)

17.56 (a) $NH_3(aq) + H_2O(l) \rightleftharpoons NH_4^+(aq) + OH^-(aq)$
NH_4NO_3 is a source of NH_4^+ (reaction product). The equilibrium shifts toward reactants, and the $[OH^-]$ decreases. The pH decreases.
(b) $HCO_3^-(aq) + H_2O(l) \rightleftharpoons H_3O^+(aq) + CO_3^{2-}(aq)$
Na_2CO_3 is a source of CO_3^{2-} (reaction product). The equilibrium shifts toward reactants, and the $[H_3O^+]$ decreases. The pH increases.
(c) Because NaOH is a strong base, addition of $NaClO_4$, a neutral salt, does not change the pH.

17.58   From $NH_4Cl(s)$, $[NH_4^+]_{initial} = \dfrac{0.10 \text{ mol}}{0.500 \text{ L}} = 0.20$ M

$$NH_3(aq) + H_2O(l) \rightleftharpoons NH_4^+(aq) + OH^-(aq)$$

initial (M)	0.40	0.20	~0
change (M)	$-x$	$+x$	$+x$
equil (M)	$0.40 - x$	$0.20 + x$	x

$K_b = \dfrac{[NH_4^+][OH^-]}{[NH_3]} = 1.8 \times 10^{-5} = \dfrac{(0.20 + x)(x)}{(0.40 - x)} \approx \dfrac{(0.20)(x)}{(0.40)}$

Solve for x. $x = [OH^-] = 3.6 \times 10^{-5}$ M

$[H_3O^+] = \dfrac{K_w}{[OH^-]} = \dfrac{1.0 \times 10^{-14}}{3.6 \times 10^{-5}} = 2.8 \times 10^{-10}$ M

$pH = -\log[H_3O^+] = -\log(2.8 \times 10^{-10}) = 9.55$

17.60 $pH = 8.90$; $[H_3O^+] = 10^{-pH} = 10^{-8.90} = 1.26 \times 10^{-9}$ M

$[OH^-] = \dfrac{K_w}{[H_3O^+]} = \dfrac{1.0 \times 10^{-14}}{1.26 \times 10^{-9}} = 7.9 \times 10^{-6}$ M

$$NH_3(aq) + H_2O(l) \rightleftharpoons NH_4^+(aq) + OH^-(aq)$$

$K_b = \dfrac{[NH_4^+][OH^-]}{[NH_3]} = 1.8 \times 10^{-5}$

$[NH_4^+] = \dfrac{(K_b)[NH_3]}{[OH^-]} = \dfrac{(1.8 \times 10^{-5})(0.016)}{(7.9 \times 10^{-6})} = 0.036$ M

17.62

$$NH_3(aq) + H_2O(l) \rightleftharpoons NH_4^+(aq) + OH^-(aq)$$

initial (M)	0.30	0	~0
change (M)	$-x$	$+x$	$+x$
equil (M)	$0.30 - x$	x	x

$K_b = \dfrac{[NH_4^+][OH^-]}{[NH_3]} = 1.8 \times 10^{-5} = \dfrac{x^2}{0.30 - x} \approx \dfrac{x^2}{0.30}$

Solve for x. $x = [OH^-] = 2.3 \times 10^{-3}$ M

$[H_3O^+] = \dfrac{K_w}{[OH^-]} = \dfrac{1.0 \times 10^{-14}}{2.3 \times 10^{-3}} = 4.3 \times 10^{-12}$ M

$pH = -\log[H_3O^+] = -\log(4.3 \times 10^{-12}) = 11.37$
Add 4.0 g of NH_4NO_3.

NH_4NO_3, 80.04; $[NH_4^+] = $ molarity of $NH_4NO_3 = \dfrac{\left(4.0 \text{ g} \times \dfrac{1 \text{ mol}}{80.04 \text{ g}}\right)}{0.100 \text{ L}} = 0.50$ M

$$NH_3(aq) + H_2O(l) \rightleftharpoons NH_4^+(aq) + OH^-(aq)$$

initial (M)	0.30	0.50	~0
change (M)	$-x$	$+x$	$+x$
equil (M)	$0.30 - x$	$0.50 + x$	x

$$K_b = \frac{[NH_4^+][OH^-]}{[NH_3]} = 1.8 \times 10^{-5} = \frac{(0.50 + x)x}{0.30 - x} \approx \frac{(0.50)x}{0.30}$$

Solve for x. $x = [OH^-] = 1.1 \times 10^{-5}$ M

$$[H_3O^+] = \frac{K_w}{[OH^-]} = \frac{1.0 \times 10^{-14}}{1.1 \times 10^{-5}} = 9.1 \times 10^{-10} \text{ M}; \quad pH = -\log[H_3O^+] = -\log(9.1 \times 10^{-10}) = 9.04$$

The % dissociation decreases because of the common ion (NH_4^+) effect.

Buffer Solutions (Sections 17.3 and 17.4)

17.64   Solutions (b), (c), and (d) are buffer solutions. Neutralization reactions for (b) and (d) result in solutions with equal concentrations of NH_3 and NH_4^+.

17.66   Both solutions buffer at the same pH because in both cases the $[NH_3]/[NH_4^+] = 1.5$. Solution (b), however, has a higher concentration of both NH_3 and NH_4^+, therefore it has the greater buffer capacity.

17.68 $NaHCO_3$, 84.01; Na_2CO_3, 105.99

$$[HCO_3^-] = \text{molarity of } NaHCO_3 = \frac{\left(4.2 \text{ g} \times \dfrac{1 \text{ mol}}{84.01 \text{ g}}\right)}{0.20 \text{ L}} = 0.25 \text{ M}$$

$$[CO_3^{2-}] = \text{molarity of } Na_2CO_3 = \frac{\left(5.3 \text{ g} \times \dfrac{1 \text{ mol}}{105.99 \text{ g}}\right)}{0.20 \text{ L}} = 0.25 \text{ M}$$

$$pH = pK_a + \log \frac{[\text{base}]}{[\text{acid}]} = pK_a + \log \frac{[CO_3^{2-}]}{[HCO_3^-]}$$

For HCO_3^-, $K_{a2} = 5.6 \times 10^{-11}$, $pK_{a2} = -\log K_{a2} = -\log(5.6 \times 10^{-11}) = 10.25$

$$pH = 10.25 + \log \frac{[0.25]}{[0.25]} = 10.25$$

The pH of a buffer solution will not change on dilution because the acid and base concentrations will change by the same amount and their ratio will remain the same.

17.70

$$CH_3CO_2H(aq) + H_2O(l) \rightleftharpoons H_3O^+(aq) + CH_3CO_2^-(aq)$$

initial (M)	0.18	~0	0.29
change (M)	−x	+x	+x
equil (M)	0.18 − x	x	0.29 + x

$$K_a = \frac{[H_3O^+][CH_3CO_2^-]}{[CH_3CO_2H]} = 1.8 \times 10^{-5} = \frac{x(0.29 + x)}{0.18 - x} \approx \frac{x(0.29)}{0.18}$$

Solve for x. $x = 1.12 \times 10^{-5}$ M $= [H_3O^+]$
For the buffer, $pH = -\log[H_3O^+] = -\log(1.12 \times 10^{-5}) = 4.95$
mol $CH_3CO_2H = (0.18 \text{ mol/L})(0.375 \text{ L}) = 0.0675$ mol CH_3CO_2H
mol $CH_3CO_2^- = (0.29 \text{ mol/L})(0.375 \text{ L}) = 0.109$ mol $CH_3CO_2^-$

(a)

$$\underset{100\%}{CH_3CO_2H(aq) \ + \ OH^-(aq) \ \rightarrow \ CH_3CO_2^-(aq) \ + \ H_2O(l)}$$

before (mol)	0.0675	0.0060	0.109
change (mol)	–0.0060	–0.0060	+0.0060
after (mol)	0.0615	0	0.115

$$[H_3O^+] = K_a \frac{[CH_3CO_2H]}{[CH_3CO_2^-]} = (1.8 \times 10^{-5})\left(\frac{0.0615}{0.115}\right) = 9.63 \times 10^{-6} \ M$$

$$pH = -\log[H_3O^+] = -\log(9.63 \times 10^{-6}) = 5.02$$

(b)

$$\underset{100\%}{CH_3CO_2^-(aq) \ + \ H_3O^+(aq) \ \rightarrow \ CH_3CO_2H(aq) \ + \ H_2O(l)}$$

before (mol)	0.109	0.0060	0.0675
change (mol)	–0.0060	–0.0060	+0.0060
after (mol)	0.103	0	0.0735

$$[H_3O^+] = K_a \frac{[CH_3CO_2H]}{[CH_3CO_2^-]} = (1.8 \times 10^{-5})\left(\frac{0.0735}{0.103}\right) = 1.28 \times 10^{-5} \ M$$

$$pH = -\log[H_3O^+] = -\log(1.28 \times 10^{-5}) = 4.89$$

17.72 $pH = pK_a + \log \dfrac{[base]}{[acid]} = pK_a + \log \dfrac{[C_3H_5O_3^-]}{[HC_3H_5O_3]}$

For $HC_3H_5O_3$, $K_a = 1.4 \times 10^{-4}$; $pK_a = -\log K_a = -\log(1.4 \times 10^{-4}) = 3.85$

$pH = 3.85 + \log \dfrac{(0.36)}{(0.58)} = 3.64$

17.74 $pH = pK_a + \log \dfrac{[base]}{[acid]} = pK_a + \log \dfrac{[HCO_3^-]}{[H_2CO_3]}$

For H_2CO_3, at 37 °C, $K_a = 7.9 \times 10^{-7}$; $pK_a = -\log K_a = -\log(7.9 \times 10^{-7}) = 6.10$
$pH = 6.10 + \log(10) = 7.10$

17.76 $pH = pK_a + \log \dfrac{[base]}{[acid]} = pK_a + \log \dfrac{[CH_3CO_2^-]}{[CH_3CO_2H]}$

For CH_3CO_2H, $K_a = 1.8 \times 10^{-5}$; $pK_a = -\log K_a = -\log(1.8 \times 10^{-5}) = 4.74$

$4.44 = 4.74 + \log \dfrac{[CH_3CO_2^-]}{[CH_3CO_2H]};$ $-0.30 = \log \dfrac{[CH_3CO_2^-]}{[CH_3CO_2H]}$

$\dfrac{[CH_3CO_2^-]}{[CH_3CO_2H]} = 10^{-0.30} = 0.50$

The solution should have 0.50 mol of $CH_3CO_2^-$ per mole of CH_3CO_2H. For example, you could dissolve 41g of CH_3CO_2Na in 1.00 L of 1.00 M CH_3CO_2H.

17.78 HSO_4^-, $K_{a2} = 1.2 \times 10^{-2}$; $pK_{a2} = -\log K_{a2} = 1.92$
$HOCl$, $K_a = 3.5 \times 10^{-8}$; $pK_a = -\log K_a = 7.46$
$C_6H_5CO_2H$, $K_a = 6.5 \times 10^{-5}$; $pK_a = -\log K_a = 4.19$
The buffer system of choice for pH = 4.50 is (c) $C_6H_5CO_2H$ - $C_6H_5CO_2^-$ because the pK_a
for $C_6H_5CO_2H$ (4.19) is closest to 4.50.

17.80   NaOH, 40.0; $20 \text{ g} \times \dfrac{1 \text{ mol}}{40.0 \text{ g}} = 0.50$ mol NaOH

(0.500 L)(1.5 mol/L) = 0.75 mol NH_4Cl

	NH_4^+(aq)	+	OH^-(aq)	$\rightleftharpoons$	NH_3(aq)	+	H_2O(l)
before reaction (mol)	0.75		0.50		0		
change (mol)	−0.50		−0.50		+0.50		
after reaction (mol)	0.25		0		0.50		

This reaction produces a buffer solution.
$[NH_4^+] = 0.25$ mol/0.500 L = 0.50 M; $[NH_3] = 0.50$ mol/0.500 L = 1.0 M

$$pH = pK_a + \log \frac{[\text{base}]}{[\text{acid}]} = pK_a + \log \frac{[NH_3]}{[NH_4^+]}$$

For NH_4^+, $K_a = \dfrac{K_w}{K_b \text{ for } NH_3} = \dfrac{1.0 \times 10^{-14}}{1.8 \times 10^{-5}} = 5.6 \times 10^{-10}$; $pK_a = -\log K_a = 9.25$

$pH = 9.25 + \log\left(\dfrac{1.0}{0.5}\right) = 9.55$

Strong Acid–Stong Base Titrations (Sections 17.5–17.6)

17.82

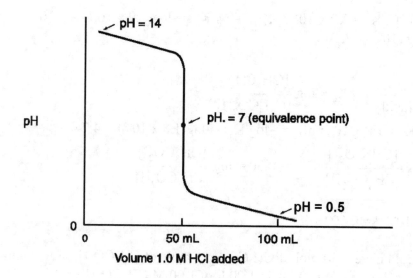

mmol NaOH = (50.0 mL)(1.0 mmol/mL) = 50 mmol
mmol HCl = mmol NaOH = 50 mmol

$$\text{vol HCl} = (50 \text{ mmol})\left(\frac{1.0 \text{ mL}}{1.0 \text{ mmol}}\right) = 50 \text{ mL}$$

50 mL of 1.0 M HCl is needed to reach the equivalence point.

17.84 $(0.0500 \text{ L})(0.116 \text{ mol/L})(1000 \text{ mmol/mol}) = 5.80 \text{ mmol NaOH}$
(a) $(0.0050 \text{ L})(0.0750 \text{ mol/L})(1000 \text{ mmol/mol}) = 0.375 \text{ mmol HCl}$

Neutralization reaction:	$H_3O^+(aq)$ +	$OH^-(aq)$ →	$H_2O(l)$
before reaction (mmol)	0.375	5.80	
change (mmol)	−0.375	−0.375	
after reaction (mmol)	0	5.425	

$$[OH^-] = \frac{5.425 \text{ mmol}}{(50.0 \text{ mL} + 5.0 \text{ mL})} = 0.0986 \text{ M}$$

$$[H_3O^+] = \frac{K_w}{[OH^-]} = \frac{1.0 \times 10^{-14}}{0.0986} = 1.01 \times 10^{-13} \text{ M}$$

$$pH = -\log[H_3O^+] = -\log(1.01 \times 10^{-13}) = 13.00$$

(b) $(0.050 \text{ L})(0.0750 \text{ mol/L})(1000 \text{ mmol/mol}) = 3.75 \text{ mmol HCl}$

Neutralization reaction:	$H_3O^+(aq)$ +	$OH^-(aq)$ →	$H_2O(l)$
before reaction (mmol)	3.75	5.80	
change (mmol)	−3.75	−3.75	
after reaction (mmol)	0	2.05	

$$[OH^-] = \frac{2.05 \text{ mmol}}{(50.0 \text{ mL} + 50 \text{ mL})} = 0.0205 \text{ M}$$

$$[H_3O^+] = \frac{K_w}{[OH^-]} = \frac{1.0 \times 10^{-14}}{0.0205} = 4.88 \times 10^{-13} \text{ M}$$

$$pH = -\log[H_3O^+] = -\log(4.88 \times 10^{-13}) = 12.31$$

(c) $(0.10 \text{ L})(0.0750 \text{ mol/L})(1000 \text{ mmol/mol}) = 7.50 \text{ mmol HCl}$

Neutralization reaction:	$H_3O^+(aq)$ +	$OH^-(aq)$ →	$H_2O(l)$
before reaction (mmol)	7.50	5.80	
change (mmol)	−5.80	−5.80	
after reaction (mmol)	1.70	0	

$$[H_3O^+] = \frac{1.70 \text{ mmol}}{(50.0 \text{ mL} + 100 \text{ mL})} = 0.0113 \text{ M}$$

$$pH = -\log[H_3O^+] = -\log(0.0113) = 1.95$$

Weak Acid–Strong Base Titrations (Section 17.7)

17.86 $\text{mmol HCO}_2\text{H} = (25.0 \text{ mL})(0.200 \text{ mmol/mL}) = 5.0 \text{ mmol}$
mmol NaOH required = mmol HCO_2H = 5.0 mmol

$$\text{mL NaOH required} = (5.0 \text{ mmol})\left(\frac{1.00 \text{ mL}}{0.250 \text{ mmol}}\right) = 20.0 \text{ mL}$$

20.0 mL of 0.250 M NaOH is required to reach the equivalence point.

For HCO_2H, $K_a = 1.8 \times 10^{-4}$; $pK_a = -\log K_a = -\log(1.8 \times 10^{-4}) = 3.74$

(a) mmol $HCO_2H = 5.0$ mmol

mmol $NaOH = (0.250$ mmol/mL$)(7.0$ mL$) = 1.75$ mmol

Neutralization reaction:

	$HCO_2H(aq)$	$+$	$OH^-(aq)$	$\rightarrow$	$HCO_2^-(aq)$	$+$	$H_2O(l)$
before reaction (mmol)	5.0		1.75		0		
change (mmol)	-1.75		-1.75		$+1.75$		
after reaction (mmol)	3.25		0		1.75		

$$[HCO_2H] = \frac{3.25 \text{ mmol}}{(25.0 \text{ mL} + 7.0 \text{ mL})} = 0.102 \text{ M};$$

$$[HCO_2^-] = \frac{1.75 \text{ mmol}}{(25.0 \text{ mL} + 7.0 \text{ mL})} = 0.0547 \text{ M}$$

	$HCO_2H(aq)$	$+$	$H_2O(l)$	$\rightleftharpoons$	$H_3O^+(aq)$	$+$	$HCO_2^-(aq)$
initial (M)	0.102				~0		0.0547
change (M)	$-x$				$+x$		$+x$
equil (M)	$0.102 - x$				x		$0.0547 + x$

$$K_a = \frac{[H_3O^+][HCO_2^-]}{[HCO_2H]} = 1.8 \times 10^{-4} = \frac{x(0.0547 + x)}{0.102 - x} \approx \frac{x(0.0547)}{0.102}$$

Solve for x. $x = [H_3O^+] = 3.4 \times 10^{-4}$ M

$pH = -\log[H_3O^+] = -\log(3.4 \times 10^{-4}) = 3.47$

(b) Halfway to the equivalence point,

$pH = pK_a = -\log K_a = -\log(1.8 \times 10^{-4}) = 3.74$

(c) At the equivalence point only the salt $NaHCO_2$ is in solution.

$$[HCO_2^-] = \frac{5.0 \text{ mmol}}{(25.0 \text{ mL} + 20.0 \text{ mL})} = 0.111 \text{ M}$$

	$HCO_2^-(aq)$	$+$	$H_2O(l)$	$\rightleftharpoons$	$HCO_2H(aq)$	$+$	$OH^-(aq)$
initial (M)	0.111				0		~0
change (M)	$-x$				$+x$		$+x$
equil (M)	$0.111 - x$				x		x

For HCO_2^-, $K_b = \dfrac{K_w}{K_a \text{ for } HCO_2H} = \dfrac{1.0 \times 10^{-14}}{1.8 \times 10^{-4}} = 5.6 \times 10^{-11}$

$$K_b = \frac{[HCO_2H][OH^-]}{[HCO_2^-]} = 5.6 \times 10^{-11} = \frac{x^2}{0.111 - x} \approx \frac{x^2}{0.111}$$

Solve for x. $x = [OH^-] = 2.5 \times 10^{-6}$ M

$$[H_3O^+] = \frac{K_w}{[OH^-]} = \frac{1.0 \times 10^{-14}}{2.5 \times 10^{-6}} = 4.0 \times 10^{-9} \text{ M}$$

$pH = -\log[H_3O^+] = -\log(4.0 \times 10^{-9}) = 8.40$

(d) mmol HCO_2H = 5.0 mmol
mmol NaOH = (0.250 mmol/mL)(25.0 mL) = 6.25 mmol
Neutralization reaction: $HCO_2H(aq)$ + $OH^-(aq)$ → $HCO_2^-(aq)$ + $H_2O(l)$
before reaction (mmol)   5.0   6.25   0
change (mmol)   −5.0   −5.0   +5.0
after reaction (mmol)   0   1.25   5.0
After the equivalence point, the pH of the solution is determined by the $[OH^-]$.

$$[OH^-] = \frac{1.25 \text{ mmol}}{(25.0 \text{ mL} + 25.0 \text{ mL})} = 2.5 \times 10^{-2} \text{ M}$$

$$[H_3O^+] = \frac{K_w}{[OH^-]} = \frac{1.0 \times 10^{-14}}{2.5 \times 10^{-2}} = 4.0 \times 10^{-13} \text{ M}$$

$$pH = -\log[H_3O^+] = -\log(4.0 \times 10^{-13}) = 12.40$$

17.88   (a)

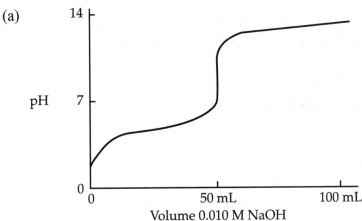

Volume 0.010 M NaOH

(b) mol NaOH required = $\left(\dfrac{0.010 \text{ mol HA}}{L}\right)(0.0500 \text{ L})\left(\dfrac{1 \text{ mol NaOH}}{1 \text{ mol HA}}\right)$ = 0.000 50 mol

vol NaOH required = (0.000 50 mol)$\left(\dfrac{1 \text{ L}}{0.010 \text{ mol}}\right)$ = 0.050 L = 50 mL

(c) A basic salt is present at the equivalence point; pH > 7.00
(d) Halfway to the equivalence point, the pH = pK_a = 4.00

Weak Base–Strong Acid Titrations (Section 17.8)

17.90   mmol NH_3 = (50.0 mL)(0.250 mmol/mL) = 12.5 mmol
mmol HNO_3 required = mmol NH_3 = 12.5 mmol

vol HNO_3 required = (12.5 mmol)$\left(\dfrac{1.00 \text{ mL}}{0.250 \text{ mmol}}\right)$ = 50.0 mL

50.0 mL of 0.250 M HNO_3 are required to reach the equivalence point.
(a) $NH_3(aq)$ + $H_2O(l)$ ⇌ $NH_4^+(aq)$ + $OH^-(aq)$
initial (M)   0.250   0   ~0
change (M)   −x   +x   +x
equil (M)   0.250 − x   x   x

$$K_b = \frac{[NH_4^+][OH^-]}{[NH_3]} = 1.8 \times 10^{-5} = \frac{x^2}{0.250 - x} \approx \frac{x^2}{0.250}$$

Solve for x. $x = [OH^-] = 2.1 \times 10^{-3}$ M

$$[H_3O^+] = \frac{K_w}{[OH^-]} = \frac{1.0 \times 10^{-14}}{2.1 \times 10^{-3}} = 4.7 \times 10^{-12} \text{ M}$$

$$pH = -\log[H_3O^+] = -\log(4.7 \times 10^{-12}) = 11.33$$

(b) 25.0 mL of HNO_3 is halfway to the equivalence point.

For NH_4^+, $K_a = \dfrac{K_w}{K_b \text{ for } NH_3} = \dfrac{1.0 \times 10^{-14}}{1.8 \times 10^{-5}} = 5.6 \times 10^{-10}$

$$pH = pK_a = -\log(5.6 \times 10^{-10}) = 9.26$$

(c) At the equivalence point only the salt NH_4NO_3 is in solution.
mmol $NH_4NO_3 = (0.250 \text{ mmol/mL})(50.0 \text{ mL}) = 12.5$ mmol

$$[NH_4^+] = \frac{12.5 \text{ mmol}}{(50.0 \text{ mL} + 50.0 \text{ mL})} = 0.125 \text{ M}$$

	$NH_4^+(aq)$	$+ H_2O(l)$	$\rightleftharpoons$	$H_3O^+(aq)$	$+ NH_3(aq)$
initial (M)	0.125			~0	0
change (M)	$-x$			$+x$	$+x$
equil (M)	$0.125 - x$			x	x

$$K_a = \frac{[H_3O^+][NH_3]}{[NH_4^+]} = 5.6 \times 10^{-10} = \frac{x^2}{0.125 - x} \approx \frac{x^2}{0.125}$$

Solve for x. $x = [H_3O^+] = 8.4 \times 10^{-6}$ M
$$pH = -\log[H_3O^+] = -\log(8.4 \times 10^{-6}) = 5.08$$

(d) mmol $NH_3 = (50.0 \text{ mL})(0.250 \text{ mmol/mL}) = 12.5$ mmol
mmol $HNO_3 = (0.250 \text{ mmol/mL})(60.0 \text{ mL}) = 15.0$ mmol

Neutralization reaction:

	$NH_3(aq)$	$+ H_3O^+(aq)$	$\rightarrow$	$NH_4^+(aq)$	$+ H_2O(l)$
before reaction (mmol)	12.5	15.0		0	
change (mmol)	-12.5	-12.5		$+12.5$	
after reaction (mmol)	0	2.5		12.5	

After the equivalence point the pH of the solution is determined by the $[H_3O^+]$.

$$[H_3O^+] = \frac{2.5 \text{ mmol}}{(50.0 \text{ mL} + 60.0 \text{ mL})} = 2.3 \times 10^{-2} \text{ M}$$

$$pH = -\log[H_3O^+] = -\log(2.3 \times 10^{-2}) = 1.64$$

17.92 (a) The strongest base has the highest pH at the equivalence point, C is the strongest base.
(b) The weakest base has the lowest pH at the equivalence point, B is the weakest base.

17.94 When equal volumes of acid and base react, all concentrations are cut in half.
(a) At the equivalence point only the salt $C_5H_{11}NHNO_3$ is in solution.

$[C_5H_{11}NH^+] = 0.10$ M

For $C_5H_{11}NH^+$, $K_a = \dfrac{K_w}{K_b \text{ for } C_5H_{11}N} = \dfrac{1.0 \times 10^{-14}}{1.3 \times 10^{-3}} = 7.7 \times 10^{-12}$

$$C_5H_{11}NH^+(aq) + H_2O(l) \rightleftharpoons H_3O^+(aq) + C_5H_{11}N(aq)$$

initial (M)	0.10	~0	0
change (M)	−x	+x	+x
equil (M)	0.10 − x	x	x

$K_a = \dfrac{[H_3O^+][C_5H_{11}N]}{[C_5H_{11}NH^+]} = 7.7 \times 10^{-12} = \dfrac{(x)(x)}{0.10 - x} \approx \dfrac{x^2}{0.10}$

Solve for x. $x = [H_3O^+] = 8.8 \times 10^{-7}$ M

pH = $-\log[H_3O^+] = -\log(8.8 \times 10^{-7}) = 6.06$

Chlorphenol red would be a suitable indicator.

(b) At the equivalence point only the salt Na_2SO_3 is in solution.

$[SO_3^{2-}] = 0.10$ M

For SO_3^{2-}, $K_b = \dfrac{K_w}{K_a \text{ for } HSO_3^-} = \dfrac{1.0 \times 10^{-14}}{6.3 \times 10^{-8}} = 1.6 \times 10^{-7}$

$$SO_3^{2-}(aq) + H_2O(l) \rightleftharpoons HSO_3^-(aq) + OH^-(aq)$$

Initial (M)	0.10	0	~0
change (M)	−x	+x	+x
equil (M)	0.10 − x	x	x

$K_b = \dfrac{[HSO_3^-][OH^-]}{[SO_3^{2-}]} = 1.6 \times 10^{-7} = \dfrac{(x)(x)}{0.10 - x} \approx \dfrac{x^2}{0.10}$

Solve for x. $x = [OH^-] = 1.26 \times 10^{-4}$ M

$[H_3O^+] = \dfrac{K_w}{[OH^-]} = \dfrac{1.0 \times 10^{-14}}{1.26 \times 10^{-4}} = 7.9 \times 10^{-11}$ M

pH = $-\log[H_3O^+] = -\log(7.9 \times 10^{-11}) = 10.10$

Thymolphthalein would be a suitable indicator.

(c) The pH is 7.00 at the equivalence point for the titration of a strong acid (HBr) with a strong base ($Ba(OH)_2$).

Bromthymol blue or phenol red would be suitable indicators. (Any indicator that changes color in the pH range 4–10 is satisfactory for a strong acid–strong base titration.)

Polyprotic Acid–Strong Base Titrations (Section 17.9)

17.96 For H_2CO_3, $K_{a1} = 4.3 \times 10^{-7}$ and $K_{a2} = 5.6 \times 10^{-11}$

(a) (25.0 mL)(0.0200 mmol/mL) = 0.500 mmol H_2CO_3

(10.0 mL)(0.0250 mmol/mL) = 0.250 mmol KOH added

0.500 mmol H_2CO_3 − 0.250 mmol KOH = 0.250 mmol HCO_3^- produced

This is halfway to the first equivalence point where pH = $pK_{a1} = -\log(4.3 \times 10^{-7}) = 6.37$

(b) At the first equivalence point, $pH = \dfrac{pK_{a1} + pK_{a2}}{2} = 8.31$

(c) Halfway between the first and second equivalence points, $pH = pK_{a2} = 10.25$

(d) At the second equivalence point only the basic salt, K_2CO_3, is in solution.

$$K_b = \frac{K_w}{K_a \text{ for } HCO_3^-} = \frac{K_w}{K_{a2}} = \frac{1.0 \times 10^{-14}}{5.6 \times 10^{-11}} = 1.8 \times 10^{-4}$$

mmol CO_3^{2-} = mmol H_2CO_3 = 0.500 mmol

$$[CO_3^{2-}] = \frac{0.500 \text{ mmol}}{(25.0 \text{ mL} + 40.0 \text{ mL})} = 0.00769 \text{ M}$$

$$\begin{array}{lcccccc}
& CO_3^{2-}(aq) & + & H_2O(l) & \rightleftharpoons & HCO_3^-(aq) & + & OH^-(aq) \\
\text{initial (M)} & 0.00769 & & & & 0 & & \sim 0 \\
\text{change (M)} & -x & & & & +x & & +x \\
\text{equil (M)} & 0.00769 - x & & & & x & & x
\end{array}$$

$$K_b = \frac{[HCO_3^-][OH^-]}{[CO_3^{2-}]} = 1.8 \times 10^{-4} = \frac{(x)(x)}{0.00769 - x}; \quad x^2 + (1.8 \times 10^{-4})x - (1.4 \times 10^{-6}) = 0$$

Use the quadratic formula to solve for x.

$$x = \frac{-(1.8 \times 10^{-4}) \pm \sqrt{(1.8 \times 10^{-4})^2 - (4)(1)(-1.4 \times 10^{-6})}}{2(1)} = \frac{-(1.8 \times 10^{-4}) \pm (2.37 \times 10^{-3})}{2}$$

$x = -1.28 \times 10^{-3}$ and 1.10×10^{-3}

Of the two solutions for x, only the positive value of x has physical meaning because x is the $[OH^-]$.

$[OH^-] = x = 1.10 \times 10^{-3}$ M

$$[H_3O^+] = \frac{K_w}{[OH^-]} = \frac{1.0 \times 10^{-14}}{1.10 \times 10^{-3}} = 9.1 \times 10^{-12} \text{ M}$$

$pH = -\log[H_3O^+] = -\log(9.1 \times 10^{-12}) = 11.04$

(e) excess KOH

(50.0 mL – 40.0 mL)(0.025 mmol/mL) = 0.250 mmol KOH = 0.250 mmol OH^-

$$[OH^-] = \frac{0.250 \text{ mmol}}{(25.0 \text{ mL} + 50.0 \text{ mL})} = 3.33 \times 10^{-3} \text{ M}$$

$$[H_3O^+] = \frac{K_w}{[OH^-]} = \frac{1.0 \times 10^{-14}}{3.33 \times 10^{-3}} = 3.0 \times 10^{-12} \text{ M}$$

$pH = -\log[H_3O^+] = -\log(3.0 \times 10^{-12}) = 11.52$

17.98   mol H_2A = mol NaOH/2 = (0.1000 mol/L)(0.03472/2 L) = 0.001 736 mol H_2A

$$H_2A \text{ molar mass} = \frac{0.2015 \text{ g } H_2A}{0.001 736 \text{ mol } H_2A} = 116.1 \text{ g/mol}$$

At the second equivalence point, the dominant reaction is:

$A^{2-}(aq) + H_2O(l) \rightleftharpoons OH^-(aq) + HA^-(aq)$

$$K_{b1} = \frac{[OH^-][HA^-]}{[A^{2-}]} \text{ and } [OH^-] = [HA^-]$$

$$K_{b1} = \frac{[OH^-]^2}{[A^{2-}]}$$

$$[A^{2-}] = \frac{0.001\ 736\ mol}{(0.02500\ L + 0.03472\ L)} = 0.02907\ M$$

$$[H_3O^+] = 10^{-pH} = 10^{-9.27} = 5.37 \times 10^{-10}\ M$$

$$[OH^-] = \frac{K_w}{[H_3O^+]} = \frac{1.00 \times 10^{-14}}{5.37 \times 10^{-10}} = 1.86 \times 10^{-5}\ M$$

$$K_{b1} = \frac{[OH^-]^2}{[A^{2-}]} = \frac{(1.86 \times 10^{-5})^2}{(0.02907)} = 1.19 \times 10^{-8}$$

$$K_{a2} \cdot K_{b1} = K_w$$

$$K_{a2} = \frac{K_w}{K_{b1}} = \frac{1.00 \times 10^{-14}}{1.19 \times 10^{-8}} = 8.40 \times 10^{-7}; \quad pK_{a2} = -\log(8.40 \times 10^{-7}) = 6.08$$

At the first equivalence point, $pH = \dfrac{pK_{a1} + pK_{a2}}{2} = 3.95$

$$3.95 = \frac{pK_{a1} + 6.08}{2}$$

$$pK_{a1} = (2)(3.95) - 6.08 = 1.82$$

Solubility Equilibria (Sections 17.10 and 17.11)

17.100   (a) $K_{sp} = [Ca^{2+}][OH^-]^2$ (b) $K_{sp} = [Ag^+]^3[PO_4^{3-}]$
         (c) $K_{sp} = [Ba^{2+}][CO_3^{2-}]$ (d) $K_{sp} = [Ca^{2+}]^5[PO_4^{3-}]^3[OH^-]$

17.102   (a) $K_{sp} = [Ca^{2+}]^3[PO_4^{2-}]^2 = (2.9 \times 10^{-7})^3(2.9 \times 10^{-7})^2 = 2.1 \times 10^{-33}$

(b) $[Ca^{2+}] = \sqrt[3]{\dfrac{K_{sp}}{[PO_4^{2-}]^2}} = \sqrt[3]{\dfrac{2.1 \times 10^{-33}}{(0.010)^2}} = 2.8 \times 10^{-10}\ M$

(c) $[PO_4^{2-}] = \sqrt{\dfrac{K_{sp}}{[Ca^{2+}]^3}} = \sqrt{\dfrac{2.1 \times 10^{-33}}{(0.010)^3}} = 4.6 \times 10^{-14}\ M$

17.104 $Pb(N_3)_2(s) \rightleftarrows Pb^{2+}(aq) + 2\ N_3^-(aq)$
         equil (M)   x   2x
         $[Pb^{2+}] = x = 8.5 \times 10^{-4}\ M; \quad [N_3^-] = 2x = 2(8.5 \times 10^{-4}\ M) = 1.7 \times 10^{-3}\ M$
         $K_{sp} = [Pb^{2+}][N_3^-]^2 = (8.5 \times 10^{-4})(1.7 \times 10^{-3})^2 = 2.5 \times 10^{-9}$

17.106   (a) $CdCO_3(s) \rightleftarrows Cd^{2+}(aq) + CO_3^-(aq)$
         equil (M)   x   x
         $[Cd^{2+}] = [CO_3^{2-}] = x = 1.0 \times 10^{-6}\ M$
         $K_{sp} = [Cd^{2+}][CO_3^{2-}] = (1.0 \times 10^{-6})^2 = 1.0 \times 10^{-12}$

(b) $\qquad$ Ca(OH)$_2$(s) $\rightleftarrows$ Ca^{2+}(aq) + 2 OH$^-$(aq)

equil (M) $\qquad\qquad\qquad\qquad$ x $\qquad\qquad$ 2x

[Ca^{2+}] = x = 1.06 x 10^{-2} M; [OH$^-$] = 2x = 2(1.06 x 10^{-2} M) = 2.12 x 10^{-2} M

K$_{sp}$ = [Ca^{2+}][OH$^-$]^2 = (1.06 x 10^{-2})(2.12 x 10^{-2})^2 = 4.76 x 10^{-6}

(c) PbBr$_2$, 367.01

$$[Pb^{2+}] = \text{molarity of PbBr}_2 = \frac{\left(4.34 \text{ g} \times \dfrac{1 \text{ mol}}{367.01 \text{ g}}\right)}{1 \text{ L}} = 1.18 \times 10^{-2} \text{ M}$$

$\qquad\qquad\qquad$ PbBr$_2$(s) $\rightleftarrows$ Pb^{2+}(aq) + 2 Br$^-$(aq)

equil (M) $\qquad\qquad\qquad\qquad$ x $\qquad\qquad$ 2x

[Pb^{2+}] = x = 1.18 x 10^{-2} M; [Br$^-$] = 2x = 2(1.18 x 10^{-2} M) = 2.36 x 10^{-2} M

K$_{sp}$ = [Pb^{2+}][Br$^-$]^2 = (1.18 x 10^{-2})(2.36 x 10^{-2})^2 = 6.57 x 10^{-6}

(d) BaCrO$_4$, 253.32

$$[Ba^{2+}] = [CrO_4^{2-}] = \text{molarity of BaCrO}_4 = \frac{\left(2.8 \times 10^{-3} \text{ g} \times \dfrac{1 \text{ mol}}{253.32 \text{ g}}\right)}{1 \text{ L}} = 1.1 \times 10^{-5} \text{ M}$$

$\qquad\qquad\qquad$ BaCrO$_4$(s) $\rightleftarrows$ Ba^{2+}(aq) + CrO$_4^{2-}$(aq)

equil (M) $\qquad\qquad\qquad\qquad$ x $\qquad\qquad$ x

[Ba^{2+}] = [CrO$_4^{2-}$] = x = 1.1 x 10^{-5} M

K$_{sp}$ = [Ba^{2+}][CrO$_4^{2-}$] = (1.1 x 10^{-5})^2 = 1.2 x 10^{-10}

17.108 (a) $\qquad\qquad\qquad$ Ag$_2$CO$_3$(s) $\rightleftarrows$ 2 Ag$^+$(aq) + CO$_3^{2-}$(aq)

equil (M) $\qquad\qquad\qquad\qquad$ 2x $\qquad\qquad$ x

K$_{sp}$ = [Ag$^+$]^2[CO$_3^{2-}$] = 8.4 x 10^{-12} = (2x)^2(x) = 4x^3

$$\text{molar solubility} = x = \sqrt[3]{\frac{8.4 \times 10^{-12}}{4}} = 1.3 \times 10^{-4} \text{ M}$$

Ag$_2$CO$_3$, 275.75

solubility = (1.3 x 10^{-4} mol/L)(275.75 g/mol) = 0.036 g/L

(b) $\qquad\qquad\qquad$ CuBr(s) $\rightleftarrows$ Cu$^+$(aq) + Br$^-$(aq)

equil (M) $\qquad\qquad\qquad\qquad$ x $\qquad\qquad$ x

K$_{sp}$ = [Cu$^+$][Br$^-$] = 6.3 x 10^{-9} = (x)(x)

molar solubility = x = $\sqrt{6.3 \times 10^{-9}}$ = 7.9 x 10^{-5} M

CuBr, 143.45

solubility = (7.9 x 10^{-5} mol/L)(143.45 g/mol) = 0.011 g/L

(c) $\qquad\qquad\qquad$ Cu$_3$(PO$_4$)$_2$(s) $\rightleftarrows$ 3 Cu^{2+}(aq) + 2 PO$_4^{3-}$(aq)

equil (M) $\qquad\qquad\qquad\qquad$ 3x $\qquad\qquad$ 2x

K$_{sp}$ = [Cu^{2+}]^3[PO$_4^{3-}$]^2 = 1.4 x 10^{-37} = (3x)^3(2x)^2 = 108x^5

$$\text{molar solubility} = x = \sqrt[5]{\frac{1.4 \times 10^{-37}}{108}} = 1.7 \times 10^{-8}$$

$Cu_3(PO_4)_2$, 380.58
solubility = $(1.7 \times 10^{-8} \text{ mol/L})(380.58 \text{ g/mol}) = 6.5 \times 10^{-6}$ g/L

Factors That Affect Solubility (Section 17.12)

17.110 $BaF_2(s) \rightleftharpoons Ba^{2+}(aq) + 2 F^-(aq)$
(a) H^+ from HCl reacts with F^- forming the weak acid HF. The equilibrium shifts to the right increasing the solubility of BaF_2.
(b) KF, source of F^-; equilibrium shifts left, solubility of BaF_2 decreases.
(c) No change in solubility.
(d) $Ba(NO_3)_2$, source of Ba^{2+}; equilibrium shifts left, solubility of BaF_2 decreases.

17.112   (a)

	$SrF_2(s) \rightleftharpoons$	$Sr^{2+}(aq)$	+	$2 F^-(aq)$
initial (M)		0.010		0
equil (M)		0.010 + x		2x

$K_{sp} = [Sr^{2+}][F^-]^2 = 4.3 \times 10^{-9} = (0.010 + x)(2x)^2 \approx (0.010)(2x)^2 = 0.040 \, x^2$

$$\text{molar solubility} = x = \sqrt{\frac{4.3 \times 10^{-9}}{0.040}} = 3.3 \times 10^{-4} \text{ M}$$

(b)

	$SrF_2(s) \rightleftharpoons$	$Sr^{2+}(aq)$	+	$2 F^-(aq)$
initial (M)		0		0.010
equil (M)		x		0.010 + 2x

$K_{sp} = [Sr^{2+}][F^-]^2 = 4.3 \times 10^{-9} = (x)(0.010 + 2x)^2 \approx (x)(0.010)^2 = x(0.00010)$

$$\text{molar solubility} = x = \frac{4.3 \times 10^{-9}}{0.00010} = 4.3 \times 10^{-5} \text{ M}$$

17.114   (a), (b), and (d) are more soluble in acidic solution.
(a) $MnS(s) + 2 H_3O^+(aq) \rightleftharpoons Mn^{2+}(aq) + H_2S(aq) + 2 H_2O(l)$
(b) $Fe(OH)_3(s) + 3 H_3O^+(aq) \rightleftharpoons Fe^{3+}(aq) + 6 H_2O(l)$
(c) $AgCl(s) \rightleftharpoons Ag^+(aq) + Cl^-(aq)$
(d) $BaCO_3(s) + H_3O^+(aq) \rightleftharpoons Ba^{2+}(aq) + HCO_3^-(aq) + H_2O(l)$

17.116 $BaCO_3(s) \rightleftharpoons Ba^{2+}(aq) + CO_3^{2-}(aq)$
(a) HNO_3 reacts with CO_3^{2-}, removing it from the solution. The solubility of $BaCO_3$ is increased.
(b) $Ba(NO_3)_2$ is a source of Ba^{2+} (reaction product). The solubility of $BaCO_3$ is decreased.
(c) Na_2CO_3 is a source of CO_3^{2-} (reaction product). The solubility of $BaCO_3$ is decreased.
(d) CH_3CO_2H reacts with CO_3^{2-}, removing it from the solution. The solubility of $BaCO_3$ is increased

17.118 (a) Because $Fe(OH)_3$ contains a basic anion, it becomes more soluble as the acidity of the solution increases. $Fe(OH)_3(s) + 3 H_3O^+(aq) \rightleftharpoons Fe^{3+}(aq) + 6 H_2O(l)$
(b) $Fe(OH)_3$ does not form a complex anion with OH^-; the solubility of $Fe(OH)_3$ is decreased by the presence of a common ion (OH^-) in the solution.
$Fe(OH)_3(s) \rightleftharpoons Fe^{3+}(aq) + 3 OH^-(aq)$
(c) Because $Fe(OH)_3$ forms the complex anion, $Fe(CN)_6^{3-}$, $Fe(OH)_3$ becomes more soluble in the presence of CN^-.
$Fe(OH)_3(s) + 6 CN^-(aq) \rightleftharpoons Fe(CN)_6^{3-}(aq) + 3 OH^-(aq)$

17.120

	$Cr^{3+}(aq)$	+	$4 OH^-(aq)$	$\rightleftharpoons$	$Cr(OH)_4^-(aq)$
before reaction (M)	0.0050		1.0		0
assume 100% reaction	−0.0050		−(4)(0.0050)		+0.0050
after reaction(M)	0		0.98		0.0050
assume small back rxn	+x		+4x		−x
equil (M)	x		0.98 + 4x		0.0050 − x

$$K_f = \frac{[Cr(OH)_4^-]}{[Cr^{3+}][OH^-]^4} = 8 \times 10^{29} = \frac{(0.0050 - x)}{(x)(0.98 + 4x)^4} \approx \frac{(0.0050)}{(x)(0.98)^4}$$

Solve for x. $x = [Cr^{3+}] = 6.8 \times 10^{-33} M = 7 \times 10^{-33} M$

$$\text{fraction uncomplexed } Cr^{3+} = \frac{[Cr^{3+}]}{[Cr(OH)_4^-]} = \frac{7 \times 10^{-33} M}{0.0050 M} = 1.4 \times 10^{-30} = 1 \times 10^{-30}$$

17.122 (a)

$Zn(OH)_2(s) \rightleftharpoons Zn^{2+}(aq) + 2 OH^-(aq)$ $\qquad K_{sp} = 4.1 \times 10^{-17}$
$Zn^{2+}(aq) + 4 OH^-(aq) \rightarrow Zn(OH)_4^{2-}(aq)$ $\qquad K_f = 3 \times 10^{15}$

dissolution rxn $Zn(OH)_2(s) + 2 OH^-(aq) \rightleftharpoons Zn(OH)_4^{2-}(aq)$
$K = (K_{sp})(K_f) = (4.1 \times 10^{-17})(3 \times 10^{15}) = 0.1$

(b)

$Cu(OH)_2(s) \rightleftharpoons Cu^{2+}(aq) + 2 OH^-(aq)$ $\qquad K_{sp} = 1.6 \times 10^{-19}$
$Cu^{2+}(aq) + 4 NH_3(aq) \rightarrow Cu(NH_3)_4^{2+}(aq)$ $\qquad K_f = 5.6 \times 10^{11}$

dissolution rxn $Cu(OH)_2(s) + 4 NH_3(aq) \rightleftharpoons Cu(NH_3)_4^{2+}(aq) + 2 OH^-(aq)$
$K = (K_{sp})(K_f) = (1.6 \times 10^{-19})(5.6 \times 10^{11}) = 9.0 \times 10^{-8}$

(c)

$AgBr(s) \rightleftharpoons Ag^+(aq) + Br^-(aq)$ $\qquad K_{sp} = 5.4 \times 10^{-13}$
$Ag^+(aq) + 2 NH_3(aq) \rightarrow Ag(NH_3)_2^+(aq)$ $\qquad K_f = 1.7 \times 10^7$

dissolution rxn $AgBr(s) + 2 NH_3(aq) \rightleftharpoons Ag(NH_3)_2^+(aq) + Br^-(aq)$
$K = (K_{sp})(K_f) = (5.4 \times 10^{-13})(1.7 \times 10^7) = 9.2 \times 10^{-6}$

17.124

	$Cr(OH)_3(s)$	+	$OH^-(aq)$	$\rightleftharpoons$	$Cr(OH)_4^-(aq)$
initial (M)			0.50		0
change (M)			−x		+x
equil (M)			0.50 − x		x

$K = (K_{sp})(K_f) = (6.7 \times 10^{-31})(8 \times 10^{29}) = 0.54$

$$K = 0.54 = \frac{[Cr(OH)_4^-]}{[OH^-]} = \frac{x}{0.50 - x}; \quad 0.27 - 0.54x = x; \quad 0.27 = 1.54x$$

molar solubility $= x = \dfrac{0.27}{1.54} = 0.2$ M

17.126 (a) $Fe(OH)_3(s) \rightleftharpoons Fe^{3+}(aq) + 3\ OH^-(aq)$ $\qquad K_{sp} = 2.6 \times 10^{-39}$

$H_3Cit(aq) + H_2O(l) \rightleftharpoons H_3O^+(aq) + H_2Cit^-(aq)$ $\qquad K_{a1} = 7.1 \times 10^{-4}$

$H_2Cit^-(aq) + H_2O(l) \rightleftharpoons H_3O^+(aq) + HCit^{2-}(aq)$ $\qquad K_{a2} = 1.7 \times 10^{-5}$

$HCit^{2-}(aq) + H_2O(l) \rightleftharpoons H_3O^+(aq) + Cit^{3-}(aq)$ $\qquad K_{a3} = 4.1 \times 10^{-7}$

$Fe^{3+}(aq) + Cit^{3-}(aq) \rightleftharpoons Fe(Cit)(aq)$ $\qquad K_f = 6.3 \times 10^{11}$

$\underline{3\ [H_3O^+(aq) + OH^-(aq) \rightleftharpoons 2\ H_2O(l)]}$ $\qquad (1/K_w)^3 = 1.0 \times 10^{42}$

$Fe(OH)_3(s) + H_3Cit(aq) \rightleftharpoons Fe(Cit)(aq) + 3\ H_2O(l)$

$K = K_{sp}K_{a1}K_{a2}K_{a3}K_f(1/K_w)^3 = 8.1$

(b) $\qquad Fe(OH)_3(s) + H_3Cit(aq) \rightleftharpoons Fe(Cit)(aq) + 3\ H_2O(l)$

initial (M) $\qquad\qquad\qquad 0.500 \qquad\qquad 0$

change (M) $\qquad\qquad\qquad -x \qquad\qquad +x$

equil (M) $\qquad\qquad\qquad 0.500 - x \qquad\quad x$

$K = \dfrac{[Fe(Cit)]}{[H_3Cit]} = 8.1 = \dfrac{x}{0.500 - x}$

$8.1(0.500 - x) = x$

$4.05 - 8.1x = x$

$4.05 = 9.1x$

$x =$ molar solubility $= 4.05/9.1 = 0.45$ M

Precipitation; Qualitative Analysis (Sections 17.13–17.15)

17.128 (a) $[CO_3^{2-}] = \dfrac{(2.0 \times 10^{-3}\ M)(0.10\ mL)}{(250\ mL)} = 8.0 \times 10^{-7}$ M

$K_{sp} = 5.0 \times 10^{-9} = [Ca^{2+}][CO_3^{2-}]$

$IP = [Ca^{2+}][CO_3^{2-}] = (8.0 \times 10^{-4})(8.0 \times 10^{-7}) = 6.4 \times 10^{-10}$

$IP < K_{sp}$; no precipitate will form.

(b) Na_2CO_3, 106; 10 mg = 0.010 g

$[CO_3^{2-}] = \dfrac{\left(0.010\ g \times \dfrac{1\ mol}{106\ g}\right)}{0.250\ L} = 3.8 \times 10^{-4}$ M

$IP = [Ca^{2+}][CO_3^{2-}] = (8.0 \times 10^{-4})(3.8 \times 10^{-4}) = 3.0 \times 10^{-7}$

$IP > K_{sp}$; $CaCO_3(s)$ will precipitate.

17.130 Assume 1.00 L of solution with a density of 1.00 g/mL. The solution weighs 1000 g.

1 ppm $F^- = \dfrac{mass\ F^-}{1000\ g} \times 10^6$

mass $F^- = \dfrac{1000\ g}{10^6} = 0.0010\ g\ F^-$

$$[F^-] = \frac{\left(0.0010 \text{ g} \times \dfrac{1 \text{ mol}}{19.0 \text{ g}}\right)}{1.00 \text{ L}} = 5.3 \times 10^{-5} \text{ M}$$

$[Ca^{2+}] = 5.0 \times 10^{-4} \text{ M}$

For CaF_2, $K_{sp} = [Ca^{2+}][F^-]^2 = 3.5 \times 10^{-11}$

For CaF_2, IP $= [Ca^{2+}][F^-]^2 = (5.0 \times 10^{-4})(5.3 \times 10^{-5})^2 = 1.4 \times 10^{-12}$

IP $< K_{sp}$; CaF_2 will not precipitate.

17.132 $K_{spa} = \dfrac{[M^{2+}][H_2S]}{[H_3O^+]^2}$;   FeS, $K_{spa} = 6 \times 10^2$;   SnS, $K_{spa} = 1 \times 10^{-5}$

Fe^{2+} and Sn^{2+} can be separated by bubbling H_2S through an acidic solution containing the two cations because their K_{spa} values are so different.

For FeS and SnS, $Q_c = \dfrac{(0.01)(0.10)}{(0.3)^2} = 1.1 \times 10^{-2}$

For FeS, $Q_c < K_{spa}$, and FeS will not precipitate.

For SnS, $Q_c > K_{spa}$, and SnS will precipitate.

17.134   FeS, $K_{spa} = \dfrac{[Fe^{2+}][H_2S]}{[H_3O^+]^2} = 6 \times 10^2$

(i) In 0.4 M HCl, $[H_3O^+] = 0.4$ M

$Q_c = \dfrac{[Fe^{2+}]_t[H_2S]_t}{[H_3O^+]_t^2} = \dfrac{(0.10)(0.10)}{(0.4)^2} = 0.0625$; $Q_c < K_{spa}$; FeS will not precipitate

(ii) pH $= 8$; $[H_3O^+] = 10^{-pH} = 10^{-8} = 1 \times 10^{-8}$ M

$Q_c = \dfrac{[Fe^{2+}]_t[H_2S]_t}{[H_3O^+]_t^2} = \dfrac{(0.10)(0.10)}{(1 \times 10^{-8})^2} = 1 \times 10^{14}$; $Q_c > K_{spa}$; FeS(s) will precipitate

17.136   (a) add Cl^- to precipitate AgCl

(b) add CO_3^{2-} to precipitate $CaCO_3$

(c) add H_2S to precipitate MnS

(d) add NH_3 and NH_4Cl to precipitate $Cr(OH)_3$

(Need buffer to control $[OH^-]$; excess OH^- produces the soluble $Cr(OH)_4^-$.)

17.138   Prepare aqueous solutions of the three salts. Add a solution of $(NH_4)_2HPO_4$. If a white precipitate forms, the solution contains Mg^{2+}. Perform flame test on the other two solutions. A yellow flame test indicates Na^+. A violet flame test indicates K^+.

Multiconcept Problems

17.140   (1) $Ca_5(PO_4)_3(OH)(s) \;\rightleftharpoons\; 5\, Ca^{2+}(aq) + 3\, PO_4^{3-}(aq) + OH^-(aq)$ $K_{sp1} = 2.3 \times 10^{-59}$

(2) $Ca_5(PO_4)_3(F)(s) \;\rightleftharpoons\; 5\, Ca^{2+}(aq) + 3\, PO_4^{3-}(aq) + F^-(aq)$ $K_{sp2} = 3.2 \times 10^{-60}$

Reverse reaction (2) to get reaction (3)

(3) $5 Ca^{2+}(aq) + 3 PO_4^{3-}(aq) + F^-(aq) \rightleftarrows Ca_5(PO_4)_3(F)(s)$ $K' = 1/K_{sp2}$

Combine reactions (1) and (3) to get reaction (4).

(4) $Ca_5(PO_4)_3(OH)(s) + F^-(aq) \rightleftarrows Ca_5(PO_4)_3(F)(s) + OH^-(aq)$

$$K(4) = (K_{sp1})(K') = \frac{K_{sp1}}{K_{sp2}} = \frac{2.3 \times 10^{-59}}{3.2 \times 10^{-60}} = 7.2$$

17.142 $[H_3O^+] = 10^{-pH} = 10^{-2.37} = 0.004\ 27$ M

$H_3Cit(aq) + H_2O(l) \rightleftarrows H_3O^+(aq) + H_2Cit^-(aq)$

$$K_{a1} = 7.1 \times 10^{-4} = \frac{[H_3O^+][H_2Cit^-]}{[H_3Cit]}$$

$(7.1 \times 10^{-4})[H_3Cit] = (0.004\ 27)[H_2Cit^-]$

$[H_3Cit] = (0.004\ 27)[H_2Cit^-]/(7.1 \times 10^{-4}) = (6.01)[H_2Cit^-]$

$H_2Cit^-(aq) + H_2O(l) \rightleftarrows H_3O^+(aq) + HCit^{2-}(aq)$

$$K_{a2} = 1.7 \times 10^{-5} = \frac{[H_3O^+][HCit^{2-}]}{[H_2Cit^-]}$$

$(1.7 \times 10^{-5})[H_2Cit^-] = (0.004\ 27)[HCit^{2-}]$

$[HCit^{2-}] = (1.7 \times 10^{-5})[H_2Cit^-]/(0.004\ 27) = (0.003\ 98)[H_2Cit^-]$

$[H_3Cit] + [H_2Cit^-] + [HCit^{2-}] + [Cit^{3-}] = 0.350$ M

Now assume $[Cit^{3-}] \approx 0$, so $[H_3Cit] + [H_2Cit^-] + [HCit^{2-}] = 0.350$ M and then by substitution:

$(6.01)[H_2Cit^-] + [H_2Cit^-] + (0.003\ 98)[H_2Cit^-] = 0.350$ M

$(7.01)[H_2Cit^-] = 0.350$ M

$[H_2Cit^-] = 0.350$ M$/7.01 = 0.050$ M

$[H_3Cit] = (6.01)[H_2Cit^-] = (6.01)(0.050$ M$) = 0.30$ M

$[HCit^{2-}] = (0.003\ 98)[H_2Cit^-] = (0.003\ 98)(0.050$ M$) = 2.0 \times 10^{-4}$ M

$HCit^{2-}(aq) + H_2O(l) \rightleftarrows H_3O^+(aq) + Cit^{3-}(aq)$

$$K_{a3} = 4.1 \times 10^{-7} = \frac{[H_3O^+][Cit^{3-}]}{[HCit^{2-}]}$$

$$[Cit^{3-}] = \frac{(K_{a3})[HCit^{2-}]}{[H_3O^+]} = \frac{(4.1 \times 10^{-7})(2.0 \times 10^{-4})}{(0.004\ 27)} = 1.9 \times 10^{-8} \text{ M}$$

17.144   (a) $Cd(OH)_2(s) \rightleftarrows Cd^{2+}(aq) + 2 OH^-(aq)$

initial (M)   0   ~0

equil (M)   x   2x

$K_{sp} = [Cd^{2+}][OH^-]^2 = 5.3 \times 10^{-15} = (x)(2x)^2 = 4x^3$

$$\text{molar solubility} = x = \sqrt[3]{\frac{5.3 \times 10^{-15}}{4}} = 1.1 \times 10^{-5} \text{ M}$$

$[OH^-] = 2x = 2(1.1 \times 10^{-5} \, M) = 2.2 \times 10^{-5} \, M$

$[H_3O^+] = \dfrac{1.0 \times 10^{-14}}{2.2 \times 10^{-5}} = 4.5 \times 10^{-10} \, M$

$pH = -\log[H_3O^+] = -\log(4.5 \times 10^{-10}) = 9.35$

(b) 90.0 mL = 0.0900 L

mol HNO_3 = (0.100 mol/L)(0.0900 L) = 0.009 00 mol HNO_3

The addition of HNO_3 dissolves some $Cd(OH)_2(s)$.

$$Cd(OH)_2(s) \; + \; 2\,HNO_3(aq) \; \rightarrow \; Cd^{2+}(aq) \; + \; 2\,H_2O(l)$$

before (mol)	0.100	0.009 00	1.1×10^{-5}
change (mol)	−0.0045	−2(0.0045)	$1.1 \times 10^{-5} + 0.0045$
after (mol)	0.0955	0	~0.0045

total volume = 100.0 mL + 90.0 mL = 190.0 mL = 0.1900 L

$[Cd^{2+}]$ = 0.0045 mol/0.1900 L = 0.024 M

$K_{sp} = 5.3 \times 10^{-15} = [Cd^{2+}][OH^-]^2 = (0.024)[OH^-]^2$

$[OH^-] = \sqrt{\dfrac{5.3 \times 10^{-15}}{0.024}} = 4.7 \times 10^{-7} \, M; \qquad [H_3O^+] = \dfrac{1.0 \times 10^{-14}}{4.7 \times 10^{-7}} = 2.1 \times 10^{-8} \, M$

$pH = -\log[H_3O^+] = -\log(2.1 \times 10^{-8}) = 7.68$

(c) volume HNO_3 = 0.0100 mol $Cd(OH)_2$ $\times \dfrac{2 \text{ mol } HNO_3}{1 \text{ mol } Cd(OH)_2} \times \dfrac{1.00 \text{ L}}{0.100 \text{ mol}} \times \dfrac{1000 \text{ mL}}{1.00 \text{ L}} = 200 \text{ mL}$

17.146 (a) $HA^-(aq) \; + \; H_2O(l) \; \rightleftharpoons \; H_3O^+(aq) \; + \; A^{2-}(aq) \qquad K_{a2} = 10^{-10}$

$HA^-(aq) \; + \; H_2O(l) \; \rightleftharpoons \; H_2A(aq) \; + \; OH^-(aq) \qquad K_b = \dfrac{K_w}{K_{a1}} = 10^{-10}$

$2\,HA^-(aq) \; \rightleftharpoons \; H_2A(aq) \; + \; A^{2-}(aq) \qquad\qquad K = \dfrac{K_{a2}}{K_{a1}} = 10^{-6}$

$2\,H_2O(l) \; \rightleftharpoons \; H_3O^+(aq) \; + \; OH^-(aq) \qquad\qquad K_w = 1.0 \times 10^{-14}$

The principal reaction of the four is the one with the largest K, and that is the third reaction.

(b) $K_{a1} = \dfrac{[H_3O^+][HA^-]}{[H_2A]}$ and $K_{a2} = \dfrac{[H_3O^+][A^{2-}]}{[HA^-]}$

$[H_3O^+] = \dfrac{K_{a1}[H_2A]}{[HA^-]}$ and $[H_3O^+] = \dfrac{K_{a2}[HA^-]}{[A^{2-}]}$

$\dfrac{K_{a1}[H_2A]}{[HA^-]} \times \dfrac{K_{a2}[HA^-]}{[A^{2-}]} = [H_3O^+]^2; \qquad \dfrac{K_{a1}K_{a2}[H_2A]}{[A^{2-}]} = [H_3O^+]^2$

Because the principal reaction is $2\,HA^-(aq) \rightleftharpoons H_2A(aq) + A^{2-}(aq)$, $[H_2A] = [A^{2-}]$.

$K_{a1}K_{a2} = [H_3O^+]^2$

$\log K_{a1} + \log K_{a2} = 2 \log [H_3O^+]$

$$\frac{\log K_{a1} + \log K_{a2}}{2} = \log[H_3O^+]; \quad \frac{-\log K_{a1} + (-\log K_{a2})}{2} = -\log[H_3O^+]$$

$$\frac{pK_{a1} + pK_{a2}}{2} = pH$$

(c)

	2 HA⁻(aq)	⇌	H₂A(aq)	+	A²⁻(aq)

	2 HA$^-$(aq) ⇌	H$_2$A(aq) +	A^{2-}(aq)
initial (M)	1.0	0	0
change (M)	$-2x$	$+x$	$+x$
equil (M)	$1.0 - 2x$	x	x

$$K = \frac{[H_2A][A^{2-}]}{[HA^-]^2} = 1 \times 10^{-6} = \frac{x^2}{(1.0 - 2x)^2}$$

Take the square root of both sides and solve for x.
$x = [A^{2-}] = 1 \times 10^{-3}$ M
mol A^{2-} = $(1 \times 10^{-3}$ mol/L$)(0.0500$ L$) = 5 \times 10^{-5}$ mol A^{2-}
number of A^{2-} ions = $(5 \times 10^{-5}$ mol A$^{2-})(6.022 \times 10^{23}$ ions/mol$) = 3 \times 10^{19}$ A^{2-} ions

17.148 (a) The first equivalence point is reached when all the H$_3$O$^+$ from the HCl, and the H$_3$O$^+$ from the first ionization of H$_3$PO$_4$, is consumed.

At the first equivalence point pH $= \dfrac{pK_{a1} + pK_{a2}}{2} = 4.66$

$[H_3O^+] = 10^{-pH} = 10^{(-4.66)} = 2.2 \times 10^{-5}$ M
$(88.0$ mL$)(0.100$ mmol/mL$) = 8.80$ mmol NaOH are used to get to the first equivalence point
(b) mmol (HCl + H$_3$PO$_4$) = mmol NaOH = 8.8 mmol
mmol H$_3$PO$_4$ = $(126.4$ mL $- 88.0$ mL$)(0.100$ mmol/mL$) = 3.84$ mmol
mmol HCl = $(8.8 - 3.84) = 4.96$ mmol
$[HCl] = \dfrac{4.96 \text{ mmol}}{40.0 \text{ mL}} = 0.124$ M; $\quad [H_3PO_4] = \dfrac{3.84 \text{ mmol}}{40.0 \text{ mL}} = 0.0960$ M
(c) 100% of the HCl is neutralized at the first equivalence point.

(d)

	H₃PO₄(aq) +	H₂O(l) ⇌	H₃O⁺(aq) +	H₂PO₄⁻(aq)
initial (M)	0.0960		0.124	0
change (M)	$-x$		$+x$	$+x$
equil (M)	$0.0960 - x$		$0.124 + x$	x

$$K_{a1} = \frac{[H_3O^+][H_2PO_4^-]}{[H_3PO_4]} = 7.5 \times 10^{-3} = \frac{(0.124 + x)(x)}{0.0960 - x}$$

$x^2 + 0.132x - (7.2 \times 10^{-4}) = 0$
Use the quadratic formula to solve for x.

$$x = \frac{-(0.132) \pm \sqrt{(0.132)^2 - 4(1)(-7.2 \times 10^{-4})}}{2(1)} = \frac{-0.132 \pm 0.142}{2}$$

$x = -0.137$ and 0.005
Of the two solutions for x, only the positive value of x has physical meaning because the other solution would give a negative $[H_3O^+]$.
$[H_3O^+] = 0.124 + x = 0.124 + 0.005 = 0.129$ M

$pH = -\log[H_3O^+] = -\log(0.129) = 0.89$

(e)

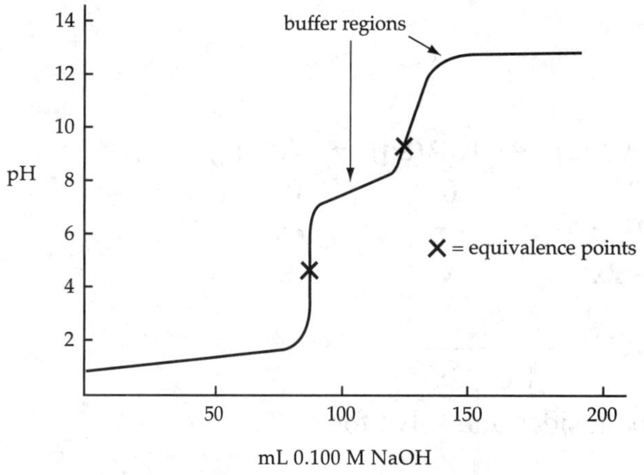

mL 0.100 M NaOH

(f) Bromcresol green or methyl orange are suitable indicators for the first equivalence point. Thymolphthalein is a suitable indicator for the second equivalence point.

17.150 $25\,^\circ C = 298\,K$

$$\Pi = 2MRT; \quad M = \frac{\Pi}{2RT} = \frac{\left(74.4\ mm\ Hg \times \dfrac{1.00\ atm}{760\ mm\ Hg}\right)}{(2)\left(0.082\ 06\ \dfrac{L \cdot atm}{K \cdot mol}\right)(298\ K)} = 0.00200\ M$$

$[M^+] = [X^-] = 0.00200\ M$

$K_{sp} = [M^+][X^-] = (0.00200)^2 = 4.00 \times 10^{-6}$

17.152 (a) species present initially:

NH_4^+	CO_3^{2-}	H_2O
acid	base	acid or base

$2H_2O(l) \rightleftarrows H_3O^+(aq) + OH^-(aq)$

$NH_4^+(aq) + H_2O(l) \rightleftarrows NH_3(aq) + H_3O^+(aq)$

$CO_3^{2-}(aq) + H_2O(l) \rightleftarrows HCO_3^-(aq) + OH^-(aq)$

$NH_3, K_b = 1.8 \times 10^{-5};$ $NH_4^+, K_a = 5.6 \times 10^{-10}$

$CO_3^{2-}, K_b = 1.8 \times 10^{-4};$ $HCO_3^-, K_a = 5.6 \times 10^{-11}$

In the mixture, proton transfer takes place from the stronger acid to the stronger base, so the principal reaction is $NH_4^+(aq) + CO_3^{2-}(aq) \rightleftarrows HCO_3^-(aq) + NH_3(aq)$

(b)

$NH_4^+(aq) + OH^-(aq) \rightleftarrows NH_3(aq) + H_2O(l)$ $K_1 = 1/K_b(NH_3)$

$CO_3^{2-}(aq) + H_2O(l) \rightleftarrows HCO_3^-(aq) + OH^-(aq)$ $K_2 = K_b(CO_3^{2-})$

$NH_4^+(aq) + CO_3^{2-}(aq) \rightleftarrows HCO_3^-(aq) + NH_3(aq)$ $K = K_1 \cdot K_2$

	NH_4^+	CO_3^{2-}	HCO_3^-	NH_3
initial (M)	0.16	0.080	0	0.16
change (M)	$-x$	$-x$	$+x$	$+x$
equil (M)	$0.16 - x$	$0.080 - x$	x	$0.16 + x$

$$K = \frac{[HCO_3^-][NH_3]}{[NH_4^+][CO_3^{2-}]} = \frac{1.8 \times 10^{-4}}{1.8 \times 10^{-5}} = 10 = \frac{x(0.16+x)}{(0.16-x)(0.080-x)}$$

$9x^2 - 2.56x + 0.128 = 0$

Use the quadratic formula to solve for x.

$$x = \frac{-(-2.56) \pm \sqrt{(-2.56)^2 - (4)(9)(0.128)}}{2(9)} = \frac{2.56 \pm 1.395}{18}$$

$x = 0.220$ and 0.0647

Of the two solutions for x, only 0.00647 has physical meaning because 0.220 leads to negative concentrations.

$[NH_4^+] = 0.16 - x = 0.16 - 0.0647 = 0.0953\ M = 0.095\ M$
$[NH_3] = 0.16 + x = 0.16 + 0.0647 = 0.225\ M = 0.23\ M$
$[CO_3^{2-}] = 0.080 - x = 0.080 - 0.0647 = 0.0153\ M = 0.015\ M$
$[HCO_3^-] = x = 0.0647\ M = 0.065\ M$

The solution is a buffer containing two different sets of conjugate acid-base pairs. Either pair can be used to calculate the pH.

For NH_4^+, $K_a = 5.6 \times 10^{-10}$ and $pK_a = 9.25$

$$pH = pK_a + \log \frac{[NH_3]}{[NH_4^+]} = 9.25 + \log \frac{(0.225)}{(0.0953)} = 9.62$$

$[H_3O^+] = 10^{-pH} = 10^{-9.62} = 2.4 \times 10^{-10}\ M$

$$[OH^-] = \frac{1.0 \times 10^{-14}}{2.4 \times 10^{-10}} = 4.2 \times 10^{-5}\ M$$

$$[H_2CO_3] = \frac{[HCO_3^-][H_3O^+]}{K_a} = \frac{(0.647)(2.4 \times 10^{-10})}{(4.3 \times 10^{-7})} = 3.6 \times 10^{-4}\ M$$

(c) For MCO_3, $IP = [M^{2+}][CO_3^{2-}] = (0.010)(0.0153) = 1.5 \times 10^{-4}$
$K_{sp}(CaCO_3) = 5.0 \times 10^{-9}$, $10^3\ K_{sp} = 5.0 \times 10^{-6}$
$K_{sp}(BaCO_3) = 2.6 \times 10^{-9}$, $10^3\ K_{sp} = 2.6 \times 10^{-6}$
$K_{sp}(MgCO_3) = 6.8 \times 10^{-6}$, $10^3\ K_{sp} = 6.8 \times 10^{-3}$
$IP > 10^3\ K_{sp}$ for $CaCO_3$ and $BaCO_3$, but $IP < 10^3\ K_{sp}$ for $MgCO_3$ so the $[CO_3^{2-}]$ is large enough to give observable precipitation of $CaCO_3$ and $BaCO_3$, but not $MgCO_3$.
(d) For $M(OH)_2$, $IP = [M^{2+}][OH^-]^2 = (0.010)(4.17 \times 10^{-5})^2 = 1.7 \times 10^{-11}$
$K_{sp}(Ca(OH)_2) = 4.7 \times 10^{-6}$, $10^3\ K_{sp} = 4.7 \times 10^{-3}$
$K_{sp}(Ba(OH)_2) = 5.0 \times 10^{-3}$, $10^3\ K_{sp} = 5.0$
$K_{sp}(Mg(OH)_2) = 5.6 \times 10^{-12}$, $10^3\ K_{sp} = 5.6 \times 10^{-9}$
$IP < 10^3\ K_{sp}$ for all three $M(OH)_2$. None precipitate.
(e)

	$CO_3^{2-}(aq)$	+ $H_2O(l)$	$\rightleftharpoons$	$HCO_3^-(aq)$	+ $OH^-(aq)$
initial (M)	0.08			0	~0
change (M)	–x			+x	+x
equil (M)	0.08 – x			x	x

$$K_b = \frac{[HCO_3^-][OH^-]}{[CO_3^{2-}]} = 1.8 \times 10^{-4} = \frac{x^2}{(0.08 - x)}$$

$x^2 + (1.8 \times 10^{-4})x - (1.44 \times 10^{-5}) = 0$

Use the quadratic formula to solve for x.

$$x = \frac{-(1.8 \times 10^{-4}) \pm \sqrt{(1.8 \times 10^{-4})^2 - (4)(1)(-1.44 \times 10^{-5})}}{2(1)} = \frac{-(1.8 \times 10^{-4}) \pm 7.59 \times 10^{-3}}{2}$$

$x = 0.0037$ and -0.0039

Of the two solutions for x, only 0.0037 has physical meaning because −0.0039 leads to negative concentrations.

$[OH^-] = x = 3.7 \times 10^{-3}$ M

For MCO_3, IP $= [M^{2+}][CO_3^{2-}] = (0.010)(0.08) = 8.0 \times 10^{-4}$

For $M(OH)_2$, IP $= [M^{2+}][OH^-]^2 = (0.010)(3.7 \times 10^{-3})^2 = 1.4 \times 10^{-7}$

Comparing IP's here and 10^3 K_{sp}'s in (c) and (d) above, Ca^{2+} and Ba^{2+} cannot be separated from Mg^{2+} using 0.08 M Na_2CO_3. Na_2CO_3 is more basic than $(NH_4)_2CO_3$ and $Mg(OH)_2$ would precipitate along with $CaCO_3$ and $BaCO_3$.

17.154 $Pb(CH_3CO_2)_2$, 325.29; PbS, 239.27

(a) mass PbS $= (2$ mL$)(1$ g/mL$)(0.003)$ x $\dfrac{1 \text{ mol } Pb(CH_3CO_2)_2}{325.29 \text{ g } Pb(CH_3CO_2)_2}$ x

$\dfrac{1 \text{ mol PbS}}{1 \text{ mol } Pb(CH_3CO_2)_2}$ x $\dfrac{239.27 \text{ g PbS}}{1 \text{ mol PbS}}$ x $(30/100) = 0.0013$ g

$= 1.3$ mg PbS per dye application

(b) $[H_3O^+] = 10^{-pH} = 10^{-5.50} = 3.16 \times 10^{-6}$ M

	PbS(s)	+	2 H$_3$O$^+$(aq)	⇌	Pb^{2+}(aq)	+	H$_2$S(aq)	+	2 H$_2$O(l)
initial (M)			3.16 x 10^{-6}		0		0		
change (M)			−2x		+x		+x		
equil (M)			3.16 x 10^{-6} − 2x		x		x		

$$K_{spa} = \frac{[Pb^{2+}][H_2S]}{[H_3O^+]^2} = \frac{x^2}{(3.16 \times 10^{-6} - 2x)^2} \approx \frac{x^2}{(3.16 \times 10^{-6})^2} = 3 \times 10^{-7}$$

$x^2 = (3.16 \times 10^{-6})^2(3 \times 10^{-7}) = 3.0 \times 10^{-18}$

$x = 1.7 \times 10^{-9}$ M $= [Pb^{2+}]$ for a saturated solution.

mass of PbS dissolved per washing =

$(3$ gal$)(3.7854$ L/1 gal$)(1.7 \times 10^{-9}$ mol/L$)$ x $\dfrac{239.27 \text{ g PbS}}{1 \text{ mol PbS}} = 4.6 \times 10^{-6}$ g PbS/washing

Number of washings required to remove 50% of the PbS from one application =

$$\frac{(0.0013 \text{ g PbS})(50/100)}{(4.6 \times 10^{-6} \text{ g PbS/washing})} = 1.4 \times 10^2 \text{ washings}$$

(c) The number of washings does not look reasonable. It seems too high considering that frequent dye application is recommended. If the PbS is located mainly on the surface of the hair, as is believed to be the case, solid particles of PbS can be lost by abrasion during shampooing.

18 Thermodynamics: Entropy, Free Energy, and Spontaneity

18.1   (a) $CO_2(s) \rightarrow CO_2(g)$
A gas has more randomness than a solid. Therefore, ΔS is positive.
(b) $I_2(g) \rightarrow 2\ I(g)$
ΔS is positive because the reaction increases the number of gaseous particles from 1 mol to 2 mol.
(c) $CaCO_3(s) \rightarrow CaO(s) + CO_2(g)$
ΔS is positive because the reaction increases the number of gaseous molecules.
(d) $Ag^+(aq) + Br^-(aq) \rightarrow AgBr(s)$
A solid has less randomness than +1 and –1 charged ions in an aqueous solution.
Therefore, ΔS is negative.

18.2   (a) $A_2(g) + 2\ B(g) \rightarrow 2\ AB(g)$.
(b) Because the reaction decreases the number of gaseous particles from 3 mol to 2 mol, the entropy change is negative.

18.3 $S = k \ln W$, $k = 1.38 \times 10^{-23}$ J/K
(a) $W = 1$; $S = (1.38 \times 10^{-23}$ J/K$) \ln (1) = 0$
(b) $W = 2^{10}$; $S = (1.38 \times 10^{-23}$ J/K$) \ln (2^{10}) = 9.57 \times 10^{-23}$ J/K
(c) $W = 3^{10}$; $S = (1.38 \times 10^{-23}$ J/K$) \ln (3^{10}) = 1.52 \times 10^{-22}$ J/K

18.4   (a) 1 mole N_2 at STP (larger volume, more randomness)
(b) $\Delta S = R \ln \left(\dfrac{V_{\text{State B}}}{V_{\text{State A}}} \right) = R \ln \left(\dfrac{11.2\ \text{L}}{22.4\ \text{L}} \right) = (8.314\ \text{J/K}) \ln(1/2) = -5.76$ J/K

18.5 $2\ H_2(g) + O_2(g) \rightarrow 2\ H_2O(l)$
$\Delta S° = [2\ S°(H_2O(l))] - [2\ S°(H_2) + S°(O_2)]$
$\Delta S° = [(2\ \text{mol})(69.9\ \text{J/(K} \cdot \text{mol}))]$
$\qquad - [(2\ \text{mol})(130.6\ \text{J/(K} \cdot \text{mol})) + (1\ \text{mol})(205.0\ \text{J/(K} \cdot \text{mol}))] = -326.4$ J/K

18.6   (a) $C_3H_8(g) + 5\ O_2(g) \rightarrow 3\ CO_2(g) + 4\ H_2O(l)$
$\Delta S°$ is negative because 6 mol of gas in the reactants are converted to 3 mol of gas in the products (less randomness).
(b) $\Delta S° = -376.6$ J/K $= [3\ S°(CO2) + 4\ S°(H_2O(l))] - [S°(C_3H_8) + 5\ S°(O_2)]$
$S°(C_3H_8) = [3\ S°(CO2) + 4\ S°(H_2O(l))] - [5\ S°(O_2)] + 376.6$ J/K
$S°(C_3H_8) = [(3\ \text{mol})(213.6\ \text{J/(K} \cdot \text{mol})) + (4\ \text{mol})(69.9\ \text{J/(K} \cdot \text{mol}))]$
$\qquad\qquad\qquad - [(5\ \text{mol})(205.0\ \text{J/(K} \cdot \text{mol}))] + 376.6$ J/K

$S°(C_3H_8) = 272.0$ J/(K $\cdot$ mol)

18.7 $CaCO_3(s) \rightarrow CaO(s) + CO_2(g)$
$\Delta H^\circ = [\Delta H^\circ_f(CaO) + \Delta H^\circ_f(CO_2)] - \Delta H^\circ_f(CaCO_3)$
$\Delta H^\circ = [(1 \text{ mol})(-634.9 \text{ kJ/mol}) + (1 \text{ mol})(-393.5 \text{ kJ/mol})]$
$\qquad\qquad\qquad - (1 \text{ mol})(-1207.6 \text{ kJ/mol}) = +179.2 \text{ kJ}$
$\Delta S^\circ = [S^\circ(CaO) + S^\circ(CO_2)] - S^\circ(CaCO_3)$
$\Delta S^\circ = [(1 \text{ mol})(38.1 \text{ J/(K} \cdot \text{mol)}) + (1 \text{ mol})(213.6 \text{ J/(K} \cdot \text{mol)})]$
$\qquad\qquad\qquad - (1 \text{ mol})(91.7 \text{ J/(K} \cdot \text{mol)}) = +160.0 \text{ J/K}$

$\Delta S_{surr} = \dfrac{-\Delta H^\circ}{T} = \dfrac{-179{,}200 \text{ J}}{298 \text{ K}} = -601 \text{ J/K}$

$\Delta S_{total} = \Delta S_{sys} + \Delta S_{surr} = 160.0 \text{ J/K} + (-601 \text{ J/K}) = -441 \text{ J/K}$
Because ΔS_{total} is negative, the reaction is not spontaneous under standard-state conditions at 25 °C.

18.8 $Br_2(l) \rightarrow Br_2(g)$
$\Delta H^\circ = \Delta H^\circ_f(Br_2(g)) = (1 \text{ mol})(30.9 \text{ kJ/mol}) = 30.9 \text{ kJ}$
$\Delta S^\circ = S^\circ(Br_2(g)) - S^\circ(Br_2(l))$
$\Delta S^\circ = (1 \text{ mol})(245.4 \text{ J/(K} \cdot \text{mol)}) - (1 \text{ mol})(152.2 \text{ J/(K} \cdot \text{mol)}) = 93.2 \text{ J/K} = 93.2 \times 10^{-3} \text{ kJ/K}$
$\Delta G = \Delta H - T\Delta S$
The boiling point (phase change) is associated with an equilibrium. Set $\Delta G = 0$ and solve for T, the boiling point.

$0 = \Delta H - T\Delta S; \qquad T_{bp} = \dfrac{\Delta H}{\Delta S} = \dfrac{30.9 \text{ kJ}}{93.2 \times 10^{-3} \text{ kJ/K}} = 331.5 \text{ K} = 58.4 \text{ °C}$

18.9 $\Delta G = \Delta H - T\Delta S = 55.3 \text{ kJ} - (298 \text{ K})(0.1757 \text{ kJ/K}) = +2.9 \text{ kJ}$
Because $\Delta G > 0$, the reaction is nonspontaneous at 25 °C (298 K)
To estimate the temperature when the reaction becomes spontaneous, set $\Delta G = 0$ and solve for T.

$0 = \Delta H - T\Delta S; \qquad T = \dfrac{\Delta H}{\Delta S} = \dfrac{55.3 \text{ kJ}}{0.1757 \text{ kJ/K}} = 315 \text{ K} = 42 \text{ °C}$

18.10 $\Delta H < 0$ (reaction involves bond making - exothermic)
$\Delta S < 0$ (the reaction has less randomness in going from reactants (2 atoms) to products (1 molecule)
$\Delta G < 0$ (the reaction is spontaneous)

18.11 (a) $CaCO_3(s) \rightarrow CaO(s) + CO_2(g)$
$\Delta H^\circ = [\Delta H^\circ_f(CaO) + \Delta H^\circ_f(CO_2)] - \Delta H^\circ_f(CaCO_3)$
$\Delta H^\circ = [(1 \text{ mol})(-634.9 \text{ kJ/mol}) + (1 \text{ mol})(-393.5 \text{ kJ/mol})]$
$\qquad\qquad\qquad - (1 \text{ mol})(-1207.6 \text{ kJ/mol}) = +179.2 \text{ kJ}$
$\Delta S^\circ = [S^\circ(CaO) + S^\circ(CO_2)] - S^\circ(CaCO_3)$
$\Delta S^\circ = [(1 \text{ mol})(38.1 \text{ J/(K} \cdot \text{mol)}) + (1 \text{ mol})(213.6 \text{ J/(K} \cdot \text{mol)})]$
$\qquad\qquad\qquad - (1 \text{ mol})(91.7 \text{ J/(K} \cdot \text{mol)}) = +160.0 \text{ J/K}$
$\Delta G^\circ = \Delta H^\circ - T\Delta S^\circ = 179.2 \text{ kJ} - (298 \text{ K})(0.1600 \text{ kJ/K}) = +131.5 \text{ kJ}$
(b) Because $\Delta G > 0$, the reaction is nonspontaneous at 25 °C (298 K).

18.12 $2 AB_2 \rightarrow A_2 + 2 B_2$
(a) $\Delta S°$ is positive because the reaction increases the number of molecules.
(b) $\Delta H°$ is positive because the reaction is endothermic.
$\Delta G° = \Delta H° - T\Delta S°$
For the reaction to be spontaneous, $\Delta G°$ must be negative. This will only occur at high temperature where $T\Delta S°$ is greater than $\Delta H°$.

18.13 $CaC_2(s) + 2 H_2O(l) \rightarrow C_2H_2(g) + Ca(OH)_2(s)$
$\Delta G° = [\Delta G°_f(C_2H_2) + \Delta G°_f(Ca(OH)_2)] - [\Delta G°_f(CaC_2) + 2 \Delta G°_f(H_2O)]$
$\Delta G° = [(1 \text{ mol})(209.9 \text{ kJ/mol}) + (1 \text{ mol})(-897.5 \text{ kJ/mol})]$
$\quad - [(1 \text{ mol})(-64.8 \text{ kJ/mol}) + (2 \text{ mol})(-237.2 \text{ kJ/mol})] = -148.4 \text{ kJ}$
This reaction can be used for the synthesis of C_2H_2 because $\Delta G < 0$.

18.14 $C_{diamond}(s) \rightarrow C_{graphite}(s)$
(a) $\Delta G° = \Delta G°_f(C_{graphite}) - \Delta G°_f(C_{diamond}) = 0 - 2.9 \text{ kJ} = -2.9 \text{ kJ}$
(b) The reaction is spontaneous because $\Delta G° < 0$.
(c) The reaction is spontaneous, but the reaction rate is extremely low.

18.15 $C(s) + 2 H_2(g) \rightarrow C_2H_4(g)$
$Q_p = \dfrac{P_{C_2H_4}}{(P_{H_2})^2} = \dfrac{(0.100)}{(100)^2} = 1.00 \times 10^{-5}$
$\Delta G = \Delta G° + RT \ln Q_p$
$\Delta G = 68.1 \text{ kJ/mol} + [8.314 \times 10^{-3} \text{ kJ/(K} \cdot \text{mol)}](298 \text{ K})\ln(1.00 \times 10^{-5}) = +39.6 \text{ kJ/mol}$
Because $\Delta G > 0$, the reaction is spontaneous in the reverse direction.

18.16 $\Delta G = \Delta G° + RT \ln Q$ and $\Delta G° = 15 \text{ kJ}$
For $A_2(g) + B_2(g) \rightleftharpoons 2 AB(g)$, $Q_p = \dfrac{(P_{AB})^2}{(P_{A_2})(P_{B_2})}$

Let the number of molecules be proportional to the partial pressure.
(1) $Q_p = 1.0$ (2) $Q_p = 0.0667$ (3) $Q_p = 18$
(a) Reaction (3) has the largest ΔG because Q_p is the largest. Reaction (2) has the smallest ΔG because Q_p is the smallest.
(b) $\Delta G = \Delta G° = 15 \text{ kJ}$ because $Q_p = 1$ and $\ln (1) = 0$.

18.17 From Problem 18.11, $\Delta G° = +131.5 \text{ kJ}$
$\Delta G° = -RT \ln K_p$
$\ln K_p = \dfrac{-\Delta G°}{RT} = \dfrac{-131.5 \text{ kJ/mol}}{[8.314 \times 10^{-3} \text{ kJ/(K} \cdot \text{mol)}](298 \text{ K})} = -53.1$
$K_p = e^{-53.1} = 9 \times 10^{-24}$

18.18 $\Delta G° = -RT \ln K = -[8.314 \times 10^{-3} \text{ kJ/(K} \cdot \text{mol)}](298 \text{ K}) \ln (1.0 \times 10^{-14}) = 80 \text{ kJ/mol}$

18.19 $H_2O(l) \rightleftharpoons H_2O(g)$

$K_p = P_{H_2O}$; K_p is equal to the vapor pressure for H_2O.

$\Delta G° = \Delta G°_f(H_2O(g)) - \Delta G°_f(H_2O(l))$

$\Delta G° = (1 \text{ mol})(-228.6 \text{ kJ/mol}) - (1 \text{ mol})(-237.2 \text{ kJ/mol}) = +8.6 \text{ kJ}$

$\Delta G° = -RT \ln K_p$

$\ln K_p = \dfrac{-\Delta G°}{RT} = \dfrac{-8.6 \text{ kJ/mol}}{[8.314 \times 10^{-3} \text{ kJ/(K} \cdot \text{mol)}](298 \text{ K})} = -3.5$

$K_p = P_{H_2O} = e^{-3.5} = 0.03 \text{ atm}$

18.20 $P_{EtOH} = 60.6 \text{ mm Hg} \times \dfrac{1.00 \text{ atm}}{760 \text{ mm Hg}} = 0.0797 \text{ atm}$

$K_p = P_{EtOH} = 0.0797 \text{ atm}$

$\Delta G° = -RT \ln K_p = -[8.314 \text{ J/K}](298 \text{ K}) \ln (0.0797) = 6267 \text{ J} = 6.27 \text{ kJ}$

18.21   (a) $\Delta S_{sys} < 0$ because there is less randomness   (b) For a spontaneous process $\Delta G < 0$
(c) As the two DNA strands come together, the formation of intermolecular forces (specifically hydrogen bonds) between DNA bases releases a substantial amount of heat. The heat released increases the entropy of the surroundings. The increase in entropy of the surroundings is greater than the decrease in entropy of the system leading to a spontaneous process which does not violate the second law of thermodynamics.

18.22   (a) $C_6H_{12}O_6(s) + 6 O_2(g) \rightleftharpoons 6 CO_2(g) + 6 H_2O(l)$ $\Delta G°' = -2870 \text{ kJ}$
$ADP^{3-}(aq) + H_2PO_4^-(aq) \rightleftharpoons ATP^{4-}(aq) + H_2O(l)$ $\Delta G°' = +30.5 \text{ kJ}$

$C_6H_{12}O_6(s) + 6 O_2(g) \rightleftharpoons 6 CO_2(g) + 6 H_2O(l)$
$\underline{32 [ADP^{3-}(aq) + H_2PO_4^-(aq) \rightleftharpoons ATP^{4-}(aq) + H_2O(l)]}$
$C_6H_{12}O_6(s) + 6 O_2(g) + 32 ADP^{3-}(aq) + 32 H_2PO_4^-(aq) \rightleftharpoons 6 CO_2(g) + 32 ATP^{4-}(aq) + 38 H_2O(l)$

$\Delta G°' = (-2870 \text{ kJ}) + 32(30.5 \text{ kJ}) = -1894 \text{ kJ}$
(b) Because $\Delta G°' < 0$, the reaction is spontaneous.

18.23   (a) Because $\Delta G°' > 0$, the reaction is not spontaneous.
(b) $\Delta G°' = (13.8 \text{ kJ}) + (-30.5 \text{ kJ}) = -16.7 \text{ kJ}$
Because the overall $\Delta G°' < 0$, the reaction is spontaneous.
(c) $\Delta G° = -RT \ln K$

$\ln K = \dfrac{-\Delta G°}{RT} = \dfrac{-(-16.7 \text{ kJ})}{[8.314 \times 10^{-3} \text{ kJ/K}](310 \text{ K})} = 6.48;$ $K = e^{6.48} = 652$

18.24   (a) $ATP^{4-}(aq) + H_2O(l) \rightleftharpoons ADP^{3-}(aq) + H_2PO_4^-(aq)$ $\Delta G°' = -30.5 \text{ kJ}$

$Q = \dfrac{[ADP^{3-}][H_2PO_4^-]}{[ATP^{4-}]} = \dfrac{(8.0 \times 10^{-3})(0.9 \times 10^{-3})}{(8.0 \times 10^{-3})}$

$$\Delta G = \Delta G^{o'} + RT \ln Q$$

$$\Delta G = -30.5 \text{ kJ} + (8.314 \times 10^{-3} \text{ kJ/K})(310 \text{ K}) \ln\left(\frac{(8.0 \times 10^{-3})(0.9 \times 10^{-3})}{(8.0 \times 10^{-3})}\right) = -48.6 \text{ kJ}$$

(b) The amount of free-energy released increases at the concentrations present in muscle cells.

18.25   creatine phosphate(aq) + H_2O(l) $\rightleftharpoons$ creatine(aq) + $H_2PO_4^-$(aq) $\Delta G^{o'} = -43.1$ kJ

<u>ADP^{3-}(aq) + $H_2PO_4^-$(aq) $\rightleftharpoons$ ATP^{4-}(aq) + H_2O(l)</u> $\Delta G^{o'} = +30.5$ kJ

creatine phosphate(aq) + ADP^{3-}(aq) $\rightleftharpoons$ creatine(aq) + ATP^{4-}(aq)

$\Delta G^{o'} = (-43.1 \text{ kJ}) + (30.5 \text{ kJ}) = -12.6$ kJ

Conceptual Problems

18.26   (a) $A_2 + AB_3 \rightarrow 3 \text{ AB}$

(b) ΔS is positive because the reaction increases the number of gaseous molecules.

18.28 $\Delta H > 0$ (heat is absorbed during sublimation)

$\Delta S > 0$ (gas has more randomness than solid)

$\Delta G < 0$ (the reaction is spontaneous)

18.30 $\Delta H = 0$ (system is an ideal gas at constant temperature)

$\Delta S < 0$ (there is less randomness in the smaller volume)

$\Delta G > 0$ (compression of a gas is not spontaneous)

18.32   (a) For <u>initial state 1</u>, $Q_p < K_p$

(more reactant (A_2) than product (A) compared to the equilibrium state)

For <u>initial state 2</u>, $Q_p > K_p$

(more product (A) than reactant (A_2) compared to the equilibrium state)

(b) $\Delta H > 0$ (reaction involves bond breaking - endothermic)

$\Delta S > 0$ (equilibrium state has more randomness than initial state 1)

$\Delta G < 0$ (reaction spontaneously proceeds toward equilibrium)

(c) $\Delta H < 0$ (reaction involves bond making - exothermic)

$\Delta S < 0$ (equilibrium state has less randomness than initial state 2)

$\Delta G < 0$ (reaction spontaneously proceeds toward equilibrium)

(d) State 1 lies to the left of the minimum in Figure 18.11. State 2 lies to the right of the minimum.

18.34   (a) Because the free energy decreases as pure reactants form products and also decreases as pure products form reactants, the free energy curve must go through a minimum somewhere between pure reactants and pure products. At the minimum point, $\Delta G = 0$ and the system is at equilibrium.

(b) The minimum in the plot is on the left side of the graph because $\Delta G^o > 0$ and the equilibrium composition is rich in reactants.

18.36   The equilibrium mixture is richer in reactant A at the higher temperature. This means the reaction is exothermic ($\Delta H < 0$). At 25 °C, $\Delta G° < 0$ because K > 1 and at 45 °C, $\Delta G° > 0$ because K < 1. Using the relationship. $\Delta G° = \Delta H° - T\Delta S°$, with $\Delta H° < 0$, $\Delta G°$ will become positive at the higher temperature only if $\Delta S°$ is negative.

Section Problems
Spontaneous Processes (Section 18.1)

18.38   (a) and (d) nonspontaneous; (b) and (c) spontaneous

18.40   (b) and (d) spontaneous (because of the large positive K_p's)

Enthalpy, Entropy, and Spontaneous Processes (Section 18.2)

18.42   Molecular randomness is called entropy. For the following reaction, the entropy increases:
$H_2O(s) \rightarrow H_2O(l)$ at 25 °C.

18.44   (a) + (solid → gas)
        (b) − (liquid → solid)
        (c) − (aqueous ions → solid)
        (d) + ($CO_2(aq) \rightarrow CO_2(g)$)

18.46   (a) − (liquid → solid)
        (b) − (decrease in number of O_2 molecules)
        (c) + (gas has more randomness in larger volume)
        (d) − (aqueous ions → solid)

Entropy (Sections 18.3–18.4)

18.48

Point Total	Possible Ways	Number of Ways
2	(1+1)	1
3	(2+1)(1+2)	2
4	(1+3)(2+2)(3+1)	3
5	(1+4)(2+3)(3+2)(4+1)	4
6	(1+5)(2+4)(3+3)(4+2)(5+1)	5
7	(1+6)(2+5)(3+4)(4+3)(5+2)(6+1)	6
8	(2+6)(3+5)(4+4)(5+3)(6+2)	5
9	(3+6)(4+5)(5+4)(6+3)	4
10	(4+6)(5+5)(6+4)	3
11	(6+5)(5+6)	2
12	(6+6)	1

Because a point total of 7 can be rolled in the most ways, it is the most probable point total.

18.50 $S = k \ln W$, $k = 1.38 \times 10^{-23}$ J/K
        (a) $W = 1$; $S = (1.38 \times 10^{-23}$ J/K$) \ln (1) = 0$
        (b) $W = 3^2$, $= 9$; $S = (1.38 \times 10^{-23}$ J/K$) \ln (3^2) = 3.03 \times 10^{-23}$ J/K

(c) $W = 1$; $S = (1.38 \times 10^{-23} \text{ J/K}) \ln (1) = 0$

(d) $W = 3^3 = 27$; $S = (1.38 \times 10^{-23} \text{ J/K}) \ln (3^3) = 4.55 \times 10^{-23}$ J/K

(e) $W = 1$; $S = (1.38 \times 10^{-23} \text{ J/K}) \ln (1) = 0$

(f) $W = 3^{6.02 \times 10^{23}}$; $S = (1.38 \times 10^{-23} \text{ J/K}) \ln (3^{6.02 \times 10^{23}}) = 9.13$ J/K

$$\Delta S = R \ln \left(\frac{V_f}{V_i} \right) = (8.314 \text{ J/K}) \ln 3 = 9.13 \text{ J/K} \qquad \text{The results are the same.}$$

18.52 $S = k \ln W$

$S_i = k \ln W_i = k \ln (1.00 \times 10^6)^{1000}$ and $S_f = k \ln W_f = k \ln (1.00 \times 10^7)^{1000}$

$$\frac{S_f}{S_i} = \frac{k \ln (1.00 \times 10^7)^{1000}}{k \ln (1.00 \times 10^6)^{1000}} = \frac{k(1000) \ln (1.00 \times 10^7)}{k(1000) \ln (1.00 \times 10^6)} = \frac{\ln(1.00 \times 10^7)}{\ln(1.00 \times 10^6)} = 1.17$$

and

$S_i = k \ln W_i = k \ln (1.00 \times 10^{16})^{1000}$ and $S_f = k \ln W_f = k \ln (1.00 \times 10^{17})^{1000}$

$$\frac{S_f}{S_i} = \frac{k \ln (1.00 \times 10^{17})^{1000}}{k \ln (1.00 \times 10^{16})^{1000}} = \frac{k(1000) \ln (1.00 \times 10^{17})}{k(1000) \ln (1.00 \times 10^{16})} = \frac{\ln(1.00 \times 10^{17})}{\ln(1.00 \times 10^{16})} = 1.06$$

18.54   (a) disordered N_2O (more randomness)

(b) quartz glass (amorphous solid, more randomness)

18.56   (a) H_2 at 25 °C in 50 L (larger volume)

(b) O_2 at 25 °C, 1 atm (larger volume)

(c) H_2 at 100 °C, 1 atm (larger volume and higher T)

(d) CO_2 at 100 °C, 0.1 atm (larger volume and higher T)

18.58 $\Delta S = nR \ln \left(\dfrac{V_f}{V_i} \right) = (0.050 \text{ mol})(8.314 \text{ J/K} \cdot \text{mol}) \ln \left(\dfrac{3.5 \text{ L}}{2.5 \text{ L}} \right) = 0.14$ J/K

18.60

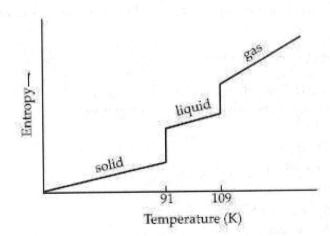

Standard Molar Entropies and Standard Entropies of Reaction (Section 18.5)

18.62 (a) $C_2H_6(g)$; more atoms/molecule
(b) $CO_2(g)$; more atoms/molecule
(c) $I_2(g)$; gas has more randomness than the solid
(d) $CH_3OH(g)$; gas has more randomness than the liquid

18.64 $2 CO(g) + O_2(g) \rightarrow 2 CO_2(g)$
$\Delta S^\circ = [2 S^\circ(CO_2)] - [2 S^\circ(CO) + S^\circ(O_2)]$
$\Delta S^\circ = [(2 \text{ mol})(213.6 \text{ J/(K} \cdot \text{mol}))]$
$\quad\quad - [(2 \text{ mol})(197.6 \text{ J/(K} \cdot \text{mol})) + (1 \text{ mol})(205.0 \text{ J/(K} \cdot \text{mol}))] = -173.0 \text{ J/K}$

18.66 (a) $2 H_2O_2(l) \rightarrow 2 H_2O(l) + O_2(g)$
$\Delta S^\circ = [2 S^\circ(H_2O(l)) + S^\circ(O_2)] - 2 S^\circ(H_2O_2)$
$\Delta S^\circ = [(2 \text{ mol})(69.9 \text{ J/(K} \cdot \text{mol})) + (1 \text{ mol})(205.0 \text{ J/(K} \cdot \text{mol}))]$
$\quad\quad - (2 \text{ mol})(110 \text{ J/(K} \cdot \text{mol})) = +125 \text{ J/K}$ (+, because moles of gas increase)
(b) $2 Na(s) + Cl_2(g) \rightarrow 2 NaCl(s)$
$\Delta S^\circ = 2 S^\circ(NaCl) - [2 S^\circ(Na) + S^\circ(Cl_2)]$
$\Delta S^\circ = (2 \text{ mol})(72.1 \text{ J/(K} \cdot \text{mol})) - [(2 \text{ mol})(51.2 \text{ J/(K} \cdot \text{mol})) + (1 \text{ mol})(223.0 \text{ J/(K} \cdot \text{mol}))]$
$\Delta S^\circ = -181.2 \text{ J/K}$ (–, because moles of gas decrease)
(c) $2 O_3(g) \rightarrow 3 O_2(g)$
$\Delta S^\circ = 3 S^\circ(O_2) - 2 S^\circ(O_3)$
$\Delta S^\circ = (3 \text{ mol})(205.0 \text{ J/(K} \cdot \text{mol})) - (2 \text{ mol})(238.8 \text{ J/(K} \cdot \text{mol}))$
$\Delta S^\circ = +137.4 \text{ J/K}$ (+, because moles of gas increase)
(d) $4 Al(s) + 3 O_2(g) \rightarrow 2 Al_2O_3(s)$
$\Delta S^\circ = 2 S^\circ(Al_2O_3) - [4 S^\circ(Al) + 3 S^\circ(O_2)]$
$\Delta S^\circ = (2 \text{ mol})(50.9 \text{ J/(K} \cdot \text{mol})) - [(4 \text{ mol})(28.3 \text{ J/(K} \cdot \text{mol})) + (3 \text{ mol})(205.0 \text{ J/(K} \cdot \text{mol}))]$
$\Delta S^\circ = -626.4 \text{ J/K}$ (–, because moles of gas decrease)

Entropy and the Second Law of Thermodynamics (Section 18.6)

18.68 In any spontaneous process, the total entropy of a system and its surroundings always increases.

18.70 $\Delta S_{surr} = \dfrac{-\Delta H}{T}$;   the temperature (T) is always positive.

(a) For an exothermic reaction, ΔH is negative and ΔS_{surr} is positive.
(b) For an endothermic reaction, ΔH is positive and ΔS_{surr} is negative.

18.72 $HgO(s) + Zn(s) \rightarrow ZnO(s) + Hg(l)$
(a) $\Delta S_{surr} = \dfrac{-\Delta H^\circ}{T} = \dfrac{-(-259.7 \times 10^3 \text{ J})}{298 \text{ K}} = +871.5 \text{ J/K}$

$\Delta S_{total} = \Delta S^\circ + \Delta S_{surr} = +7.8 \text{ J/K} + 871.5 \text{ J/K} = +879.3 \text{ J/K}$
The reaction is spontaneous because ΔS_{total} is > 0.
(b) Because $\Delta S^\circ > 0$ and $\Delta H^\circ < 0$, there is no temperature at which the reaction is not spontaneous.

18.74 $3 O_2(g) \rightarrow 2 O_3(g)$

$\Delta H^\circ = 2 \Delta H^\circ_f(O_3) = (2 \text{ mol})(143 \text{ kJ/mol}) = 286 \text{ kJ} = 286 \times 10^3 \text{ J}$

$\Delta S^\circ = 2 S^\circ(O_3) - 3 S^\circ(O_2)]$

$\Delta S^\circ = (2 \text{ mol})(238.8 \text{ J/(K} \cdot \text{mol})) - (3 \text{ mol})(205.0 \text{ J/(K} \cdot \text{mol}))]$

$\Delta S^\circ = -137.4 \text{ J/K}$

$\Delta S_{total} = \Delta S^\circ + \Delta S_{surr} = \Delta S^\circ + \dfrac{-\Delta H^\circ}{T} = -137.4 \text{ J/K} + \dfrac{-(286 \times 10^3 \text{ J})}{298 \text{ K}} = -1097 \text{ J/K}$

Because $\Delta S_{total} < 0$, the reaction is not spontaneous under standard-state conditions at 25 °C.

18.76 $2 HgO(s) \rightarrow 2 Hg(l) + O_2(g)$

$\Delta H^\circ = 0 - 2 \Delta H^\circ_f(HgO)$

$\Delta H^\circ = -(2 \text{ mol})(-90.8 \text{ kJ/mol}) = +181.6 \text{ kJ} = +181.6 \times 10^3 \text{ J}$

$\Delta S^\circ = [2 S^\circ(Hg) + S^\circ(O_2)] - 2 S^\circ(HgO)$

$\Delta S^\circ = [(2 \text{ mol})(76.0 \text{ J/(K} \cdot \text{mol})) + (1 \text{ mol})(205.0 \text{ J/(K} \cdot \text{mol}))] - (2 \text{ mol})(70.3 \text{ J/(K} \cdot \text{mol}))$

$\Delta S^\circ = \Delta S_{sys} = +216.4 \text{ J/K}$

$\Delta S_{surr} = \dfrac{-\Delta H^\circ}{T} = \dfrac{-(181.6 \times 10^3 \text{ J})}{298 \text{ K}} = -609.4 \text{ J/K}$

$\Delta S_{total} = \Delta S^\circ + \Delta S_{surr} = 216.4 \text{ J/K} + (-609.4 \text{ J/K}) = -393.0 \text{ J/K}$

Because $\Delta S_{total} < 0$, the reaction is not spontaneous under standard-state conditions at 25 °C.

(b) $\Delta S_{total} = \Delta S^\circ + \Delta S_{surr} = \Delta S^\circ + \dfrac{-\Delta H^\circ}{T}$

To find the temperature where the reaction becomes spontaneous, set $\Delta S_{total} = 0$ and solve for T.

$0 = \Delta S^\circ + \dfrac{-\Delta H^\circ}{T} = +216.4 \text{ J/K} + \dfrac{-(181.6 \times 10^3 \text{ J})}{T}$

$T = \dfrac{-(181.6 \times 10^3 \text{ J})}{-216.4 \text{ J/K}} = 839.2 \text{ K}$

18.78 (a) $\Delta S_{surr} = \dfrac{-\Delta H_{vap}}{T} = \dfrac{-30,700 \text{ J/mol}}{343 \text{ K}} = -89.5 \text{ J/(K} \cdot \text{mol})$

$\Delta S_{total} = \Delta S_{vap} + \Delta S_{surr} = 87.0 \text{ J/(K} \cdot \text{mol}) + (-89.5 \text{ J/(K} \cdot \text{mol})) = -2.5 \text{ J/(K} \cdot \text{mol})$

(b) $\Delta S_{surr} = \dfrac{-\Delta H_{vap}}{T} = \dfrac{-30,700 \text{ J/mol}}{353 \text{ K}} = -87.0 \text{ J/(K} \cdot \text{mol})$

$\Delta S_{total} = \Delta S_{vap} + \Delta S_{surr} = 87.0 \text{ J/(K} \cdot \text{mol}) + (-87.0 \text{ J/(K} \cdot \text{mol})) = 0$

(c) $\Delta S_{surr} = \dfrac{-\Delta H_{vap}}{T} = \dfrac{-30,700 \text{ J/mol}}{363 \text{ K}} = -84.6 \text{ J/(K} \cdot \text{mol})$

$\Delta S_{total} = \Delta S_{vap} + \Delta S_{surr} = 87.0 \text{ J/(K} \cdot \text{mol}) + (-84.6 \text{ J/(K} \cdot \text{mol})) = +2.4 \text{ J/(K} \cdot \text{mol})$

Benzene does not boil at 70 °C (343 K) because ΔS_{total} is negative.

The normal boiling point for benzene is 80 °C (353 K), where $\Delta S_{total} = 0$.

Free Energy (Section 18.7)

18.80

ΔH	ΔS	$\Delta G = \Delta H - T\Delta S$	Reaction Spontaneity
–	+	–	Spontaneous at all temperatures
–	–	– or +	Spontaneous at low temperatures where $|\Delta H| > |T\Delta S|$ Nonspontaneous at high temperatures where $|\Delta H| < |T\Delta S|$
+	–	+	Nonspontaneous at all temperatures
+	+	– or +	Spontaneous at high temperatures where $T\Delta S > \Delta H$ Nonspontaneous at low temperature where $T\Delta S < \Delta H$

18.82   (e) a negative free-energy change.

18.84   (a) 0 °C   (temperature is below mp); $\Delta H > 0$, $\Delta S > 0$, $\Delta G > 0$
        (b) 15 °C (temperature is above mp); $\Delta H > 0$, $\Delta S > 0$, $\Delta G < 0$

18.86 $\Delta H_{vap} = 30.7$ kJ/mol
        $\Delta S_{vap} = 87.0$ J/(K · mol) $= 87.0$ x 10^{-3} kJ/(K · mol)
        $\Delta G_{vap} = \Delta H_{vap} - T\Delta S_{vap}$
        (a) $\Delta G_{vap} = 30.7$ kJ/mol $- (343$ K$)(87.0$ x 10^{-3} kJ/(K · mol)) $= +0.9$ kJ/mol
        At 70 °C (343 K), benzene does not boil because ΔG_{vap} is positive.
        (b) $\Delta G_{vap} = 30.7$ kJ/mol $- (353$ K$)(87.0$ x 10^{-3} kJ/(K · mol)) $= 0$
        80 °C (353 K) is the boiling point for benzene because $\Delta G_{vap} = 0$
        (c) $\Delta G_{vap} = 30.7$ kJ/mol $- (363$ K$)(87.0$ x 10^{-3} kJ/(K · mol)) $= -0.9$ kJ/mol
        At 90 °C (363 K), benzene boils because ΔG_{vap} is negative.

18.88   At the melting point (phase change), $\Delta G_{fusion} = 0$.
        $\Delta G_{fusion} = \Delta H_{fusion} - T\Delta S_{fusion}$
        $0 = \Delta H_{fusion} - T\Delta S_{fusion}$
        $$T = \frac{\Delta H_{fusion}}{\Delta S_{fusion}} = \frac{18.02 \text{ kJ/mol}}{45.56 \text{ x } 10^{-3} \text{ kJ/(K · mol)}} = 395.5 \text{ K} = 122.4 \text{ °C}$$

18.90   At the normal boiling point, $\Delta G = 0$.
        $$\Delta G_{vap} = \Delta H_{vap} - T\Delta S_{vap}; \quad T = \frac{\Delta H_{vap}}{\Delta S_{vap}} = \frac{38,600 \text{ J}}{110 \text{ J/K}} = 351 \text{ K} = 78 \text{ °C}$$

Standard Free-Energy Changes and Standard Free Energies of Formation (Sections 18.8–18.9)

18.92   (a) $\Delta G°$ is the change in free energy that occurs when reactants in their standard states are converted to products in their standard states.
        (b) $\Delta G°_f$ is the free-energy change for formation of one mole of a substance in its standard state from the most stable form of the constituent elements in their standard states.

18.94 (a) $N_2(g) + 2 O_2(g) \rightarrow 2 NO_2(g)$

$\Delta H° = 2 \Delta H°_f(NO_2) = (2 \text{ mol})(33.2 \text{ kJ/mol}) = 66.4 \text{ kJ}$

$\Delta S° = 2 S°(NO_2) - [S°(N_2) + 2 S°(O_2)]$

$\Delta S° = (2 \text{ mol})(240.0 \text{ J/(K} \cdot \text{mol})) - [(1 \text{ mol})(191.5 \text{ J/(K} \cdot \text{mol})) + (2 \text{ mol})(205.0 \text{ J/(K} \cdot \text{mol}))]$

$\Delta S° = -121.5 \text{ J/K} = -121.5 \times 10^{-3} \text{ kJ/K}$

$\Delta G° = \Delta H° - T\Delta S° = 66.4 \text{ kJ} - (298 \text{ K})(-121.5 \times 10^{-3} \text{ kJ/K}) = +102.6 \text{ kJ}$

Because $\Delta G°$ is positive, the reaction is nonspontaneous under standard-state conditions at 25 °C.

(b) $CH_3CH_2OH(l) + O_2(g) \rightarrow CH_3CO_2H(l) + H_2O(l)$

$\Delta H° = [\Delta H°_f(CH_3CO_2H) + \Delta H°_f(H_2O)] - \Delta H°_f(CH_3CH_2OH)$

$\Delta H° = [(1 \text{ mol})(-484.5 \text{ kJ/mol}) + (1 \text{ mol})(-285.8 \text{ kJ/mol})] - (1 \text{ mol})(-277.7 \text{ kJ/mol}) = -492.6 \text{ kJ}$

$\Delta S° = [S°(CH_3CO_2H) + S°(H_2O)] - [S°(CH_3CH_2OH) + S°(O_2)]$

$\Delta S° = [(1 \text{ mol})(160 \text{ J/(K} \cdot \text{mol})) + (1 \text{ mol})(69.9 \text{ J/(K} \cdot \text{mol}))]$
$\qquad\qquad - [(1 \text{ mol})(161 \text{ J/(K} \cdot \text{mol})) + (1 \text{ mol})(205.0 \text{ J/(K} \cdot \text{mol}))]$

$\Delta S° = -136.1 \text{ J/K} = -136.1 \times 10^{-3} \text{ kJ/K}$

$\Delta G° = \Delta H° - T\Delta S° = -492.6 \text{ kJ} - (298 \text{ K})(-136.1 \times 10^{-3} \text{ kJ/K}) = -452.0 \text{ kJ}$

Because $\Delta G°$ is negative, the reaction is spontaneous under standard-state conditions at 25 °C.

18.96 $2 KClO_3(s) \rightarrow 2 KCl(s) + 3 O_2(g)$

$\Delta G° = 2 \Delta G°_f(KCl) - 2 \Delta G°_f(KClO_3)$

$\Delta G° = (2 \text{ mol})(-408.5 \text{ kJ/mol}) - (2 \text{ mol})(-296.3 \text{ kJ/mol}) = -224.4 \text{ kJ}$

18.98 A compound is thermodynamically stable with respect to its constituent elements at 25 °C if $\Delta G°_f$ is negative.

	$\Delta G°_f$ (kJ/mol)	Stable
(a) $BaCO_3(s)$	−1134.4	yes
(b) $HBr(g)$	−53.4	yes
(c) $N_2O(g)$	+104.2	no
(d) $C_2H_4(g)$	+68.1	no

18.100 $C_2H_4(g) + Cl_2(g) \rightarrow CH_2ClCH_2Cl(l)$

$\Delta G° = \Delta G°_f(CH_2ClCH_2Cl) - \Delta G°_f(C_2H_4)$

$\Delta G° = (1 \text{ mol})(-79.6 \text{ kJ/mol}) - (1 \text{ mol})(68.1 \text{ kJ/mol}) = -147.7 \text{ kJ}$

Because $\Delta G° < 0$, dichloroethane can be synthesized from gaseous C_2H_4 and Cl_2, each at 25 °C and 1 atm pressure.

18.102 $CH_2=CH_2(g) + H_2O(l) \rightarrow CH_3CH_2OH(l)$

$\Delta H° = \Delta H°_f(CH_3CH_2OH) - [\Delta H°_f(CH_2=CH_2) + \Delta H°_f(H_2O)]$

$\Delta H° = (1 \text{ mol})(-277.7 \text{ kJ/mol}) - [(1 \text{ mol})(52.3 \text{ kJ/mol}) + (1 \text{ mol})(-285.8 \text{ kJ/mol})]$

$\Delta H° = -44.2 \text{ kJ}$

$\Delta S° = S°(CH_3CH_2OH) - [S°(CH_2=CH_2) + S°(H_2O)]$

$\Delta S° = (1 \text{ mol})(161 \text{ J/(K} \cdot \text{mol})) - [(1 \text{ mol})(219.5 \text{ J/(K} \cdot \text{mol})) + (1 \text{ mol})(69.9 \text{ J/(K} \cdot \text{mol}))]$

$\Delta S° = -128 \text{ J/(K} \cdot \text{mol}) = -128 \times 10^{-3} \text{ kJ/(K} \cdot \text{mol})$

$\Delta G° = \Delta H° - T\Delta S° = -44.2 \text{ kJ} - (298 \text{ K})(-128 \times 10^{-3} \text{ kJ/K}) = -6.1 \text{ kJ}$

Because $\Delta G°$ is negative, the reaction is spontaneous under standard-state conditions at 25 °C. The reaction becomes nonspontaneous at high temperatures because $\Delta S°$ is negative. To find the crossover temperature, set $\Delta G = 0$ and solve for T.

$$T = \frac{\Delta H^\circ}{\Delta S^\circ} = \frac{-44,200 \text{ J}}{-128 \text{ J/K}} = 345 \text{ K} = 72 \text{ °C}$$

The reaction becomes nonspontaneous at 72 °C.

18.104 $3 \text{ C}_2\text{H}_2(g) \rightarrow \text{C}_6\text{H}_6(l)$
$\Delta G^\circ = \Delta G^\circ_f(\text{C}_6\text{H}_6) - 3 \ \Delta G^\circ_f(\text{C}_2\text{H}_2)$
$\Delta G^\circ = (1 \text{ mol})(124.5 \text{ kJ/mol}) - (3 \text{ mol})(209.9 \text{ kJ/mol}) = -505.2 \text{ kJ}$
Because ΔG° is negative, the reaction is possible. Look for a catalyst.
Because ΔG°_f for benzene is positive (+124.5 kJ/mol), the synthesis of benzene from graphite and gaseous H_2 at 25 °C and 1 atm pressure is not possible.

Free Energy for Reactions under Nonstandard-State Conditions (Section 18.10)

18.106 $\Delta G = \Delta G^\circ + RT \ln Q$

18.108 $2 \text{ NO}(g) + \text{Cl}_2(g) \rightarrow 2 \text{ NOCl}(g)$
$\Delta G^\circ = 2 \ \Delta G^\circ_f(\text{NOCl}) - 2 \ \Delta G^\circ_f(\text{NO})$
$\Delta G^\circ = (2 \text{ mol})(66.1 \text{ kJ/mol}) - (2 \text{ mol})(87.6 \text{ kJ/mol}) = -43.0 \text{ kJ}$

$$\Delta G = \Delta G^\circ + RT \ln \left[\frac{(P_{\text{NOCl}})^2}{(P_{\text{NO}})^2(P_{\text{Cl}_2})} \right]$$

$$\Delta G = (-43.0 \text{ kJ/mol}) + [8.314 \times 10^{-3} \text{ kJ/(K} \cdot \text{mol})](298 \text{ K})\ln \left[\frac{(2.00)^2}{(1.00 \times 10^{-3})^2(1.00 \times 10^{-3})} \right]$$

$\Delta G = +11.8 \text{ kJ/mol};$ The reaction is spontaneous in the reverse direction.

18.110 $\Delta G = \Delta G^\circ + RT \ln \left[\dfrac{(P_{\text{SO}_3})^2}{(P_{\text{SO}_2})^2(P_{\text{O}_2})} \right]$

(a) $\Delta G = (-141.8 \text{ kJ/mol}) + [8.314 \times 10^{-3} \text{ kJ/(K} \cdot \text{mol})](298 \text{ K})\ln \left[\dfrac{(1.0)^2}{(100)^2(100)} \right] = -176.0 \text{ kJ/mol}$

(b) $\Delta G = (-141.8 \text{ kJ/mol}) + [8.314 \times 10^{-3} \text{ kJ/(K} \cdot \text{mol})](298 \text{ K})\ln \left[\dfrac{(10)^2}{(2.0)^2(1.0)} \right] = -133.8 \text{ kJ/mol}$

(c) $Q = 1, \ln Q = 0, \Delta G = \Delta G^\circ = -141.8 \text{ kJ/mol}$

Free Energy and Chemical Equilibrium (Section 18.11)

18.112 $\Delta G^\circ = -RT \ln K$
(a) If $K > 1$, ΔG° is negative.   (b) If $K = 1$, $\Delta G^\circ = 0$.   (c) If $K < 1$, ΔG° is positive.

18.114 $\text{C}_2\text{H}_5\text{OH}(l) \rightleftharpoons \text{C}_2\text{H}_5\text{OH}(g)$
$\Delta G^\circ = \Delta G^\circ_f(\text{C}_2\text{H}_5\text{OH}(g)) - \Delta G^\circ_f(\text{C}_2\text{H}_5\text{OH}(l))$
$\Delta G^\circ = (1 \text{ mol})(-167.9 \text{ kJ/mol}) - (1 \text{ mol})(-174.9 \text{ kJ/mol}) = +7.0 \text{ kJ}$

$$\Delta G° = -RT \ln K$$

$$\ln K = \frac{-\Delta G°}{RT} = \frac{-(7.0 \text{ kJ/mol})}{[8.314 \times 10^{-3} \text{ kJ/(K} \cdot \text{mol)}](298 \text{ K})} = -2.83$$

$$K = e^{-2.83} = 0.059; \quad K = K_p = P_{C_2H_5OH} = 0.059 \text{ atm}$$

18.116 $Br_2(l) \rightleftarrows Br_2(g)$

$\Delta G° = \Delta G°_f(Br_2(g)) = 3.14 \text{ kJ/mol}$

$\Delta G° = -RT \ln K$

$$\ln K = \frac{-\Delta G°}{RT} = \frac{-(3.14 \text{ kJ/mol})}{[8.314 \times 10^{-3} \text{ kJ/(K} \cdot \text{mol)}](298 \text{ K})} = -1.267$$

$$K = e^{-1.267} = 0.282; \quad K = K_p = P_{Br_2} = 0.282 \text{ atm} = 0.28 \text{ atm}$$

18.118 $2 \ CH_2{=}CH_2(g) + O_2(g) \rightarrow 2 \ C_2H_4O(g)$

$\Delta G° = 2 \ \Delta G°_f(C_2H_4O) - 2 \ \Delta G°_f(CH_2{=}CH_2)$

$\Delta G° = (2 \text{ mol})(-13.1 \text{ kJ/mol}) - (2 \text{ mol})(68.1 \text{ kJ/mol}) = -162.4 \text{ kJ}$

$\Delta G° = -RT \ln K$

$$\ln K = \frac{-\Delta G°}{RT} = \frac{-(-162.4 \text{ kJ/mol})}{[8.314 \times 10^{-3} \text{ kJ/(K} \cdot \text{mol)}](298 \text{ K})} = 65.55$$

$$K = K_p = e^{65.55} = 2.9 \times 10^{28}$$

18.120 $NH_4NO_3(s) \rightarrow N_2O(g) + 2 \ H_2O(g)$

(a) $\Delta G° = [\Delta G°_f(N_2O) + 2 \ \Delta G°_f(H_2O)] - \Delta G°_f(NH_4NO_3)$

$\Delta G° = [(1 \text{ mol})(104.2 \text{ kJ/mol}) + (2 \text{ mol})(-228.6 \text{ kJ/mol})] - (1 \text{ mol})(-184.0 \text{ kJ/mol})$

$\Delta G° = -169.0 \text{ kJ}$

Because $\Delta G°$ is negative, the reaction is spontaneous.

(b) Because the reaction increases the number of moles of gas, $\Delta S°$ is positive.

$\Delta G° = \Delta H° - T\Delta S°$

As the temperature is raised, $\Delta G°$ becomes more negative.

(c) $\Delta G° = -RT \ln K$

$$\ln K = \frac{-\Delta G°}{RT} = \frac{-(-169.0 \text{ kJ/mol})}{[8.314 \times 10^{-3} \text{ kJ/(K} \cdot \text{mol)}](298 \text{ K})} = 68.21$$

$$K = K_p = e^{68.21} = 4.2 \times 10^{29}$$

(d) $Q = (P_{N_2O})(P_{H_2O})^2 = (30)(30)^2 = (30)^3$

$\Delta G = \Delta G° + RT \ln Q$

$\Delta G = -169.0 \text{ kJ/mol} + [8.314 \times 10^{-3} \text{ kJ/(K} \cdot \text{mol)}](298 \text{ K}) \ln[(30)^3] = -143.7 \text{ kJ/mol}$

18.122 $MgCO_3(s) \rightarrow MgO(s) + CO_2(g)$

$\Delta H° = [\Delta H°_f(MgO) + \Delta H°_f(CO_2)] - \Delta H°_f(MgCO_3)$

$\Delta H° = [(1 \text{ mol})(-601.1 \text{ kJ/mol}) + (1 \text{ mol})(-393.5 \text{ kJ/mol})] - (1 \text{ mol})(-1096 \text{ kJ/mol}) = +101 \text{ kJ}$

$\Delta S° = [S°(MgO) + S°(CO_2)] - S°(MgCO_3)$

$\Delta S° = [(1 \text{ mol})(26.9 \text{ J/(K} \cdot \text{mol)}) + (1 \text{ mol})(213.6 \text{ J/(K} \cdot \text{mol)})] - (1 \text{ mol})(65.7 \text{ J/(K} \cdot \text{mol)})$

$\Delta S° = 174.8 \text{ J/K} = 174.8 \times 10^{-3} \text{ kJ/K}$

The equilibrium pressure of CO_2 is equal to $K_p = P_{CO_2}$. K_p is not affected by the quantities of $MgCO_3$ and MgO present. K_p can be calculated from $\Delta G°$.

$\Delta G° = \Delta H° - T\Delta S°$

$\Delta G° = -RT \ln K_p$

(a) $\Delta G° = 101 \text{ kJ} - (298 \text{ K})(174.8 \times 10^{-3} \text{ kJ/K}) = +49 \text{ kJ}$

$$\ln K_p = \frac{-\Delta G°}{RT} = \frac{-49 \text{ kJ/mol}}{[8.314 \times 10^{-3} \text{ kJ/(K} \cdot \text{mol)}](298 \text{ K})} = -19.8$$

$K_p = P_{CO_2} = e^{-19.8} = 3 \times 10^{-9} \text{ atm}$

(b) $\Delta G° = 101 \text{ kJ} - (553 \text{ K})(174.8 \times 10^{-3} \text{ kJ/K}) = 4.3 \text{ kJ}$

$$\ln K_p = \frac{-\Delta G°}{RT} = \frac{-4.3 \text{ kJ/mol}}{[8.314 \times 10^{-3} \text{ kJ/(K} \cdot \text{mol)}](553 \text{ K})} = -0.94$$

$K_p = P_{CO_2} = e^{-0.94} = 0.39 \text{ atm}$

(c) $P_{CO_2} = 0.39 \text{ atm}$ because the temperature is the same as in (b).

18.124   (a) $\Delta H° = 2 \Delta H°_f(NH_3) = (2 \text{ mol})(-46.1 \text{ kJ/mol}) = -92.2 \text{ kJ}$

$\Delta G° = 2 \Delta G°_f(NH_3) = (2 \text{ mol})(-16.5 \text{ kJ/mol}) = -33.0 \text{ kJ}$

$\Delta G° = \Delta H° - T\Delta S°$

$\Delta H° - \Delta G° = T\Delta S°$

$$\Delta S° = \frac{\Delta H° - \Delta G°}{T} = \frac{-92.2 \text{ kJ} - (-33.0 \text{ kJ})}{298 \text{ K}} = -0.199 \text{ kJ/K} = -199 \text{ J/K}$$

(b) $\Delta S°$ is negative because the number of mol of gas molecules decreases from 4 mol to 2 mol on going from reactants to products.

(c) The reaction is spontaneous because $\Delta G°$ is negative.

(d) $\Delta G° = \Delta H° - T\Delta S° = -92.2 \text{ kJ} - (350 \text{ K})(-0.199 \text{ kJ/K}) = -22.55 \text{ kJ}$

$\Delta G° = -RT \ln K_p$

$$\ln K_p = \frac{-\Delta G°}{RT} = \frac{-(-22.55 \text{ kJ/mol})}{[8.314 \times 10^{-3} \text{ kJ/(K} \cdot \text{mol)}](350 \text{ K})} = 7.749$$

$K_p = e^{7.749} = 2.3 \times 10^3$

$\Delta n = 2 - (1 + 3) = -2$

$K_c = K_p \left(\dfrac{1}{RT}\right)^{\Delta n} = (2.3 \times 10^3)\left(\dfrac{1}{RT}\right)^{-2} = (2.3 \times 10^3)(RT)^2$

$K_c = (2.3 \times 10^3)[(0.082\ 06)(350)]^2 = 1.9 \times 10^6$

Multiconcept Problems

18.126 (a)

	$\Delta H_{vap}/T_{bp}$
ammonia	98 J/K
benzene	87 J/K
carbon tetrachloride	85 J/K
chloroform	87 J/K
mercury	94 J/K

(b) All processes are the conversion of a liquid to a gas at the boiling point. They should all have similar ΔS values. $\Delta H_{vap}/T_{bp}$ is equal to ΔS_{vap}.

(c) NH_3 deviates from Trouton's rule because of hydrogen bonding. $NH_3(l)$ has less randomness and ΔS_{vap} is larger. Hg has metallic bonding which also leads to less randomness of the liquid.

18.128 $Br_2(l) \;\rightleftarrows\; Br_2(g)$

$\Delta S^\circ = S^\circ(Br_2(g)) - S^\circ(Br_2(l))$

$\Delta S^\circ = (1\ mol)(245.4\ J/(K \cdot mol)) - (1\ mol)(152.2\ J/(K \cdot mol)) = 93.2\ J/K = 93.2 \times 10^{-3}\ kJ/K$

$\Delta G = \Delta H^\circ - T\Delta S^\circ$

At the boiling point, $\Delta G = 0$.

$0 = \Delta H^\circ - T_{bp}\Delta S^\circ$

$T_{bp} = \dfrac{\Delta H^\circ}{\Delta S^\circ}$

$\Delta H^\circ = T_{bp}\ \Delta S^\circ = (332\ K)(93.2 \times 10^{-3}\ kJ/K) = 30.9\ kJ$

$K_p = P_{Br_2} = \left(227\ mm\ Hg\ x\ \dfrac{1\ atm}{760\ mm\ Hg} \right) = 0.299\ atm$

$\Delta G^\circ = -RT \ln K_p$ and $\Delta G^\circ = \Delta H^\circ - T\Delta S^\circ$ (set equations equal to each other)

$\Delta H^\circ - T\Delta S^\circ = -RT \ln K_p$ (rearrange)

$\ln K_p = \dfrac{-\Delta H^\circ}{R}\dfrac{1}{T} + \dfrac{\Delta S^\circ}{R}$ (solve for T)

$T = \dfrac{\left(\dfrac{-\Delta H^\circ}{R} \right)}{\left(\ln K_p - \dfrac{\Delta S^\circ}{R} \right)} = \dfrac{\left(\dfrac{-30.9\ kJ/mol}{8.314 \times 10^{-3}\ kJ/(K \cdot mol)} \right)}{\left(\ln(0.299) - \dfrac{93.2 \times 10^{-3}\ kJ/(K \cdot mol)}{8.314 \times 10^{-3}\ kJ/(K \cdot mol)} \right)} = 299\ K = 26\ ^\circ C$

$Br_2(l)$ has a vapor pressure of 227 mm Hg at 26 °C.

18.130 $\Delta H^\circ = [2\ \Delta H^\circ_f(Cl^-(aq))] - [2\ \Delta H^\circ_f(Br^-(aq))]$

$\Delta H^\circ = [(2\ mol)(-167.2\ kJ/mol)] - [(2\ mol)(-121.5\ kJ/mol)] = -91.4\ kJ$

$\Delta S^\circ = [S^\circ(Br_2(l)) + 2\ S^\circ(Cl^-(aq))] - [2\ S^\circ(Br^-(aq)) + S^\circ(Cl_2(g))]$

$\Delta S^\circ = [(1\ mol)(152.2\ J/(K \cdot mol)) + (2\ mol)(56.5\ J/(K \cdot mol))]$
$\qquad\qquad - [(2\ mol)(82.4\ J/(K \cdot mol)) + (1\ mol)(223.0\ J/(K \cdot mol))] = -122.6\ J/K$

$80\ ^\circ C = 80 + 273 = 353\ K$

$\Delta G^\circ = \Delta H^\circ - T\Delta S^\circ = -91.4\ kJ - (353\ K)(-122.6 \times 10^{-3}\ kJ/K) = -48.1\ kJ$

$\Delta G° = -RT \ln K$

$$\ln K = \frac{-\Delta G°}{RT} = \frac{-(-48.1 \text{ kJmol})}{[8.314 \times 10^{-3} \text{ kJ/(K·mol)}](353 \text{ K})} = 16.4$$

$K = e^{16.4} = 1.3 \times 10^7$

18.132 $35 \text{ °C} = 35 + 273 = 308 \text{ K}$

$\Delta G° = \Delta H° - T\Delta S° = -352 \text{ kJ} - (308 \text{ K})(-899 \times 10^{-3} \text{ kJ/K}) = -75.1 \text{ kJ}$

$\Delta G° = -RT \ln K_p$

$$\ln K_p = \frac{-\Delta G°}{RT} = \frac{-(-75.1 \text{ kJ/mol})}{[8.314 \times 10^{-3} \text{ kJ/(K·mol)}](308 \text{ K})} = 29.33$$

$K_p = e^{29.33} = 5.5 \times 10^{12}$

$$K_p = \frac{1}{(P_{H_2O})^6} = 5.5 \times 10^{12}$$

$$P_{H_2O} = \sqrt[6]{\frac{1}{5.5 \times 10^{12}}} = 0.0075 \text{ atm}$$

$$P_{H_2O} = 0.0075 \text{ atm} \times \frac{760 \text{ mm Hg}}{1 \text{ atm}} = 5.7 \text{ mm Hg}$$

18.134 $N_2O_4(g) \rightleftharpoons 2 NO_2(g)$

$\Delta H° = 2 \Delta H°_f(NO_2) - \Delta H°_f(N_2O_4) = (2 \text{ mol})(33.2 \text{ kJ}) - (1 \text{ mol})(11.1 \text{ kJ}) = 55.3 \text{ kJ}$

$\Delta S° = 2 S°(NO_2) - S°(N_2O_4) = (2 \text{ mol})(240.0 \text{ J/(K·mol)}) - (1 \text{ mol})(304.3 \text{ J/(K·mol)})$

$\Delta S° = 175.7 \text{ J/K} = 175.7 \times 10^{-3} \text{ kJ/K}$

$\Delta G° = \Delta H° - T\Delta S°$ and $\Delta G° = -RT \ln K_p$; Set these two equations equal to each other and solve for T.

$\Delta H° - T\Delta S° = -RT \ln K_p$

$\Delta H° = T\Delta S° - RT \ln K_p = T(\Delta S° - R \ln K_p)$

$$T = \frac{\Delta H°}{\Delta S° - R \ln K_p}$$

(a) $P_{N_2O_4} + P_{NO_2} = 1.00 \text{ atm}$ and $P_{NO_2} = 2 P_{N_2O_4}$

$P_{N_2O_4} + 2 P_{N_2O_4} = 3 P_{N_2O_4} = 1.00 \text{ atm}$

$P_{N_2O_4} = 1.00 \text{ atm}/3 = 0.333 \text{ atm}$

$P_{NO_2} = 1.00 \text{ atm} - P_{N_2O_4} = 1.00 - 0.333 = 0.667 \text{ atm}$

$$K_p = \frac{(P_{NO_2})^2}{P_{N_2O_4}} = \frac{(0.667)^2}{(0.333)} = 1.34$$

$$T = \frac{\Delta H°}{\Delta S° - R \ln K_p}$$

$$T = \frac{55.3 \text{ kJ/mol}}{[175.7 \times 10^{-3} \text{ kJ/(K·mol)}] - [8.314 \times 10^{-3} \text{ kJ/(K·mol)}] \ln(1.34)} = 319 \text{ K}$$

$T = 319 \text{ K} = 319 - 273 = 46 \,^\circ\text{C}$

(b) $P_{N_2O_4} + P_{NO_2} = 1.00$ atm and $P_{NO_2} = P_{N_2O_4}$ so $P_{NO_2} = P_{N_2O_4} = 0.50$ atm

$$K_p = \frac{(P_{NO_2})^2}{P_{N_2O_4}} = \frac{(0.500)^2}{(0.500)} = 0.500$$

$$T = \frac{\Delta H^\circ}{\Delta S^\circ - R \ln K_p}$$

$$T = \frac{55.3 \text{ kJ/mol}}{[175.7 \times 10^{-3} \text{ kJ/(K} \cdot \text{mol)}] - [8.314 \times 10^{-3} \text{ kJ/(K} \cdot \text{mol)}] \ln(0.500)} = 305 \text{ K}$$

$T = 305 \text{ K} = 305 - 273 = 32 \,^\circ\text{C}$

18.136 $N_2(g) + 3 H_2(g) \rightleftarrows 2 NH_3(g)$

$\Delta H^\circ = 2\,\Delta H^\circ_f(NH_3) - [\Delta H^\circ_f(N_2) + 3\,\Delta H^\circ_f(H_2)] = (2 \text{ mol})(-46.1 \text{ kJ}) - [0] = -92.2 \text{ kJ}$

$\Delta S^\circ = 2\,S^\circ(NH_3) - [S^\circ(N_2) + 3\,S^\circ(H_2)]$

$\Delta S^\circ = (2 \text{ mol})(192.3 \text{ J/(K} \cdot \text{mol)})$
$\qquad\qquad - [(1 \text{ mol})(191.5 \text{ J/(K} \cdot \text{mol)}) + (3 \text{ mol})(130.6 \text{ J/(K} \cdot \text{mol)})] = -198.7 \text{ J/K}$

$\Delta G^\circ = \Delta H^\circ - T\Delta S^\circ = -92.2 \text{ kJ} - (673 \text{ K})(-198.7 \times 10^{-3} \text{ kJ/K}) = 41.5 \text{ kJ}$

$\Delta G^\circ = -RT \ln K_p$

$$\ln K_p = \frac{-\Delta G^\circ}{RT} = \frac{-41.5 \text{ kJ/mol}}{[8.314 \times 10^{-3} \text{ kJ/(K} \cdot \text{mol)}](673 \text{ K})} = -7.42$$

$K_p = e^{-7.42} = 6.0 \times 10^{-4}$

Because $K_p = K_c(RT)^{\Delta n}$, $K_c = K_p(RT)^{-\Delta n}$

$K_c = K_p(RT)^2 = (6.0 \times 10^{-4})[(0.082\,06)(673)]^2 = 1.83$

N_2, 28.01; H_2, 2.016

Initial concentrations:

$$[N_2] = \frac{(14.0 \text{ g})\left(\dfrac{1 \text{ mol}}{28.01 \text{ g}}\right)}{5.00 \text{ L}} = 0.100 \text{ M} \quad \text{and} \quad [H_2] = \frac{(3.024 \text{ g})\left(\dfrac{1 \text{ mol}}{2.016 \text{ g}}\right)}{5.00 \text{ L}} = 0.300 \text{ M}$$

	$N_2(g)$	$+$	$3 H_2(g)$	$\rightleftarrows$	$2 NH_3(g)$
initial (M)	0.100		0.300		0
change (M)	$-x$		$-3x$		$+2x$
equil (M)	$0.100 - x$		$0.300 - 3x$		$2x$

$$K_c = \frac{[NH_3]^2}{[N_2][H_2]^3} = \frac{(2x)^2}{(0.100 - x)(0.300 - 3x)^3} = \frac{4x^2}{27(0.100 - x)^4} = 1.83$$

$$\left(\frac{x}{(0.100 - x)^2}\right)^2 = \frac{(27)(1.83)}{4} = 12.35; \qquad \frac{x}{(0.100 - x)^2} = \sqrt{12.35} = 3.514$$

$3.514x^2 - 1.703x + 0.03514 = 0$

Use the quadratic formula to solve for x.

$$x = \frac{-(-1.703) \pm \sqrt{(-1.703)^2 - (4)(3.514)(0.03514)}}{2(3.514)} = \frac{1.703 \pm 1.551}{7.028}$$

$x = 0.463$ and 0.0216

Of the two solutions for x, only 0.0216 has physical meaning because 0.463 would lead to negative concentrations of N_2 and H_2.

$[N_2] = 0.100 - x = 0.100 - 0.0216 = 0.078$ M

$[H_2] = 0.300 - 3x = 0.300 - 3(0.0216) = 0.235$ M; $[NH_3] = 2x = 2(0.0216) = 0.043$ M

18.138 $Pb(s) + PbO_2(s) + 2 H^+(aq) + 2 HSO_4^-(aq) \rightarrow 2 PbSO_4(s) + 2 H_2O(l)$

(a) $\Delta G° = [2 \Delta G°_f(PbSO_4) + 2 \Delta G°_f(H_2O)] - [\Delta G°_f(PbO_2) + 2 \Delta G°_f(HSO_4^-)]$

$\Delta G° = (2 \text{ mol})(-813.2 \text{ kJ/mol}) + (2 \text{ mol})(-237.2 \text{ kJ/mol})]$

$\qquad - [(1 \text{ mol})(-217.4 \text{ kJ/mol}) + (2 \text{ mol})(-756.0 \text{ kJ/mol})] = -371.4$ kJ

(b) $°C = 5/9(°F - 32) = 5/9(10 - 32) = -12.2 \ °C;$ $-12.2 \ °C = 261$ K

$\Delta H° = [2 \Delta H°_f(PbSO_4) + 2 \Delta H°_f(H_2O)] - [\Delta H°_f(PbO_2) + 2 \Delta H°_f(HSO_4^-)]$

$\Delta H° = [(2 \text{ mol})(-919.9 \text{ kJ/mol}) + (2 \text{ mol})(-285.8 \text{ kJ/mol})]$

$\qquad - [(1 \text{ mol})(-277 \text{ kJ/mol}) + (2 \text{ mol})(-887.3 \text{ kJ/mol})] = -359.8$ kJ

$\Delta S° = [2 S°(PbSO_4) + 2 S°(H_2O)] - [S°(Pb) + S°(PbO_2) + 2 S°(H^+) + 2 S°(HSO_4^-)]$

$\Delta S° = [(2 \text{ mol})(148.6 \text{ J/(K} \cdot \text{mol)}) + (2 \text{ mol})(69.9 \text{ J/(K} \cdot \text{mol)})]$

$\qquad - [(1 \text{ mol})(64.8 \text{ J/(K} \cdot \text{mol)}) + (1 \text{ mol})(68.6 \text{ J/(K} \cdot \text{mol)})$

$\qquad + (2 \text{ mol})(132 \text{ J/(K} \cdot \text{mol)})] = 39.6 \text{ J/K} = 39.6 \times 10^{-3}$ kJ/K

$\Delta G° = \Delta H° - T\Delta S° = -359.8 \text{ kJ} - (261 \text{ K})(39.6 \times 10^{-3} \text{ kJ/K}) = -370.1$ kJ at 261 K

	$HSO_4^-(aq)$	$+ \ H_2O(l)$	$\rightleftharpoons$	$H_3O^+(aq)$	$+ \ SO_4^{2-}(aq)$
initial (M)	0.100			0.100	0
change (M)	$-x$			$+x$	$+x$
equil (M)	$0.100 - x$			$0.100 + x$	x

$$K_{a2} = \frac{[H_3O^+][SO_4^{2-}]}{[HSO_4^-]} = 1.2 \times 10^{-2} = \frac{(0.100 + x)x}{0.100 - x}$$

$x^2 + 0.112x - (1.2 \times 10^{-3}) = 0$

Use the quadratic formula to solve for x.

$$x = \frac{-(0.112) \pm \sqrt{(0.112)^2 - (4)(1)(-1.2 \times 10^{-3})}}{2(1)} = \frac{-0.112 \pm 0.132}{2}$$

$x = -0.122$ and 0.010

Of the two solutions for x, only 0.010 has physical meaning because -0.122 would lead to negative concentrations of H_3O^+ and SO_4^{2-}.

$[H^+] = 0.100 + x = 0.100 + 0.010 = 0.110$ M

$[HSO_4^-] = 0.100 - x = 0.100 - 0.010 = 0.090$ M

$$\Delta G = \Delta G° + RT \ln \frac{1}{[H^+]^2[HSO_4^-]^2}$$

$$\Delta G = (-370.1 \text{ kJ/mol}) + [8.314 \times 10^{-3} \text{ kJ/(K} \cdot \text{mol)}](261 \text{ K}) \ln \frac{1}{(0.110)^2(0.090)^2}$$

$\Delta G = -350.1$ kJ/mol

18.140 $PV = nRT$

$$n_{NH_3} = \frac{PV}{RT} = \frac{\left(744 \text{ mm Hg } \times \dfrac{1.00 \text{ atm}}{760 \text{ mm Hg}}\right)(1.00 \text{ L})}{\left(0.082\ 06 \dfrac{\text{L} \cdot \text{atm}}{\text{K} \cdot \text{mol}}\right)(298.1 \text{ K})} = 0.0400 \text{ mol NH}_3$$

500.0 mL = 0.5000 L

$[NH_3] = 0.0400$ mol/0.5000 L = 0.0800 M

$NH_3(aq) + H_2O(l) \rightleftharpoons NH_4^+(aq) + OH^-(aq)$

$\Delta H° = [\Delta H°_f(NH_4^+) + \Delta H°_f(OH^-)] - [\Delta H°_f(NH_3) + \Delta H°_f(H_2O)]$

$\Delta H° = [(1 \text{ mol})(-132.5 \text{ kJ/mol}) + (1 \text{ mol})(-230.0 \text{ kJ/mol})]$
$\qquad\qquad - [(1 \text{ mol})(-80.3 \text{ kJ/mol}) + (1 \text{ mol})(-285.8 \text{ kJ/mol})] = +3.6 \text{ kJ}$

$\Delta S° = [S°(NH_4^+) + S°(OH^-)] - [S°(NH_3) + S°(H_2O)]$

$\Delta S° = [(1 \text{ mol})(113 \text{ J/(K} \cdot \text{mol)}) + (1 \text{ mol})(-10.8 \text{ J/(K} \cdot \text{mol)})]$
$\qquad\qquad - [(1 \text{ mol})(111 \text{ J/(K} \cdot \text{mol)}) + (1 \text{ mol})(69.9 \text{ J/(K} \cdot \text{mol)})] = -78.7 \text{ J/K}$

$T = 2.0 \text{ °C} = 2.0 + 273.1 = 275.1 \text{ K}$

$\Delta G° = \Delta H° - T\Delta S° = 3.6 \text{ kJ} - (275.1 \text{ K})(-78.7 \times 10^{-3} \text{ kJ/K}) = 25.3 \text{ kJ}$

$\Delta G° = -RT \ln K_b$

$$\ln K_b = \frac{-\Delta G°}{RT} = \frac{-25.3 \text{ kJ/mol}}{[8.314 \times 10^{-3} \text{ kJ/(K} \cdot \text{mol)}](275.1 \text{ K})} = -11.06$$

$K_b = e^{-11.06} = 1.6 \times 10^{-5}$

	$NH_3(aq)$	$+ H_2O(l)$	$\rightleftharpoons$	$NH_4^+(aq)$	$+ OH^-(aq)$
initial (M)	0.0800			0	~0
change (M)	−x			+x	+x
equil (M)	0.0800 − x			x	x

$$\text{at 2 °C, } K_b = \frac{[NH_4^+][OH^-]}{[NH_3]} = 1.6 \times 10^{-5} = \frac{x^2}{0.0800 - x} \approx \frac{x^2}{0.0800}$$

$x^2 = (1.6 \times 10^{-5})(0.0800)$

$x = [OH^-] = \sqrt{(1.6 \times 10^{-5})(0.0800)} = 1.13 \times 10^{-3} \text{ M}$

$$[H_3O^+] = \frac{1.0 \times 10^{-14}}{1.13 \times 10^{-3}} = 8.85 \times 10^{-12} \text{ M}$$

$pH = -\log[H_3O^+] = -\log(8.85 \times 10^{-12}) = 11.05$

18.142 (a) $N_2O_4(g) \rightleftharpoons 2 NO_2(g)$

$\Delta H° = 2 \Delta H°_f(NO_2) - \Delta H°_f(N_2O_4) = (2 \text{ mol})(33.2 \text{ kJ/mol}) - (1 \text{ mol})(11.1 \text{ kJ/mol}) = 55.3 \text{ kJ}$

$\Delta S° = 2 S°(NO_2) - S°(N_2O_4) = (2 \text{ mol})(240.0 \text{ J/(K} \cdot \text{mol)}) - (1 \text{ mol})(304.3 \text{ J/(K} \cdot \text{mol)})$

$\Delta S° = 175.7 \text{ J/K} = 175.7 \times 10^{-3} \text{ kJ/K}$

$\Delta G° = \Delta H° - T\Delta S° = 55.3 \text{ kJ} - (373 \text{ K})(175.7 \times 10^{-3} \text{ kJ/K}) = -10.2 \text{ kJ}$

$$K_p = \frac{(P_{NO_2})^2}{P_{N_2O_4}}$$

$$\Delta G^\circ = -RT \ln K_p; \qquad \ln K_p = \frac{-\Delta G^\circ}{RT} = \frac{-(-10.2 \text{ kJ/mol})}{[8.314 \times 10^{-3} \text{ kJ/(K} \cdot \text{mol)}](373 \text{ K})} = 3.29$$

$$K_p = e^{3.29} = 27$$

	$N_2O_4(g)$	$\rightleftharpoons$	$2 \ NO_2(g)$
initial (atm)	1.00		1.00
change (atm)	$-x$		$+2x$
equil (atm)	$1.00 - x$		$1.00 + 2x$

$$K_p = \frac{(P_{NO_2})^2}{P_{N_2O_4}} = 27 = \frac{(1.00 + 2x)^2}{(1.00 - x)}$$

$$4x^2 + 31x - 26 = 0$$

Use the quadratic formula to solve for x.

$$x = \frac{-(31) \pm \sqrt{(31)^2 - (4)(4)(-26)}}{2(4)} = \frac{-31 \pm 37.1}{8}$$

$$x = 0.76 \text{ and } -8.5$$

Of the two solutions for x, only 0.76 has physical meaning because -8.5 would lead to a negative partial pressure for NO_2.

$$P_{N_2O_4} = 1.00 - x = 1.00 - 0.76 = 0.24 \text{ atm}; \quad P_{NO_2} = 1.00 + 2x = 1.00 + 2(0.76) = 2.52 \text{ atm}$$

(b) One resonance structure is shown here.

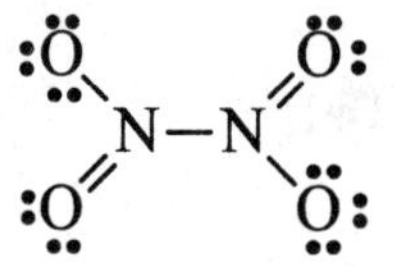

Each N is sp^2 hybridized.
There is a trigonal planar geometry about each N.

19 Electrochemistry

19.1 $NO_3^-(aq) + Cu(s) \rightarrow NO(g) + Cu^{2+}(aq)$
[$Cu(s) \rightarrow Cu^{2+}(aq) + 2 e^-$] x 3   (oxidation half reaction)

$NO_3^-(aq) \rightarrow NO(g)$
$NO_3^-(aq) \rightarrow NO(g) + 2 H_2O(l)$
$4 H^+(aq) + NO_3^-(aq) \rightarrow NO(g) + 2 H_2O(l)$
[$3 e^- + 4 H^+(aq) + NO_3^-(aq) \rightarrow NO(g) + 2 H_2O(l)$] x 2   (reduction half reaction)

Combine the two half reactions.
$2 NO_3^-(aq) + 8 H^+(aq) + 3 Cu(s) \rightarrow 3 Cu^{2+}(aq) + 2 NO(g) + 4 H_2O(l)$

19.2 $I^-(aq) \rightarrow I_3^-(aq)$
$3 I^-(aq) \rightarrow I_3^-(aq)$
[$3 I^-(aq) \rightarrow I_3^-(aq) + 2 e^-$] x 8   (oxidation half reaction)

$IO_3^-(aq) \rightarrow I_3^-(aq)$
$3 IO_3^-(aq) \rightarrow I_3^-(aq)$
$3 IO_3^-(aq) \rightarrow I_3^-(aq) + 9 H_2O(l)$
$18 H^+(aq) + 3 IO_3^-(aq) \rightarrow I_3^-(aq) + 9 H_2O(l)$
$18 H^+(aq) + 3 IO_3^-(aq) + 16 e^- \rightarrow I_3^-(aq) + 9 H_2O(l)$ (reduction half reaction)

Combine the two half reactions.
$18 H^+(aq) + 3 IO_3^-(aq) + 24 I^-(aq) \rightarrow 9 I_3^-(aq) + 9 H_2O(l)$
Divide each coefficient by 3.
$6 H^+(aq) + IO_3^-(aq) + 8 I^-(aq) \rightarrow 3 I_3^-(aq) + 3 H_2O(l)$

19.3 $I^-(aq) \rightarrow I_2(s)$
$2 I^-(aq) \rightarrow I_2(s)$
[$2 I^-(aq) \rightarrow I_2(s) + 2 e^-$] x 3 (oxidation half reaction)

$MnO_4^-(aq) \rightarrow MnO_2(s)$
$MnO_4^-(aq) \rightarrow MnO_2(s) + 2 H_2O(l)$
$4 H^+(aq) + MnO_4^-(aq) \rightarrow MnO_2(s) + 2 H_2O(l)$
[$4 H^+(aq) + MnO_4^-(aq) + 3 e^- \rightarrow MnO_2(s) + 2 H_2O(l)$] x 2   (reduction half reaction)

Combine the two half reactions.
$6 I^-(aq) + 2 MnO_4^-(aq) + 8 H^+(aq) \rightarrow 3 I_2(s) + 2 MnO_2(s) + 4 H_2O(l)$
$6 I^-(aq) + 2 MnO_4^-(aq) + 8 H^+(aq) + 8 OH^-(aq) \rightarrow$
                         $3 I_2(s) + 2 MnO_2(s) + 4 H_2O(l) + 8 OH^-(aq)$
$6 I^-(aq) + 2 MnO_4^-(aq) + 4 H_2O(l) \rightarrow 3 I_2(s) + 2 MnO_2(s) + 8 OH^-(aq)$

19.4 $Fe(OH)_2(s) \rightarrow Fe(OH)_3(s)$
$[Fe(OH)_2(s) + OH^-(aq) \rightarrow Fe(OH)_3(s) + e^-] \times 4$ (oxidation half reaction)

$O_2(g) \rightarrow 2 H_2O(l)$
$4 H^+(aq) + O_2(g) \rightarrow 2 H_2O(l)$
$4 e^- + 4 H^+(aq) + O_2(g) \rightarrow 2 H_2O(l)$
$4 e^- + 4 H^+(aq) + 4 OH^-(aq) + O_2(g) \rightarrow 2 H_2O(l) + 4 OH^-(aq)$
$4 e^- + 4 H_2O(l) + O_2(g) \rightarrow 2 H_2O(l) + 4 OH^-(aq)$
$4 e^- + 2 H_2O(l) + O_2(g) \rightarrow 4 OH^-(aq)$ (reduction half reaction)

Combine the two half reactions.
$4 Fe(OH)_2(s) + 4 OH^-(aq) + 2 H_2O(l) + O_2(g) \rightarrow 4 Fe(OH)_3(s) + 4 OH^-(aq)$
$4 Fe(OH)_2(s) + 2 H_2O(l) + O_2(g) \rightarrow 4 Fe(OH)_3(s)$

19.5 $2 Ag^+(aq) + Ni(s) \rightarrow 2 Ag(s) + Ni^{2+}(aq)$
There is a Ni anode in an aqueous solution of Ni^{2+}, and a Ag cathode in an aqueous solution of Ag^+. A salt bridge connects the anode and cathode compartment. The electrodes are connected through an external circuit.

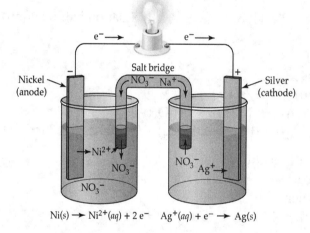

19.6

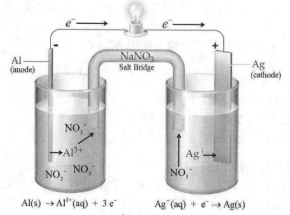

anode reaction $Al(s) \rightarrow Al^{3+}(aq) + 3 e^-$
cathode reaction $\underline{3 Ag^+(aq) + 3 e^- \rightarrow 3 Ag(s)}$
overall reaction $Al(s) + 3 Ag^+(aq) \rightarrow Al^{3+}(aq) + 3 Ag(s)$

19.7 $Pb(s) + Br_2(l) \rightarrow Pb^{2+}(aq) + 2 Br^-(aq)$

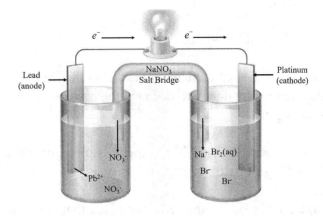

19.8   (a) and (b)

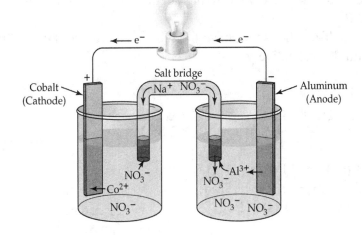

    (c) $2 Al(s) + 3 Co^{2+}(aq) \rightarrow 2 Al^{3+}(aq) + 3 Co(s)$
    (d) $Al(s)|Al^{3+}(aq)||Co^{2+}(aq)|Co(s)$

19.9 $2 Ag^+(aq) + Cd(s) \rightarrow 2 Ag(s) + Cd^{2+}(aq)$
    $n = 2$ mol e$^-$

$$\Delta G^\circ = -nFE^\circ = -(2 \text{ mol e}^-)\left(\frac{96,500 \text{ C}}{1 \text{ mol e}^-}\right)(1.20 \text{ V})\left(\frac{1 \text{ J}}{1 \text{ C} \cdot \text{V}}\right) = -231,600 \text{ J} = -232 \text{ kJ}$$

19.10 $Hg(l) + I_2(s) \rightarrow Hg^{2+}(aq) + 2 I^-(aq)$
    $n = 2$ mol e$^-$ and $1 \text{ J/C} = 1\text{V}$
    $\Delta G^\circ = 59.8 \text{ kJ} = 59,800 \text{ J} = -nFE^\circ$

$$E^\circ = \frac{-\Delta G^\circ}{nF} = \frac{-(59,800 \text{ J})}{(2 \text{ mol e}^-)\left(\dfrac{96,500 \text{ C}}{1 \text{ mol e}^-}\right)} = -0.310 \text{ J/C} = -0.310 \text{ V}$$

Because $E^\circ < 0$, the reaction is nonspontaneous.

19.11 $Cl_2(g) + 2 e^- \rightarrow 2 Cl^-(aq)$ $E° = 1.36$ V
       $Cu^{2+}(aq) + 2 e^- \rightarrow Cu(s)$ $E° = 0.34$ V
       $Zn^{2+}(aq) + 2 e^- \rightarrow Zn(s)$ $E° = -0.76$ V
       $Al^{3+}(aq) + 3 e^- \rightarrow Al(s)$ $E° = -1.66$ V

The last half-reaction has the least tendency to occur in the forward direction (most negative E°) and the greatest tendency to occur in the reverse direction. Therefore, Al is the strongest reducing agent.

19.12 (a) D is the strongest reducing agent. D^+ has the most negative standard reduction potential. A^{3+} is the strongest oxidizing agent. It has the most positive standard reduction potential.
(b) An oxidizing agent can oxidize any reducing agent that is below it in the table. B^{2+} can oxidize C and D.
A reducing agent can reduce any oxidizing agent that is above it in the table. C can reduce A^{3+} and B^{2+}.
(c) Use the two half-reactions that have the most positive and the most negative standard reduction potentials, respectively.

$$A^{3+} + 2 e^- \rightarrow A^+ \qquad\qquad 1.47 \text{ V}$$
$$\underline{2 \times (D \rightarrow D^+ + e^-)} \qquad\quad \underline{1.38 \text{ V}}$$
$$A^{3+} + 2 D \rightarrow A^+ + 2 D^+ \qquad 2.85 \text{ V}$$

19.13 $2 Fe^{3+}(aq) + 2 I^-(aq) \rightarrow 2 Fe^{2+}(aq) + I_2(s)$
       reduction: $Fe^{3+}(aq) + e^- \rightarrow Fe^{2+}(aq)$ $E° = 0.77$ V
       oxidation: $2 I^-(aq) \rightarrow I_2(s) + 2 e^-$ $\underline{E° = -0.54 \text{ V}}$
                                                              overall $E° = 0.23$ V
Because E° for the overall reaction is positive, this reaction can occur under standard-state conditions.

19.14 An oxidizing agent can oxidize any reducing agent that lies below it in Table 19.1 but can't oxidize one that lies above it. $Mg^{2+}(aq)$ is below Ag(s) in the table so $Mg^{2+}(aq)$ can't oxidize Ag(s), no reaction occurs.

19.15 (a) $Ni(s) + 2 Ag^+(aq) \rightarrow Ni^{2+}(aq) + 2 Ag(s)$
       oxidation: $Ni(s) \rightarrow Ni^{2+}(aq) + 2 e^-$ $E° = 0.26$ V
       reduction: $Ag^+(aq) + e^- \rightarrow Ag(s)$ $\underline{E° = 0.80 \text{ V}}$
                                                          overall $E° = 1.06$ V
(b) $Ni(s) | Ni^{2+}(aq)(1.0 \text{ M}) \| Ag^+(aq)(1.0 \text{ M}) | Ag(s)$
(c) Ni(s) is the anode and Ag(s) is the cathode.

19.16 oxidation: $Al(s) \rightarrow Al^{3+}(aq) + 3 e^-$ $E° = 1.66$ V
       reduction: $\underline{Cr^{3+}(aq) + 3 e^- \rightarrow Cr(s)}$ $\underline{E° = ?}$
       overall $Al(s) + Cr^{3+}(aq) \rightarrow Al^{3+}(aq) + Cr(s)$ $E° = 0.92$ V
The standard reduction potential for the Cr^{3+}/Cr half cell is:
$E° = 0.92 - 1.66 = -0.74$ V

19.17 $Cu(s) + 2 Fe^{3+}(aq) \rightarrow Cu^{2+}(aq) + 2 Fe^{2+}(aq)$

$E^\circ = E^\circ_{Cu \rightarrow Cu^{2+}} + E^\circ_{Fe^{3+} \rightarrow Fe^{2+}} = -0.34 \text{ V} + 0.77 \text{ V} = 0.43 \text{ V}; \quad n = 2 \text{ mol e}^-$

$E = E^\circ - \dfrac{0.0592 \text{ V}}{n} \log \dfrac{[Cu^{2+}][Fe^{2+}]^2}{[Fe^{3+}]^2} = 0.43 \text{ V} - \dfrac{(0.0592 \text{ V})}{2} \log \dfrac{(0.25)(0.20)^2}{(1.0 \times 10^{-4})^2} = 0.25 \text{ V}$

19.18   (a) anode: $4[Al(s) \rightarrow Al^{3+}(aq) + 3 e^-]$ $\qquad\qquad\qquad\qquad$ $E^\circ = 1.66 \text{ V}$
$\qquad$ cathode: $\underline{3[O_2(g) + 4 H^+(aq) + 4 e^- \rightarrow 2 H_2O(l)]}$ $\qquad$ $E^\circ = 1.23 \text{ V}$
$\qquad$ overall: $4 Al(s) + 3 O_2(g) + 12 H^+(aq) \rightarrow 4 Al^{3+}(aq) + 6 H_2O(l)$ $\quad$ $E^\circ = 2.89 \text{ V}$

(b) & (c) $E = E^\circ - \dfrac{2.303 RT}{nF} \log \dfrac{[Al^{3+}]^4}{(P_{O_2})^3 [H^+]^{12}}$

$E = 2.89 \text{ V} - \dfrac{(2.303)\left(8.314 \dfrac{J}{K \cdot mol}\right)(310 \text{ K})}{(12 \text{ mol e}^-)(96{,}500 \text{ C/mol e}^-)} \log \left(\dfrac{(1.0 \times 10^{-9})^4}{(0.20)^3 (1.0 \times 10^{-7})^{12}} \right)$

$E = 2.89 \text{ V} - 0.257 \text{ V} = 2.63 \text{ V}$

19.19 $5 [Cu(s) \rightarrow Cu^{2+}(aq) + 2 e^-]$ $\qquad$ (oxidation half reaction)
$\qquad$ $2 [5 e^- + 8 H^+(aq) + MnO_4^-(aq) \rightarrow Mn^{2+}(aq) + 4 H_2O(l)]$ $\quad$ (reduction half reaction)

$\qquad$ $5 Cu(s) + 16 H^+(aq) + 2 MnO_4^-(aq) \rightarrow 5 Cu^{2+}(aq) + 2 Mn^{2+}(aq) + 8 H_2O(l)$

$\qquad$ $\Delta E = -\dfrac{0.0592 \text{ V}}{n} \log \dfrac{[Cu^{2+}]^5 [Mn^{2+}]^2}{[MnO_4^-]^2 [H^+]^{16}}$

The anode compartment contains Cu^{2+}.

$\qquad$ $\Delta E = -\dfrac{0.0592 \text{ V}}{10} \log \dfrac{(100)^5 (1)^2}{(1)^2 (1)^{16}} = -0.059 \text{ V}$

19.20 $Zn(s) + Cu^{2+}(aq) \rightarrow Zn^{2+}(aq) + Cu(s)$
$\qquad$ oxidation: $Zn(s) \rightarrow Zn^{2+}(aq) + 2 e^-$ $\qquad\qquad$ $E^\circ = 0.76 \text{ V}$
$\qquad$ reduction: $Cu^{2+}(aq) + 2 e^- \rightarrow Cu(s)$ $\qquad\qquad$ $\underline{E^\circ = 0.34 \text{ V}}$
$\qquad\qquad\qquad\qquad\qquad\qquad\qquad\qquad\qquad$ overall $E^\circ = 1.10 \text{ V}$

$\qquad$ $E = 1.16 \text{ V} = E^\circ - \dfrac{0.0592 \text{ V}}{n} \log \dfrac{[Zn^{2+}]}{[Cu^{2+}]} = 1.10 \text{ V} - \dfrac{(0.0592 \text{ V})}{2} \log \dfrac{[Zn^{2+}]}{[Cu^{2+}]}$

$\qquad$ $\dfrac{(1.16 \text{ V} - 1.10 \text{ V})}{(-0.0592 \text{ V}/2)} = \log \dfrac{[Zn^{2+}]}{[Cu^{2+}]}$

$\qquad$ $\log \dfrac{[Zn^{2+}]}{[Cu^{2+}]} = -2.03$ $\quad$ and $\quad$ $\dfrac{[Zn^{2+}]}{[Cu^{2+}]} = 10^{-2.03} = 9.3 \times 10^{-3}$

19.21 $H_2(g) + Pb^{2+}(aq) \rightarrow 2 H^+(aq) + Pb(s)$
$\qquad$ $E^\circ = E^\circ_{H_2 \rightarrow H^+} + E^\circ_{Pb^{2+} \rightarrow Pb} = 0 \text{ V} + (-0.13 \text{ V}) = -0.13 \text{ V}; \quad n = 2 \text{ mol e}^-$

$$E = E^\circ - \frac{0.0592\,V}{n} \log \frac{[H_3O^+]^2}{[Pb^{2+}](P_{H_2})}$$

$$0.28\,V = -0.13\,V - \frac{(0.0592\,V)}{2} \log \frac{[H_3O^+]^2}{(1)(1)} = -0.13\,V - (0.0592\,V) \log [H_3O^+]$$

$pH = -\log[H_3O^+]$ therefore $0.28\,V = -0.13\,V + (0.0592\,V)\,pH$

$$pH = \frac{(0.28\,V + 0.13\,V)}{0.0592\,V} = 6.9$$

19.22 $H_2(g) + Hg_2Cl_2(s) \rightarrow 2\,H^+(aq) + 2\,Hg(l) + 2Cl^-(aq)$

$E^\circ = E^\circ_{H_2 \rightarrow H^+} + E^\circ_{Hg_2Cl_2 \rightarrow Hg} = 0\,V + 0.28\,V = 0.28\,V;$ $n = 2$ mol e^-

$$E = E^\circ - \frac{0.0592\,V}{n} \log \frac{[H_3O^+]^2[Cl^-]^2}{(P_{H_2})}$$

$$E = 0.28\,V - \frac{0.0592\,V}{2} \log \frac{(1.0 \times 10^{-7})^2(1.0)^2}{(1.0)} = 0.69\,V$$

19.23 $4\,Fe^{2+}(aq) + O_2(g) + 4\,H^+(aq) \rightarrow 4\,Fe^{3+}(aq) + 2\,H_2O(l)$

$E^\circ = E^\circ_{Fe^{2+} \rightarrow Fe^{3+}} + E^\circ_{O_2 \rightarrow H_2O} = -0.77\,V + 1.23\,V = 0.46\,V;$ $n = 4$ mol e^-

$$E^\circ = \frac{0.0592\,V}{n} \log K; \quad \log K = \frac{nE^\circ}{0.0592\,V} = \frac{(4)(0.46\,V)}{0.0592\,V} = 31; \quad K = 10^{31} \text{ at } 25\,^\circ C$$

19.24 $E^\circ = \dfrac{0.0592\,V}{n} \log K = \dfrac{0.0592\,V}{2} \log(1.8 \times 10^{-5}) = -0.140\,V$

19.25 $\text{Charge} = \left(1.00 \times 10^5\,\dfrac{C}{s}\right)(8.00\,h)\left(\dfrac{60\,\min}{h}\right)\left(\dfrac{60\,s}{\min}\right) = 2.88 \times 10^9\,C$

$\text{Moles of } e^- = (2.88 \times 10^9\,C)\left(\dfrac{1\,\text{mol } e^-}{96,500\,C}\right) = 2.98 \times 10^4\,\text{mol } e^-$

cathode reaction: $Al^{3+} + 3\,e^- \rightarrow Al$

$\text{mass Al} = (2.98 \times 10^4\,\text{mol } e^-) \times \dfrac{1\,\text{mol Al}}{3\,\text{mol } e^-} \times \dfrac{26.98\,\text{g Al}}{1\,\text{mol Al}} \times \dfrac{1\,kg}{1000\,g} = 268\,\text{kg Al}$

19.26 $3.00\,\text{g Ag} \times \dfrac{1\,\text{mol Ag}}{107.9\,\text{g Ag}} = 0.0278\,\text{mol Ag}$

cathode reaction: $Ag^+(aq) + e^- \rightarrow Ag(s)$

$\text{Charge} = (0.0278\,\text{mol Ag})\left(\dfrac{1\,\text{mol } e^-}{1\,\text{mol Ag}}\right)\left(\dfrac{96,500\,C}{1\,\text{mol } e^-}\right) = 2682.7\,C$

$\text{Time} = \dfrac{C}{A} = \left(\dfrac{2682.7\,C}{0.100\,C/s} \times \dfrac{1\,h}{3600\,s}\right) = 7.45\,h$

19.27   A fuel cell and a battery are both galvanic cells that convert chemical energy into electrical energy utilizing a spontaneous redox reaction. A fuel cell differs from an ordinary battery in that the reactants are not contained within the cell but instead are continuously supplied from an external reservoir.

19.28   The main obstacles to large-scale commercialization of the fuel-cell-powered vehicles (FCVs) are the relatively high cost of the FCVs, the lack of low cost, nonpolluting sources of hydrogen, and development of a hydrogen-fuel infrastructure.

19.29   (a)   anode reaction $2 H_2(g) \rightarrow 4 H^+(aq) + 4 e^-$ $E° = 0.00$ V
                cathode reaction $O_2(g) + 4 H^+(aq) + 4 e^- \rightarrow 2 H_2O(l)$ $E° = 1.23$ V
                overall reaction $2 H_2(g) + O_2(g) \rightarrow 2 H_2O(l)$ $E° = 1.23$ V

            (b) $\Delta G° = -nFE° = -(4 \text{ mol } e^-)\left(\dfrac{96,500 \text{ C}}{1 \text{ mol } e^-}\right)(1.23 \text{ V})\left(\dfrac{1 \text{ J}}{1 \text{ C} \cdot \text{V}}\right) = -474,780 \text{ J} = -475 \text{ kJ}$

        $E° = \dfrac{0.0592 \text{ V}}{n} \log K; \quad \log K = \dfrac{n E°}{0.0592 \text{ V}} = \dfrac{(4)(1.23 \text{ V})}{0.0592 \text{ V}} = 83.1$

        $K = 10^{83.1} = 1.28 \times 10^{83}$

            (c) $E = E° - \dfrac{0.0592 \text{ V}}{n} \log \dfrac{1}{(P_{H_2})^2 (P_{O_2})} = 1.23 \text{ V} - \dfrac{0.0592 \text{ V}}{4} \log \dfrac{1}{(25)^2(25)} = 1.29 \text{ V}$

19.30   (a)   anode reaction $2 H_2(g) \rightarrow 4 H^+(aq) + 4 e^-$ $E° = 0.00$ V
                cathode reaction $O_2(g) + 4 H^+(aq) + 4 e^- \rightarrow 2 H_2O(l)$ $E° = 1.23$ V
                overall reaction $2 H_2(g) + O_2(g) \rightarrow 2 H_2O(l)$ $E° = 1.23$ V

            (b) $E = E° - \dfrac{0.0592 \text{ V}}{n} \log \dfrac{1}{(P_{H_2})^2 (P_{O_2})} = 1.23 \text{ V} - \dfrac{0.0592 \text{ V}}{4} \log \dfrac{1}{(6)^2(0.2)} = 1.24 \text{ V}$

19.31 $2 CH_3OH(l) + 3 O_2(g) \rightarrow 2 CO_2(g) + 4 H_2O(l)$
        $\Delta G° = [2 \Delta G°_f(CO_2) + 4 \Delta G°_f(H_2O)] - [2 \Delta G°_f(CH_3OH)]$
        $\Delta G° = [(2 \text{ mol})(-394.4 \text{ kJ/mol}) + (4 \text{ mol})(-237.2 \text{ kJ/mol})] - (2 \text{ mol})(-166.6 \text{ kJ/mol})$
        $\Delta G° = -1404 \text{ kJ}$
        anode: $2 CH_3OH(l) + 2 H_2O(l) \rightarrow 2 CO_2(g) + 12 H^+(aq) + 12 e^-$
        cathode: $3 O_2(g) + 12 H^+(aq) + 12 e^- \rightarrow 6 H_2O(l)$
        $n = 12 \text{ mol } e^-$ and $1 \text{ J} = 1 \text{ C} \times 1 \text{ V}$

        $\Delta G° = -nFE° \qquad E° = \dfrac{-\Delta G°}{nF} = \dfrac{-(-1,404,000 \text{ J})}{(12 \text{ mol } e^-)\left(\dfrac{96,500 \text{ C}}{1 \text{ mol } e^-}\right)} = +1.21 \text{ J/C} = +1.21 \text{ V}$

        $E° = \dfrac{0.0592 \text{ V}}{n} \log K; \quad \log K = \dfrac{n E°}{0.0592 \text{ V}} = \dfrac{(12)(1.21 \text{ V})}{0.0592 \text{ V}} = 245; \quad K = 10^{245} = 1 \times 10^{245}$

19.32   (a) H is reduced and C is oxidized.
(b) H is reduced and C is oxidized. H_2O is the oxidizing agent and CO is the reducing agent.
(c) The drawback of the process is the high reaction temperatures, which require a lot of energy and the greenhouse gas CO_2 is also produced.

19.33 $6 H_2O(l) \rightarrow 2 H_2(g) + O_2(g) + 4 H^+(aq) + 4 OH^-(aq)$
1 A = 1 C/s

(a) $\text{mol e}^- = 250.0 \dfrac{C}{s} \times 30 \text{ min} \times \dfrac{60 \text{ s}}{\text{min}} \times \dfrac{1 \text{ mol e}^-}{96{,}500 \text{ C}} = 4.66 \text{ mol e}^-$

$\text{mass } H_2 = 4.66 \text{ mol e}^- \times \dfrac{2 \text{ mol } H_2}{4 \text{ mol e}^-} \times \dfrac{2.02 \text{ g } H_2}{1 \text{ mol } H_2} = 4.71 \text{ g } H_2$

(b) $\text{charge} = 25 \text{ mol } O_2 \times \dfrac{4 \text{ mol e}^-}{1 \text{ mol } O_2} \times \dfrac{96{,}500 \text{ C}}{1 \text{ mol e}^-} = 9.65 \times 10^6 \text{ C}$

$\text{time} = \dfrac{9.65 \times 10^6 \text{ C}}{500.0 \text{ C/s}} \times \dfrac{1 \text{ min}}{60 \text{ s}} \times \dfrac{1 \text{ h}}{60 \text{ min}} = 5.36 \text{ h}$

Conceptual Problems

19.34   (a) - (d)

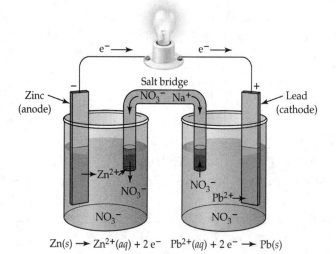

$Zn(s) \rightarrow Zn^{2+}(aq) + 2 e^- \quad Pb^{2+}(aq) + 2 e^- \rightarrow Pb(s)$

(e) anode reaction $Zn(s) \rightarrow Zn^{2+}(aq) + 2 e^-$
cathode reaction $\underline{Pb^{2+}(aq) + 2 e^- \rightarrow Pb(s)}$
overall reaction $Zn(s) + Pb^{2+}(aq) \rightarrow Zn^{2+}(aq) + Pb(s)$

19.36 $Zn(s) + Cu^{2+}(aq) \rightarrow Zn^{2+}(aq) + Cu(s); \quad E = E^\circ - \dfrac{0.0592 \text{ V}}{2} \log \dfrac{[Zn^{2+}]}{[Cu^{2+}]}$

(a) E increases because increasing $[Cu^{2+}]$ decreases $\log \dfrac{[Zn^{2+}]}{[Cu^{2+}]}$.

(b) E will decrease because addition of H_2SO_4 increases the volume which, decreases $[Cu^{2+}]$ and increases $\log \dfrac{[Zn^{2+}]}{[Cu^{2+}]}$.

(c) E decreases because increasing $[Zn^{2+}]$ increases $\log \dfrac{[Zn^{2+}]}{[Cu^{2+}]}$.

(d) Because there is no change in $[Zn^{2+}]$, there is no change in E.

19.38   (a) - (b)

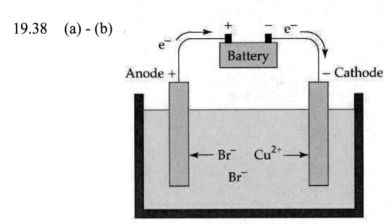

      (c) anode reaction $2\,Br^-(aq) \rightarrow Br_2(aq) + 2\,e^-$
            cathode reaction $\underline{Cu^{2+}(aq) + 2\,e^- \rightarrow Cu(s)}$
            overall reaction $Cu^{2+}(aq) + 2\,Br^-(aq) \rightarrow Cu(s) + Br_2(aq)$

19.40   (a) & (b)

Graphite or titanium container (Cathode)

Ca^{2+}

Cl^- Cl^-

Graphite (Anode)

Pellet of TiO_2

      (c)   anode reaction $2\,O^{2-} \rightarrow O_2(g) + 4\,e^-$
             cathode reaction $\underline{TiO_2(s) + 4\,e^- \rightarrow Ti(s) + 2\,O^{2-}}$
             overall reaction $TiO_2(s) \rightarrow Ti(s) + O_2(g)$

Section Problems
Balancing Redox Reactions (Section 19.1)

19.42   (a) Cl oxidation number decreases from +1 to 0; reduction.
       (b) Pt oxidation number decreases from +2 to 0; reduction.
       (c) Cr oxidation number increases from 0 to +3; oxidation.
       (d) Sb oxidation number increases from +3 to +5; oxidation

19.44 (a) $HClO(aq) \rightarrow Cl_2(g)$
$$2 HClO(aq) \rightarrow Cl_2(g)$$
$$2 HClO(aq) \rightarrow Cl_2(g) + H_2O(l)$$
$$2 HClO(aq) + 2 H^+(aq) \rightarrow Cl_2(g) + H_2O(l)$$
$$2 HClO(aq) + 2 H^+(aq) + 2 e^- \rightarrow Cl_2(g) + H_2O(l)$$

(b) $Pt^{2+}(aq) \rightarrow Pt(s)$
$$Pt^{2+}(aq) + 2 e^- \rightarrow Pt(s)$$

(c) $Cr(s) \rightarrow Cr^{3+}(aq)$
$$Cr(s) \rightarrow Cr^{3+}(aq) + 3 e^-$$

(d) $SbCl_4^-(aq) \rightarrow SbCl_6^-(aq)$
$$6 SbCl_4^-(aq) \rightarrow 4 SbCl_6^-(aq)$$
$$6 SbCl_4^-(aq) \rightarrow 4 SbCl_6^-(aq) + 2 Sb^{3+}(aq)$$
$$6 SbCl_4^-(aq) \rightarrow 4 SbCl_6^-(aq) + 2 Sb^{3+}(aq) + 8 e^-$$

19.46 (a) $Te(s) + NO_3^-(aq) \rightarrow TeO_2(s) + NO(g)$
oxidation: $Te(s) \rightarrow TeO_2(s)$
reduction: $NO_3^-(aq) \rightarrow NO(g)$

(b) $H_2O_2(aq) + Fe^{2+}(aq) \rightarrow Fe^{3+}(aq) + H_2O(l)$
oxidation: $Fe^{2+}(aq) \rightarrow Fe^{3+}(aq)$
reduction: $H_2O_2(aq) \rightarrow H_2O(l)$

19.48 (a) $Cr_2O_7^{2-}(aq) \rightarrow Cr^{3+}(aq)$
$$Cr_2O_7^{2-}(aq) \rightarrow 2 Cr^{3+}(aq)$$
$$Cr_2O_7^{2-}(aq) \rightarrow 2 Cr^{3+}(aq) + 7 H_2O(l)$$
$$14 H^+(aq) + Cr_2O_7^{2-}(aq) \rightarrow 2 Cr^{3+}(aq) + 7 H_2O(l)$$
$$14 H^+(aq) + Cr_2O_7^{2-}(aq) + 6 e^- \rightarrow 2 Cr^{3+}(aq) + 7 H_2O(l)$$

(b) $CrO_4^{2-}(aq) \rightarrow Cr(OH)_4^-(aq)$
$$4 H^+(aq) + CrO_4^{2-}(aq) \rightarrow Cr(OH)_4^-(aq)$$
$$4 H^+(aq) + 4 OH^-(aq) + CrO_4^{2-}(aq) \rightarrow Cr(OH)_4^-(aq) + 4 OH^-(aq)$$
$$4 H_2O(l) + CrO_4^{2-}(aq) \rightarrow Cr(OH)_4^-(aq) + 4 OH^-(aq)$$
$$4 H_2O(l) + CrO_4^{2-}(aq) + 3 e^- \rightarrow Cr(OH)_4^-(aq) + 4 OH^-(aq)$$

(c) $Bi^{3+}(aq) \rightarrow BiO_3^-(aq)$
$$Bi^{3+}(aq) + 3 H_2O(l) \rightarrow BiO_3^-(aq)$$
$$Bi^{3+}(aq) + 3 H_2O(l) \rightarrow BiO_3^-(aq) + 6 H^+(aq)$$
$$Bi^{3+}(aq) + 3 H_2O(l) + 6 OH^-(aq) \rightarrow BiO_3^-(aq) + 6 H^+(aq) + 6 OH^-(aq)$$
$$Bi^{3+}(aq) + 3 H_2O(l) + 6 OH^-(aq) \rightarrow BiO_3^-(aq) + 6 H_2O(l)$$
$$Bi^{3+}(aq) + 6 OH^-(aq) \rightarrow BiO_3^-(aq) + 3 H_2O(l)$$
$$Bi^{3+}(aq) + 6 OH^-(aq) \rightarrow BiO_3^-(aq) + 3 H_2O(l) + 2 e^-$$

(d) $ClO^-(aq) \rightarrow Cl^-(aq)$
$$ClO^-(aq) \rightarrow Cl^-(aq) + H_2O(l)$$
$$2 H^+(aq) + ClO^-(aq) \rightarrow Cl^-(aq) + H_2O(l)$$
$$2 H^+(aq) + 2 OH^-(aq) + ClO^-(aq) \rightarrow Cl^-(aq) + H_2O(l) + 2 OH^-(aq)$$
$$2 H_2O(l) + ClO^-(aq) \rightarrow Cl^-(aq) + H_2O(l) + 2 OH^-(aq)$$
$$H_2O(l) + ClO^-(aq) \rightarrow Cl^-(aq) + 2 OH^-(aq)$$
$$H_2O(l) + ClO^-(aq) + 2 e^- \rightarrow Cl^-(aq) + 2 OH^-(aq)$$

19.50 (a) $Zn(s) \rightarrow Zn^{2+}(aq)$

$Zn(s) \rightarrow Zn^{2+}(aq) + 2\ e^-$　　　　　(oxidation half reaction)

$VO^{2+}(aq) \rightarrow V^{3+}(aq)$

$VO^{2+}(aq) \rightarrow V^{3+}(aq) + H_2O(l)$

$2\ H^+(aq) + VO^{2+}(aq) \rightarrow V^{3+}(aq) + H_2O(l)$

$[2\ H^+(aq) + VO^{2+}(aq) + e^- \rightarrow V^{3+}(aq) + H_2O(l)] \times 2$　　(reduction half reaction)

Combine the two half reactions.

$Zn(s) + 2\ VO^{2+}(aq) + 4\ H^+(aq) \rightarrow Zn^{2+}(aq) + 2\ V^{3+}(aq) + 2\ H_2O(l)$

(b) $Cr^{2+}(aq) \rightarrow Cr^{3+}(aq)$

$[Cr^{2+}(aq) \rightarrow Cr^{3+}(aq) + e^-] \times 4$　　　(oxidation half reaction)

$TeO_2(s) \rightarrow Te(s)$

$TeO_2(s) \rightarrow Te(s) + 2\ H_2O(l)$

$TeO_2(s) + 4\ H^+(aq) \rightarrow Te(s) + 2\ H_2O(l)$

$TeO_2(s) + 4\ H^+(aq) + 4\ e^- \rightarrow Te(s) + 2\ H_2O(l)$　　(reduction half reaction)

Combine the two half reactions.

$4\ Cr^{2+}(aq) + TeO_2(s) + 4\ H^+(aq) \rightarrow 4\ Cr^{3+}(aq) + Te(s) + 2\ H_2O(l)$

(c) $I^-(aq) \rightarrow I_3^-(aq)$

$3\ I^-(aq) \rightarrow I_3^-(aq)$

$[3\ I^-(aq) \rightarrow I_3^-(aq) + 2\ e^-] \times 8$　　　(oxidation half reaction)

$IO_3^-(aq) \rightarrow I_3^-(aq)$

$3\ IO_3^-(aq) \rightarrow I_3^-(aq)$

$3\ IO_3^-(aq) \rightarrow I_3^-(aq) + 9\ H_2O(l)$

$18\ H^+(aq) + 3\ IO_3^-(aq) \rightarrow I_3^-(aq) + 9\ H_2O(l)$

$18\ H^+(aq) + 3\ IO_3^-(aq) + 16\ e^- \rightarrow I_3^-(aq) + 9\ H_2O(l)$　(reduction half reaction)

Combine the two half reactions.

$18\ H^+(aq) + 3\ IO_3^-(aq) + 24\ I^-(aq) \rightarrow 9\ I_3^-(aq) + 9\ H_2O(l)$

Divide each coefficient by 3.

$6\ H^+(aq) + IO_3^-(aq) + 8\ I^-(aq) \rightarrow 3\ I_3^-(aq) + 3\ H_2O(l)$

19.52 (a) $MnO_4^-(aq) \rightarrow MnO_2(s)$

$MnO_4^-(aq) \rightarrow MnO_2(s) + 2\ H_2O(l)$

$4\ H^+(aq) + MnO_4^-(aq) \rightarrow MnO_2(s) + 2\ H_2O(l)$

$[4\ H^+(aq) + MnO_4^-(aq) + 3\ e^- \rightarrow MnO_2(s) + 2\ H_2O(l)] \times 2$　(reduction half reaction)

$IO_3^-(aq) \rightarrow IO_4^-(aq)$

$H_2O(l) + IO_3^-(aq) \rightarrow IO_4^-(aq)$

$H_2O(l) + IO_3^-(aq) \rightarrow IO_4^-(aq) + 2\ H^+(aq)$

$[H_2O(l) + IO_3^-(aq) \rightarrow IO_4^-(aq) + 2\ H^+(aq) + 2\ e^-] \times 3$　　(oxidation half reaction)

Combine the two half reactions.

$8 H^+(aq) + 3 H_2O(l) + 2 MnO_4^-(aq) + 3 IO_3^-(aq) \rightarrow$
$\qquad\qquad 6 H^+(aq) + 4 H_2O(l) + 2 MnO_2(s) + 3 IO_4^-(aq)$

$2 H^+(aq) + 2 MnO_4^-(aq) + 3 IO_3^-(aq) \rightarrow 2 MnO_2(s) + 3 IO_4^-(aq) + H_2O(l)$

$2 H^+(aq) + 2 OH^-(aq) + 2 MnO_4^-(aq) + 3 IO_3^-(aq) \rightarrow$
$\qquad\qquad 2 MnO_2(s) + 3 IO_4^-(aq) + H_2O(l) + 2 OH^-(aq)$

$2 H_2O(l) + 2 MnO_4^-(aq) + 3 IO_3^-(aq) \rightarrow$
$\qquad\qquad 2 MnO_2(s) + 3 IO_4^-(aq) + H_2O(l) + 2 OH^-(aq)$

$H_2O(l) + 2 MnO_4^-(aq) + 3 IO_3^-(aq) \rightarrow 2 MnO_2(s) + 3 IO_4^-(aq) + 2 OH^-(aq)$

(b) $N_2(g) \rightarrow N_2H_4(aq)$
$N_2(g) + 4 H^+(aq) \rightarrow N_2H_4(aq)$
$N_2(g) + 4 H^+(aq) + 4 e^- \rightarrow N_2H_4(aq)$ (reduction half reaction)

$NO_2^-(aq) \rightarrow NO_3^-(aq)$
$NO_2^-(aq) + H_2O(l) \rightarrow NO_3^-(aq)$
$[NO_2^-(aq) + H_2O(l) \rightarrow NO_3^-(aq) + 2 H^+(aq) + 2 e^-] \times 2$ (oxidation half reaction)

Combine the two half reactions.
$N_2(g) + 2 NO_2^-(aq) + 2 H_2O(l) \rightarrow N_2H_4(aq) + 2 NO_3^-(aq)$

(c) $ClO_4^-(aq) \rightarrow ClO_2^-(aq)$
$ClO_4^-(aq) \rightarrow ClO_2^-(aq) + 2 H_2O(l)$
$4 H^+(aq) + ClO_4^-(aq) \rightarrow ClO_2^-(aq) + 2 H_2O(l)$
$4 H^+(aq) + ClO_4^-(aq) + 4 e^- \rightarrow ClO_2^-(aq) + 2 H_2O(l)$ (reduction half reaction)

$H_2O_2(aq) \rightarrow O_2(g)$
$H_2O_2(aq) \rightarrow O_2(g) + 2 H^+(aq)$
$[H_2O_2(aq) \rightarrow O_2(g) + 2 H^+(aq) + 2 e^-] \times 2$ (oxidation half reaction)

Combine the two half reactions.
$4 H^+(aq) + ClO_4^-(aq) + 2 H_2O_2(aq) \rightarrow ClO_2^-(aq) + 2 H_2O(l) + 2 O_2(g) + 4 H^+(aq)$
$ClO_4^-(aq) + 2 H_2O_2(aq) \rightarrow ClO_2^-(aq) + 2 H_2O(l) + 2 O_2(g)$

Galvanic Cells (Sections 19.2 and 19.3)

19.54   The cathode of a galvanic cell is considered to be the positive electrode because electrons flow through the external circuit toward the positive electrode (the cathode).

19.56   (a) $Cd(s) + Sn^{2+}(aq) \rightarrow Cd^{2+}(aq) + Sn(s)$

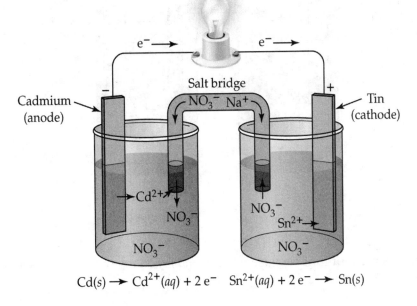

$$Cd(s) \rightarrow Cd^{2+}(aq) + 2\,e^- \qquad Sn^{2+}(aq) + 2\,e^- \rightarrow Sn(s)$$

(b) $2\,Al(s) + 3\,Cd^{2+}(aq) \rightarrow 2\,Al^{3+}(aq) + 3\,Cd(s)$

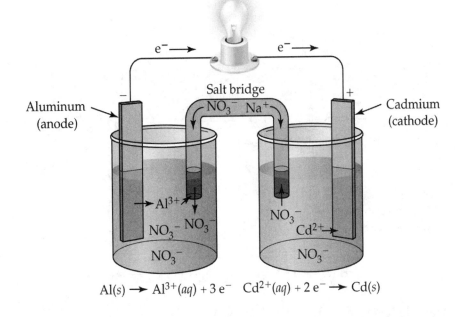

$$Al(s) \rightarrow Al^{3+}(aq) + 3\,e^- \qquad Cd^{2+}(aq) + 2\,e^- \rightarrow Cd(s)$$

(c) $6\,Fe^{2+}(aq) + Cr_2O_7^{2-}(aq) + 14\,H^+(aq) \rightarrow 6\,Fe^{3+}(aq) + 2\,Cr^{3+}(aq) + 7\,H_2O(l)$

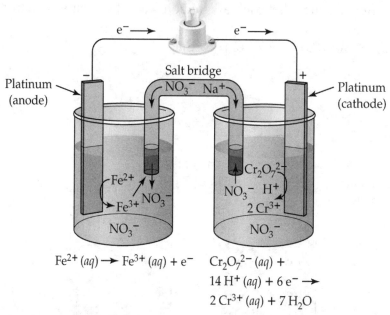

$Fe^{2+}(aq) \rightarrow Fe^{3+}(aq) + e^-$

$Cr_2O_7^{2-}(aq) +$
$14\,H^+(aq) + 6\,e^- \rightarrow$
$2\,Cr^{3+}(aq) + 7\,H_2O$

19.58 $2\,Br^-(aq) + Cl_2(g) \rightarrow Br_2(l) + 2\,Cl^-(aq)$
Inert electrodes are required because none of the reactants or products is an electrical conductor.

19.60 $Al(s)\big|Al^{3+}(aq)\big\|Cd^{2+}\big|Cd(s)$

19.62 (a)

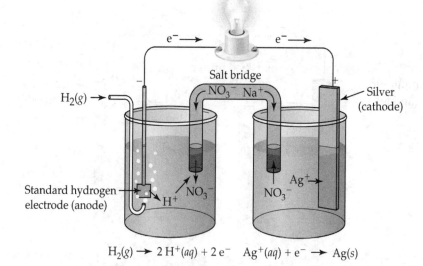

$H_2(g) \rightarrow 2\,H^+(aq) + 2\,e^-$ $Ag^+(aq) + e^- \rightarrow Ag(s)$

(b) anode reaction $H_2(g) \rightarrow 2\,H^+(aq) + 2\,e^-$
    cathode reaction $\underline{2\,Ag^+(aq) + 2\,e^- \rightarrow 2\,Ag(s)}$
    overall reaction $H_2(g) + 2\,Ag^+(aq) \rightarrow 2\,H^+(aq) + 2\,Ag(s)$
(c) $Pt(s)\big|H_2(g)\big|H^+(aq)\big\|Ag^+(aq)\big|Ag(s)$

19.64 (a) anode reaction $Co(s) \rightarrow Co^{2+}(aq) + 2\,e^-$
       cathode reaction $\underline{Cu^{2+}(aq) + 2\,e^- \rightarrow Cu(s)}$
       overall reaction $Co(s) + Cu^{2+}(aq) \rightarrow Co^{2+}(aq) + Cu(s)$

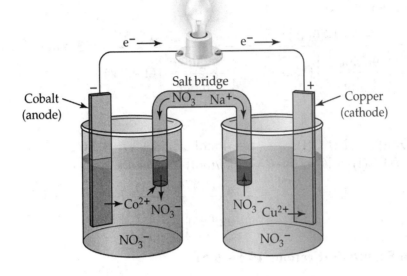

(b) anode reaction $2\,Fe(s) \rightarrow 2\,Fe^{2+}(aq) + 4\,e^-$
    cathode reaction $\underline{O_2(g) + 4\,H^+(aq) + 4\,e^- \rightarrow 2\,H_2O(l)}$
    overall reaction $2\,Fe(s) + O_2(g) + 4\,H^+(aq) \rightarrow 2\,Fe^{2+}(aq) + 2\,H_2O(l)$

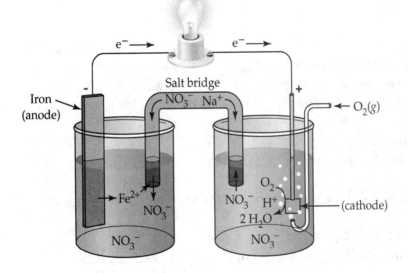

Cell Potentials and Free-Energy Changes (Section 19.4)

19.66   E is the standard cell potential ($E°$) when all reactants and products are in their standard states–solutes at 1 M concentrations, gases at a partial pressure of 1 atm, solids and liquids in pure form, all at 25 °C.

19.68 $Zn(s) + Ag_2O(s) \rightarrow ZnO(s) + 2\,Ag(s);$ $n = 2\ mol\ e^-$

$$\Delta G = -nFE = -(2\ mol\ e^-)\left(\frac{96{,}500\ C}{1\ mol\ e^-}\right)(1.60\ V)\left(\frac{1\ J}{1\ C \cdot V}\right) = -308{,}800\ J = -309\ kJ$$

19.70 $\Delta G^\circ = -nFE^\circ;\ \ 1\ J = C \cdot V$

$$n = \frac{-\Delta G^\circ}{FE^\circ} = \frac{-(-414{,}000\ J)}{\left(\dfrac{96{,}500\ C}{1\ mol\ e^-}\right)(1.43\ V)} = \frac{-(-414{,}000\ C \cdot V)}{\left(\dfrac{96{,}500\ C}{1\ mol\ e^-}\right)(1.43\ V)} = 3\ mol\ e^-$$

$x = 3$

19.72 $2\,H_2(g) + O_2(g) \rightarrow 2\,H_2O(l);$ $n = 4\ mol\ e^-$ and $1\ V = 1\ J/C$
$\Delta G^\circ = 2\,\Delta G^\circ_f(H_2O(l)) = (2\ mol)(-237.2\ kJ/mol) = -474.4\ kJ$

$$\Delta G^\circ = -nFE^\circ \qquad E^\circ = \frac{-\Delta G^\circ}{nF} = \frac{-(-474{,}400\ J)}{(4\ mol\ e^-)\left(\dfrac{96{,}500\ C}{1\ mol\ e^-}\right)} = +1.23\ J/C = +1.23\ V$$

Standard Reduction Potentials (Sections 19.5–19.6)

19.74 $Sn^{4+}(aq) < Br_2(aq) < MnO_4^-$

19.76 $Cr_2O_7^{2-}(aq)$ is highest in the table of standard reduction potentials, therefore it is the strongest oxidizing agent.
$Fe^{2+}(aq)$ is lowest in the table of standard reduction potentials, therefore it is the weakest oxidizing agent.

19.78   (a) oxidizing agents: $PbO_2, H^+, Cr_2O_7^{2-}$;   reducing agents: Al, Fe, Ag
        (b) PbO_2 is the strongest oxidizing agent. H^+ is the weakest oxidizing agent.
        (c) Al is the strongest reducing agent.   Ag is the weakest reducing agent.
        (d) oxidized by Cu^{2+}: Fe and Al;   reduced by H_2O_2: PbO_2 and $Cr_2O_7^{2-}$

19.80   (a) oxidation: $Sn^{2+}(aq) \rightarrow Sn^{4+}(aq) + 2\,e^-$ $E^\circ = -0.15\ V$
           reduction: $Br_2(aq) + 2\,e^- \rightarrow 2\,Br^-(aq)$ $\underline{E^\circ = \ \ 1.09\ V}$
                                              overall $E^\circ = +0.94\ V$
Because the overall E° is positive, $Sn^{2+}(aq)$ can be oxidized by $Br_2(aq)$.

        (b) oxidation: $Sn^{2+}(aq) \rightarrow Sn^{4+}(aq) + 2\,e^-$ $E^\circ = -0.15\ V$
           reduction: $Ni^{2+}(aq) + 2\,e^- \rightarrow Ni(s)$ $\underline{E^\circ = -0.26\ V}$
                                         overall $E^\circ = -0.41\ V$
Because the overall E° is negative, $Ni^{2+}(aq)$ cannot be reduced by $Sn^{2+}(aq)$.

        (c) oxidation: $2\,Ag(s) \rightarrow 2\,Ag^+(aq) + 2\,e^-$ $E^\circ = -0.80\ V$
           reduction: $Pb^{2+}(aq) + 2\,e^- \rightarrow Pb(s)$ $\underline{E^\circ = -0.13\ V}$
                                         overall $E^\circ = -0.93\ V$
Because the overall E° is negative, $Ag(s)$ cannot be oxidized by $Pb^{2+}(aq)$.

(d) oxidation: $H_2SO_3(aq) + H_2O(l) \rightarrow SO_4^{2-}(aq) + 4\,H^+(aq) + 2\,e^-$ $E^\circ = -0.17$ V

reduction: $I_2(s) + 2\,e^- \rightarrow 2\,I^-(aq)$ $E^\circ = \underline{\ \ 0.54}$ V

overall $E^\circ = +0.37$ V

Because the overall E° is positive, $I_2(s)$ can be reduced by H_2SO_3.

19.82 (a) oxidation: $2\,Cr^{3+}(aq) + 7\,H_2O(l) \rightarrow Cr_2O_7^{2-}(aq) + 14\,H^+(aq) + 6\,e^-$ $E^\circ = -1.36$ V

reduction: $O_2(g) + 4\,H^+(aq) + 4\,e^- \rightarrow 2\,H_2O(l)$ $E^\circ = \underline{\ \ 1.23}$ V

overall $E^\circ = -0.13$ V

There is no reaction because the overall E° is negative.

(b) oxidation: $Pb(s) \rightarrow Pb^{2+}(aq) + 2\,e^-$ $E^\circ = \ \ 0.13$ V

reduction: $2\,Ag^+(aq) + 2\,e^- \rightarrow 2\,Ag(s)$ $E^\circ = \underline{\ \ 0.80}$ V

overall $E^\circ = +0.93$ V

$Pb(s) + 2\,Ag^+(aq) \rightarrow Pb^{2+}(aq) + 2\,Ag(s)$

The reaction is spontaneous because the overall E° is positive.

(c) oxidation: $H_2C_2O_4(aq) \rightarrow 2\,CO_2(g) + 2\,H^+(aq) + 2\,e^-$ $E^\circ = \ \ 0.49$ V

reduction: $Cl_2(g) + 2\,e^- \rightarrow 2\,Cl^-(aq)$ $E^\circ = \underline{\ \ 1.36}$ V

overall $E^\circ = +1.85$ V

$Cl_2(g) + H_2C_2O_4(aq) \rightarrow 2\,Cl^-(aq) + 2\,CO_2(g) + 2\,H^+(aq)$

The reaction is spontaneous because the overall E° is positive.

(d) oxidation: $Ni(s) \rightarrow Ni^{2+}(aq) + 2\,e^-$ $E^\circ = \ \ 0.26$ V

reduction: $2\,HClO(aq) + 2\,H^+(aq) + 2\,e^- \rightarrow Cl_2(g) + H_2O(l)$ $E^\circ = \underline{\ \ 1.61}$ V

overall $E^\circ = +1.87$ V

$Ni(s) + 2\,HClO(aq) + 2\,H^+(aq) \rightarrow Ni^{2+}(aq) + Cl_2(g) + H_2O(l)$

The reaction is spontaneous because the overall E° is positive.

19.84 oxidation: $Zn(s) \rightarrow Zn^{2+}(aq) + 2\,e^-$ $E^\circ = 0.76$ V

reduction: $2[Eu^{3+}(aq) + e^- \rightarrow Eu^{2+}(aq)]$ $E^\circ = \underline{\ ?\ }$

overall $Zn(s) + 2\,Eu^{3+}(aq) \rightarrow Zn^{2+}(aq) + 2\,Eu^{2+}(aq)$ $E^\circ = 0.40$ V

The standard reduction potential for the Eu^{3+}/Eu^{2+} half cell is:

$E^\circ = 0.40 - 0.76 = -0.36$ V

19.86 oxidation: $Co(s) \rightarrow Co^{2+}(aq) + 2\,e^-$ $E^\circ = 0.28$ V

reduction: $I_2(s) + 2\,e^- \rightarrow 2\,I^-(aq)$ $E^\circ = \underline{0.54}$ V

overall $I_2(s) + Co(s) \rightarrow Co^{2+}(aq) + 2\,I^-(aq)$ $E^\circ = 0.82$ V

$n = 2$ mol e^-

$$\Delta G^\circ = -nFE^\circ = -(2 \text{ mol } e^-)\left(\frac{96{,}500 \text{ C}}{1 \text{ mol } e^-}\right)(0.82 \text{ V})\left(\frac{1 \text{ J}}{1 \text{ C} \cdot \text{V}}\right) = -158{,}260 \text{ J} = -1.6 \times 10^2 \text{ kJ}$$

19.88 oxidation: $2\,Al(s) \rightarrow 2\,Al^{3+}(aq) + 6\,e^-$ $E^\circ = \ \ 1.66$ V

reduction: $3\,Cd^{2+}(aq) + 6\,e^- \rightarrow 3\,Cd(s)$ $E^\circ = -0.40$ V

overall $2\,Al(s) + 3\,Cd^{2+}(aq) \rightarrow Al^{3+}(aq) + 3\,Cd(s)$ $E^\circ = \ \ 1.26$ V

$n = 6$ mol e^-

$$\Delta G^\circ = -nFE^\circ = -(6 \text{ mol e}^-)\left(\frac{96,500 \text{ C}}{1 \text{ mol e}^-}\right)(1.26 \text{ V})\left(\frac{1 \text{ J}}{1 \text{ C} \cdot \text{V}}\right) = -729,540 \text{ J} = -730 \text{ kJ}$$

19.90 (a) $2 \text{ Fe}^{2+}(aq) + \text{Pb}^{2+}(aq) \rightarrow 2 \text{ Fe}^{3+}(aq) + \text{Pb}(s)$

        oxidation: $2 \text{ Fe}^{2+}(aq) \rightarrow 2 \text{ Fe}^{3+}(aq) + 2 \text{ e}^-$ $E^\circ = -0.77 \text{ V}$

        reduction: $\text{Pb}^{2+}(aq) + 2 \text{ e}^- \rightarrow \text{Pb}(s)$ $\underline{E^\circ = -0.13 \text{ V}}$

                                       overall $E^\circ = -0.90 \text{ V}$

    Because the overall E° is negative, this reaction is nonspontaneous.

    (b) $\text{Mg}(s) + \text{Ni}^{2+}(aq) \rightarrow \text{Mg}^{2+}(aq) + \text{Ni}(s)$

    oxidation: $\text{Mg}(s) \rightarrow \text{Mg}^{2+}(aq) + 2 \text{ e}^-$ $E^\circ = 2.37 \text{ V}$

    reduction: $\text{Ni}^{2+}(aq) + 2 \text{ e}^- \rightarrow \text{Ni}(s)$ $\underline{E^\circ = -0.26 \text{ V}}$

                                       overall $E^\circ = 2.11 \text{ V}$

    Because the overall E° is positive, this reaction is spontaneous.

19.92 (a) $2 \text{ MnO}_4^-(aq) + 16 \text{ H}^+(aq) + 5 \text{ Sn}^{2+}(aq) \rightarrow 2 \text{ Mn}^{2+}(aq) + 5 \text{ Sn}^{4+}(aq) + 8 \text{ H}_2\text{O}(l)$

    (b) MnO_4^- is the oxidizing agent; Sn^{2+} is the reducing agent.

    (c) $E^\circ = 1.51 \text{ V} + (-0.15 \text{ V}) = 1.36 \text{ V}$

The Nernst Equation (Sections 19.7 and 19.8)

19.94 $2 \text{ Ag}^+(aq) + \text{Ni}(s) \rightarrow 2 \text{ Ag}(s) + \text{Ni}^{2+}(aq)$

    oxidation: $\text{Ni}(s) \rightarrow \text{Ni}^{2+}(aq) + 2 \text{ e}^-$ $E^\circ = 0.26 \text{ V}$

    reduction: $2 \text{ Ag}^+(aq) + 2 \text{ e}^- \rightarrow 2 \text{ Ag}(s)$ $\underline{E^\circ = 0.80 \text{ V}}$

                                       overall $E^\circ = 1.06 \text{ V}$

$$E = E^\circ - \frac{0.0592 \text{ V}}{n} \log \frac{[\text{Ni}^{2+}]}{[\text{Ag}^+]^2} = 1.06 \text{ V} - \frac{(0.0592 \text{ V})}{2} \log \frac{(0.100)}{(0.010)^2} = 0.97 \text{ V}$$

19.96 $\text{Pb}(s) + \text{Cu}^{2+}(aq) \rightarrow \text{Pb}^{2+}(aq) + \text{Cu}(s)$

    oxidation: $\text{Pb}(s) \rightarrow \text{Pb}^{2+}(aq) + 2 \text{ e}^-$ $E^\circ = 0.13 \text{ V}$

    reduction: $\text{Cu}^{2+}(aq) + 2 \text{ e}^- \rightarrow \text{Cu}(s)$ $\underline{E^\circ = 0.34 \text{ V}}$

                                       overall $E^\circ = 0.47 \text{ V}$

$$E = E^\circ - \frac{0.0592 \text{ V}}{n} \log \frac{[\text{Pb}^{2+}]}{[\text{Cu}^{2+}]} = 0.47 \text{ V} - \frac{(0.0592 \text{ V})}{2} \log \frac{1.0}{(1.0 \times 10^{-4})} = 0.35 \text{ V}$$

When $E = 0$, $\quad 0 = E^\circ - \dfrac{0.0592 \text{ V}}{n} \log \dfrac{[\text{Pb}^{2+}]}{[\text{Cu}^{2+}]} = 0.47 \text{ V} - \dfrac{(0.0592 \text{ V})}{2} \log \dfrac{1.0}{[\text{Cu}^{2+}]}$

$$0 = 0.47 \text{ V} + \frac{(0.0592 \text{ V})}{2} \log [\text{Cu}^{2+}]$$

$$\log [\text{Cu}^{2+}] = (-0.47 \text{ V})\left(\frac{2}{0.0592 \text{ V}}\right) = -15.88$$

$$[\text{Cu}^{2+}] = 10^{-15.88} = 1 \times 10^{-16} \text{ M}$$

19.98 $Zn(s) + Cu^{2+}(aq) \rightarrow Zn^{2+}(aq) + Cu(s)$

oxidation: $Zn(s) \rightarrow Zn^{2+}(aq) + 2\,e^-$ $E^\circ = 0.76$ V

reduction: $Cu^{2+}(aq) + 2\,e^- \rightarrow Cu(s)$ $\underline{E^\circ = 0.34 \text{ V}}$

overall $E^\circ = 1.10$ V

$$E = 1.07 \text{ V} = E^\circ - \frac{0.0592\text{ V}}{n}\log\frac{[Zn^{2+}]}{[Cu^{2+}]} = 1.10 \text{ V} - \frac{(0.0592\text{ V})}{2}\log\left(\frac{[Zn^{2+}]}{[Cu^{2+}]}\right)$$

$$1.07 \text{ V} - 1.10 \text{ V} = -\frac{(0.0592\text{ V})}{2}\log\left(\frac{[Zn^{2+}]}{[Cu^{2+}]}\right)$$

$$\frac{0.03 \text{ V}}{\dfrac{(0.0592\text{ V})}{2}} = \log\left(\frac{[Zn^{2+}]}{[Cu^{2+}]}\right) = 1; \quad \frac{[Zn^{2+}]}{[Cu^{2+}]} = 10^1 = 10$$

19.100 (a) $E = E^\circ - \dfrac{0.0592\text{ V}}{n}\log[I^-]^2 = 0.54 \text{ V} - \dfrac{(0.0592\text{ V})}{2}\log(0.020)^2 = 0.64$ V

(b) $E = E^\circ - \dfrac{0.0592\text{ V}}{n}\log\dfrac{[Fe^{2+}]}{[Fe^{3+}]} = 0.77 \text{ V} - \dfrac{(0.0592\text{ V})}{1}\log\left(\dfrac{0.10}{0.10}\right) = 0.77$ V

(c) $E = E^\circ - \dfrac{0.0592\text{ V}}{n}\log\dfrac{[Sn^{4+}]}{[Sn^{2+}]} = -0.15 \text{ V} - \dfrac{(0.0592\text{ V})}{2}\log\left(\dfrac{0.40}{0.0010}\right) = -0.23$ V

(d) $E = E^\circ - \dfrac{0.0592\text{ V}}{n}\log\dfrac{[Cr_2O_7^{2-}][H^+]^{14}}{[Cr^{3+}]^2} = -1.36 \text{ V} - \dfrac{(0.0592\text{ V})}{6}\log\left(\dfrac{(1.0)(0.010)^{14}}{1.0}\right)$

$$E = -1.36 \text{ V} - \frac{(0.0592\text{ V})}{6}(14)\log(0.010) = -1.08 \text{ V}$$

19.102 $E = E^\circ - \dfrac{0.0592\text{ V}}{n}\log\dfrac{P_{H_2}}{[H_3O^+]^2}$; $E^\circ = 0$, $n = 2$ mol e^-, and $P_{H_2} = 1$ atm

(a) $[H_3O^+] = 1.0$ M; $E = -\dfrac{0.0592\text{ V}}{2}\log\dfrac{1}{(1.0)^2} = 0$

(b) pH = 4.00, $[H_3O^+] = 10^{-4.00} = 1.0 \times 10^{-4}$ M

$$E = -\frac{0.0592\text{ V}}{2}\log\frac{1}{(1.0 \times 10^{-4})^2} = -0.24 \text{ V}$$

(c) $[H_3O^+] = 1.0 \times 10^{-7}$ M; $E = -\dfrac{0.0592\text{ V}}{2}\log\dfrac{1}{(1.0 \times 10^{-7})^2} = -0.41$ V

(d) $[OH^-] = 1.0$ M; $[H_3O^+] = \dfrac{K_w}{[OH^-]} = \dfrac{1.0 \times 10^{-14}}{1.0} = 1.0 \times 10^{-14}$ M

$$E = -\frac{0.0592\text{ V}}{2}\log\frac{1}{(1.0 \times 10^{-14})^2} = -0.83 \text{ V}$$

19.104 For Pb^{2+}, $E = -0.13 - \dfrac{0.0592\,V}{2} \log \dfrac{1}{[Pb^{2+}]}$

For Cd^{2+}, $E = -0.40 - \dfrac{0.0592\,V}{2} \log \dfrac{1}{[Cd^{2+}]}$

Set these two equations for E equal to each other and solve for $[Cd^{2+}]/[Pb^{2+}]$.

$$-0.13 - \dfrac{0.0592\,V}{2} \log \dfrac{1}{[Pb^{2+}]} = -0.40 - \dfrac{0.0592\,V}{2} \log \dfrac{1}{[Cd^{2+}]}$$

$$0.27 = \dfrac{0.0592\,V}{2}(\log[Cd^{2+}] - \log[Pb^{2+}]) = \dfrac{0.0592\,V}{2} \log \dfrac{[Cd^{2+}]}{[Pb^{2+}]}$$

$$\log \dfrac{[Cd^{2+}]}{[Pb^{2+}]} = \dfrac{(0.27)(2)}{0.0592} = 9.1; \qquad \dfrac{[Cd^{2+}]}{[Pb^{2+}]} = 10^{9.1} = 1 \times 10^9$$

19.106 $H_2(g) + Ni^{2+}(aq) \rightarrow 2\,H^+(aq) + Ni(s)$

$E° = E°_{H_2 \rightarrow H^+} + E°_{Ni^{2+} \rightarrow Ni} = 0\,V + (-0.26\,V) = -0.26\,V$

$$E = E° - \dfrac{0.0592\,V}{n} \log \dfrac{[H_3O^+]^2}{[Ni^{2+}](P_{H_2})}$$

$$0.15\,V = -0.26\,V - \dfrac{(0.0592\,V)}{2} \log \dfrac{[H_3O^+]^2}{(1)(1)}$$

$0.15\,V = -0.26\,V - (0.0592\,V)\log[H_3O^+]$

$pH = -\log[H_3O^+]$ therefore $0.15\,V = -0.26\,V + (0.0592\,V)\,pH$

$pH = \dfrac{(0.15\,V + 0.26\,V)}{0.0592\,V} = 6.9$

Standard Cell Potentials and Equilibrium Constants (Section 19.9)

19.108 $\Delta G° = -nFE°$

Because n and F are always positive, $\Delta G°$ is negative when E° is positive because of the negative sign in the equation.

$$E° = \dfrac{0.0592\,V}{n} \log K; \quad \log K = \dfrac{nE°}{0.0592\,V}; \qquad K = 10^{\frac{nE°}{0.0592}}$$

If E° is positive, the exponent is positive (because n is positive), and K is greater than 1.

19.110 $Ni(s) + 2\,Ag^+(aq) \rightarrow Ni^{2+}(aq) + 2\,Ag(s)$

oxidation: $Ni(s) \rightarrow Ni^{2+}(aq) + 2\,e^-$ $E° = 0.26\,V$

reduction: $2\,Ag^+(aq) + 2\,e^- \rightarrow 2\,Ag(s)$ $\underline{E° = 0.80\,V}$

overall $E° = 1.06\,V$

$$E° = \dfrac{0.0592\,V}{n} \log K; \quad \log K = \dfrac{nE°}{0.0592\,V} = \dfrac{(2)(1.06\,V)}{0.0592\,V} = 35.8; \quad K = 10^{35.8} = 6 \times 10^{35}$$

19.112 $Cd(s) + Sn^{2+}(aq) \rightarrow Cd^{2+}(aq) + Sn(s)$

oxidation $Cd(s) \rightarrow Cd^{2+}(aq) + 2\ e^-$ $E° = 0.40\ V$

reduction: $Sn^{2+}(aq) + 2\ e^- \rightarrow Sn(s)$ $\underline{E° = -0.14\ V}$

overall $E° = 0.26\ V$

$$E° = \frac{0.0592\ V}{n} \log K; \quad \log K = \frac{nE°}{0.0592\ V} = \frac{(2)(0.26\ V)}{0.0592\ V} = 8.8; \quad K = 10^{8.8} = 6.3 \times 10^8$$

19.114 $Hg_2^{2+}(aq) \rightarrow Hg(l) + Hg^{2+}(aq)$

oxidation: $\frac{1}{2}[Hg_2^{2+}(aq) \rightarrow 2\ Hg^{2+}(aq) + 2\ e^-]$ $E° = -0.92\ V$

reduction: $\frac{1}{2}[Hg_2^{2+}(aq) + 2\ e^- \rightarrow 2\ Hg(l)]$ $\underline{E° = 0.80\ V}$

overall $E° = -0.12\ V$

$$E° = \frac{0.0592\ V}{n} \log K$$

$$\log K = \frac{nE°}{0.0592\ V} = \frac{(1)(-0.12\ V)}{0.0592\ V} = -2.027; \quad K = 10^{-2.027} = 9 \times 10^{-3}$$

19.116 First calculate $E°$ for the galvanic cell in order to determine $E°_1$.

anode: $5\ [2\ Hg(l) + 2\ Br^-(aq) \rightarrow Hg_2Br_2(s) + 2\ e^-]$ $E°_1 = ?$

cathode: $\underline{2\ [MnO_4^-(aq) + 8\ H^+(aq) + 5\ e^- \rightarrow Mn^{2+}(aq) + 4\ H_2O(l)]}$ $E°_2 = 1.51\ V$

overall: $2\ MnO_4^-(aq) + 10\ Hg(l) + 10\ Br^-(aq) + 16\ H^+(aq) \rightarrow$
                $2\ Mn^{2+}(aq) + 5\ Hg_2Br_2(s) + 8\ H_2O(l)$

$n = 10$ mol e^-

$$E = E° - \frac{0.0592\ V}{n} \log \frac{[Mn^{2+}]^2}{[Br^-]^{10}[MnO_4^-]^2[H^+]^{16}}$$

$$1.214\ V = E° - \frac{(0.0592\ V)}{10} \log \left(\frac{(0.10)^2}{(0.10)^{10}(0.10)^2(0.10)^{16}} \right)$$

$$1.214\ V = E° - \frac{(0.0592\ V)}{10} \log \frac{1}{(0.10)^{26}} = E° - 0.154\ V$$

$E° = 1.214 + 0.154 = 1.368\ V$

$E°_1 + E°_2 = 1.368\ V; \quad E°_1 + 1.51\ V = 1.368\ V; \quad E°_1 = 1.368\ V - 1.51\ V = -0.142\ V$

oxidation: $2\ Hg(l) \rightarrow Hg_2^{2+}(aq) + 2\ e^-$ $E° = -0.80\ V$ (Appendix D)

reduction: $\underline{Hg_2Br_2(s) + 2\ e^- \rightarrow 2\ Hg(l) + 2\ Br^-(aq)}$ $E° = +0.142\ V$ (from $E°_1$)

overall: $Hg_2Br_2(s) \rightarrow Hg_2^{2+}(aq) + 2\ Br^-(aq)$ $E° = -0.658\ V$

$$E° = \frac{0.0592\ V}{n} \log K; \quad \log K = \frac{nE°}{0.0592\ V} = \frac{(2)(-0.658\ V)}{0.0592\ V} = -22.2$$

$K = K_{sp} = 10^{-22.2} = 6 \times 10^{-23}$

Batteries (Section 19.10)

19.118 $2\ PbSO_4(s) + 2\ H_2O(l) \rightarrow Pb(s) + PbO_2(s) + 2\ H^+(aq) + 2\ HSO_4^-(aq)$

19.120 (a) &(b)

oxidation:	$2[Al(s) \rightarrow Al^{3+}(aq) + 3\ e^-]$		$E° = 1.66$ V
reduction:	$3[Cu^{2+}(aq) + 2\ e^- \rightarrow Cu(s)]$		$E° = 0.34$ V
overall:	$2\ Al(s) + 3\ Cu^{2+}(aq) \rightarrow 2\ Al^{3+}(aq) + 3Cu(s)$		$E° = 2.00$ V

(c)

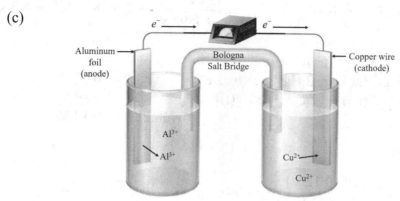

(d) This battery will only deliver 2.00 V, which is not enough to charge an iPhone.

19.122 (a)

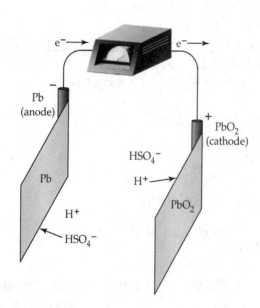

(b)   anode: $Pb(s) + HSO_4^-(aq) \rightarrow PbSO_4(s) + H^+(aq) + 2\ e^-$ $E° = 0.296$ V
        cathode: $PbO_2(s) + 3\ H^+(aq) + HSO_4^-(aq) + 2\ e^- \rightarrow PbSO_4(s) + 2\ H_2O(l)$ $E° = 1.628$ V
     overall $Pb(s) + PbO_2(s) + 2\ H^+(aq) + 2\ HSO_4^-(aq) \rightarrow 2\ PbSO_4(s) + 2\ H_2O(l)$ $E° = 1.924$ V

(c) $E° = \dfrac{0.0592\ V}{n} \log K;\quad \log K = \dfrac{n\,E°}{0.0592\ V} = \dfrac{(2)(1.924\ V)}{0.0592\ V} = 65.0;\quad K = 1 \times 10^{65}$

(d) When the cell reaction reaches equilibrium the cell voltage = 0.

Corrosion (Section 19.11)

19.124 Rust is a hydrated form of iron(III) oxide ($Fe_2O_3 \cdot H_2O$). Rust forms from the oxidation
        of Fe in the presence of O_2 and H_2O. Rust can be prevented by coating Fe with Zn
        (galvanizing).

19.126 Cr forms a protective oxide coating similar to Al.

19.128 (d) A strip of magnesium is attached to steel because the magnesium is more easily oxidized than iron.

19.130 Mn and Al

Electrolysis (Sections 19.12–19.14)

19.132 (a)

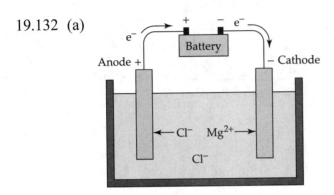

(b) anode: $2\ Cl^-(l)\ \rightarrow\ Cl_2(g)\ +\ 2\ e^-$
  cathode: $\underline{Mg^{2+}(l)\ +\ 2\ e^-\ \rightarrow\ Mg(l)}$
  overall: $Mg^{2+}(l)\ +\ 2\ Cl^-(l)\ \rightarrow\ Mg(l)\ +\ Cl_2(g)$

19.134 Possible anode reactions:
    $2\ Cl^-(aq)\ \rightarrow\ Cl_2(g)\ +\ 2\ e^-$
    $2\ H_2O(l)\ \rightarrow\ O_2(g)\ +\ 4\ H^+(aq)\ +\ 4\ e^-$

  Possible cathode reactions:
    $2\ H_2O(l)\ +\ 2\ e^-\ \rightarrow\ H_2(g)\ +\ 2\ OH^-(aq)$
    $Mg^{2+}(aq)\ +\ 2\ e^-\ \rightarrow\ Mg(s)$

  Actual reactions:
  anode: $2\ Cl^-(aq)\ \rightarrow\ Cl_2(g)\ +\ 2\ e^-$
  cathode: $2\ H_2O(l)\ +\ 2\ e^-\ \rightarrow\ H_2(g)\ +\ 2\ OH^-(aq)$

  This anode reaction takes place instead of $2\ H_2O(l) \rightarrow O_2(g) + 4\ H^+(aq) + 4\ e^-$ because of a high overvoltage for formation of gaseous O_2.
  This cathode reaction takes place instead of $Mg^{2+}(aq) + 2\ e^- \rightarrow Mg(s)$ because H_2O is easier to reduce than Mg^{2+}.

19.136 (a) NaBr
  anode: $2\ Br^-(aq)\ \rightarrow\ Br_2(l)\ +\ 2\ e^-$
  cathode: $\underline{2\ H_2O(l)\ +\ 2\ e^-\ \rightarrow\ H_2(g)\ +\ 2\ OH^-(aq)}$
  overall: $2\ H_2O(l) + 2\ Br^-(aq) \rightarrow Br_2(l) + H_2(g) + 2\ OH^-(aq)$

  (b) $CuCl_2$
  anode: $2\ Cl^-(aq)\ \rightarrow\ Cl_2(g)\ +\ 2\ e^-$
  cathode: $\underline{Cu^{2+}(aq)\ +\ 2\ e^-\ \rightarrow\ Cu(s)}$
  overall: $Cu^{2+}(aq) + 2\ Cl^-(aq) \rightarrow Cu(s) + Cl_2(g)$

(c) LiOH

anode: $4 \; OH^-(aq) \; \rightarrow \; O_2(g) \; + \; 2 \; H_2O(l) \; + \; 4 \; e^-$

cathode: $\underline{4 \; H_2O(l) \; + \; 4 \; e^- \; \rightarrow \; 2 \; H_2(g) \; + \; 4 \; OH^-(aq)}$

overall: $2 \; H_2O(l) \; \rightarrow \; O_2(g) \; + \; 2 \; H_2(g)$

19.138 $Ag^+(aq) \; + \; e^- \; \rightarrow \; Ag(s)$; $1 \; A = 1 \; C/s$

$$\text{mass Ag} = 2.40 \; \frac{C}{s} \times 20.0 \; \text{min} \times \frac{60 \; s}{1 \; \text{min}} \times \frac{1 \; \text{mol} \; e^-}{96{,}500 \; C} \times \frac{1 \; \text{mol Ag}}{1 \; \text{mol} \; e^-} \times \frac{107.87 \; \text{g Ag}}{1 \; \text{mol Ag}} = 3.22 \; g$$

19.140 $2 \; Na^+(l) \; + \; 2 \; Cl^-(l) \; \rightarrow \; 2 \; Na(l) \; + \; Cl_2(g)$

$Na^+(l) \; + \; e^- \; \rightarrow \; Na(l)$; $1 \; A = 1 \; C/s$; $1.00 \times 10^3 \; kg = 1.00 \times 10^6 \; g$

$$\text{Charge} = 1.00 \times 10^6 \; \text{g Na} \times \frac{1 \; \text{mol Na}}{22.99 \; \text{g Na}} \times \frac{1 \; \text{mol} \; e^-}{1 \; \text{mol Na}} \times \frac{96{,}500 \; C}{1 \; \text{mol} \; e^-} = 4.20 \times 10^9 \; C$$

$$\text{Time} = \frac{4.20 \times 10^9 \; C}{30{,}000 \; C/s} \times \frac{1 \; h}{3600 \; s} = 38.9 \; h$$

$$1.00 \times 10^6 \; \text{g Na} \times \frac{1 \; \text{mol Na}}{22.99 \; \text{g Na}} \times \frac{1 \; \text{mol Cl}_2}{2 \; \text{mol Na}} = 21{,}748.6 \; \text{mol Cl}_2$$

$PV = nRT$

$$V = \frac{nRT}{P} = \frac{(21{,}748.6 \; \text{mol})\left(0.082 \; 06 \; \dfrac{L \cdot atm}{K \cdot mol}\right)(273.15 \; K)}{1.00 \; atm} = 4.87 \times 10^5 \; L \; Cl_2$$

19.142 $M^{2+} \; + \; 2 \; e^- \; \rightarrow \; M$; $20.0 \; A = 20.0 \; C/s$

$$\text{mol} \; e^- = 20.0 \; \frac{C}{s} \times 325 \; \text{min} \times \frac{60 \; s}{\text{min}} \times \frac{1 \; \text{mol} \; e^-}{96{,}500 \; C} = 4.04 \; \text{mol} \; e^-$$

$$4.04 \; \text{mol} \; e^- \times \frac{1 \; \text{mol M}}{2 \; \text{mol} \; e^-} = 2.02 \; \text{mol M}$$

$$\text{molar mass} = \frac{111 \; \text{g M}}{2.02 \; \text{mol M}} = 54.9 \; \text{g/mol}, \; M^{2+} = Mn^{2+}$$

19.144 $\text{volume} = \left(0.0100 \; \text{mm} \times \dfrac{1 \; cm}{10 \; mm}\right)(10.0 \; cm)^2 = 0.100 \; cm^3$

$$\text{mol} \; Al_2O_3 = (0.100 \; cm^3)(3.97 \; g/cm^3) \frac{1 \; \text{mol} \; Al_2O_3}{102.0 \; \text{g} \; Al_2O_3} = 3.892 \times 10^{-3} \; \text{mol} \; Al_2O_3$$

$$\text{mole} \; e^- = 3.892 \times 10^{-3} \; \text{mol} \; Al_2O_3 \times \frac{6 \; \text{mol} \; e^-}{1 \; \text{mol} \; Al_2O_3} = 0.02335 \; \text{mol} \; e^-$$

$$\text{coulombs} = 0.02335 \; \text{mol} \; e^- \times \frac{96{,}500 \; C}{1 \; \text{mol} \; e^-} = 2253 \; C$$

$$\text{time} = \frac{C}{A} = \frac{2253 \; C}{0.600 \; C/s} \times \frac{1 \; \text{min}}{60 \; s} = 62.6 \; \text{min}$$

19.146 $PbSO_4(s) + 2\,H_2O(l) \rightarrow PbO_2(s) + 4\,H^+(aq) + SO_4^-(aq) + 2\,e^-$
$\underline{PbSO_4(s) + 2\,e^- \rightarrow Pb(s) + SO_4^-(aq)}$
$2\,PbSO_4(s) + 2\,H_2O(l) \rightarrow Pb(s) + PbO_2(s) + 2\,H^+(aq) + 2\,HSO_4^-(aq)$

(a) The reaction represents an electrolytic cell.
(b) During the recharging (electrolysis) 250.0 g of $PbSO_4$ are oxidized to PbO_2 and 250.0 g of $PbSO_4$ are reduced to Pb.

$$\text{mol } PbSO_4 = 250.0 \text{ g } PbSO_4 \times \frac{1 \text{ mol } PbSO_4}{303.3 \text{ g } PbSO_4} = 0.824 \text{ mol } PbSO_4$$

$$\text{mole } e^- = 0.824 \text{ mol } PbSO_4 \times \frac{2 \text{ mol } e^-}{1 \text{ mol } PbSO_4} = 1.648 \text{ mol } e^-$$

$$\text{coulombs} = 1.648 \text{ mol } e^- \times \frac{96{,}500 \text{ C}}{1 \text{ mol } e^-} = 1.591 \times 10^5 \text{ C}$$

(c) $\text{time} = \dfrac{C}{A} = \dfrac{1.591 \times 10^5 \text{ C}}{500 \text{ C/s}} \times \dfrac{1 \text{ min}}{60 \text{ s}} = 5.3 \text{ min}$

Multiconcept Problems

19.148 (a) Ag^+ is the strongest oxidizing agent because Ag^+ has the most positive standard reduction potential.
Pb is the strongest reducing agent because Pb^{2+} has the most negative standard reduction potential.

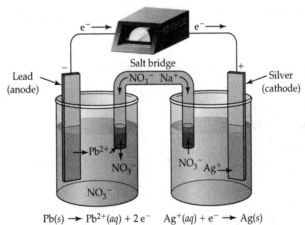

(b) $Pb(s) \rightarrow Pb^{2+}(aq) + 2\,e^-$ $Ag^+(aq) + e^- \rightarrow Ag(s)$

(c) $Pb(s) + 2\,Ag^+(aq) \rightarrow Pb^{2+}(aq) + 2\,Ag(s)$; $n = 2 \text{ mol } e^-$
$E^\circ = E^\circ_{ox} + E^\circ_{red} = 0.13 \text{ V} + 0.80 \text{ V} = 0.93 \text{ V}$

$$\Delta G^\circ = -nFE^\circ = -(2 \text{ mol } e^-)\left(\frac{96{,}500 \text{ C}}{1 \text{ mol } e^-}\right)(0.93 \text{ V})\left(\frac{1 \text{ J}}{1 \text{ C} \cdot \text{V}}\right) = -179{,}490 \text{ J} = -180 \text{ kJ}$$

$E^\circ = \dfrac{0.0592 \text{ V}}{n} \log K$; $\log K = \dfrac{nE^\circ}{0.0592 \text{ V}} = \dfrac{(2)(0.93 \text{ V})}{0.0592 \text{ V}} = 31$; $K = 10^{31}$

(d) $E = E^\circ - \dfrac{0.0592 \text{ V}}{n} \log \dfrac{[Pb^{2+}]}{[Ag^+]^2} = 0.93 \text{ V} - \dfrac{0.0592 \text{ V}}{2} \log\left(\dfrac{0.01}{(0.01)^2}\right) = 0.87 \text{ V}$

19.150 (a) $5 \times [2\ Cl^-(aq) \rightarrow Cl_2(g) + 2\ e^-]$ $\hspace{1cm}$ $E° = -1.36$ V
$\hspace{1.6cm}$ $2 \times [MnO_4^-(aq) + 8\ H^+(aq) + 5\ e^- \rightarrow Mn^{2+}(aq) + 4\ H_2O(l)]$ $\hspace{1cm}$ $E° = 1.51$ V
$\hspace{1.6cm}$ $10\ Cl^-(aq) + 2\ MnO_4^-(aq) + 16\ H^+(aq) \rightarrow 5\ Cl_2(g) + 2\ Mn^{2+}(aq) + 8\ H_2O(l)$ $E° = 0.15$ V

$\hspace{0.8cm}$ (b) $E° = 0.15$ V

$$\Delta G° = -nFE° = -(10\ \text{mol e}^-)\left(\frac{96{,}500\ C}{1\ \text{mol e}^-}\right)(0.15\ V)\left(\frac{1\ J}{1\ C \cdot V}\right) = -144{,}750\ J = -1.4 \times 10^2\ kJ$$

$\hspace{0.8cm}$ (c) $KMnO_4$, 158.0

$$179\ g\ KMnO_4 \times \frac{1\ \text{mol}\ KMnO_4}{158.0\ g\ KMnO_4} \times \frac{5\ \text{mol}\ Cl_2}{2\ \text{mol}\ KMnO_4} = 2.83\ \text{mol}\ Cl_2$$

$$PV = nRT; \quad V = \frac{nRT}{P} = \frac{(2.83\ \text{mol})\left(0.082\ 06\ \dfrac{L \cdot atm}{K \cdot mol}\right)(298\ K)}{1.0\ atm} = 69\ L$$

19.152 (a) $\Delta G° = -nFE°$
$\hspace{1.2cm}$ $\Delta G°_3 = \Delta G°_1 + \Delta G°_2$ therefore $-n_3FE°_3 = -n_1FE°_1 + (-n_2FE°_2)$
$\hspace{1.2cm}$ $n_3E°_3 = n_1E°_1 + n_2E°_2$

$$E°_3 = \frac{n_1 E°_1 + n_2 E°_2}{n_3}$$

$\hspace{0.8cm}$ (b) $E°_3 = \dfrac{(3)(-0.04\ V) + (2)(0.45\ V)}{1} = 0.78$ V

$\hspace{0.8cm}$ (c) $E°$ values would be additive ($E°_3 = E°_1 + E°_2$) if reaction (3) is an overall cell reaction because the electrons in the two half reactions, (1) and (2), cancel. That is, $n_1 = n_2 = n_3$ in the equation for $E°_3$.

19.154 $H_2MoO_4(aq) + As(s) \rightarrow Mo^{3+}(aq) + H_3AsO_4(aq)$

$\hspace{1cm}$ $H_2MoO_4(aq) \rightarrow Mo^{3+}(aq)$
$\hspace{1cm}$ $H_2MoO_4(aq) \rightarrow Mo^{3+}(aq) + 4\ H_2O(l)$
$\hspace{1cm}$ $6\ H^+(aq) + H_2MoO_4(aq) \rightarrow Mo^{3+}(aq) + 4\ H_2O(l)$
$\hspace{1cm}$ $[3\ e^- + 6\ H^+(aq) + H_2MoO_4(aq) \rightarrow Mo^{3+}(aq) + 4\ H_2O(l)] \times 5$
$\hspace{3.2cm}$ (reduction half reaction)

$\hspace{1cm}$ $As(s) \rightarrow H_3AsO_4(aq)$
$\hspace{1cm}$ $As(s) + 4\ H_2O(l) \rightarrow H_3AsO_4(aq)$
$\hspace{1cm}$ $As(s) + 4\ H_2O(l) \rightarrow H_3AsO_4(aq) + 5\ H^+(aq)$
$\hspace{1cm}$ $[As(s) + 4\ H_2O(l) \rightarrow H_3AsO_4(aq) + 5\ H^+(aq) + 5\ e^-] \times 3$ (oxidation half reaction)

$\hspace{0.6cm}$ Combine the two half reactions.
$\hspace{1cm}$ $30\ H^+(aq) + 5\ H_2MoO_4(aq) + 3\ As(s) + 12\ H_2O(l) \rightarrow$
$\hspace{3cm}$ $5\ Mo^{3+}(aq) + 3\ H_3AsO_4(aq) + 15\ H^+(aq) + 20\ H_2O(l)$
$\hspace{1cm}$ $15\ H^+(aq) + 5\ H_2MoO_4(aq) + 3\ As(s) \rightarrow 5\ Mo^{3+}(aq) + 3\ H_3AsO_4(aq) + 8\ H_2O(l)$

$\hspace{1cm}$ $5 \times [H_2MoO_4(aq) + 2\ H^+(aq) + 2\ e^- \rightarrow MoO_2(s) + 2\ H_2O(l)]$ $\hspace{0.6cm}$ $E° = +0.646$ V
$\hspace{1cm}$ $5 \times [MoO_2(s) + 4\ H^+(aq) + e^- \rightarrow Mo^{3+}(aq) + 2\ H_2O(l)]$ $\hspace{0.6cm}$ $E° = -0.008$ V

$$3 \times [As(s) + 3\,H_2O(l) \rightarrow H_3AsO_3(aq) + 3\,H^+(aq) + 3\,e^-] \qquad E° = -0.240\ V$$
$$3 \times [H_3AsO_3(aq) + H_2O(l) \rightarrow H_3AsO_4(aq) + 2\,H^+(aq) + 2\,e^-] \qquad E° = -0.560\ V$$

$$15\,H^+(aq) + 5\,H_2MoO_4(aq) + 3\,As(s) \rightarrow 5\,Mo^{3+}(aq) + 3\,H_3AsO_4(aq) + 8\,H_2O(l)$$

$$\Delta G° = -nFE° = -(10\ \text{mol e}^-)\left(\frac{96{,}500\ C}{1\ \text{mol e}^-}\right)(0.646\ V)\left(\frac{1\ J}{1\ C \cdot V}\right) = -623{,}390\ J = -623.4\ kJ$$

$$\Delta G° = -nFE° = -(5\ \text{mol e}^-)\left(\frac{96{,}500\ C}{1\ \text{mol e}^-}\right)(-0.008\ V)\left(\frac{1\ J}{1\ C \cdot V}\right) = 3{,}860\ J = +3.9\ kJ$$

$$\Delta G° = -nFE° = -(9\ \text{mol e}^-)\left(\frac{96{,}500\ C}{1\ \text{mol e}^-}\right)(-0.240\ V)\left(\frac{1\ J}{1\ C \cdot V}\right) = 208{,}440\ J = +208.4\ kJ$$

$$\Delta G° = -nFE° = -(6\ \text{mol e}^-)\left(\frac{96{,}500\ C}{1\ \text{mol e}^-}\right)(-0.560\ V)\left(\frac{1\ J}{1\ C \cdot V}\right) = 324{,}240\ J = +324.2\ kJ$$

$$\Delta G°(\text{total}) = -623.4\ kJ + 3.9\ kJ + 208.4\ kJ + 324.2\ kJ = -86.9\ kJ = -86{,}900\ J$$

$$1\ V = 1\ J/C$$

$$\Delta G° = -nFE°;\quad E° = \frac{-\Delta G°}{nF} = \frac{-(-86{,}900\ J)}{(15\ \text{mol e}^-)\left(\dfrac{96{,}500\ C}{1\ \text{mol e}^-}\right)} = +0.060\ J/C = +0.060\ V$$

19.156 $\quad Ba(s) + Cl_2(g) \rightarrow BaCl_2(s) \qquad\qquad \Delta G°_f = -806.7\ kJ/mol$

$\qquad\quad \underline{BaCl_2(s) \rightarrow Ba^{2+}(aq) + 2\,Cl^-(aq)} \qquad \underline{\Delta G°_1 = -16.7\ kJ/mol}$

$\qquad\quad Ba(s) + Cl_2(g) \rightarrow Ba^{2+}(aq) + 2\,Cl^-(aq) \quad \Delta G°_2 = \Delta G°_f + \Delta G°_1 = -823.4\ kJ/mol$

$$E° = \frac{-\Delta G°}{nF} = \frac{-(-823{,}400\ J)}{(2\ \text{mol e}^-)\left(\dfrac{96{,}500\ C}{1\ \text{mol e}^-}\right)} = +4.266\ J/C = +4.266\ V$$

$\qquad\quad Ba(s) + Cl_2(g) \rightarrow Ba^{2+}(aq) + 2\,Cl^-(aq) \quad E° = +4.266\ V$

$\qquad\quad \underline{2\,Cl^-(aq) \rightarrow Cl_2(g) + 2\,e^-} \qquad\qquad\quad \underline{E° = -1.36\ V}$

$\qquad\quad Ba(s) \rightarrow Ba^{2+}(aq) + 2\,e^- \qquad\qquad\qquad E° = \ \ \ 2.91\ V$

$\qquad\quad Ba^{2+}(aq) + 2\,e^- \rightarrow Ba(s) \qquad\qquad\quad E° = -2.91\ V$

19.158 $\quad$ (a) $\ Cr_2O_7^{2-}(aq) + 6\,Fe^{2+}(aq) + 14\,H^+(aq) \rightarrow 2\,Cr^{3+}(aq) + 6\,Fe^{3+}(aq) + 7\,H_2O(l)$

$\qquad\quad$ (b) The two half reactions are:

$\qquad\quad$ oxidation: $\quad Fe^{2+}(aq) \rightarrow Fe^{3+}(aq) + e^- \qquad\qquad\qquad\qquad E° = -0.77\ V$

$\qquad\quad$ reduction: $\quad Cr_2O_7^{2-}(aq) + 14\,H^+(aq) + 6\,e^- \rightarrow 2\,Cr^{3+}(aq) + 7\,H_2O(l)\ \ E° = \ 1.36\ V$

$\qquad\quad$ At the equivalence point the potential is given by either of the following expressions:

$$(1)\quad E = 1.36\ V - \frac{0.0592\ V}{6}\log\frac{[Cr^{3+}]^2}{[Cr_2O_7^{2-}][H^+]^{14}}$$

$$(2)\quad E = 0.77\ V - \frac{0.0592\ V}{1}\log\frac{[Fe^{2+}]}{[Fe^{3+}]}$$

where E is the same in both because equilibrium is reached and the solution can have only one potential. Multiplying (1) by 6, adding it to (2), and using some stoichiometric relationships at the equivalence point will simplify the log term.

$$7E = [(6 \times 1.36 \text{ V}) + 0.77 \text{ V}] - (0.0592 \text{ V})\log \frac{[Fe^{2+}][Cr^{3+}]^2}{[Fe^{3+}][Cr_2O_7^{2-}][H^+]^{14}}$$

At the equivalence point, $[Fe^{2+}] = 6[Cr_2O_7^{2-}]$ and $[Fe^{3+}] = 3[Cr^{3+}]$. Substitute these equalities into the previous equation.

$$7E = [(6 \times 1.36 \text{ V}) + 0.77 \text{ V}] - (0.0592 \text{ V})\log \frac{6[Cr_2O_7^{2-}][Cr^{3+}]^2}{3[Cr^{3+}][Cr_2O_7^{2-}][H^+]^{14}}$$

Cancel identical terms.

$$7E = [(6 \times 1.36 \text{ V}) + 0.77 \text{ V}] - (0.0592 \text{ V})\log \frac{6[Cr^{3+}]}{3[H^+]^{14}}$$

$$\text{mol Fe}^{2+} = (0.120 \text{ L})(0.100 \text{ mol/L}) = 0.0120 \text{ mol Fe}^{2+}$$

$$\text{mol Cr}_2O_7^{2-} = 0.0120 \text{ mol Fe}^{2+} \times \frac{1 \text{ mol Cr}_2O_7^{2-}}{6 \text{ mol Fe}^{2+}} = 0.002\ 00 \text{ mol Cr}_2O_7^{2-}$$

$$\text{volume Cr}_2O_7^{2-} = 0.002\ 00 \text{ mol} \times \frac{1 \text{ L}}{0.120 \text{ mol}} = 0.0167 \text{ L}$$

At the equivalence point assume mol Fe^{3+} = initial mol Fe^{2+} = 0.0120 mol
Total volume at the equivalence point is 0.120 L + 0.0167 L = 0.1367 L

$$[Fe^{3+}] = \frac{0.0120 \text{ mol}}{0.1367 \text{ L}} = 0.0878 \text{ M}; \quad [Cr^{3+}] = [Fe^{3+}]/3 = (0.0878 \text{ M})/3 = 0.0293 \text{ M}$$

$$[H^+] = 10^{-pH} = 10^{-2.00} = 0.010 \text{ M}$$

$$7E = [(6 \times 1.36 \text{ V}) + 0.77 \text{ V}] - (0.0592 \text{ V})\log \frac{6(0.0293)}{3(0.010)^{14}} = 8.93 - 1.585 = 7.345 \text{ V}$$

$$E = \frac{7.345 \text{ V}}{7} = 1.05 \text{ V at the equivalence point.}$$

19.160 (a) anode: $Fe(s) + 2 OH^-(aq) \rightarrow Fe(OH)_2(s) + 2 e^-$
      cathode: $\underline{2 \times [NiO(OH)(s) + H_2O(l) + e^- \rightarrow Ni(OH)_2(s) + OH^-(aq)]}$
      overall: $Fe(s) + 2 NiO(OH)(s) + 2 H_2O(l) \rightarrow Fe(OH)_2(s) + 2 Ni(OH)_2(s)$

(b)

$$\Delta G° = -nFE° = -(2 \text{ mol e}^-)\left(\frac{96,500 \text{ C}}{1 \text{ mol e}^-}\right)(1.37 \text{ V})\left(\frac{1 \text{ J}}{1 \text{ C} \cdot \text{V}}\right) = -264,410 \text{ J} = -264 \text{ kJ}$$

$$E° = \frac{0.0592 \text{ V}}{n} \log K; \quad \log K = \frac{nE°}{0.0592 \text{ V}} = \frac{(2)(1.37 \text{ V})}{0.0592 \text{ V}} = 46.3$$

$$K = 10^{46.3} = 2 \times 10^{46}$$

(c) It would still be 1.37 V because OH^- does not appear in the overall cell reaction. The overall cell reaction contains only solids and one liquid, therefore the cell voltage does not change because there are no concentration changes.

(d) $Fe(OH)_2$, 89.86; 1 A = 1 C/s

$$\text{mol e}^- = (0.250 \text{ C/s})(40.0 \text{ min})\left(\frac{60 \text{ s}}{1 \text{ min}}\right)\left(\frac{1 \text{ mol e}^-}{96,500 \text{ C}}\right) = 6.22 \times 10^{-3} \text{ mol e}^-$$

$$\text{mass Fe(OH)}_2 = (6.22 \times 10^{-3} \text{ mol e}^-) \times \frac{1 \text{ mol Fe(OH)}_2}{2 \text{ mol e}^-} \times \frac{89.86 \text{ g Fe(OH)}_2}{1 \text{ mol Fe(OH)}_2} = 0.279 \text{ g}$$

H_2O molecules consumed =

$$(6.22 \times 10^{-3} \text{ mol e}^-) \times \frac{2 \text{ mol H}_2\text{O}}{2 \text{ mol e}^-} \times \frac{6.022 \times 10^{23} \text{ H}_2\text{O molecules}}{1 \text{ mol H}_2\text{O}} = 3.75 \times 10^{21} \text{ H}_2\text{O molecules}$$

19.162 (a) cathode:

(1) $\text{MnO}_2(s) + 4 \text{ H}^+(aq) + 2 \text{ e}^- \rightarrow \text{Mn}^{2+}(aq) + 2 \text{ H}_2\text{O}(l)$ $\quad\quad$ E° = +1.22 V

(2) $\text{Mn(OH)}_2(s) + \text{OH}^-(aq) \rightarrow \text{MnO(OH)}(s) + \text{H}_2\text{O}(l) + \text{e}^-$ $\quad$ E° = +0.380 V

(3) $\text{Mn}^{2+}(aq) + 2 \text{ OH}^-(aq) \rightarrow \text{Mn(OH)}_2(s)$ $\quad$ $K = 1/K_{sp} = 1/(2.1 \times 10^{-13}) = 4.8 \times 10^{12}$

(4) $\underline{4 \times [\text{H}_2\text{O}(l) \rightarrow \text{H}^+(aq) + \text{OH}^-(aq)]}$ $\quad$ $K = (K_w)^4 = (1.0 \times 10^{-14})^4 = 1.0 \times 10^{-56}$

$\quad\quad$ $\text{MnO}_2(s) + \text{H}_2\text{O}(l) + \text{e}^- \rightarrow \text{MnO(OH)}(s) + \text{OH}^-(aq)$

$$\Delta G°_1 = -nFE° = -(2 \text{ mol e}^-)\left(\frac{96,500 \text{ C}}{1 \text{ mol e}^-}\right)(1.22 \text{ V})\left(\frac{1 \text{ J}}{1 \text{ C} \cdot \text{V}}\right) = -235,460 \text{ J} = -235.5 \text{ kJ}$$

$$\Delta G°_2 = -nFE° = -(1 \text{ mol e}^-)\left(\frac{96,500 \text{ C}}{1 \text{ mol e}^-}\right)(0.380 \text{ V})\left(\frac{1 \text{ J}}{1 \text{ C} \cdot \text{V}}\right) = -36,670 \text{ J} = -36.7 \text{ kJ}$$

$$\Delta G°_3 = -RT \ln K = -(8.314 \times 10^{-3} \text{ kJ/K})(298 \text{ K}) \ln (4.8 \times 10^{12}) = -72.3 \text{ kJ}$$

$$\Delta G°_4 = -RT \ln K = -(8.314 \times 10^{-3} \text{ kJ/K})(298 \text{ K}) \ln (1.0 \times 10^{-56}) = +319.5 \text{ kJ}$$

$\Delta G°(\text{total}) = -235.5 \text{ kJ} - 36.7 \text{ kJ} - 72.3 \text{ kJ} + 319.5 \text{ kJ} = -25.0 \text{ kJ} = -25,000 \text{ J}$

1 V = 1 J/C

$$\Delta G° = -nFE°; \quad E° = \frac{-\Delta G°}{nF} = \frac{-(-25,000 \text{ J})}{(1 \text{ mol e}^-)\left(\frac{96,500 \text{ C}}{1 \text{ mol e}^-}\right)} = +0.259 \text{ J/C} = +0.259 \text{ V}$$

$E°_{cathode} = +0.259 \text{ V}$

anode:

(1) $\text{Zn}(s) \rightarrow \text{Zn}^{2+}(aq) + 2 \text{ e}^-$ $\quad\quad\quad\quad$ E° = +0.76 V

(2) $\underline{\text{Zn}^{2+}(aq) + 2 \text{ OH}^-(aq) \rightarrow \text{Zn(OH)}_2(s)}$ $\quad$ $K = 1/K_{sp} = 1/(4.1 \times 10^{-17}) = 2.4 \times 10^{16}$

$\quad\quad$ $\text{Zn}(s) + 2 \text{ OH}^-(aq) \rightarrow \text{Zn(OH)}_2(s) + 2 \text{ e}^-$

$$\Delta G°_1 = -nFE° = -(2 \text{ mol e}^-)\left(\frac{96,500 \text{ C}}{1 \text{ mol e}^-}\right)(0.76 \text{ V})\left(\frac{1 \text{ J}}{1 \text{ C} \cdot \text{V}}\right) = -146,680 \text{ J} = -146.7 \text{ kJ}$$

$$\Delta G°_2 = -RT \ln K = -(8.314 \times 10^{-3} \text{ kJ/K})(298 \text{ K}) \ln (2.4 \times 10^{16}) = -93.4 \text{ kJ}$$

$\Delta G°(\text{total}) = -146.7 \text{ kJ} - 93.4 \text{ kJ} = -240.1 \text{ kJ} = -240,100 \text{ J}$

1 V = 1 J/C

$$\Delta G^\circ = -nFE^\circ; \quad E^\circ = \frac{-\Delta G^\circ}{nF} = \frac{-(-240{,}100 \text{ J})}{(2 \text{ mol e}^-)\left(\dfrac{96{,}500 \text{ C}}{1 \text{ mol e}^-}\right)} = +1.24 \text{ J/C} = +1.24 \text{ V}$$

$E^\circ_{\text{anode}} = +1.24 \text{ V}$

$E^\circ_{\text{cell}} = E^\circ_{\text{cathode}} + E^\circ_{\text{anode}} = 0.259 \text{ V} + 1.24 \text{ V} = 1.50 \text{ V}$

(b) $FeO_4^{2-}(aq) \rightarrow Fe(OH)_3(s)$

$FeO_4^{2-}(aq) \rightarrow Fe(OH)_3(s) + H_2O(l)$

$FeO_4^{2-}(aq) + 5 H^+(aq) \rightarrow Fe(OH)_3(s) + H_2O(l)$

$FeO_4^{2-}(aq) + 5 H^+(aq) + 3 e^- \rightarrow Fe(OH)_3(s) + H_2O(l)$

$FeO_4^{2-}(aq) + 5 H^+(aq) + 5 OH^-(aq) + 3 e^- \rightarrow Fe(OH)_3(s) + H_2O(l) + 5 OH^-(aq)$

$FeO_4^{2-}(aq) + 5 H_2O(l) + 3 e^- \rightarrow Fe(OH)_3(s) + H_2O(l) + 5 OH^-(aq)$

$FeO_4^{2-}(aq) + 4 H_2O(l) + 3 e^- \rightarrow Fe(OH)_3(s) + 5 OH^-(aq)$

(c) K_2FeO_4, 198.04; MnO_2, 86.94

$$\text{coulombs} = 10.00 \text{ g } K_2FeO_4 \times \frac{1 \text{ mol } K_2FeO_4}{198.04 \text{ g } K_2FeO_4} \times \frac{3 \text{ mol e}^-}{1 \text{ mol } K_2FeO_4} \times \frac{96{,}500 \text{ C}}{1 \text{ mol e}^-} =$$

1.46×10^4 C from 10.00 g K_2FeO_4

$$\text{coulombs} = 10.00 \text{ g } MnO_2 \times \frac{1 \text{ mol } MnO_2}{86.94 \text{ g } MnO_2} \times \frac{1 \text{ mol e}^-}{1 \text{ mol } MnO_2} \times \frac{96{,}500 \text{ C}}{1 \text{ mol e}^-} =$$

1.11×10^4 C from 10.00 g MnO_2

19.164 The overall cell reaction is:

$2 Fe^{3+}(aq) + 2 Hg(l) + 2 Cl^-(aq) \rightarrow 2 Fe^{2+}(aq) + Hg_2Cl_2(s)$

The Nernst equation can be applied to separate half reactions.

One half reaction is for the calomel reference electrode.

$2 Hg(l) + 2 Cl^-(aq) \rightarrow Hg_2Cl_2(s) + 2 e^- \qquad E^\circ = -0.28 \text{ V}$

When $[Cl^-] = 2.9$ M,

$$E_{\text{calomel}} = E^\circ - \frac{0.0592 \text{ V}}{n} \log \frac{1}{[Cl^-]^2} = -0.28 \text{ V} - \frac{0.0592 \text{ V}}{2} \log \frac{1}{(2.9)^2} = -0.25 \text{ V}$$

Balance the titration redox reaction: $MnO_4^-(aq) + Fe^{2+}(aq) \rightarrow Mn^{2+}(aq) + Fe^{3+}(aq)$

$[Fe^{2+}(aq) \rightarrow Fe^{3+}(aq) + e^-] \times 5$

$MnO_4^-(aq) \rightarrow Mn^{2+}(aq)$

$MnO_4^-(aq) \rightarrow Mn^{2+}(aq) + 4 H_2O(l)$

$MnO_4^-(aq) + 8 H^+(aq) \rightarrow Mn^{2+}(aq) + 4 H_2O(l)$

$MnO_4^-(aq) + 8 H^+(aq) + 5 e^- \rightarrow Mn^{2+}(aq) + 4 H_2O(l)$

Combine the two half reactions.

$MnO_4^-(aq) + 5 Fe^{2+}(aq) + 8 H^+(aq) \rightarrow Mn^{2+}(aq) + 5 Fe^{3+}(aq) + 4 H_2O(l)$

initial mol $Fe^{2+} = (0.010 \text{ mol/L})(0.1000 \text{ L}) = 0.0010 \text{ mol } Fe^{2+}$

mL MnO_4^- needed to reach endpoint =

$$0.0010 \text{ mol Fe}^{2+} \text{ x } \frac{1 \text{ mol MnO}_4^-}{5 \text{ mol Fe}^{2+}} \text{ x } \frac{1.00 \text{ L}}{0.010 \text{ mol MnO}_4^-} \text{ x } \frac{1000 \text{ mL}}{1.00 \text{ L}} = 20.0 \text{ mL}$$

(a) initial mol Fe^{2+} = (0.010 mol/L)(0.1000 L) = 0.0010 mol Fe^{2+}

mol MnO_4^- in 5.0 mL = (0.010 mol/L)(0.0050 L) = 0.000 050 mol MnO_4^-

$$MnO_4^-(aq) + 5 \text{ Fe}^{2+}(aq) + 8 \text{ H}^+(aq) \rightarrow Mn^{2+}(aq) + 5 \text{ Fe}^{3+}(aq) + 4 \text{ H}_2O(l)$$

before (mol)	0.000 050	0.0010		0
change (mol)	−0.000 050	−5(0.000 050)		+5(0.000 050)
after (mol)	0	0.000 75		0.000 25

Again, the Nernst equation can be applied to separate half reactions.

The other half reaction is: $Fe^{3+}(aq) + e^- \rightarrow 2 \text{ Fe}^{2+}(aq)$ $E° = +0.77$ V

E for the half reaction after adding 5.0 mL of MnO_4^- is

$$E_{Fe^{3+}/Fe^{2+}} = E° - \frac{0.0592 \text{ V}}{n} \log \frac{[Fe^{2+}]}{[Fe^{3+}]} = 0.77 \text{ V} - \frac{0.0592 \text{ V}}{1} \log \frac{(0.000 \text{ } 75)}{(0.000 \text{ } 25)} = 0.74 \text{ V}$$

(Note in the Nernst equation above, we are taking a ratio of Fe^{2+} to Fe^{3+} so we can ignore volumes and just use moles instead of molarity.)

$E_{cell} = E_{Fe^{3+}/Fe^{2+}} + E_{calomel} = 0.74 \text{ V} + (-0.25 \text{ V}) = 0.49 \text{ V}$

(b) initial mol Fe^{2+} = (0.010 mol/L)(0.1000 L) = 0.0010 mol Fe^{2+}

mol MnO_4^- in 10.0 mL = (0.010 mol/L)(0.0100 L) = 0.000 10 mol MnO_4^-

$$MnO_4^-(aq) + 5 \text{ Fe}^{2+}(aq) + 8 \text{ H}^+(aq) \rightarrow Mn^{2+}(aq) + 5 \text{ Fe}^{3+}(aq) + 4 \text{ H}_2O(l)$$

before (mol)	0.000 10	0.0010		0
change (mol)	−0.000 10	−5(0.000 10)		+5(0.000 10)
after (mol)	0	0.000 50		0.000 50

E for the half reaction after adding 10.0 mL of MnO_4^- is

$$E_{Fe^{3+}/Fe^{2+}} = E° - \frac{0.0592 \text{ V}}{n} \log \frac{[Fe^{2+}]}{[Fe^{3+}]} = 0.77 \text{ V} - \frac{0.0592 \text{ V}}{1} \log \frac{(0.000 \text{ } 50)}{(0.000 \text{ } 50)} = 0.77 \text{ V}$$

$E_{cell} = E_{Fe^{3+}/Fe^{2+}} + E_{calomel} = 0.77 \text{ V} + (-0.25 \text{ V}) = 0.52 \text{ V}$

(c) initial mol Fe^{2+} = (0.010 mol/L)(0.1000 L) = 0.0010 mol Fe^{2+}

mol MnO_4^- in 19.0 mL = (0.010 mol/L)(0.0190 L) = 0.000 19 mol MnO_4^-

$$MnO_4^-(aq) + 5 \text{ Fe}^{2+}(aq) + 8 \text{ H}^+(aq) \rightarrow Mn^{2+}(aq) + 5 \text{ Fe}^{3+}(aq) + 4 \text{ H}_2O(l)$$

before (mol)	0.000 19	0.0010		0
change (mol)	−0.000 19	−5(0.000 19)		+5(0.000 19)
after (mol)	0	0.000 05		0.000 95

E for the half reaction after adding 19.0 mL of MnO_4^- is

$$E_{Fe^{3+}/Fe^{2+}} = E° - \frac{0.0592 \text{ V}}{n} \log \frac{[Fe^{2+}]}{[Fe^{3+}]} = 0.77 \text{ V} - \frac{0.0592 \text{ V}}{1} \log \frac{(0.000 \text{ } 05)}{(0.000 \text{ } 95)} = 0.85 \text{ V}$$

$E_{cell} = E_{Fe^{3+}/Fe^{2+}} + E_{calomel} = 0.85 \text{ V} + (-0.25 \text{ V}) = 0.60 \text{ V}$

(d) 21.0 mL is past the endpoint so the MnO_4^- is in excess and all of the Fe^{2+} is consumed.

initial mol Fe^{2+} = (0.010 mol/L)(0.1000 L) = 0.0010 mol Fe^{2+}

mol MnO_4^- in 21.0 mL = (0.010 mol/L)(0.0210 L) = 0.000 21 mol MnO_4^-

$$MnO_4^-(aq) + 5\ Fe^{2+}(aq) + 8\ H^+(aq) \rightarrow Mn^{2+}(aq) + 5\ Fe^{3+}(aq) + 4\ H_2O(l)$$

before (mol)	0.000 21	0.0010		0	0
change (mol)	−0.000 20	−5(0.000 20)		+0.000 20	+5(0.000 20)
after (mol)	0.000 01	0		0.000 20	0.0010

Because the Fe^{2+} is totally consumed, there is a new half reaction:

$$MnO_4^-(aq) + 8\ H^+(aq) + 5\ e^- \rightarrow Mn^{2+}(aq) + 4\ H_2O(l) \qquad E° = 1.51\ V$$

The total volume = 100.0 mL + 21.0 mL = 121.0 mL = 0.1210 L

$[MnO_4^-]$ = 0.000 01 mol/0.1210 L = 0.000 083 M

$[Mn^{2+}]$ = 0.000 20 mol/0.1210 L = 0.001 65 M

We need to determine $[H^+]$ in order to determine the half reaction potential.

$[H_2SO_4]_{dil} \cdot 121.0\ mL = [H_2SO_4]_{conc} \cdot 100.0\ mL$

$[H_2SO_4]_{dil}$ = [(1.50 M)(100.0 mL)]/121.0 mL = 1.24 M

We can ignore the small amount of H^+ consumed by the titration itself, because the H_2SO_4 concentration is so large.

Consider the dissociation of H_2SO_4. From the complete dissociation of the first proton, $[H^+] = [HSO_4^-]$ = 1.24 M.

For the dissociation of the second proton, the following equilibrium must be considered:

$$HSO_4^-(aq) \rightleftharpoons H^+(aq) + SO_4^{2-}(aq)$$

initial (M)	1.24	1.24	0
change (M)	−x	+x	+x
equil (M)	1.24 − x	1.24 + x	x

$$K_{a2} = \frac{[H^+][SO_4^{2-}]}{[HSO_4^-]} = 1.2 \times 10^{-2} = \frac{(1.24 + x)(x)}{1.24 - x}$$

$x^2 + 1.252x - 0.0149 = 0$

Use the quadratic formula to solve for x.

$$x = \frac{-(1.252) \pm \sqrt{(1.252)^2 - 4(1)(-0.0149)}}{2(1)} = \frac{-1.252 \pm 1.276}{2}$$

x = −1.264 and 0.012

Of the two solutions for x, only the positive value of x has physical meaning, since x is the $[SO_4^{2-}]$.

$[H^+]$ = 1.24 + x = 1.24 + 0.012 = 1.25 M

$$E_{MnO_4^-/Mn^{2+}} = E° - \frac{0.0592\ V}{n} \log \frac{[Mn^{2+}]}{[MnO_4^-][H^+]^8}$$

$$= 1.51\ V - \frac{0.0592\ V}{5} \log \frac{(0.001\ 65)}{(0.000\ 083)(1.25)^8} = 1.50\ V$$

$E_{cell} = E_{MnO_4^-/Mn^{2+}} + E_{calomel}$ = 1.50 V + (−0.25 V) = 1.25 V

Notice that there is a dramatic change in the potential at the equivalence point.

20 Nuclear Chemistry

20.1   (a) In beta emission, the mass number is unchanged, and the atomic number increases by one. $^{106}_{44}Ru \rightarrow \ ^{0}_{-1}e \ + \ ^{106}_{45}Rh$

      (b) In alpha emission, the mass number decreases by four, and the atomic number decreases by two. $^{189}_{83}Bi \rightarrow \ ^{4}_{2}He \ + \ ^{185}_{81}Tl$

      (c) In electron capture, the mass number is unchanged, and the atomic number decreases by one. $^{204}_{84}Po \ + \ ^{0}_{-1}e \rightarrow \ ^{204}_{83}Bi$

20.2 $^{148}_{69}Tm$ decays to $^{148}_{68}Er$ by either positron emission or electron capture.

20.3 $t_{1/2} = \dfrac{0.693}{k} = \dfrac{0.693}{1.08 \times 10^{-2} \ h^{-1}} = 64.2 \ h$

20.4 $k = \dfrac{0.693}{t_{1/2}} = \dfrac{0.693}{3.82 \ d} = 0.181 \ d^{-1}$

20.5 $\ln\left(\dfrac{N}{N_0}\right) = -0.693\left(\dfrac{t}{t_{1/2}}\right) = -0.693\left(\dfrac{16{,}230 \ y}{5715 \ y}\right) = -1.968$

      $\dfrac{N}{N_0} = e^{-1.968} = 0.140; \quad \dfrac{N}{100\%} = 0.140; \quad N = 14.0\%$

20.6   Assume $N_0 = 100\%$ and $N = 89.2\%$ at $t = 5.00 \ y$

      $\ln\left(\dfrac{N}{N_0}\right) = (-0.693)\left(\dfrac{t}{t_{1/2}}\right); \qquad \dfrac{N}{N_0} = \dfrac{\text{Decay rate at time } t}{\text{Decay rate at time } t = 0}$

      $\ln\left(\dfrac{89.2}{100}\right) = (-0.693)\left(\dfrac{5.00 \ y}{t_{1/2}}\right); \qquad t_{1/2} = 30.3 \ y$

20.7 $\ln\left(\dfrac{N}{N_0}\right) = (-0.693)\left(\dfrac{t}{t_{1/2}}\right); \qquad \dfrac{N}{N_0} = \dfrac{\text{Decay rate at time } t}{\text{Decay rate at } t = 0}$

      $\ln\left(\dfrac{10{,}860}{16{,}800}\right) = (-0.693)\left(\dfrac{28.0 \ d}{t_{1/2}}\right); \qquad t_{1/2} = 44.5 \ d$

20.8 $\ln\left(\dfrac{N}{N_0}\right) = (-0.693)\left(\dfrac{t}{t_{1/2}}\right)$; $\qquad \dfrac{N}{N_0} = \dfrac{\text{Decay rate at time t}}{\text{Decay rate at } t = 0}$

$t_{1/2} = 44.5 \text{ d}$

$\ln\left(\dfrac{N}{16{,}800}\right) = (-0.693)\left(\dfrac{40.0 \text{ d}}{44.5 \text{ d}}\right)$

$\ln N - \ln(16{,}800) = (-0.693)\left(\dfrac{40.0 \text{ d}}{44.5 \text{ d}}\right)$

$\ln N = (-0.693)\left(\dfrac{40.0 \text{ d}}{44.5 \text{ d}}\right) + \ln(16{,}800) = 9.106$

$N = e^{9.106} = 9011 \text{ disintegrations/min}$

20.9 $\ln\left(\dfrac{N}{N_0}\right) = (-0.693)\left(\dfrac{t}{t_{1/2}}\right)$; $\qquad \dfrac{N}{N_0} = \dfrac{\text{Decay rate at time t}}{\text{Decay rate at } t = 0}$

$\ln\left(\dfrac{2.4}{15.3}\right) = (-0.693)\left(\dfrac{t}{5715 \text{ y}}\right)$; $\qquad t = 1.53 \times 10^4 \text{ y}$

20.10

	^{40}K	$\rightarrow$	^{40}Ar
then (mmol)	N_0		0
change (mmol)	-0.95		$+0.95$
now (mmol)	$N = 1.20$		0.95

$N_0 - 0.95 = 1.20 \text{ mmol therefore, } N_0 = 1.20 + 0.95 = 2.15 \text{ mmol and } N = 1.20 \text{ mmol}$

$\ln\left(\dfrac{N}{N_0}\right) = (-0.693)\left(\dfrac{t}{t_{1/2}}\right)$

$\ln\left(\dfrac{1.20}{2.15}\right) = (-0.693)\left(\dfrac{t}{1.25 \times 10^9 \text{ y}}\right)$; $\qquad t = 1.05 \times 10^9 \text{ y}$

20.11 For $^{16}_{8}O$:

First, calculate the total mass of the nucleons (8 n + 8 p)

Mass of 8 neutrons = (8)(1.008 66) = 8.069 28

Mass of 8 protons = (8)(1.007 28) = 8.058 24

Mass of 8 n + 8 p   = 16.127 52

Next, calculate the mass of a ^{16}O nucleus by subtracting the mass of 8 electrons from the mass of a ^{16}O atom.

Mass of ^{16}O atom   = 15.994 91

−Mass of 8 electrons = −(8)(5.486 x 10^{-4})   = −0.004 39

Mass of ^{16}O nucleus   = 15.990 52

Then subtract the mass of the ^{16}O nucleus from the mass of the nucleons to find the mass defect:

Mass defect = mass of nucleons – mass of nucleus

$\qquad$ = (16.127 52) – (15.990 52) = 0.137 00 u

Mass defect in grams = (0.137 00 u)(1.660 54 x 10^{-24} g/u) = 2.2749 x 10^{-25} g

Mass defect in g/mol = (2.2749 x 10^{-25} g)(6.022 x 10^{23} mol^{-1}) = 0.136 99 g/mol

Now, use the Einstein equation to convert the mass defect into the binding energy.

$\Delta E = \Delta mc^2$ = (0.136 99 g/mol)(10^{-3} kg/g)(3.00 x 10^8 m/s)^2

ΔE = 1.233 x 10^{13} J/mol = 1.233 x 10^{10} kJ/mol

20.12 ^{6}Li atomic mass = (3 proton mass) + (3 neutron mass) + (3 electron mass) – (mass defect)

$\qquad$ = (3 x 1.007 28) + (3 x 1.008 66) + (3 x 5.486 x 10^{-4}) – (0.034 37) = 6.0151 u

Mass defect in g/mol = 0.034 37 g/mol

Now, use the Einstein equation to convert the mass defect into the binding energy.

$\Delta E = \Delta mc^2$ = (0.034 37 g/mol)(10^{-3} kg/g)(3.00 x 10^8 m/s)^2

ΔE = 3.093 x 10^{12} J/mol = 3.093 x 10^9 kJ/mol

$$\Delta E = \frac{3.093 \times 10^{12} \text{ J/mol}}{6.022 \times 10^{23} \text{ nuclei/mol}} \times \frac{1 \text{ MeV}}{1.60 \times 10^{-13} \text{ J}} \times \frac{1 \text{ nucleus}}{6 \text{ nucleons}} = 5.35 \frac{\text{MeV}}{\text{nucleon}}$$

20.13 $$\Delta m = \frac{\Delta E}{c^2} = \frac{-820 \times 10^3 \text{ J}}{(3.00 \times 10^8 \text{ m/s})^2} = \frac{-820 \times 10^3 \text{ kg·m}^2/\text{s}^2}{(3.00 \times 10^8 \text{ m/s})^2} = -9.11 \times 10^{-12} \text{ kg}$$

$$\Delta m = \frac{-9.11 \times 10^{-9} \text{ g}}{2 \text{ mol NaCl}} = 4.56 \times 10^{-9} \text{ g/mol}$$

20.14 $$\Delta m = \frac{\Delta E}{c^2} = \frac{-3.9 \times 10^{10} \text{ J}}{(3.00 \times 10^8 \text{ m/s})^2} = \frac{-3.9 \times 10^{10} \text{ kg·m}^2/\text{s}^2}{(3.00 \times 10^8 \text{ m/s})^2} = -4.3 \times 10^{-7} \text{ kg}$$

Δm = –4.3 x 10^{-4} g

20.15 $^1_0$n + $^{235}_{92}$U $\rightarrow$ $^{137}_{52}$Te + $^{97}_{40}$Zr + 2 $^1_0$n

mass $^{235}_{92}$U	235.0439
mass $^1_0$n	1.008 66
–mass $^{137}_{52}$Te	–136.9254
–mass $^{97}_{40}$Zr	–96.9110
–mass 2 $^1_0$n	–(2)(1.008 66)
mass change	0.1988 u

(0.1988 u)(1.660 54 x 10^{-24} g/u)(6.022 x 10^{23} mol^{-1}) = 0.1988 g/mol

$\Delta E = \Delta mc^2$ = (0.1988 g/mol)(10^{-3} kg/g)(3.00 x 10^8 m/s)^2

ΔE = 1.79 x 10^{13} J/mol = 1.79 x 10^{10} kJ/mol

20.16   (a) $^{238}_{92}U \rightarrow {}^{232}_{90}Th + {}^{4}_{2}He$

(b)

mass ^{238}U	238.0508
mass ^{232}Th	-234.0436
mass ^{4}He	-4.0026
mass change	0.0046 u

$(0.0046 \text{ u/atom})(1.660\ 54 \times 10^{-24} \text{ g/u}) = 7.6 \times 10^{-27} \text{ g/atom}$

$(0.0046 \text{ u})(1.660\ 54 \times 10^{-24} \text{ g/u})(6.022 \times 10^{23} \text{ mol}^{-1}) = 0.0046 \text{ g/mol}$
$\Delta E = \Delta mc^2 = (0.0046 \text{ g/mol})(10^{-3} \text{ kg/g})(3.00 \times 10^8 \text{ m/s})^2$
$\Delta E = 4.1 \times 10^{11} \text{ J/mol} = 4.1 \times 10^8 \text{ kJ/mol}$
(c) Mass is lost and energy is released.

20.17 $^{70}_{30}Zn + {}^{208}_{82}Pb \rightarrow ? + {}^{1}_{0}n; \qquad ? = {}^{70+208-1}_{30+82}X = {}^{277}_{112}Cn$

20.18 $^{238}_{92}U + {}^{12}_{6}C \rightarrow {}^{246}_{98}Cf + 4\,{}^{1}_{0}n$

20.19 $^{24}_{11}Na \rightarrow {}^{24}_{12}Mg + {}^{0}_{-1}e$

20.20   (a) $^{18}_{8}O + {}^{1}_{1}H \rightarrow {}^{18}_{9}F + ?; \quad ? = {}^{18+1-18}_{8+1-9}X = {}^{1}_{0}n$

(b) $^{16}_{8}O + {}^{3}_{2}He \rightarrow {}^{18}_{9}F + ?; \quad ? = {}^{16+3-18}_{8+2-9}X = {}^{0}_{+1}e + {}^{1}_{0}n$

20.21 ^{125}I, $t_{1/2} = 59.5$ days

$k = \dfrac{0.693}{t_{1/2}} = \dfrac{0.693}{59.5 \text{ d}} = 0.0116 \text{ d}^{-1}$

$\ln\left(\dfrac{N}{N_0}\right) = -kt; \quad \dfrac{\ln\left(\dfrac{N}{N_0}\right)}{-k} = t; \quad \text{Let } N_0 = 1 \text{ and } N = 0.1$

$\dfrac{\ln\left(\dfrac{0.1}{1}\right)}{-0.0116 \text{ d}^{-1}} = t = 198.5 \text{ days}$

20.22   First find the activity of the ^{51}Cr after 17.0 days.

$\ln\left(\dfrac{N}{N_0}\right) = (-0.693)\left(\dfrac{t}{t_{1/2}}\right); \qquad \dfrac{N}{N_0} = \dfrac{\text{Decay rate at time t}}{\text{Decay rate at time t} = 0}$

$$\ln\left(\frac{N}{4.10}\right) = (-0.693)\left(\frac{17.0 \text{ d}}{27.7 \text{ d}}\right)$$

$\ln N - \ln(4.10) = -0.4253$

$\ln N = -0.4253 + \ln(4.10) = 0.9857$

$N = e^{0.9857} = 2.68 \ \mu\text{Ci/mL}$

$(20.0 \text{ mL})(2.68 \ \mu\text{Ci/mL}) = (\text{total blood volume})(0.009\ 35 \ \mu\text{Ci/mL})$

total blood volume = 5732 mL = 5.73 L

20.23 $^{10}\text{B} + {}^{1}\text{n} \rightarrow {}^{4}\text{He} + {}^{7}\text{Li} + \gamma$

mass ^{10}B	10.012 937
mass ^{1}n	1.008 665
–mass ^{4}He	– 4.002 603
–mass ^{7}Li	–7.016 004
mass change	0.002 995 u

$(0.002\ 995 \text{ u})(1.660\ 54 \times 10^{-24} \text{ g/u}) = 4.973 \times 10^{-27} \text{ g}$

$\Delta E = \Delta mc^2 = (4.973 \times 10^{-27} \text{ g})(10^{-3} \text{ kg/g})(3.00 \times 10^{8} \text{ m/s})^2 = 4.476 \times 10^{-13} \text{ J}$

$\text{Kinetic energy} = 2.31 \text{ MeV} \times \dfrac{1.60 \times 10^{-13} \text{ J}}{1 \text{ MeV}} = 3.696 \times 10^{-13} \text{ J}$

$\gamma \text{ photon energy} = \Delta E - KE = 4.476 \times 10^{-13} \text{ J} - 3.696 \times 10^{-13} \text{ J} = 7.80 \times 10^{-14} \text{ J}$

$= 7.80 \times 10^{-14} \text{ J} \times \dfrac{1 \text{ MeV}}{1.60 \times 10^{-13} \text{ J}} = 0.488 \text{ MeV}$

20.24   1 Ci $= 3.7 \times 10^{10}$ disintegrations/s and 1.0 rad $= 2.2 \times 10^{11}$ disintegrations of ^{60}Co

$1800 \text{ rad} \times \dfrac{2.2 \times 10^{11} \text{ disintegrations}}{1 \text{ rad}} = 3.96 \times 10^{14} \text{ disintegrations}$

$30 \text{ Ci} = (30)(3.7 \times 10^{10} \text{ disintegrations/s}) = 1.11 \times 10^{12} \text{ disintegrations/s}$

$\dfrac{1.11 \times 10^{12} \text{ disintegrations/s}}{1 \text{ source}} \times 201 \text{ sources} = 2.23 \times 10^{14} \text{ disintegrations/s}$

$\text{time} = \dfrac{3.96 \times 10^{14} \text{ disintegrations}}{2.23 \times 10^{14} \text{ disintegrations/s}} = 1.8 \text{ s}$

Section Problems
Nuclear Reactions and Radioactivity (Sections 20.1–20.2)

20.26   Positron emission is the conversion of a proton in the nucleus into a neutron plus an ejected positron.
Electron capture is the process in which a proton in the nucleus captures an inner-shell electron and is thereby converted into a neutron.

20.28   In beta emission a neutron is converted to a proton and the atomic number increases. In positron emission a proton is converted to a neutron and the atomic number decreases.

20.30   (a) $^{126}_{50}\text{Sn} \rightarrow \, ^{0}_{-1}\text{e} + \, ^{126}_{51}\text{Sb}$ (b) $^{210}_{88}\text{Ra} \rightarrow \, ^{4}_{2}\text{He} + \, ^{206}_{86}\text{Rn}$

(c) $^{77}_{37}\text{Rb} \rightarrow \, ^{0}_{1}\text{e} + \, ^{77}_{36}\text{Kr}$ (d) $^{76}_{36}\text{Kr} + \, ^{0}_{-1}\text{e} \rightarrow \, ^{76}_{35}\text{Br}$

20.32   The mass number decreases by four, and the atomic number decreases by two. This is characteristic of alpha emission. $^{214}_{90}\text{Th} \rightarrow \, ^{210}_{88}\text{Ra} + \, ^{4}_{2}\text{He}$

20.34   (a) $^{188}_{80}\text{Hg} \rightarrow \, ^{188}_{79}\text{Au} + \, ^{0}_{1}\text{e}$ (b) $^{218}_{85}\text{At} \rightarrow \, ^{214}_{83}\text{Bi} + \, ^{4}_{2}\text{He}$

(c) $^{234}_{90}\text{Th} \rightarrow \, ^{234}_{91}\text{Pa} + \, ^{0}_{-1}\text{e}$

20.36   (a) $^{162}_{75}\text{Re} \rightarrow \, ^{158}_{73}\text{Ta} + \, ^{4}_{2}\text{He}$ (b) $^{138}_{62}\text{Sm} + \, ^{0}_{-1}\text{e} \rightarrow \, ^{138}_{61}\text{Pm}$

(c) $^{188}_{74}\text{W} \rightarrow \, ^{188}_{75}\text{Re} + \, ^{0}_{-1}\text{e}$ (d) $^{165}_{73}\text{Ta} \rightarrow \, ^{165}_{72}\text{Hf} + \, ^{0}_{1}\text{e}$

20.38 $^{100}_{43}\text{Tc} \rightarrow \, ^{0}_{1}\text{e} + \, ^{100}_{42}\text{Mo}$ (positron emission)

$^{100}_{43}\text{Tc} + \, ^{0}_{-1}\text{e} \rightarrow \, ^{100}_{42}\text{Mo}$ (electron capture)

Nuclear Stability (Section 20.3)

20.40 ^{160}W is neutron poor and decays by alpha emission. ^{185}W is neutron rich and decays by beta emission.

20.42 $^{196}_{82}\text{Pb}$ is neutron poor and decays by positron emission. $^{206}_{82}\text{Pb}$ is stable.

20.44 $^{241}_{95}\text{Am} \rightarrow \, ^{237}_{93}\text{Np} + \, ^{4}_{2}\text{He}$

$^{237}_{93}\text{Np} \rightarrow \, ^{233}_{91}\text{Pa} + \, ^{4}_{2}\text{He}$

$^{233}_{91}\text{Pa} \rightarrow \, ^{233}_{92}\text{U} + \, ^{0}_{-1}\text{e}$

$^{233}_{92}\text{U} \rightarrow \, ^{229}_{90}\text{Th} + \, ^{4}_{2}\text{He}$

$^{229}_{90}\text{Th} \rightarrow \, ^{225}_{88}\text{Ra} + \, ^{4}_{2}\text{He}$

$^{225}_{88}\text{Ra} \rightarrow \, ^{225}_{89}\text{Ac} + \, ^{0}_{-1}\text{e}$

$^{225}_{89}\text{Ac} \rightarrow \, ^{221}_{87}\text{Fr} + \, ^{4}_{2}\text{He}$

$^{221}_{87}\text{Fr} \rightarrow \, ^{217}_{85}\text{At} + \, ^{4}_{2}\text{He}$

$^{217}_{85}\text{At} \rightarrow \, ^{213}_{83}\text{Bi} + \, ^{4}_{2}\text{He}$

$^{213}_{83}\text{Bi} \rightarrow \, ^{213}_{84}\text{Po} + \, ^{0}_{-1}\text{e}$

$$^{213}_{84}\text{Po} \rightarrow \ ^{209}_{82}\text{Pb} + \ ^{4}_{2}\text{He}$$

$$^{209}_{82}\text{Pb} \rightarrow \ ^{209}_{83}\text{Bi} + \ ^{0}_{-1}\text{e}$$

20.46   Each alpha emission decreases the mass number by four and the atomic number by two. Each beta emission increases the atomic number by one.

$$^{232}_{90}\text{Th} \rightarrow \ ^{208}_{82}\text{Pb}$$

Number of α emissions $= \dfrac{\text{Th mass number} - \text{Pb mass number}}{4}$

$$= \dfrac{232 - 208}{4} = 6 \ \alpha \ \text{emissions}$$

The atomic number decreases by 12 as a result of 6 alpha emissions. The resulting atomic number is $(90 - 12) = 78$.
Number of β emissions = Pb atomic number $- 78 = 82 - 78 = 4$ β emissions

20.48   Each alpha emission decreases the mass number by four and the atomic number by two. Each beta emission increases the atomic number by one.

$$^{237}_{93}\text{Np} \rightarrow \ ^{209}_{83}\text{Bi}$$

Number of α emissions $= \dfrac{\text{Np mass number} - \text{Bi mass number}}{4}$

$$= \dfrac{237 - 209}{4} = 7 \ \alpha \ \text{emissions}$$

The atomic number decreases by 14 as a result of 7 alpha emissions. The resulting atomic number is $(93 - 14) = 79$.
Number of β emissions = Bi atomic number $- 79 = 83 - 79 = 4$ β emissions

Radioactive Decay Rates (Section 20.4)

20.50 $k = \dfrac{0.693}{t_{1/2}} = \dfrac{0.693}{2.805 \ \text{d}} = 0.247 \ \text{d}^{-1}$

20.52 $t_{1/2} = \dfrac{0.693}{k} = \dfrac{0.693}{7.95 \times 10^{-3} \ \text{d}^{-1}} = 87.17 \text{d}$

$$\ln\left(\dfrac{N}{N_0}\right) = (-0.693)\left(\dfrac{t}{t_{1/2}}\right) = (-0.693)\left(\dfrac{185 \ \text{d}}{87.17 \ \text{d}}\right) = -1.4707$$

$\dfrac{N}{N_0} = e^{-1.4707} = 0.2298; \qquad \dfrac{N}{100\%} = 0.2298; \qquad N = 23.0\%$

20.54 $t_{1/2} = (102 \ \text{y})(365 \ \text{d/y})(24 \ \text{h/d})(3600 \ \text{s/h}) = 3.2167 \times 10^{9} \ \text{s}$

$k = \dfrac{0.693}{t_{1/2}} = \dfrac{0.693}{3.2167 \times 10^{9} \ \text{s}} = 2.1544 \times 10^{-10} \ \text{s}^{-1}$

$$N = (1.0 \times 10^{-9} \text{ g})\left(\frac{1 \text{ mol Po}}{209 \text{ g Po}}\right)(6.022 \times 10^{23} \text{ atoms/mol}) = 2.881 \times 10^{12} \text{ atoms}$$

Decay rate = kN = $(2.1544 \times 10^{-10} \text{ s}^{-1})(2.881 \times 10^{12} \text{ atoms}) = 6.21 \times 10^{2} \text{ s}^{-1}$
621 α particles are emitted in 1.0 s.

20.56   Decay rate = kN

$$N = (1.0 \times 10^{-3} \text{ g})\left(\frac{1 \text{ mol } ^{79}\text{Se}}{79 \text{ g}}\right)(6.022 \times 10^{23} \text{ atoms/mol}) = 7.6 \times 10^{18} \text{ atoms}$$

$$k = \frac{\text{Decay rate}}{N} = \frac{1.5 \times 10^{5}/\text{s}}{7.6 \times 10^{18}} = 2.0 \times 10^{-14} \text{ s}^{-1}$$

$$t_{1/2} = \frac{0.693}{k} = \frac{0.693}{2.0 \times 10^{-14} \text{ s}^{-1}} = 3.5 \times 10^{13} \text{ s}$$

$$t_{1/2} = (3.5 \times 10^{13} \text{ s})\left(\frac{1 \text{ h}}{3600 \text{ s}}\right)\left(\frac{1 \text{ d}}{24 \text{ h}}\right)\left(\frac{1 \text{ y}}{365 \text{ d}}\right) = 1.1 \times 10^{6} \text{ y}$$

20.58 $\ln\left(\dfrac{N}{N_0}\right) = (-0.693)\left(\dfrac{t}{t_{1/2}}\right)$; $\dfrac{N}{N_0} = \dfrac{\text{Decay rate at time t}}{\text{Decay rate at time t} = 0}$

$\ln\left(\dfrac{6990}{8540}\right) = (-0.693)\left(\dfrac{10.0 \text{ d}}{t_{1/2}}\right)$; $t_{1/2} = 34.6 \text{ d}$

20.60   (a) $1 \rightarrow 1/2 \rightarrow 1/4 \rightarrow 1/8$
After three half-lives, 1/8 of the strontium-90 will remain.

(b) $k = \dfrac{0.693}{t_{1/2}} = \dfrac{0.693}{29 \text{ y}} = 0.0239 \text{ y}^{-1} = 0.024 \text{ y}^{-1}$

(c) $t = \dfrac{\ln\dfrac{N}{N_o}}{-k} = \dfrac{\ln\dfrac{(\text{Sr–90})_t}{(\text{Sr–90})_o}}{-k} = \dfrac{\ln\dfrac{(0.01)}{(1)}}{-0.0239\text{y}^{-1}} = 193 \text{ y}$

20.62 $\ln\left(\dfrac{N}{N_0}\right) = -0.693\left(\dfrac{t}{t_{1/2}}\right) = -0.693\left(\dfrac{9.5 \text{ y}}{87.7 \text{ y}}\right) = -0.0751$

$\dfrac{N}{N_0} = e^{-0.0751} = 0.928$

power output = 240 W x 0.928 = 223 W = 220 W

Radiocarbon Dating (Section 20.5)

20.64 $\ln\left(\dfrac{N}{N_0}\right) = (-0.693)\left(\dfrac{t}{t_{1/2}}\right)$; $\dfrac{N}{N_0} = \dfrac{\text{Decay rate at time t}}{\text{Decay rate at t} = 0}$

$t_{1/2} = 5715 \text{ y}$

$$\ln\left(\frac{2.3}{15.3}\right) = (-0.693)\left(\frac{t}{5715\ y}\right)$$

age of bone $= t = 1.6 \times 10^4$ y

20.66 $\quad \ln\left(\dfrac{N}{N_0}\right) = (-0.693)\left(\dfrac{t}{t_{1/2}}\right); \qquad \dfrac{N}{N_0} = \dfrac{\text{Decay rate at time t}}{\text{Decay rate at t} = 0}$

$t_{1/2} = 5715$ y

$$\ln\left(\frac{14.4}{15.3}\right) = (-0.693)\left(\frac{t}{5715\ y}\right)$$

manuscript age $= t = 500$ y

20.68 $\quad \ln\left(\dfrac{N}{N_0}\right) = (-0.693)\left(\dfrac{t}{t_{1/2}}\right); \qquad \dfrac{N}{N_0} = \dfrac{\text{Decay rate at time t}}{\text{Decay rate at t} = 0}$

$t_{1/2} = 5715$ y and $t = 46,000$ y; Let $N_0 = 100$

$$\ln\left(\frac{N}{100}\right) = (-0.693)\left(\frac{46,000\ y}{5715\ y}\right)$$

$$\ln N - \ln(100) = (-0.693)\left(\frac{46,000\ y}{5715\ y}\right)$$

$$\ln N = (-0.693)\left(\frac{46,000\ y}{5715\ y}\right) + \ln(100) = -0.97; \quad N = e^{-0.97} = 0.38\% \text{ left}$$

20.70 $\quad$ U-238, $t_{1/2} = 4.47 \times 10^9$ yr

At time t, U-238 $= 105$ μmol and Pb-206 $= 33$ μmol

U-238 at time $t_0 = (105 + 33)$ μmol $= 138$ μmol

$$\ln\left(\frac{N}{N_0}\right) = (-0.693)\left(\frac{t}{t_{1/2}}\right)$$

$$\ln\left(\frac{^{238}U_t}{^{238}U_0}\right) = \ln\left(\frac{105}{138}\right) = (-0.693)\left(\frac{t}{4.47 \times 10^9\ yr}\right)$$

age of rock $= t = 1.8 \times 10^9$ yr

20.72

	^{40}K	$\rightarrow$	^{40}Ar
then (mmol)	N_0		0
change (mmol)	-0.25		$+0.25$
now (mmol)	$N = 3.35$		0.25

$N_0 - 0.25 = 3.35$ mmol therefore, $N_0 = 3.35 + 0.25 = 3.60$ mmol and $N = 3.35$ mmol

$$\ln\left(\frac{N}{N_0}\right) = (-0.693)\left(\frac{t}{t_{1/2}}\right)$$

$$\ln\left(\frac{3.35}{3.60}\right) = (-0.693)\left(\frac{t}{1.25 \times 10^9 \text{ y}}\right)$$

age of rock = t = 1.30×10^8 y

Energy Changes during Nuclear Reactions (Section 20.6)

20.74   The loss in mass that occurs when protons and neutrons combine to form a nucleus is called the mass defect. The lost mass is converted into the binding energy that is used to hold the nucleons together.

20.76   (a) For $_{26}^{52}\text{Fe}$:

First, calculate the total mass of the nucleons (26 n + 26 p)

Mass of 26 neutrons = (26)(1.008 66) = 26.225 16
Mass of 26 protons = (26)(1.007 28) = 26.189 28
Mass of 26 n + 26 p   = 52.414 44

Next, calculate the mass of a ^{52}Fe nucleus by subtracting the mass of 26 electrons from the mass of a ^{52}Fe atom.

 Mass of ^{52}Fe atom   = 51.948 11
−Mass of 26 electrons = −(26)(5.486 x 10^{-4}) = −0.014 26
Mass of ^{52}Fe nucleus   = 51.933 85

Then subtract the mass of the ^{52}Fe nucleus from the mass of the nucleons to find the mass defect:

Mass defect = mass of nucleons − mass of nucleus
              = (52.414 44) − (51.933 85) = 0.480 59 u

Mass defect in g/mol:
(0.480 59 u)(1.660 54 x 10^{-24} g/u)(6.022 x 10^{23} mol^{-1}) = 0.480 58 g/mol

(b) For $_{42}^{92}\text{Mo}$:

First, calculate the total mass of the nucleons (50 n + 42 p)

Mass of 50 neutrons = (50)(1.008 66) = 50.433 00
Mass of 42 protons = (42)(1.007 28) = 42.305 76
Mass of 50 n + 42 p   = 92.738 76

Next, calculate the mass of a ^{92}Mo nucleus by subtracting the mass of 42 electrons from the mass of a ^{92}Mo atom.

 Mass of ^{92}Mo atom   = 91.906 81
−Mass of 42 electrons = −(42)(5.486 x 10^{-4}) = −0.023 04
Mass of ^{92}Mo nucleus   = 91.883 77

Then subtract the mass of the ^{92}Mo nucleus from the mass of the nucleons to find the mass defect:

Mass defect = mass of nucleons − mass of nucleus
              = (92.738 76) − (91.883 77) = 0.854 99 u

Mass defect in g/mol:
(0.854 99 u)(1.660 54 x 10^{-24} g/u)(6.022 x 10^{23} mol^{-1}) = 0.854 99 g/mol

20.78   (a) For $^{50}_{24}$Cr:

First, calculate the total mass of the nucleons (26 n + 24 p)

Mass of 26 neutrons = (26)(1.008 66) = 26.225 16

Mass of 24 protons = (24)(1.007 28) = 24.174 72

Mass of 26 n + 24 p   = 50.399 88

Next, calculate the mass of a ^{50}Cr nucleus by subtracting the mass of 24 electrons from the mass of a ^{50}Cr atom.

Mass of ^{50}Cr atom   = 49.946 05

−Mass of 24 electrons = −(24)(5.486 x 10^{-4}) = −0.013 17

Mass of ^{50}Cr nucleus   = 49.932 88

Then subtract the mass of the ^{50}Cr nucleus from the mass of the nucleons to find the mass defect:

Mass defect = mass of nucleons − mass of nucleus

= (50.399 88) − (49.932 88) = 0.467 00 u

Mass defect in g/mol:

(0.467 00 u)(1.660 54 x 10^{-24} g/u)(6.022 x 10^{23} mol^{-1}) = 0.466 99 g/mol

Now, use the Einstein equation to convert the mass defect into the binding energy.

$\Delta E = \Delta mc^2$ = (0.466 99 g/mol)(10^{-3} kg/g)(3.00 x 10^8 m/s)^2

ΔE = 4.203 x 10^{13} J/mol = 4.203 x 10^{10} kJ/mol

$$\Delta E = \frac{4.203 \times 10^{13} \text{ J/mol}}{6.022 \times 10^{23} \text{ nuclei/mol}} \times \frac{1 \text{ MeV}}{1.60 \times 10^{-13} \text{ J}} \times \frac{1 \text{ nucleus}}{50 \text{ nucleons}} = 8.72 \text{ MeV/nucleon}$$

(b) For $^{64}_{30}$Zn:

First, calculate the total mass of the nucleons (34 n + 30 p)

Mass of 34 neutrons = (34)(1.008 66) = 34.294 44

Mass of 30 protons = (30)(1.007 28) = 30.218 40

Mass of 34 n + 30 p   = 64.512 84

Next, calculate the mass of a ^{64}Zn nucleus by subtracting the mass of 30 electrons from the mass of a ^{64}Zn atom.

Mass of ^{64}Zn atom   = 63.929 15

−Mass of 30 electrons = −(30)(5.486 x 10^{-4}) = −0.016 46

Mass of ^{64}Zn nucleus   = 63.912 69

Then subtract the mass of the ^{64}Zn nucleus from the mass of the nucleons to find the mass defect:

Mass defect = mass of nucleons − mass of nucleus

= (64.512 84) − (63.912 69) = 0.600 15 u

Mass defect in g/mol:

(0.600 15 u)(1.660 54 x 10^{-24} g/u)(6.022 x 10^{23} mol^{-1}) = 0.600 14 g/mol

Now, use the Einstein equation to convert the mass defect into the binding energy.

$\Delta E = \Delta mc^2$ = (0.600 14 g/mol)(10^{-3} kg/g)(3.00 x 10^8 m/s)^2

ΔE = 5.401 x 10^{13} J/mol = 5.401 x 10^{10} kJ/mol

$$\Delta E = \frac{5.401 \times 10^{13} \text{ J/mol}}{6.022 \times 10^{23} \text{ nuclei/mol}} \times \frac{1 \text{ MeV}}{1.60 \times 10^{-13} \text{ J}} \times \frac{1 \text{ nucleus}}{64 \text{ nucleons}} = 8.76 \text{ MeV/nucleon}$$

The ^{64}Zn is more stable because ΔE is larger.

20.80   (a) For $^{58}_{28}$Ni:

First, calculate the total mass of the nucleons (30 n + 28 p)

Mass of 30 neutrons = (30)(1.008 66) = 30.259 80
Mass of 28 protons = (28)(1.007 28) = 28.203 84
Mass of 30 n + 28 p   = 58.463 64

Next, calculate the mass of a ^{58}Ni nucleus by subtracting the mass of 28 electrons from the mass of a ^{58}Ni atom.

 Mass of ^{58}Ni atom   = 57.935 35
−Mass of 28 electrons = −(28)(5.486 x 10^{-4})   = −0.015 36
Mass of ^{58}Ni nucleus   = 57.919 99

Then subtract the mass of the ^{58}Ni nucleus from the mass of the nucleons to find the mass defect:

Mass defect = mass of nucleons − mass of nucleus
         = (58.463 64) − (57.919 99) = 0.543 65 u

Mass defect in g/mol:

(0.543 65 u)(1.660 54 x 10^{-24} g/u)(6.022 x 10^{23} mol^{-1}) = 0.543 64 g/mol

Now, use the Einstein equation to convert the mass defect into the binding energy.

$\Delta E = \Delta mc^2$ = (0.543 64 g/mol)(10^{-3} kg/g)(3.00 x 10^8 m/s)^2

ΔE = 4.893 x 10^{13} J/mol = 4.893 x 10^{10} kJ/mol

$$\Delta E = \frac{4.893 \times 10^{13} \text{ J/mol}}{6.022 \times 10^{23} \text{ nuclei/mol}} \times \frac{1 \text{ MeV}}{1.60 \times 10^{-13} \text{ J}} \times \frac{1 \text{ nucleus}}{58 \text{ nucleons}} = 8.76 \text{ MeV/nucleon}$$

(b) For $^{84}_{36}$Kr:

First, calculate the total mass of the nucleons (48 n + 36 p)

Mass of 48 neutrons = (48)(1.008 66) = 48.415 68
Mass of 36 protons = (36)(1.007 28) = 36.262 08
Mass of 48 n + 36 p   = 84.677 76

Next, calculate the mass of a ^{84}Kr nucleus by subtracting the mass of 36 electrons from the mass of a ^{84}Kr atom.

 Mass of ^{84}Kr atom   = 83.911 51
−Mass of 36 electrons = −(36)(5.486 x 10^{-4})   = −0.019 75
Mass of ^{84}Kr nucleus   = 83.891 76

Then subtract the mass of the ^{84}Kr nucleus from the mass of the nucleons to find the mass defect:

Mass defect = mass of nucleons − mass of nucleus
         = (84.677 76) − (83.891 76) = 0.786 00 u

Mass defect in g/mol:

(0.786 00 u)(1.660 54 x 10^{-24} g/u)(6.022 x 10^{23} mol^{-1}) = 0.785 98 g/mol

Now, use the Einstein equation to convert the mass defect into the binding energy.

$\Delta E = \Delta mc^2$ = (0.785 98 g/mol)(10^{-3} kg/g)(3.00 x 10^8 m/s)^2

ΔE = 7.074 x 10^{13} J/mol = 7.074 x 10^{10} kJ/mol

$$\Delta E = \frac{7.074 \times 10^{13} \text{ J/mol}}{6.022 \times 10^{23} \text{ nuclei/mol}} \times \frac{1 \text{ MeV}}{1.60 \times 10^{-13} \text{ J}} \times \frac{1 \text{ nucleus}}{84 \text{ nucleons}} = 8.74 \text{ MeV/nucleon}$$

20.82　$^{174}_{77}\text{Ir} \rightarrow {}^{170}_{75}\text{Re} + {}^{4}_{2}\text{He}$

mass $^{174}_{77}\text{Ir}$ 　　　　　　　　173.966 66

−mass $^{170}_{75}\text{Re}$ 　　　　　　　−169.958 04

−mass $^{4}_{2}\text{He}$ 　　　　　　　　　− 4.002 60

mass change 　　　　　　　　　0.006 02 u

$(0.006\ 02\ \text{u})(1.660\ 54 \times 10^{-24}\ \text{g/u})(6.022 \times 10^{23}\ \text{mol}^{-1}) = 0.006\ 02\ \text{g/mol}$
$\Delta E = \Delta mc^2 = (0.006\ 02\ \text{g/mol})(10^{-3}\ \text{kg/g})(3.00 \times 10^8\ \text{m/s})^2$
$\Delta E = 5.42 \times 10^{11}\ \text{J/mol} = 5.42 \times 10^8\ \text{kJ/mol}$

20.84　$\Delta m = \dfrac{\Delta E}{c^2} = \dfrac{-92.2 \times 10^3\ \text{J}}{(3.00 \times 10^8\ \text{m/s})^2} = \dfrac{-92.2 \times 10^3\ \text{kg·m}^2/\text{s}^2}{(3.00 \times 10^8\ \text{m/s})^2} = -1.02 \times 10^{-12}\ \text{kg}$

$\Delta m = \dfrac{-1.02 \times 10^{-9}\ \text{g}}{2\ \text{mol NH}_3} = 5.10 \times 10^{-10}\ \text{g/mol}$

20.86　$^{232}_{90}\text{Th} \rightarrow {}^{208}_{82}\text{Pb} + 6\ {}^{4}_{2}\text{He} + 4\ {}^{0}_{-1}\text{e}$

Reactant: $^{232}_{90}\text{Th}$ nucleus = $^{232}_{90}\text{Th}$ atom − 90 e⁻

Product: $^{208}_{82}\text{Pb}$ nucleus + (6)($^{4}_{2}\text{He}$ nucleus) + 4 e⁻

　　　= ($^{208}_{82}\text{Pb}$ atom − 82 e⁻) + (6)($^{4}_{2}\text{He}$ atom − 2 e⁻) + 4 e⁻

　　　= $^{208}_{82}\text{Pb}$ atom + (6)($^{4}_{2}\text{He}$ atom) − 90 e⁻

Change: ($^{232}_{90}\text{Th}$ atom − 90 e⁻) − [$^{208}_{82}\text{Pb}$ atom + (6)($^{4}_{2}\text{He}$ atom) − 90 e⁻]

　　　= $^{232}_{90}\text{Th}$ atom − [$^{208}_{82}\text{Pb}$ atom + (6)($^{4}_{2}\text{He}$ atom)]　　(electrons cancel)

Mass change = (232.038 054) − [(207.976 627) + (6)(4.002 603)]
　　　　　　= 0.045 809 u

$(0.045\ 809\ \text{u})(1.660\ 54 \times 10^{-24}\ \text{g/u})(6.022 \times 10^{23}\ \text{mol}^{-1}) = 0.045\ 808\ \text{g/mol}$
$\Delta E = \Delta mc^2 = (0.045\ 808\ \text{g/mol})(10^{-3}\ \text{kg/g})(3.00 \times 10^8\ \text{m/s})^2$
$\Delta E = 4.12 \times 10^{12}\ \text{J/mol} = 4.12 \times 10^9\ \text{kJ/mol}$

Fission and Fusion (Section 20.7)

20.88　(a) Nuclear fission is induced by bombarding a U-235 sample with beta particles.

20.90　$^{10}_{5}\text{B} + {}^{1}_{0}\text{n} \rightarrow {}^{7}_{3}\text{Li} + {}^{4}_{2}\text{He}$

20.92　Fuel rods in a power plant cannot be used to make an atomic weapon unless the fuel rod is processed and significantly enriched in the fissionable U-235.

20.94 $2\,{}^{2}_{1}\text{H}\;\rightarrow\;{}^{3}_{2}\text{He}\;+\;{}^{1}_{0}\text{n}$

   mass $2\,{}^{2}_{1}\text{H}$ $2(2.0141)$

   $-\text{mass}\;{}^{3}_{2}\text{He}$ -3.0160

   $-\text{mass}\;{}^{1}_{0}\text{n}$ $-1.008\,66$

   mass change $0.003\,54\ \text{u}$

   $(0.003\,54\ \text{u})(1.660\,54 \times 10^{-24}\ \text{g/u})(6.022 \times 10^{23}\ \text{mol}^{-1}) = 0.003\,54\ \text{g/mol}$
   $\Delta E = \Delta mc^2 = (0.003\,54\ \text{g/mol})(10^{-3}\ \text{kg/g})(3.00 \times 10^{8}\ \text{m/s})^2$
   $\Delta E = 3.2 \times 10^{11}\ \text{J/mol} = 3.2 \times 10^{8}\ \text{kJ/mol}$

20.96 ${}^{2}_{1}\text{H}\;+\;{}^{3}_{2}\text{He}\;\rightarrow\;{}^{4}_{2}\text{He}\;+\;{}^{1}_{1}\text{H}$

   mass ${}^{2}_{1}\text{H}$ 2.0141

   mass ${}^{3}_{2}\text{He}$ 3.0160

   $-\text{mass}\;{}^{4}_{2}\text{He}$ -4.0026

   $-\text{mass}\;{}^{1}_{1}\text{H}$ -1.0078

   mass change $0.0197\ \text{u}$

   $(0.0197\ \text{u})(1.660\,54 \times 10^{-24}\ \text{g/u})(6.022 \times 10^{23}\ \text{mol}^{-1}) = 0.0197\ \text{g/mol}$
   $\Delta E = \Delta mc^2 = (0.0197\ \text{g/mol})(10^{-3}\ \text{kg/g})(3.00 \times 10^{8}\ \text{m/s})^2$
   $\Delta E = 1.77 \times 10^{12}\ \text{J/mol} = 1.77 \times 10^{9}\ \text{kJ/mol}$

Nuclear Transmutation (Section 20.8)

20.98   (a) ${}^{109}_{47}\text{Ag}\;+\;{}^{4}_{2}\text{He}\;\rightarrow\;{}^{113}_{49}\text{In}$

   (b) ${}^{10}_{5}\text{B}\;+\;{}^{4}_{2}\text{He}\;\rightarrow\;{}^{13}_{7}\text{N}\;+\;{}^{1}_{0}\text{n}$

20.100 ${}^{209}_{83}\text{Bi}\;+\;{}^{58}_{26}\text{Fe}\;\rightarrow\;{}^{266}_{109}\text{Mt}\;+\;{}^{1}_{0}\text{n}$

20.102 ${}^{238}_{92}\text{U}\;+\;{}^{2}_{1}\text{H}\;\rightarrow\;{}^{238}_{93}\text{Np}\;+\;2\,{}^{1}_{0}\text{n}$

20.104   (a) ${}^{249}_{98}\text{Cf}\;+\;{}^{48}_{20}\text{Ca}\;\rightarrow\;?\;+\;3\,{}^{1}_{0}\text{n}\,;\qquad ? = {}^{249+48-3}_{98+20}\text{X} = {}^{294}_{118}\text{Og}$

   ${}^{249}_{98}\text{Cf}\;+\;{}^{48}_{20}\text{Ca}\;\rightarrow\;{}^{294}_{118}\text{Og}\;+\;3\,{}^{1}_{0}\text{n}$

   (b) ${}^{294}_{118}\text{Og}\;\rightarrow\;?\;+\;4\,{}^{4}_{2}\text{He}\,;\qquad ? = {}^{294-16}_{118-8}\text{X} = {}^{278}_{110}\text{Ds}$

20.106 ${}^{293}_{118}\text{X}$, ${}^{289}_{116}\text{Y}$, and ${}^{285}_{114}\text{Z}$

Detecting and Measuring Radioactivity (Section 20.9)

20.108   1 Sv = 1 Gy and 1 Gy = 1 J/kg

$5000 \ \mu Sv = 5000 \times 10^{-6} \ Sv = 5000 \times 10^{-6} \ Gy = 5000 \times 10^{-6} \ J/kg$

joules absorbed = $(5000 \times 10^{-6} \ J/kg)(60 \ kg) = 0.3 \ J$

20.110 $4.0 \ pCi = 4.0 \times 10^{-12} \ Ci$

$1 \ Ci = 3.7 \times 10^{10} \ Bq = 3.7 \times 10^{10} \ disintegrations/s$

(a) $4.0 \times 10^{-12} \ Ci \times \dfrac{3.7 \times 10^{10} \ disintegrations/s}{1 \ Ci} \times \dfrac{60 \ s}{1 \ min} = 8.9 \ disintegrations/min$

(b) $Rate = \dfrac{8.9 \ dis}{1 \ min} \times \dfrac{60 \ min}{1 \ h} \times \dfrac{24 \ h}{1 \ d} = 1.3 \times 10^{4} \ dis/d$

$Rate = 1.3 \times 10^{4} \ dis/d = kN = \left(\dfrac{0.693}{3.8 \ d} \right) N; \ \text{solve for N}$

$N = 7.1 \times 10^{4} \ ^{222}Rn$

20.112   For ^{16}N, $t_{1/2} = 7.13 \ s$; $\ k = \dfrac{0.693}{t_{1/2}} = \dfrac{0.693}{7.13 \ s} = 0.0972 \ s^{-1}$

Decay rate = kN

$N = (50.0 \times 10^{-3} \ g) \left(\dfrac{1 \ mol \ N-16}{16 \ g \ N} \right) (6.022 \times 10^{23} \ atoms/mol) = 1.88 \times 10^{21} \ atoms$

activity = decay rate = $kN = (0.0972 \ s^{-1})(1.88 \times 10^{21} \ atoms) = 1.83 \times 10^{20} \ dis/s$

decay rate = $1.83 \times 10^{20} \ Bq$

decay rate = $1.83 \times 10^{20} \ Bq \times \dfrac{1 \ Ci}{3.7 \times 10^{10} \ Bq} = 4.94 \times 10^{9} \ Ci$

Multiconcept Problems

20.114   For radioactive decay, $\ln \dfrac{N}{N_o} = -kt$

For ^{235}U, $\quad k_1 = \dfrac{0.693}{t_{1/2}} = \dfrac{0.693}{7.04 \times 10^{8} \ y} = 9.84 \times 10^{-10} \ y^{-1}$

For ^{238}U, $\quad k_2 = \dfrac{0.693}{t_{1/2}} = \dfrac{0.693}{4.47 \times 10^{9} \ y} = 1.55 \times 10^{-10} \ y^{-1}$

For ^{235}U, $\quad \ln \dfrac{N_1}{N_{o1}} = -k_1 t \ \text{and} \ \ln \dfrac{N_1}{N_{o1}} + k_1 t = 0$

For ^{238}U, $\quad \ln \dfrac{N_2}{N_{o2}} = -k_2 t \ \text{and} \ \ln \dfrac{N_2}{N_{o2}} + k_2 t = 0$

Set the two equations that are equal to zero equal to each other and solve for t.

$$\ln \frac{N_1}{N_{o1}} + k_1 t = \ln \frac{N_2}{N_{o2}} + k_2 t;$$

$$\ln \frac{N_1}{N_{o1}} - \ln \frac{N_2}{N_{o2}} = k_2 t - k_1 t = (k_2 - k_1)t$$

$$\ln \frac{\left(\dfrac{N_1}{N_{o1}}\right)}{\left(\dfrac{N_2}{N_{o2}}\right)} = (k_2 - k_1)t, \ \text{now} \ N_{o1} = N_{o2}, \text{ so } \ln \frac{N_1}{N_2} = (k_2 - k_1)t$$

$$\frac{N_1}{N_2} = 7.25 \times 10^{-3}, \text{ so } \ln(7.25 \times 10^{-3}) = (1.55 \times 10^{-10} \ y^{-1} - 9.84 \times 10^{-10} \ y^{-1})t$$

$$t = \frac{-4.93}{-8.29 \times 10^{-10} \ y^{-1}} = 5.9 \times 10^9 \ y$$

The age of the elements is 5.9×10^9 y (6 billion years).

20.116 $\quad E = (1.50 \ \text{MeV})\left(\dfrac{1.60 \times 10^{-13} \ J}{1 \ \text{MeV}}\right) = 2.40 \times 10^{-13} \ J$

$$\lambda = \frac{hc}{E} = \frac{(6.626 \times 10^{-34} \ J \cdot s)(3.00 \times 10^8 \ m/s)}{2.40 \times 10^{-13} \ J} = 8.28 \times 10^{-13} \ m = 0.000\ 828 \ nm$$

20.118 $\quad BaCO_3$, 197.34

$$1.000 \ g \ BaCO_3 \times \frac{1 \ mol \ BaCO_3}{197.34 \ g \ BaCO_3} \times \frac{1 \ mol \ C}{1 \ mol \ BaCO_3} \times \frac{12.011 \ g \ C}{1 \ mol \ C} = 0.060\ 86 \ g \ C$$

4.0×10^{-3} Bq $= 4.0 \times 10^{-3}$ disintegrations/s
$(4.0 \times 10^{-3}$ Bq $= 4.0 \times 10^{-3}$ disintegrations/s$)(60$ s/min$) = 0.24$ disintegrations/min

$$\text{sample radioactivity} = \frac{0.24 \ \text{disintegrations/min}}{0.060\ 86 \ g \ C} = 3.94 \ \text{disintegrations/min per gram of C}$$

$$\ln\left(\frac{N}{N_0}\right) = (-0.693)\left(\frac{t}{t_{1/2}}\right); \qquad \frac{N}{N_0} = \frac{\text{Decay rate at time } t}{\text{Decay rate at time } t = 0}$$

$$\ln\left(\frac{3.94}{15.3}\right) = (-0.693)\left(\frac{t}{5715 \ y}\right); \qquad t = 11{,}200 \ y$$

20.120 First find the activity (N) that the ^{28}Mg would have after 2.4 hours assuming that none of it was removed by precipitation as $MgCO_3$.

$$\ln\left(\frac{N}{N_0}\right) = (-0.693)\left(\frac{t}{t_{1/2}}\right)$$

$$\frac{N}{N_0} = \frac{\text{Decay rate at time t}}{\text{Decay rate at time t} = 0}$$

$$\ln\left(\frac{N}{0.112}\right) = (-0.693)\left(\frac{2.40 \text{ h}}{20.91 \text{ h}}\right)$$

$\ln N - \ln(0.112) = -0.0795$

$\ln N = -0.0795 + \ln(0.112) = -2.27$

$N = e^{-2.27} = 0.103 \text{ } \mu\text{Ci/mL}$

$20.00 \text{ mL} = 0.020 \text{ } 00 \text{ L}$ and $15.00 \text{ mL} = 0.015 \text{ } 00 \text{ L}$

mol $MgCl_2 = (0.007 \text{ } 50 \text{ mol/L})(0.020 \text{ } 00 \text{ L}) = 1.50 \times 10^{-4}$ mol $MgCl_2$

mol $Na_2CO_3 = (0.012 \text{ } 50 \text{ mol/L})(0.015 \text{ } 00 \text{ L}) = 1.88 \times 10^{-4}$ mol Na_2CO_3

The mol of CO_3^{2-} are in excess, so assume that all of the Mg^{2+} precipitates as $MgCO_3$ according to the reaction:

	$Mg^{2+}(aq)$	+	$CO_3^{2-}(aq)$	$\rightarrow$	$MgCO_3(s)$
initial (mol)	0.000 150		0.000 188		0
change (mol)	– 0.000 150		– 0.000 150		+ 0.000 150
final (mol)	0		0.000 038		0.000 150

$[CO_3^{2-}] = 0.000 \text{ } 038 \text{ mol}/(0.020 \text{ } 00 \text{ L} + 0.015 \text{ } 00 \text{ L}) = 0.001 \text{ } 09 \text{ M}$

Now consider the dissolution of $MgCO_3$ in the presence of CO_3^{2-}.

	$MgCO_3(s)$	$\rightleftharpoons$	$Mg^{2+}(aq)$	+	$CO_3^{2-}(aq)$
initial (M)			0		0.001 09
change (M)			+x		+x
equil (M)			x		0.001 09 + x

The $[Mg^{2+}]$ in the filtrate is proportional to its activity after 2.40 h.

$$x = [Mg^{2+}] = 0.029 \text{ } \mu\text{Ci/mL} \times \frac{0.007 \text{ } 50 \text{ M}}{0.103 \text{ } \mu\text{Ci/mL}} = 0.002 \text{ } 11 \text{ M}$$

$[CO_3^{2-}] = 0.001 \text{ } 09 + x = 0.001 \text{ } 09 + 0.002 \text{ } 11 = 0.003 \text{ } 20 \text{ M}$

$K_{sp} = [Mg^{2+}][CO_3^{2-}] = (0.002 \text{ } 11)(0.003 \text{ } 20) = 6.8 \times 10^{-6}$

21 | Transition Elements and Coordination Chemistry

21.1   Mn   [Ar] ↿ ↿ ↿ ↿ ↿   ⇵   __ __ __
                    3d   4s   4p

Mn^{4+} in MnO$_2$, [Ar] 3d^3;   [Ar] ↿ ↿ ↿ __ __   __   __ __ __
                                            3d   4s   4p

There are 3 unpaired electrons in MnO$_2$.

21.2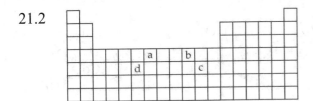

        (a) Mn   (b) Ni^{2+}

        (c) Ag   (d) Mo^{3+}

21.3   In Na$_4$[Fe(CN)$_6$] each sodium is in the +1 oxidation state (+4 total); each cyanide (CN$^-$) has a −1 charge (−6 total). The compound is neutral; therefore, the oxidation state of the iron is +2.

21.4   In [Cr(NH$_3$)$_2$(SCN)$_4$]$^-$ each NH$_3$ has no charge; each thiocyanate (SCN$^-$) has a −1 charge (−4 total). The compound has a −1 charge; therefore, the oxidation state of the chromium is +3.

21.5   diamminediaquacopper(II) chloride

21.6   triamminetrichlorocobalt(III)

21.7   [Pt(NH$_3$)$_2$(OH)$_2$]Cl$_2$

21.8   [Pt(NH$_3$)$_3$Cl]Cl, triamminechloroplatinum(II)chloride

21.9   Structures (1) and (4) are identical and are the cis isomer. Structures (2) and (3) are identical and are the trans isomer.

21.10 (1) and (2) are the same. (3) and (4) are the same. (1) and (2) are different from (3) and (4).

21.11 Two diastereoisomers are possible for [Cr(en)$_2$Cl$_2$]$^+$.

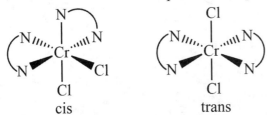

21.12 (a) One diastereoisomer for $[CoCl_2Br_2]^{2-}$;

$$\left[\begin{array}{c} Cl \\ | \\ Cl \diagdown Co \cdots\!\!\textit{''}Br \\ | \\ Br \end{array} \right]^{2-}$$

(b) One diastereoisomer for $Pt(en)Cl_2$;

(c) Two diastereoisomer for $Pt(NH_3)_2(SCN)_2$;

cis   trans

(d) Two diastereoisomer for $Co(NH_3)_3(NO_2)_3$;

21.13 $[Rh(en)Cl_2Br_2]$ is chiral.

21.14 (a) (2) is chiral and (1) and (3) are achiral.
  (b) enantiomer of (2)

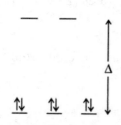

21.15

$[Fe(CN)_6]^{4-}$
no unpaired e^-

21.16 Cl^- is a weak field ligand and therefore $[FeCl_6]^{3-}$ is a high-spin complex with 5 unpaired electrons. CN^- is a strong field ligand and therefore $[Fe(CN)_6]^{3-}$ is a low-spin complex with 1 unpaired electron. $[FeCl_6]^{3-}$ is more paramagnetic because it has more unpaired electrons.

21.17 (a) $[MnCl_4]^{2-}$, Mn^{2+}, [Ar] $3d^5$

↑ ↑

↑ ↑ ↑

5 unpaired e^-

21.18 A diamagnetic four coordinate d^8 complex is most likely square planar.

21.19 Pt^{2+} [Xe] ↑↓ ↑↓ ↑↓ ↑ ↑ (5d) __ (6s) __ __ __ (6p)

$[PtCl_4]^{2-}$ [Xe] ↑↓ ↑↓ ↑↓ ↑↓ (5d) | ↑↓ (5d) ↑↓ (6s) ↑↓ ↑↓ (6p) | __

dsp^2 0 unpaired e^-

21.20 (a) Fe^{3+} [Ar] ↑ ↑ ↑ ↑ ↑ (3d) __ (4s) __ __ __ (4p)

$[Fe(CN)_6]^{3-}$ [Ar] ↑↓ ↑↓ ↑ (3d) | ↑↓ ↑↓ (3d) ↑↓ (4s) ↑↓ ↑↓ ↑↓ (4p) |

d^2sp^3 1 unpaired e^-

(b) Co^{2+} [Ar] ↑↓ ↑↓ ↑ ↑ ↑ (3d) __ (4s) __ __ __ (4p)

$[Co(H_2O)_6]^{2+}$ [Ar] ↑↓ ↑↓ ↑ ↑ ↑ (3d) | ↑↓ (4s) ↑↓ ↑↓ ↑↓ (4p) ↑↓ ↑↓ (4d) | __ __ __ (4d)

sp^3d^2 3 unpaired e^-

21.21 The ligand donor atom is N. It can serve as a ligand donor atom because it possesses a lone pair of electrons.

21.22 (a) diamminedichloroplatinum(II)
(b) oxidation state = +2 and coordination number is 4
(c) Lewis acid is Pt^{2+} and Lewis bases are Cl^- and NH_3
(d) [Xe] $4f^{14}$ $5d^8$

21.23 $Pt(NH_3)_2Cl_2$

$\overline{x^2-y^2}$

$\underset{xy}{\underline{\uparrow\downarrow}}$

$\underset{z^2}{\underline{\uparrow\downarrow}}$

$\underset{xz}{\underline{\uparrow\downarrow}}\quad\underset{yz}{\underline{\uparrow\downarrow}}$

no unpaired electrons

21.24 (a) The chloride concentration is relatively high in blood plasma and, according to Le Chatelier's Principle, a high concentration of product shifts the equilibrium position toward the reactants. Inside the cell, the chloride concentration is lower, thus shifting the equilibrium positions toward the products.
(b) diammineaquachloroplatinum(II)
(c) According to the spectrochemical series, H_2O is a stronger field ligand than Cl^-. The crystal field splitting energy is larger, which corresponds to shorter wavelength of maximum absorption.

21.25 (a) +4
(b) diamminetetrachloroplatinum(IV)
(c) cis and trans isomers.

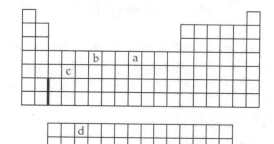

cis   trans

(d) Both the cis and trans isomers have symmetry planes and are achiral.

Conceptual Problems

21.26

(a) Co   (b) Cr

(c) Zr   (d) Pr

21.28   (a) The atomic radii decrease, at first markedly and then more gradually. Toward the end of the series, the radii increase again. The decrease in atomic radii is a result of an increase in Z_{eff}. The increase is due to electron-electron repulsions in doubly occupied d orbitals.
(b) The densities of the transition metals are inversely related to their atomic radii. The densities initially increase from left to right and then decrease toward the end of the series.
(c) Ionization energies generally increase from left to right across the series. The general trend correlates with an increase in Z_{eff} and a decrease in atomic radii.
(d) The standard oxidation potentials generally decrease from left to right across the first transition series. This correlates with the general trend in ionization energies.

21.30   (a) tridentate   (b) bidentate   (c) bidentate   (d) tetradentate
All four ligands can form chelate rings.

21.32   (1) dichloroethylenediamineplatinum(II)
(2) trans-diammineaquachloroplatinate(II) ion
(3) amminepentachloroplatinate(IV) ion
(4) cis-diaquabis(ethylenediamine)platinum(IV) ion

21.34   (a) (1) chiral;   (2) achiral;   (3) chiral;   (4) chiral
(b)

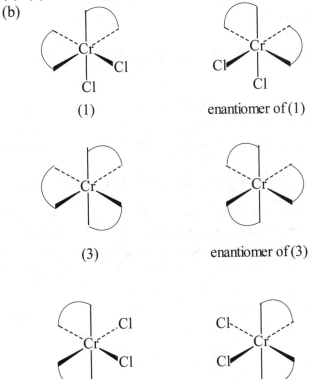

(1)   enantiomer of (1)

(3)   enantiomer of (3)

(4)   enantiomer of (4)

(c) (1) and (4) are enantiomers.

Section Problems
Electron Configurations (Section 21.1)

21.36 (a) Co $[Ar]\ 3d^7\ 4s^2$ ⇅ ⇅ ↑ ↑ ↑   ⇅
                                                   3d   4s

        (b) Co^{2+} $[Ar]\ 3d^7$ ⇅ ⇅ ↑ ↑ ↑   __
                                                   3d   4s

        (c) Co(V) in CoO_4^{3-} $[Ar]\ 3d^4$ ↑ ↑ ↑ ↑ __   __
                                                   3d   4s

        (d) Co(IV) in CoF_6^{2-} $[Ar]\ 3d^5$ ↑ ↑ ↑ ↑ ↑   __
                                                   3d   4s

21.38 (a) Cu^{2+}, $[Ar]\ 3d^9$ ⇅ ⇅ ⇅ ⇅ ↑   1 unpaired e^-
                                       3d

        (b) Ti^{2+}, $[Ar]\ 3d^2$ ↑ ↑ __ __ __   2 unpaired e^-
                                       3d

        (c) Zn^{2+}, $[Ar]\ 3d^{10}$ ⇅ ⇅ ⇅ ⇅ ⇅   0 unpaired e^-
                                       3d

        (d) Cr^{3+}, $[Ar]\ 3d^3$ ↑ ↑ ↑ __ __   3 unpaired e^-
                                       3d

Properties of Transition Elements (Section 21.2)

21.40 Ti is harder than K and Ca largely because the sharing of d, as well as s, electrons results in stronger metallic bonding.

21.42 (a) The decrease in radii with increasing atomic number is expected because the added d electrons only partially shield the added nuclear charge. As a result, Z_{eff} increases. With increasing Z_{eff}, the electrons are more strongly attracted to the nucleus, and atomic size decreases.
        (b) The densities of the transition metals are inversely related to their atomic radii.

21.44 The smaller than expected sizes of the third-transition series atoms are associated with what is called the lanthanide contraction, the general decrease in atomic radii of the f-block lanthanide elements. The lanthanide contraction is due to the increase in Z_{eff} as the 4f subshell is filled.

21.46   Sc $(631 + 1235) = 1866$ kJ/mol
        Ti $(659 + 1310) = 1969$ kJ/mol
        V $(651 + 1410) = 2061$ kJ/mol
        Cr $(653 + 1591) = 2224$ kJ/mol
        Mn $(717 + 1509) = 2226$ kJ/mol
        Fe $(762 + 1562) = 2324$ kJ/mol
        Co $(760 + 1648) = 2408$ kJ/mol
        Ni $(737 + 1753) = 2490$ kJ/mol
        Cu $(745 + 1958) = 2703$ kJ/mol
        Zn $(906 + 1733) = 2639$ kJ/mol

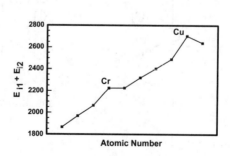

Across the first transition element series, Z_{eff} increases and there is an almost linear increase in the sum of the first two ionization energies. This is what is expected if the two electrons are removed from the 4s orbital. Higher than expected values for the sum of the first two ionization energies are observed for Cr and Cu because of their anomalous electron configurations (Cr $3d^5\ 4s^1$; Cu $3d^{10}\ 4s^1$). An increasing Z_{eff} affects 3d orbitals more than the 4s orbital and the second ionization energy for an electron from the 3d orbital is higher than expected.

21.48 Ti is more easily oxidized than is Zn because of a smaller Z_{eff}.

21.50 (a) $Cr(s) + 2\ H^+(aq) \rightarrow Cr^{2+}(aq) + H_2(g)$ (b) $Zn(s) + 2\ H^+(aq) \rightarrow Zn^{2+}(aq) + H_2(g)$
       (c) N.R.   (d) $Fe(s) + 2\ H^+(aq) \rightarrow Fe^{2+}(aq) + H_2(g)$

Oxidation States (Section 21.3)

21.52 (a) Ti, (b) V, and (c) Cr all have more than one oxidation state.

21.54 Sc(III), Ti(IV), V(V), Cr(VI), Mn(VII), Fe(VI), Co(III), Ni(II), Cu(II), Zn(II)

21.56 Cu^{2+} is a stronger oxidizing agent than Cr^{2+} because of a higher Z_{eff}.

21.58 A compound with vanadium in the +2 oxidation state is expected to be a reducing agent, because early transition metal atoms have a relatively low effective nuclear charge and are easily oxidized to higher oxidation states.

21.60 $Mn^{2+} < MnO_2 < MnO_4^-$ because of increasing oxidation state of Mn.

Coordination Compounds (Section 21.4)

21.62   Coordination Number
       (a) $[AgCl_2]^-$ 2
       (b) $[Cr(H_2O)_5Cl]^{2+}$ 6
       (c) $[Co(NCS)_4]^{2-}$ 4
       (d) $[ZrF_8]^{4-}$ 8
       (e) $[Fe(EDTA)(H_2O)]^-$ 7

21.64 (a) $AgCl_2^-$
       $2\ Cl^-$
       The oxidation state of the Ag is +1.
       (b) $[Cr(H_2O)_5Cl]^{2+}$
       $4\ H_2O$ (no charge) $1\ Cl^-$
       The oxidation state of the Cr is +3.
       (c) $[Co(NCS)_4]^{2-}$
       $4\ NCS^-$
       The oxidation state of the Co is +2.

(d) $[ZrF_8]^{4-}$

8 F^-

The oxidation state of the Zr is +4.

(e) $[Fe(EDTA)(H_2O)]^-$

H_2O (no charge) $EDTA^{4-}$

The oxidation state of the Fe is +3.

21.66 (a) $Co(NH_3)_3(NO_2)_3$

3 NH_3 (no charge)   3 NO_2^-

The oxidation state of the Co is +3.

(b) $[Ag(NH_3)_2]NO_3$

2 NH_3 (no charge)   1 NO_3^-

The oxidation state of the Ag is +1.

(c) $K_3[Cr(C_2O_4)_2Cl_2]$

3 K^+ 2 $C_2O_4^{2-}$ 2 Cl^-

The oxidation state of the Cr is +3.

(d) $Cs[CuCl_2]$

1 Cs^+ 2 Cl^-

The oxidation state of the Cu is +1.

21.68 (a) $Ir(NH_3)_3Cl_3$

(b) $Ni(en)_2Br_2$

(c) $[Pt(en)_2(SCN)_2]^{2+}$

21.70

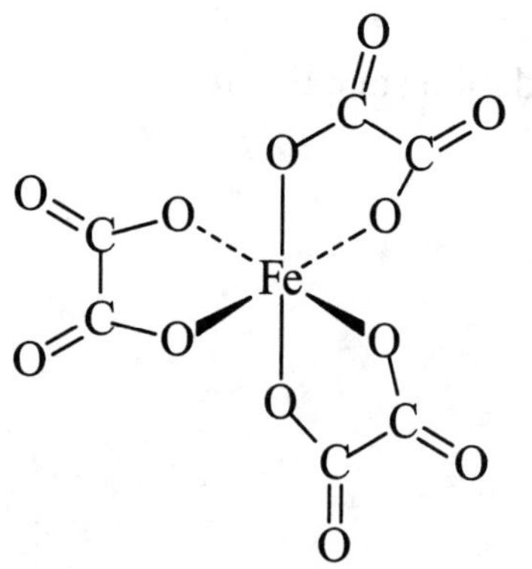

The iron is in the +3 oxidation state, and the coordination number is six. The geometry about the Fe is octahedral. The oxalate ligand is behaving as a bidentate chelating ligand. There are three chelate rings, one formed by each oxalate ligand.

21.72 (a) $[NC{-}Au{-}CN]^-$

$Na[Au(CN)_2]$ is linear. Each CN has a –1 charge. The Na has a +1 charge. The oxidation state of the Au is +1. The coordination number for Au is 2.

(b)

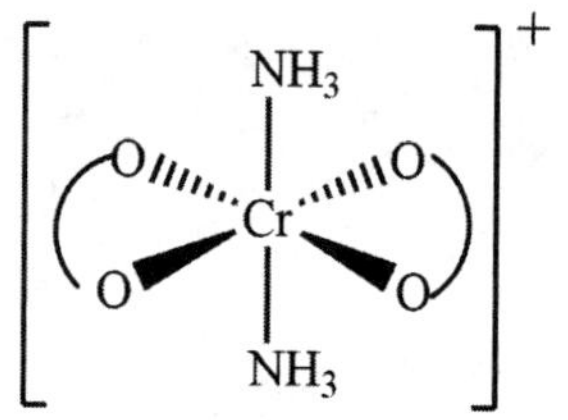

$[Cr(NH_3)_2(C_2O_4)_2]NO_2$ is octahedral. Each NH_3 has no charge. Each $C_2O_4^{2-}$ has a –2 charge. NO_2^- has a –1 charge. The oxidation state of the Cr is +5. The coordination number for Cr is 6.

Ligands (Section 21.5)

21.74 (a) Ni^{2+} is the Lewis acid. Ethylenediamine is the Lewis base.
   (b) Ethylenediamine is the ligand and the two N's are the donor atoms.
   (c) $[Ni(en)_3]^{2+}$ is octahedral with a coordination number of 6.

21.76 $EDTA^{4-}$ in mayonnaise will complex any metal cations that are present in trace amounts. Free metal ions can catalyze the oxidation of oils, causing the mayonnaise to become rancid. The bidentate ligand $H_2NCH_2CO_2^-$ will not bind to metal ions as strongly as does the hexadentate $EDTA^{4-}$ and so would not be an effective substitute for $EDTA^{4-}$.

Naming Coordination Compounds (Section 21.6)

21.78 (a) tetrachloromanganate(II)   (b) hexaamminenickel(II)
   (c) tricarbonatocobaltate(III)   (d) bis(ethylenediamine)dithiocyanatoplatinum(IV)

21.80 (a) cesium tetrachloroferrate(III)   (b) hexaaquavanadium(III) nitrate
   (c) tetraamminedibromocobalt(III) bromide   (d) diglycinatocopper(II)

21.82 (a) $[Pt(NH_3)_4]Cl_2$ (b) $Na_3[Fe(CN)_6]$
   (c) $[Pt(en)_3](SO_4)_2$ (d) $Rh(NH_3)_3(SCN)_3$

Isomers (Sections 21.7–21.8)

21.84 (a) (1) $[Ru(NH_3)_5(NO_2)]Cl$, tetraamminenitroruthenium(II) chloride
       (2) $[Ru(NH_3)_5(ONO)]Cl$, tetraamminenitritoruthenium(II) chloride
       (3) $[Ru(NH_3)_5Cl]NO_2$, tetraamminechlororuthenium(II) nitrite
   (b) (1) and (2) are linkage isomers.
   (c) (1) and (2) are ionization isomers with (3).

21.86 (a) $[Cr(NH_3)_2Cl_4]^-$ can exist as cis and trans diastereoisomers.

cis   trans

   (b) $[Co(NH_3)_5Br]^{2+}$ cannot exist as diastereoisomers.
   (c) $[MnCl_2Br_2]^{2-}$ (tetrahedral) cannot exist as diastereoisomers.
   (d) $[Pt(NH_3)_2Br_2]^{2-}$ (square planar) can exist as cis and trans diastereoisomers.

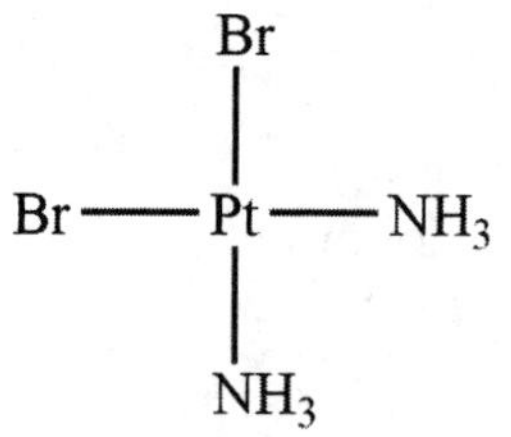

cis 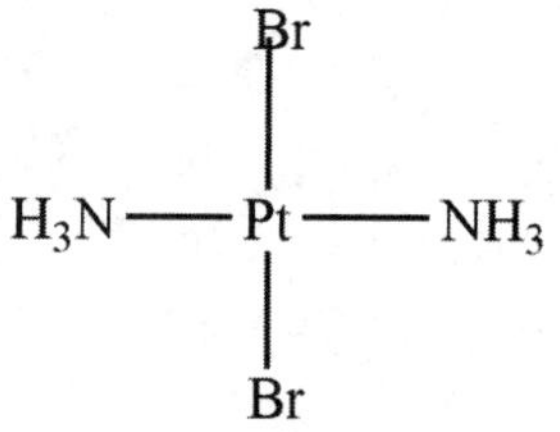   trans

21.88 (c) cis-$[Cr(en)_2(H_2O)_2]^{3+}$ (d) $[Cr(C_2O_4)_3]^{3-}$

21.90

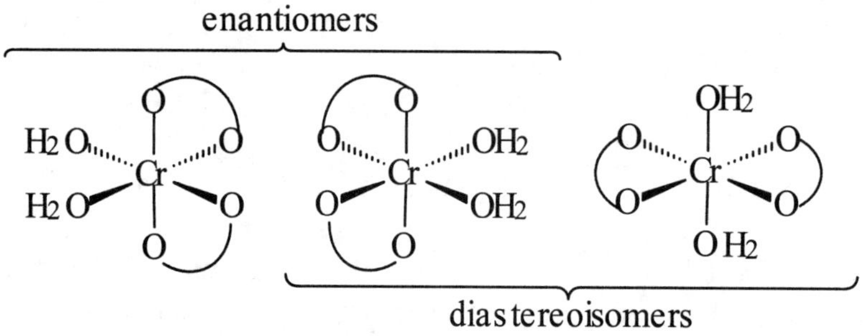

enantiomers

diastereoisomers

21.92 Plane-polarized light is light in which the electric vibrations of the light wave are restricted to a single plane. The following chromium complex can rotate the plane of plane-polarized light.

$[Cr(en)_3]^{3+}$

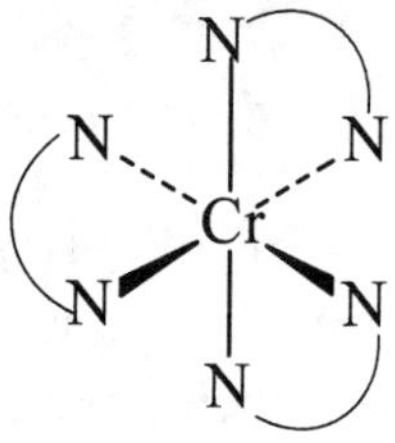

21.94   Linkage isomers:

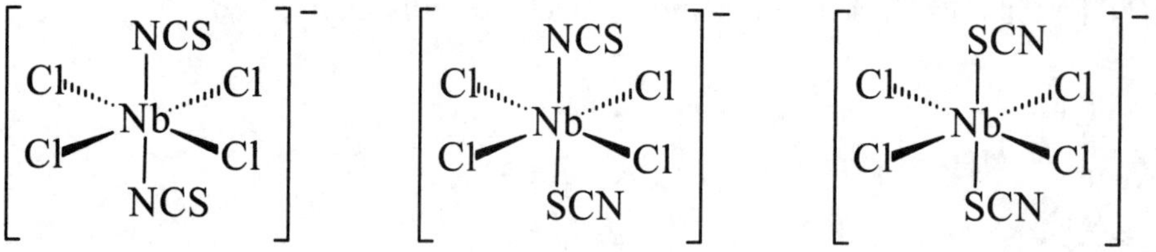

Linkage isomers:

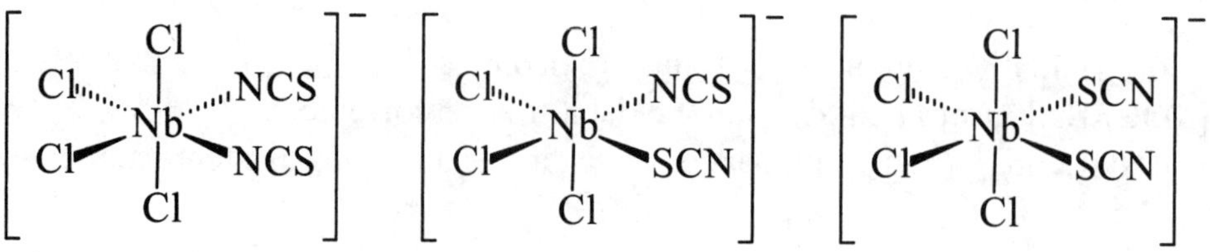

21.96

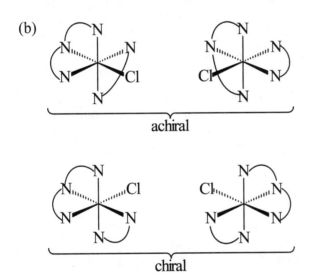

1 can exist as enantiomers.

21.98 (a) A tridentate ligand bonds to a metal using electron pairs on three donor atoms. Diethylenetriamine (dien) has three nitrogens that it can use to bond to a metal.

(b)

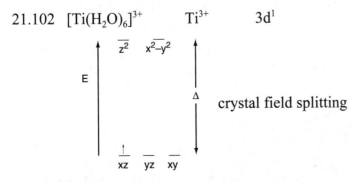

Color of Complexes and Crystal Field Theory (Sections 21.9–21.10)

21.100 The measure of the amount of light absorbed by a substance is called the absorbance, and a graph of absorbance versus wavelength is called an absorption spectrum. If a complex absorbs at 455 nm, its color is orange (use the color wheel in Figure 21.20).

21.102 $[Ti(H_2O)_6]^{3+}$ Ti^{3+} $3d^1$

$[Ti(H_2O)_6]^{3+}$ is colored because it can absorb light in the visible region, exciting the electron to the higher-energy set of orbitals.

21.104 $\lambda = 544$ nm $= 544 \times 10^{-9}$ m

$$\Delta = \frac{hc}{\lambda} = \frac{(6.626 \times 10^{-34} \text{ J} \cdot \text{s})(3.00 \times 10^8 \text{ m/s})}{(544 \times 10^{-9} \text{ m})} = 3.65 \times 10^{-19} \text{ J}$$

$\Delta = (3.65 \times 10^{-19} \text{ J/ion})(6.022 \times 10^{23} \text{ ion/mol}) = 219{,}803$ J/mol $= 220$ kJ/mol

For $[Ti(H_2O)_6]^{3+}$, $\Delta = 240$ kJ/mol

Because $\Delta_{NCS^-} < \Delta_{H_2O}$ for the Ti complex, NCS^- is a weaker-field ligand than H_2O. If $[Ti(NCS)_6]^{3-}$ absorbs at 544 nm, its color should be red (use the color wheel in Figure 21.20).

21.106   (a) $[CrF_6]^{3-}$ (b) $[V(H_2O)_6]^{3+}$ (c) $[Fe(CN)_6]^{3-}$

$\underline{}$ $\underline{}$ $\underline{}$ $\underline{}$ $\underline{}$ $\underline{}$

$\underline{\uparrow}$ $\underline{\uparrow}$ $\underline{\uparrow}$
3 unpaired e$^-$

$\underline{\uparrow}$ $\underline{\uparrow}$ $\underline{}$
2 unpaired e$^-$

$\underline{\uparrow\downarrow}$ $\underline{\uparrow\downarrow}$ $\underline{\uparrow}$
1 unpaired e$^-$

21.108 $Ni^{2+}(aq)$ $Zn^{2+}(aq)$

$\underline{\uparrow}$ $\underline{\uparrow}$ $\underline{\uparrow\downarrow}$ $\underline{\uparrow\downarrow}$

$\underline{\uparrow\downarrow}$ $\underline{\uparrow\downarrow}$ $\underline{\uparrow\downarrow}$ $\underline{\uparrow\downarrow}$ $\underline{\uparrow\downarrow}$ $\underline{\uparrow\downarrow}$

$Ni^{2+}(aq)$ is green because the Ni^{2+} ion can absorb light, which promotes electrons from the filled d orbitals to the higher energy half-filled d orbitals. $Zn^{2+}(aq)$ is colorless because the d orbitals are completely filled and no electrons can be promoted, so no light is absorbed.

21.110   Weak-field ligands produce a small Δ. Strong-field ligands produce a large Δ. For a metal complex with weak-field ligands, $\Delta < P$, where P is the pairing energy, and it is easier to place an electron in either d_{z^2} or $d_{x^2-y^2}$ than to pair up electrons; high-spin complexes result. For a metal complex with strong-field ligands, $\Delta > P$ and it is easier to pair up electrons than to place them in either d_{z^2} or $d_{x^2-y^2}$; low-spin complexes result.

21.112 ___ x^2-y^2

⇅ xy

⇅ z^2

⇅ ⇅ xz, yz

Square planar geometry is most common for metal ions with d^8 configurations because this configuration favors low-spin complexes in which all four lower energy d orbitals are filled, and the higher energy $d_{x^2-y^2}$ orbital is vacant.

21.114 (a) $[Mn(CN)_6]^{3-}$ Mn^{3+} [Ar] $3d^4$
CN$^-$ is a strong-field ligand. The Mn^{3+} complex is low-spin.

— —

⇅ ↑ ↑   2 unpaired e$^-$, paramagnetic

(b) $[Zn(NH_3)_4]^{2+}$ Zn^{2+} [Ar] $3d^{10}$
$[Zn(NH_3)_4]^{2+}$ is tetrahedral.

⇅ ⇅ ⇅

⇅ ⇅   no unpaired e$^-$, diamagnetic

(c) $[Fe(CN)_6]^{4-}$ Fe^{2+} [Ar] $3d^6$
CN$^-$ is a strong-field ligand. The Fe^{2+} complex is low-spin.

— —

⇅ ⇅ ⇅   no unpaired e$^-$, diamagnetic

(d) $[FeF_6]^{4-}$ Fe^{2+} [Ar] $3d^6$
F$^-$ is a weak-field ligand. The Fe^{2+} complex is high-spin.

↑ ↑

⇅ ↑ ↑   4 unpaired e$^-$, paramagnetic

21.116 Cr^{3+} is a $3d^3$ ion. Regardless of the crystal field splitting energy, the three electrons singly occupy the three lower energy d orbitals.

21.118   (a) $[Mn(H_2O)_6]^{2+}$ high-spin Mn^{2+}, $3d^5$

↿ ↿

↿ ↿ ↿
5 unpaired e⁻

(b) $Pt(NH_3)_2Cl_2$ square-planar Pt^{2+}, $5d^8$

—

⇅

⇅

⇅ ⇅
no unpaired e⁻

(c) $[FeO_4]^{2-}$ tetrahedral Fe(VI), $3d^2$

— — —

↿ ↿
2 unpaired e⁻

(d) $[Ru(NH_3)_6]^{2+}$ low-spin Ru^{2+}, $4d^6$

— —

⇅ ⇅ ⇅
no unpaired e⁻

21.120   The nitro ($-NO_2$) complex is orange, which means it absorbs in the blue region (see color wheel) of the visible spectrum. The nitrito ($-ONO$) is red, which means it absorbs in the green region. The energy of the absorbed light is related to ligand field strength. Blue is higher energy than green, therefore nitro ($-NO_2$) is the stronger field ligand.

21.122 ML_5 (square pyramid)

— x^2-y^2

— z^2

—   xy

— —   xz, yz

21.124 (a) Fe^{2+}, sodium pentacyanonitrosylferrate(II)

(b) — —

$$\underline{\uparrow\downarrow}\quad\underline{\uparrow\downarrow}\quad\underline{\uparrow\downarrow}$$

low-spin, no unpaired e^-

Valence Bond Theory (Section 21.11)

21.126 (a) $[Ti(H_2O)_6]^{3+}$

Ti^{3+} [Ar] $\underline{\uparrow}\;\underline{\;}\;\underline{\;}\;\underline{\;}\;\underline{\;}$ $\underline{\;}$ $\underline{\;}\;\underline{\;}\;\underline{\;}$

                 3d   4s   4p

$[Ti(H_2O)_6]^{3+}$ [Ar] $\underline{\uparrow}\;\underline{\;}\;\underline{\;}$ | $\underline{\uparrow\downarrow}\;\underline{\uparrow\downarrow}$ $\underline{\uparrow\downarrow}$ $\underline{\uparrow\downarrow}\;\underline{\uparrow\downarrow}\;\underline{\uparrow\downarrow}$ |

                 3d   4s   4p

                       d^2sp^3 1 unpaired e^-

(b) $[NiBr_4]^{2-}$

Ni^{2+} [Ar] $\underline{\uparrow\downarrow}\;\underline{\uparrow\downarrow}\;\underline{\uparrow\downarrow}\;\underline{\uparrow}\;\underline{\uparrow}$ $\underline{\;}$ $\underline{\;}\;\underline{\;}\;\underline{\;}$

                 3d   4s   4p

$[NiBr_4]^{2-}$ [Ar] $\underline{\uparrow\downarrow}\;\underline{\uparrow\downarrow}\;\underline{\uparrow\downarrow}\;\underline{\uparrow}\;\underline{\uparrow}$ | $\underline{\uparrow\downarrow}$ $\underline{\uparrow\downarrow}\;\underline{\uparrow\downarrow}\;\underline{\uparrow\downarrow}$ |

                 3d   4s   4p

                       sp^3 2 unpaired e^-

(c) $[Fe(CN)_6]^{3-}$ (low-spin)

Fe^{3+} [Ar] $\underline{\uparrow}\;\underline{\uparrow}\;\underline{\uparrow}\;\underline{\uparrow}\;\underline{\uparrow}$ $\underline{\;}$ $\underline{\;}\;\underline{\;}\;\underline{\;}$

                 3d   4s   4p

$[Fe(CN)_6]^{3-}$ [Ar] $\underline{\uparrow\downarrow}\;\underline{\uparrow\downarrow}\;\underline{\uparrow}$ | $\underline{\uparrow\downarrow}\;\underline{\uparrow\downarrow}$ $\underline{\uparrow\downarrow}$ $\underline{\uparrow\downarrow}\;\underline{\uparrow\downarrow}\;\underline{\uparrow\downarrow}$ |

                 3d   4s   4p

                       d^2sp^3 1 unpaired e^-

(d) $[MnCl_6]^{3-}$ (high-spin)

Mn^{3+} [Ar] $\underline{\uparrow}\;\underline{\uparrow}\;\underline{\uparrow}\;\underline{\uparrow}\;\underline{\;}$ $\underline{\;}$ $\underline{\;}\;\underline{\;}\;\underline{\;}$

                 3d   4s   4p

$[MnCl_6]^{3-}$ [Ar] $\underline{\uparrow}\;\underline{\uparrow}\;\underline{\uparrow}\;\underline{\uparrow}\;\underline{\;}$ | $\underline{\uparrow\downarrow}$ $\underline{\uparrow\downarrow}\;\underline{\uparrow\downarrow}\;\underline{\uparrow\downarrow}$ $\underline{\uparrow\downarrow}\;\underline{\uparrow\downarrow}$ | $\underline{\;}\;\underline{\;}\;\underline{\;}$

                 3d   4s   4p   4d

                       sp^3d^2 4 unpaired e^-

21.128 (a) +3, M = Cr or Ni

(b)

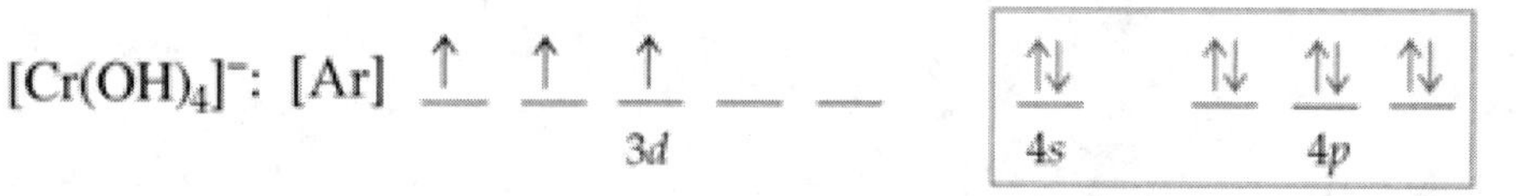

$[Cr(OH)_4]^-$: [Ar] ↑ ↑ ↑ __ __ | ↑↓ | ↑↓ ↑↓ ↑↓ |
           3d   4s   4p

Four sp^3 bonds to the ligands

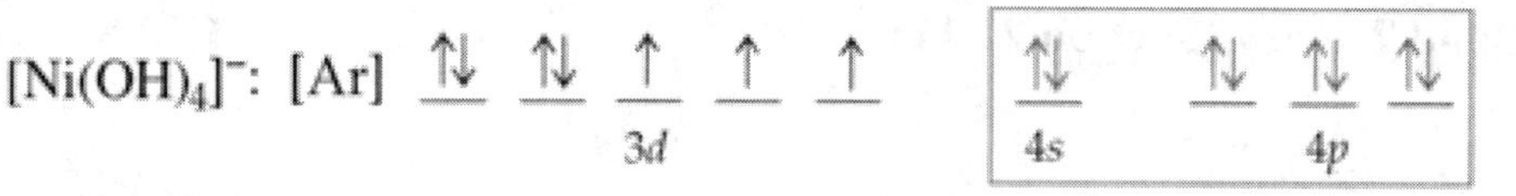

$[Ni(OH)_4]^-$: [Ar] ↑↓ ↑↓ ↑ ↑ ↑ | ↑↓ | ↑↓ ↑↓ ↑↓ |
             3d   4s   4p

(c) $[Cr(OH)_4]^-$

Four sp^3 bonds to the ligands

Multiconcept Problems

21.130 (a)

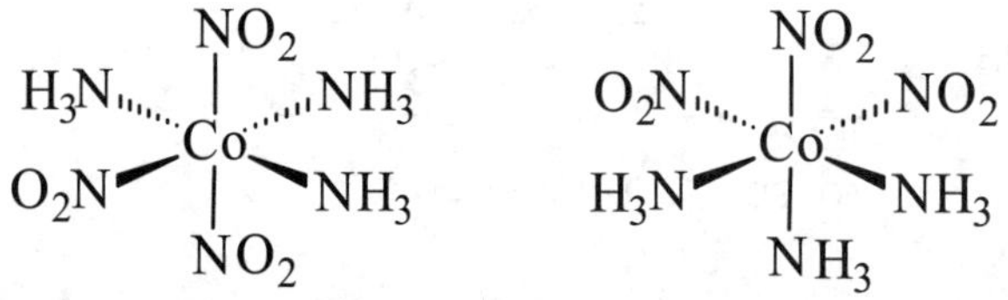

(b) $[Co(NH_3)_4(NO_2)_2][Co(NH_3)_2(NO_2)_4]$

21.132 $[Mn(CN)_6]^{3-}$ Mn^{3+} $3d^4$

valence bond theory

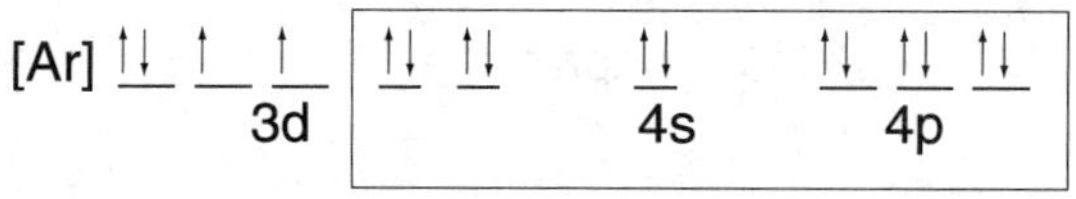

[Ar] ↑↓ ↑ ↑ | ↑↓ ↑↓ | ↑↓ | ↑↓ ↑↓ ↑↓ | d^2sp^3
      3d   4s   4p

crystal field theory

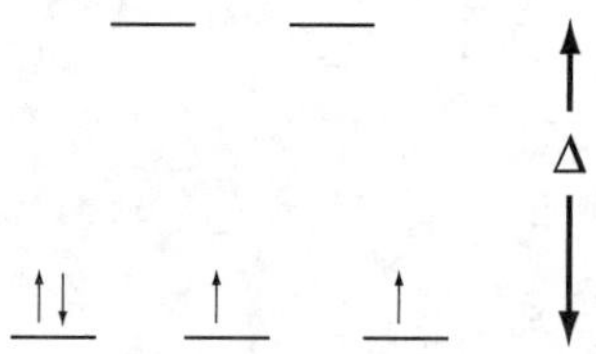

Crystal field model predicts 2 unpaired electrons.

21.134 (a) $Fe^{3+}(aq) + 3\,C_2O_4^{2-}(aq) \rightleftharpoons [Fe(C_2O_4)_3]^{3-}(aq)$ $K_f = 3.3 \times 10^{20}$

$[Fe(C_2O_4)_3]^{3-}(aq) \rightleftharpoons Fe^{3+}(aq) + 3\,C_2O_4^{2-}(aq)$ $K = 1/K_f = 3.0 \times 10^{-21}$

	$[Fe(C_2O_4)_3]^{3-}(aq)$	$Fe^{3+}(aq)$	$3\,C_2O_4^{2-}(aq)$
initial (M)	0.100	0	0
change (M)	−x	+x	+3x
equil (M)	0.100 − x	x	3x

$$K = \frac{[Fe^{3+}][C_2O_4^{2-}]^3}{[[Fe(C_2O_4)_3]^{3-}]} = 3.0 \times 10^{-21} = \frac{x(3x)^3}{0.100 - x} \approx \frac{x(3x)^3}{0.100}$$

$$3.0 \times 10^{-22} = 27x^4$$

$$x = [Fe^{3+}] = \sqrt[4]{3.0 \times 10^{-22}/27} = 1.8 \times 10^{-6} \text{ M}$$

(b) $[Fe(C_2O_4)_3]^{3-}(aq) \rightleftharpoons Fe^{3+}(aq) + 3 C_2O_4^{2-}(aq)$ $\qquad K_1 = 3.0 \times 10^{-21}$

$\quad 3 H_3O^+(aq) + 3 C_2O_4^{2-}(aq) \rightleftharpoons 3 HC_2O_4^-(aq) + 3 H_2O(l)$ $\qquad K_2 = (1/K_{a2})^3$

$\underline{\quad 3 H_3O^+(aq) + 3 HC_2O_4^-(aq) \rightleftharpoons 3 H_2C_2O_4(aq) + 3 H_2O(l) \qquad K_3 = (1/K_{a1})^3}$

$\quad [Fe(C_2O_4)_3]^{3-}(aq) + 6 H_3O^+(aq) \rightleftharpoons Fe^{3+}(aq) + 3 H_2C_2O_4(aq) + 6 H_2O(l)$

$$K_{overall} = K_1K_2K_3 = (3.0 \times 10^{-21})\left(\frac{1}{6.4 \times 10^{-5}}\right)^3\left(\frac{1}{5.9 \times 10^{-2}}\right)^3 = 5.6 \times 10^{-5}$$

$\Delta G = -RT\ln K = -[8.314 \times 10^{-3} \text{ kJ/(K} \cdot \text{mol)}](298 \text{ K}) \ln (5.6 \times 10^{-5}) = 24.3 \text{ kJ/mol}$
The reaction is nonspontaneous because ΔG is positive.

(c) $\uparrow \ \uparrow$

$\underline{\uparrow} \ \ \underline{\uparrow} \ \ \underline{\uparrow}$
5 unpaired e$^-$

(d)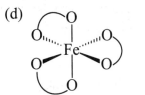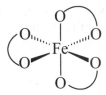

The complex is chiral. Enantiomers are shown.

21.136 (a) $[Fe^{2+}(aq) \rightarrow Fe^{3+}(aq) + e^-] \times 5$ $\qquad\qquad\qquad E^° = -0.77$ V
(oxidation half reaction)

$MnO_4^-(aq) \rightarrow Mn^{2+}(aq)$
$MnO_4^-(aq) \rightarrow Mn^{2+}(aq) + 4 H_2O(l)$
$8 H^+(aq) + MnO_4^-(aq) \rightarrow Mn^{2+}(aq) + 4 H_2O(l)$
$8 H^+(aq) + MnO_4^-(aq) + 5 e^- \rightarrow Mn^{2+}(aq) + 4 H_2O(l)$ $\qquad E^° = 1.51$ V
(reduction half reaction)

Combine the two half reactions.
$5 Fe^{2+}(aq) + MnO_4^-(aq) + 8 H^+(aq) \rightarrow 5 Fe^{3+}(aq) + Mn^{2+}(aq) + 4 H_2O(l)$

(b) $E^° = -0.77$ V + 1.51 V = 0.74 V

$$\Delta G^° = -nFE^° = -(5 \text{ mol e}^-)\left(\frac{96,500 \text{ C}}{\text{mol e}^-}\right)(0.74 \text{ V})\left(\frac{1 \text{ J}}{1 \text{ C}\cdot\text{V}}\right)\left(\frac{1 \text{ kJ}}{1000 \text{ J}}\right) = -3.6 \times 10^2 \text{ kJ}$$

$$E° = \frac{0.0592 \text{ V}}{n} \log K$$

$$\log K = \frac{nE°}{0.0592 \text{ V}} = \frac{(5)(0.74 \text{ V})}{0.0592 \text{ V}} = 62.5; \quad K = 10^{62.5} = 3 \times 10^{62}$$

(c)

(d) The paramagnetism of the solution increases as the reaction proceeds. When $Fe^{2+}(aq)$ is oxidized to $Fe^{3+}(aq)$, the number of unpaired electrons goes from 4 to 5. In addition, the $Mn^{2+}(aq)$ produced has 5 unpaired electrons.

(e) $34.83 \text{ mL} = 0.03483 \text{ L}$

$\text{mol MnO}_4^-(aq) = (0.051\ 32 \text{ mol/L})(0.034\ 83 \text{ L}) = 1.787 \times 10^{-3} \text{ mol MnO}_4^-$

$$\text{mass Fe} = 1.787 \times 10^{-3} \text{ mol MnO}_4^- \times \frac{5 \text{ mol Fe}^{2+}}{1 \text{ mol MnO}_4^-} \times \frac{55.847 \text{ g Fe}}{1 \text{ mol Fe}^{2+}} = 0.4990 \text{ g Fe}$$

$$\text{mass \% Fe} = \frac{0.4990 \text{ g Fe}}{1.265 \text{ g sample}} \times 100\% = 39.45\% \text{ Fe}$$

21.138 (a) $t_{1/2} = \dfrac{0.693}{k} = \dfrac{0.693}{3.2 \times 10^{-5} \text{ s}^{-1}} = 2.16 \times 10^4 \text{ s}$

$t_{1/2} = (2.16 \times 10^4 \text{ s})(1 \text{ h}/3600 \text{ s}) = 6.0 \text{ h}$

(b) Let $A = \text{trans-[Co(en)}_2\text{Cl}_2]^+$

$$\ln \frac{[A]_t}{[A]_o} = -kt; \qquad \ln [A]_t - \ln [A]_o = -kt; \qquad \ln [A]_t = \ln [A]_o - kt$$

$\ln [A]_t = \ln(0.138) - (3.2 \times 10^{-5} \text{ s}^{-1})(16.5 \text{ h})(3600 \text{ s/h}) = -3.88$

$[A]_t = e^{-3.88} = 0.021 \text{ M}$

(c) $\text{trans-[Co(en)}_2\text{Cl}_2]^+(aq) \rightarrow \text{trans-[Co(en)}_2\text{Cl]}^{2+}(aq) + \text{Cl}^-(aq) \qquad \text{(slow)}$

$\text{trans-[Co(en)}_2\text{Cl]}^{2+}(aq) + H_2O(l) \rightarrow \text{trans-[Co(en)}_2(H_2O)\text{Cl]}^{2+}(aq) \quad \text{(fast)}$

(d)

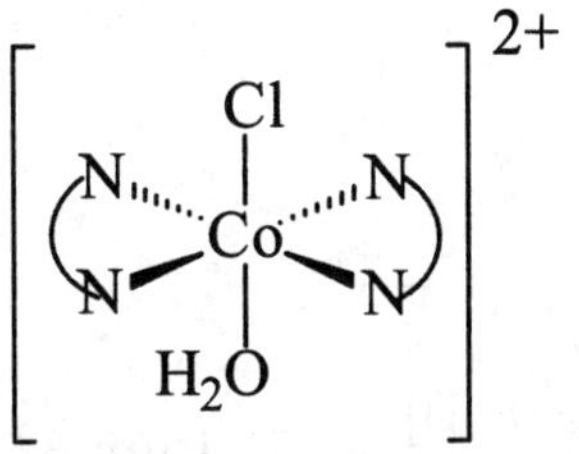

2+

The reaction product is achiral because it has several mirror planes.

(e) ___ x^2-y^2

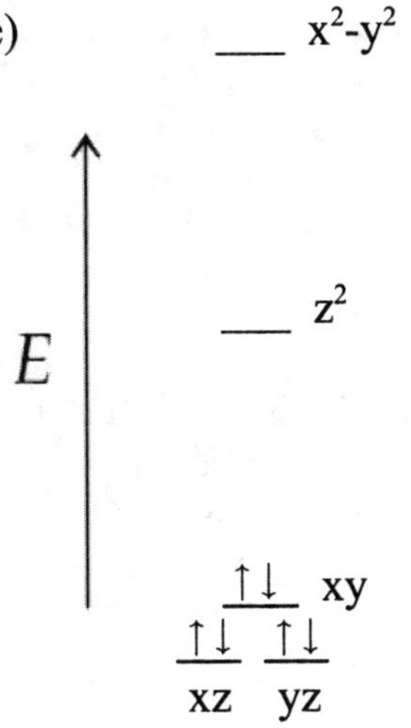

21.140 (a) & (b) Let  represent the tfac⁻ ligand.

$A =$

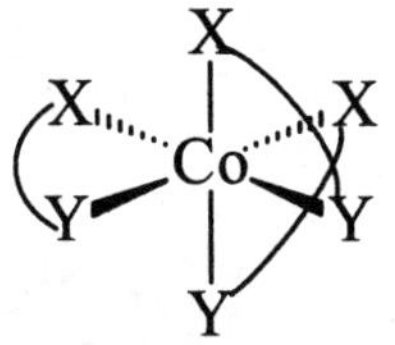

$B =$

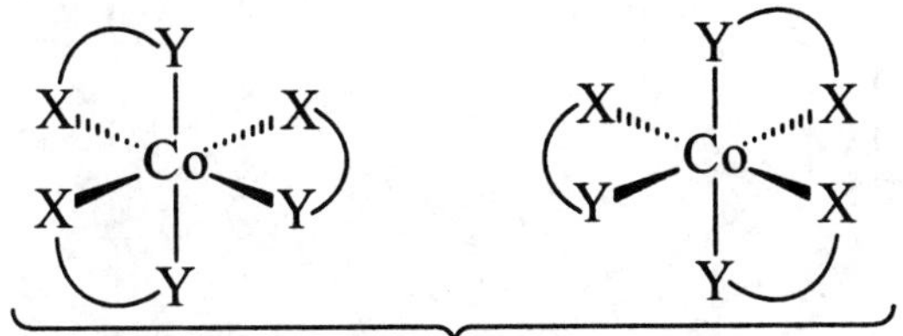

enantiomers

(c) $k_1 = 0.0889 \text{ h}^{-1}$, $T_1 = 66.1 \text{ °C} = 66.1 + 273.15 = 339.2 \text{ K}$
$k_2 = (0.0870 \text{ min}^{-1})(60 \text{ min}/1 \text{ h}) = 5.22 \text{ h}^{-1}$, $T_2 = 99.2 \text{ °C} = 99.2 + 273.15 = 372.3 \text{ K}$

$$\ln\left(\frac{k_2}{k_1}\right) = \left(\frac{-E_a}{R}\right)\left(\frac{1}{T_2} - \frac{1}{T_1}\right)$$

$$E_a = -\frac{[\ln k_2 - \ln k_1](R)}{\left(\dfrac{1}{T_2} - \dfrac{1}{T_1}\right)}$$

$$E_a = -\frac{[\ln(5.22) - \ln(0.0889)][8.314 \times 10^{-3} \text{ kJ/(K} \cdot \text{mol)}]}{\left(\dfrac{1}{372.3 \text{ K}} - \dfrac{1}{339.2 \text{ K}}\right)} = 129 \text{ kJ/mol}$$

(d)   ⟍ ⟍

$$\underset{\text{no unpaired e}^-\text{, diamagnetic}}{\underline{\uparrow\downarrow}\quad\underline{\uparrow\downarrow}\quad\underline{\uparrow\downarrow}}$$

21.142   (a)

	$Au^{3+}(aq)$	$+$	$4\ SCN^-(aq)$	$\rightleftarrows$	$Au(SCN)_4^-(aq)$
assume 100% reaction (M)	0		0		0.050
assume small back reaction (M)	+x		+4x		−x
equil (M)	x		4x		0.050 − x

$$K_f = \frac{[Au(SCN)_4^-]}{[Au^{3+}][SCN^-]^4} = 10^{37} = \frac{(0.050 - x)}{(x)(4x)^4} \approx \frac{0.050}{x(4x)^4}$$

Solve for x. $x = [Au^{3+}] = \sqrt[5]{\dfrac{0.05}{(256)(10^{37})}} = 7 \times 10^{-9} \text{ M}$

(b)

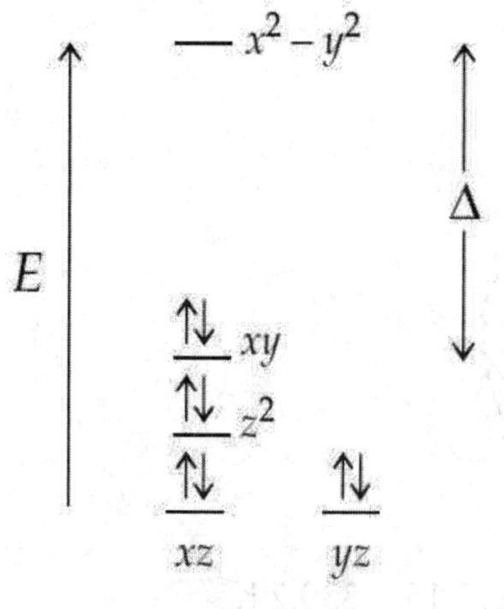

$$\underset{\substack{Au(SCN)_4^- \\ \text{no unpaired e}^-}}{}$$

The Main-Group Elements

22.1   Cl (group 7A) is more nonmetallic because it lies to the right and above Sb (group 5A).

22.2   Element A

22.3   Carbon forms strong π bonds with oxygen. Silicon does not form strong π bonds with oxygen, and what results are chains of alternating silicon and oxygen singly bonded to each other.

22.4   Because of the small sizes of their atoms, the second-row elements generally form a maximum of four covalent bonds. By contrast, the larger atoms of the third-row elements can accommodate more than four nearest neighbors and can therefore form more than four bonds. Thus, nitrogen forms only NCl_3, but phosphorus forms both PCl_3 and PCl_5.

22.5   (a) (1) ZrH_x, interstitial   (2) PH_3, covalent   (3) HBr, covalent   (4) LiH, ionic
        (b) (1) and (4) are likely to be solids at 25 °C. (2) and (3) are likely to be gases at 25 °C. Covalent hydrides, like (2) and (3), form discrete molecules and have only relatively weak intermolecular forces, resulting in gases. (4) is an ionic metal hydride with strong ion-ion forces holding the 3-dimensional lattice together in the solid state. (1) is an interstitial hydride with the metal atoms in a solid crystal lattice and H's occupying holes.
        (c) $LiH(s) + H_2O(l) \rightarrow H_2(g) + Li^+(aq) + OH^-(aq)$

22.6   (a) A, NaH; B, PdH_x; C, H_2S; D, HI
        (b) NaH (ionic); PdH_x (interstitial); H_2S and HI (covalent)
        (c) H_2S and HI (molecular); NaH and PdH_x (3-dimensional crystal)
        (d) NaH: Na +1, H −1
             H_2S: S −2, H +1
             HI: I −1, H +1

22.7   (a) $Si_8O_{24}^{16-}$

22.8

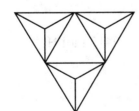

$Si_3O_9^{6-}$

415

22.9   A is Li; B is Ga; C is C
(a) Li_2O, Ga_2O_3, CO_2
(b) Li_2O is the most ionic. CO_2 is the most covalent.
(c) CO_2 is the most acidic. Li_2O is the most basic.
(d) Ga_2O_3 is amphoteric and can react with both $H^+(aq)$ and $OH^-(aq)$.

22.10   (a) $Li_2O(s) + H_2O(l) \rightarrow 2\ Li^+(aq) + 2\ OH^-(aq)$
(b) $SO_3(l) + H_2O(l) \rightarrow H^+(aq) + HSO_4^-(aq)$
(c) $Cr_2O_3(s) + 6\ H^+(aq) \rightarrow 2\ Cr^{3+}(aq) + 3\ H_2O(l)$
(d) $Cr_2O_3(s) + 2\ OH^-(aq) + 3\ H_2O(l) \rightarrow 2\ Cr(OH)_4^-(aq)$

22.11 $H_2(g) + 1/2\ O_2(g) \rightarrow H_2O(g)$ $\Delta H° = -242$ kJ

$$\text{mol } H_2 = 1.45 \times 10^6\ \text{L} \times \frac{0.088\ \text{kg}}{1\ \text{L}} \times \frac{1000\ \text{g}}{1\ \text{kg}} \times \frac{1\ \text{mol } H_2}{2.016\ \text{g } H_2} = 6.33 \times 10^7\ \text{mol } H_2$$

$$q = 6.33 \times 10^7\ \text{mol } H_2 \times \frac{242\ \text{kJ}}{1\ \text{mol } H_2} = 1.5 \times 10^{10}\ \text{kJ}$$

$$\text{mass } O_2 = 6.33 \times 10^7\ \text{mol } H_2 \times \frac{0.5\ \text{mol } O_2}{1\ \text{mol } H_2} \times \frac{32.00\ \text{g } O_2}{1\ \text{mol } O_2} \times \frac{1\ \text{kg}}{1000\ \text{g}} = 1.0 \times 10^6\ \text{kg } O_2$$

22.12   (a) $CH_4(g) + H_2O(g) \xrightarrow[\text{Ni catalyst}]{} CO(g) + 3\ H_2(g)$

$CO(g) + H_2O(g) \xrightarrow{400\ °C} CO_2(g) + H_2(g)$

(b) The production of hydrogen from methane contributes to climate change because it produces CO_2 as a byproduct.

22.13   (a)
$H_2O(g) + C(s) \xrightarrow{1000\ °C} CO(g) + H_2(g)$
(b) $C_3H_8(g) + 3\ H_2O(g) \rightarrow 7\ H_2(g) + 3\ CO(g)$

22.14   Hydrogen can be stored as a solid in the form of solid interstitial hydrides or in the recently discovered tube-shaped molecules called carbon nanotubes.

22.15   Assume 12.0 g of Pd with a volume of 1.0 cm^3.
$V_{H_2} = 935\ cm^3 = 935\ mL = 0.935\ L$

$$PV = nRT; \quad n_{H_2} = \frac{PV}{RT} = \frac{(1.00\ \text{atm})(0.935\ \text{L})}{\left(0.082\ 06\ \dfrac{\text{L} \cdot \text{atm}}{\text{K} \cdot \text{mol}}\right)(273\ \text{K})} = 0.0417\ \text{mol } H_2$$

$n_H = 2\,n_{H_2} = 0.0834\ \text{mol } H$

$$12.0\ \text{g Pd} \times \frac{1\ \text{mol Pd}}{106.42\ \text{g Pd}} = 0.113\ \text{mol Pd}$$

$Pd_{0.113}H_{0.0834}$

$Pd_{0.113/0.113}H_{0.0834/0.113}$

$PdH_{0.74}$

g H = (0.0834 mol H)(1.008 g/mol) = 0.0841 g H

$d_H = 0.0841$ g/cm^3; $M_H = \dfrac{0.0834 \text{ mol}}{0.001 \text{ L}} = 83.4$ M

22.16   (a) TiH_2, 49.88; Assume 1.0 cm^3 of TiH_2, which has a mass of 3.75 g.

3.75 g TiH_2 x $\dfrac{1 \text{ mol } TiH_2}{49.88 \text{ g } TiH_2}$ = 0.0752 mol TiH_2

0.0752 mol TiH_2 x $\dfrac{2 \text{ mol H}}{1 \text{ mol } TiH_2}$ = 0.150 mol H

0.150 mol H x $\dfrac{1.008 \text{ g H}}{1 \text{ mol H}}$ = 0.151 g H

$d_H = 0.15$ g/cm^3; the density of H in TiH_2 is about 2.1 times the density of liquid H_2.

(b)

PV = nRT;   V = $\dfrac{nRT}{P}$ = $\dfrac{\left(0.15 \text{ g x }\dfrac{1 \text{ mol}}{2.016 \text{ g}}\right)\left(0.082\ 06\ \dfrac{\text{L}\cdot\text{atm}}{\text{K}\cdot\text{mol}}\right)(273 \text{ K})}{1.00 \text{ atm}}$ = 1.7 L H_2

1.7 L = 1.7 x 10^3 mL = 1.7 x 10^3 cm^3

22.17   (a) NH_3 is the Lewis base and BH_3 is the Lewis acid.
(b) Nitrogen and boron both have sp^3 hybrid orbitals with bond angles of close to 109.5°.

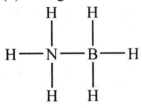

Conceptual Problems

22.18   (a)   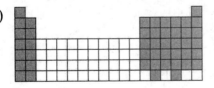   main-group elements

(b)   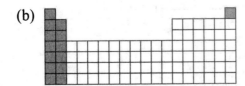   s-block elements

(c) 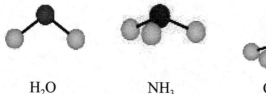 p-block elements

(d) main-group metals

(e) nonmetals

(f) semimetals

22.20   (a) A, KH; B, MgH$_2$; C, H$_2$O; D, HCl
        (b) HCl
        (c) KH(s) + H$_2$O(l) → H$_2$(g) + K$^+$(aq) + OH$^-$(aq)
             MgH$_2$(s) + 2 H$_2$O(l) → 2 H$_2$(g) + Mg^{2+}(aq) + 2 OH$^-$(aq)
        (d) HCl reacts with water to give an acidic solution. KH and MgH$_2$ react with water to give a basic solution.

22.22   (a)

H$_2$O   NH$_3$ CH$_4$

2nd row elements cannot form expanded octets.

(b)

22.24   (a) (1) –2, +4;   (2) –2, +6;   (3) –2, +2
        (b) (1) covalent;   (2) covalent;   (3) ionic

(c) (1) acidic;   (2) acidic;   (3) basic
(d) (1) carbon;   (2) sulfur

22.26   (a) A, CaO; B, Al_2O_3; C, SO_3; D, SeO_3
(b) CaO (basic); Al_2O_3 (amphoteric); SO_3 and SeO_3 (acidic)
(c) CaO (most ionic); SO_3 (most covalent)
(d) CaO and Al_2O_3 (3-dimensional crystal); SO_3 and SeO_3 (molecular)
(e) CaO (highest melting point); SO_3 (lowest melting point)

22.28   (a) N_2, O_2, F_2, P_4 (tetrahedral), S_8 (crown-shaped ring), Cl_2
(b) $:N≡N:$ $\ddot{O}=\ddot{O}$ $:\ddot{F}-\ddot{F}:$ $:\ddot{C}l—\ddot{C}l:$

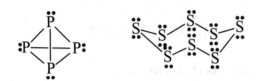

(c) The smaller N and O can form strong π bonds, whereas P and S cannot. In both F_2 and Cl_2, the atoms are joined by a single bond.

22.30   (a) CO_2, Cl_2O_7, SO_3, N_2O_5
(b) $:\ddot{O}=C=\ddot{O}:$

(resonance structures are needed)

(resonance structures are needed)

Section Problems
Periodic Trends (Section 22.1)

22.32   (a) Cl (group 7A) is above and to the right of As (group 5A) in the periodic table, therefore Cl has the higher ionization energy.
(b) Mg is above Ba in group 2A, therefore Mg has the higher ionization energy.

22.34   (a) P (group 5A) is to the right and above of Sn (group 4A), therefore Sn has the larger atomic radius.
(b) Br (group 7A) is to the right of Ge (group 4A) in the same row, therefore Ge has the larger atomic radius.

22.36 (a) I (group 7A) is to the right of Sb (group 5A) in the same row, therefore I has the higher electronegativity.
(b) P (group 5A) is above Sb (group 5A), therefore P has the higher electronegativity.

22.38 (a) Bi (group 5A) is below As (group 5A), therefore Bi has more metallic character.
(b) Ge (group 4A) is below Si (group 5A), therefore Ge has more metallic character.

22.40 In each case the more ionic compound is the one formed between a metal and nonmetal.
(a) CaH_2 (b) Ga_2O_3 (c) KCl   (d) $AlCl_3$

22.42 Molecular   (a) B_2H_6 (c) SO_3 (d) $GeCl_4$
Extended three-dimensional structure   (b) $KAlSi_3O_8$

22.44 (a) Sn   (b) Cl   (c) Sn   (d) Se   (e) B

Distinctive Properties of the Second-Row Elements (Section 22.2)

22.46 The smaller B atom can bond to a maximum of four nearest neighbors, whereas the larger Al atom can accommodate more than four nearest neighbors.

22.48 In O_2 a π bond is formed by 2p orbitals on each O. S does not form strong π bonds with its 3p orbitals, which leads to the S_8 ring structure with single bonds.

22.50 In N_2 π bonds are formed by 2p orbitals on each N. P does not form strong π bonds with its 3p orbitals, which leads to the P_4 tetrahedral structure with single bonds.

Group 1A: Hydogen (Section 22.3)

22.52 (a) $Zn(s) + 2 H^+(aq) \rightarrow H_2(g) + Zn^{2+}(aq)$
(b) at 1000 °C, $H_2O(g) + C(s) \rightarrow CO(g) + H_2(g)$
(c) at 1100 °C with a Ni catalyst, $H_2O(g) + CH_4(g) \rightarrow CO(g) + 3 H_2(g)$
(d) There are a number of possibilities. (b) and (c) above are two; electrolysis is another:
  $2 H_2O(l) \rightarrow 2 H_2(g) + O_2(g)$

22.54 $CaH_2(s) + 2 H_2O(l) \rightarrow 2 H_2(g) + Ca^{2+}(aq) + 2 OH^-(aq)$
CaH_2, 42.09; 25 °C = 298 K

$$PV = nRT; \quad n_{H_2} = \frac{PV}{RT} = \frac{(1.00 \text{ atm})(2.0 \times 10^5 \text{ L})}{\left(0.082\ 06 \frac{L \cdot atm}{K \cdot mol}\right)(298 \text{ K})} = 8.18 \times 10^3 \text{ mol } H_2$$

$$8.18 \times 10^3 \text{ mol } H_2 \times \frac{1 \text{ mol } CaH_2}{2 \text{ mol } H_2} \times \frac{42.09 \text{ g } CaH_2}{1 \text{ mol } CaH_2} \times \frac{1 \text{ kg}}{1000 \text{ g}} = 1.7 \times 10^2 \text{ kg } CaH_2$$

22.56 (a) NaH, ionic   (b) CH_4, covalent

22.58   (a) MgH_2, H^- (b) PH_3, covalent   (c) KH, H^- (d) HBr, covalent

22.60 H_2S – covalent hydride, gas, weak acid in H_2O
          NaH – ionic hydride, solid (salt like), reacts with H_2O to produce H_2
          PdH_x – metallic (interstitial) hydride, solid, stores hydrogen

22.62   (a)   H—S̈e—H , bent   (b) H—Äs—H, trigonal pyramidal
                                                                    |
                                                                    H

          (c)   H   , tetrahedral
                        |
                 H—Si—H
                        |
                        H

22.64   A nonstoichiometric compound is a compound whose atomic composition cannot be
          expressed as a ratio of small whole numbers. An example is PdH_x. The lack of stoichiometry
          results from the hydrogen occupying holes in the solid state structure.

22.66   (a) $SrH_2(s)$ + 2 $H_2O(l)$ → 2 $H_2(g)$ + $Sr^{2+}(aq)$ + 2 $OH^-(aq)$
          (b) $NH_3(g)$ + $H_2O(l)$ → $NH_4^+(aq)$ + $OH^-(aq)$

Group 1A and 2A: Alkali and Alkaline Earth Metals (Section 22.4)

22.68   Predicted for Fr: melting point ≈ 23 °C   boiling point ≈ 650 °C
                                          density ≈ 2 g/cm³   atomic radius ≈ 275 pm

22.70   (a) 2 K(s) + 2 $H_2O(l)$ → 2 $K^+(aq)$ + 2 $OH^-(aq)$ + $H_2(g)$
          (b) 2 K(s) + $Br_2(l)$ → 2 KBr(s)
          (c) K(s) + $O_2(g)$ → $KO_2(s)$

22.72   (a) 2 Cs(s) + 2 $H_2O(l)$ → 2 $Cs^+(aq)$ + 2 $OH^-(aq)$ + $H_2(g)$
          (b) Rb(s) + $O_2(g)$ → $RbO_2(s)$

22.74   2 Mg(s) + $O_2(g)$ → 2 MgO(s)
          MgO(s) + $H_2O(l)$ → $Mg(OH)_2(aq)$

22.76   anode $Mg^{2+}(l)$ + 2 e^- → Mg(l)
          cathode $\underline{2\ Cl^-(l)\ →\ Cl_2(g)\ +\ 2\ e^-}$
          overall $Mg^{2+}(l)$ + 2 $Cl^-(l)$ → Mg(l) + $Cl_2(g)$

22.78   (a) O^{2-} (b) O_2^{2-}

Group 3A Elements (Section 22.5)

22.80   (a) Al   (b) Tl   (c) B

421

22.82 +3 for B, Al, Ga and In;   +1 for Tl

22.84 Boron is a hard semiconductor with a high melting point. Boron forms only molecular compounds and does not form an aqueous B^{3+} ion. $B(OH)_3$ is an acid.

22.86 (a) An electron-deficient molecule does not have enough electrons to form a two-center, two-electron bond between each pair of bonded atoms.
(b) A three-center, two-electron bond has two electrons in an orbital involving three atoms, such as the two B-H-B bonds in diborane.
(c)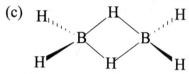

The terminal B–H bonds in diborane are ordinary two-center, two-electron bonds. The bridging B–H–B bonds in diborane are three-center, two-electron bonds. A three-center, two-electron bond is longer than a two-center, two-electron bond because it has less electron density between each pair of adjacent atoms.

Group 4A Elements (Section 22.6)

22.88 (a) Pb   (b) C   (c) Si   (d) C

22.90 (a) $GeBr_4$, tetrahedral; Ge is sp^3 hybridized.
(b) CO_2, linear; C is sp hybridized.
(c) CO_3^{2-}, trigonal planar; C is sp^2 hybridized.
(d) $SnCl_3^-$, trigonal pyramidal; Sn is sp^3 hybridized.

22.92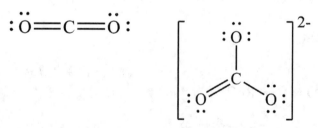

Carbon monoxide will have the strongest carbon-oxygen bond because it is a triple bond.

22.94 C, Si, Ge and Sn have allotropes with the diamond structure.
Sn and Pb have metallic allotropes.
C (nonmetal), Si (semimetal), Ge (semimetal), Sn (semimetal and metal), Pb (metal)

22.96 CO bonds to hemoglobin and prevents it from carrying O_2. CN^- bonds to cytochrome oxidase and interferes with the electron transfer associated with oxidative phosphorylation.

22.98 Silicon and germanium are semimetals, and tin and lead are metals. Silicon is a hard, gray, semiconducting solid that melts at 1414 °C. It crystallizes in a diamondlike structure but does not form a graphitelike allotrope because of the relatively poor overlap of silicon p orbitals. Germanium is a relatively high-melting, brittle semiconductor that has the same crystal structure as diamond and silicon. Tin exists in two allotropic forms: the usual silvery white metallic form called white tin and a brittle, semiconducting form with the diamond structure called gray tin. Both white tin and lead are soft, malleable, low-melting metals. Only the metallic form occurs for lead.

22.100

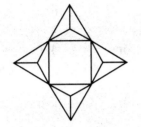

$Si_4O_{12}^{8-}$

22.102 Carbon is a versatile element that can form millions of very stable compounds with elements such as N, O, and H. Biomolecules contain chains and rings with many C–C bonds. Si–Si bonds are much less stable and chains of Si atoms are uncommon. In addition, carbon can form very stable $p\pi$-$p\pi$ multiple bonds. On the other hand, the chemistry of silicon (which cannot form stable $p\pi$-$p\pi$ bonds) is dominated by structures based on the SiO_4^{4-} anion.

Group 5A elements (Section 22.7)

22.104 (a) P   (b) Sb and Bi   (c) N   (d) Bi

22.106 (a) N_2O, +1   (b) N_2H_4, −2   (c) Ca_3P_2, −3
        (d) H_3PO_3, +3   (e) H_3AsO_4, +5

22.108 :N≡N:
        N_2 is unreactive because of the large amount of energy necessary to break the N≡N triple bond.

22.110 (a) NO_2^-, bent   (b) PH_3, trigonal pyramidal
        (c) PF_5, trigonal bipyramidal   (d) PCl_4^+, tetrahedral

22.112 (a) Nitric acid is a strong oxidizing agent, but phosphoric acid is not because the nitrogen atom is smaller and more electronegative than the phosphorus atom. This favors its reduction.
        (b) P, As, and Sb can use d orbitals to become five-coordinate. Nitrogen cannot.

22.114   N═N═O

For left nitrogen:	Isolated nitrogen valence electrons	5
	Bound nitrogen bonding electrons	4
	Bound nitrogen nonbonding electrons	4
	Formal charge = 5 − ½(4) − 4 = −1	

For center nitrogen:	Isolated nitrogen valence electrons	5
	Bound nitrogen bonding electrons	8
	Bound nitrogen nonbonding electrons	0
	Formal charge = 5 – ½(8) – 0 = +1	

For oxygen:	Isolated oxygen valence electrons	6
	Bound oxygen bonding electrons	4
	Bound oxygen nonbonding electrons	4
	Formal charge = 6 – ½(4) – 4 = 0	

$$:N\equiv N - \overset{..}{\underset{..}{O}}:$$

For left nitrogen:	Isolated nitrogen valence electrons	5
	Bound nitrogen bonding electrons	6
	Bound nitrogen nonbonding electrons	2
	Formal charge = 5 – ½(6) – 2 = 0	

For center nitrogen	Isolated nitrogen valence electrons	5
	Bound nitrogen bonding electrons	8
	Bound nitrogen nonbonding electrons	0
	Formal charge = 5 – ½(8) – 0 = +1	

For oxygen:	Isolated oxygen valence electrons	6
	Bound oxygen bonding electrons	2
	Bound oxygen nonbonding electrons	6
	Formal charge = 6 – ½(2) – 6 = –1	

$$:\overset{..}{\underset{..}{N}} - N \equiv O:$$

For left nitrogen:	Isolated nitrogen valence electrons	5
	Bound nitrogen bonding electrons	2
	Bound nitrogen nonbonding electrons	6
	Formal charge = 5 – ½(2) – 6 = –2	

For center nitrogen:	Isolated nitrogen valence electrons	5
	Bound nitrogen bonding electrons	8
	Bound nitrogen nonbonding electrons	0
	Formal charge = 5 – ½(8) – 0 = +1	

For oxygen:	Isolated oxygen valence electrons	6
	Bound oxygen bonding electrons	6
	Bound oxygen nonbonding electrons	2
	Formal charge = 6 – ½(6) – 2 = +1	

$$\overset{-1}{\underset{\cdot\cdot}{\overset{\cdot\cdot}{N}}}=\overset{+1}{N}=\overset{\cdot\cdot}{\underset{\cdot\cdot}{\overset{\cdot\cdot}{O}}} \longleftrightarrow \overset{+1}{:N}\equiv\overset{-1}{N}-\overset{\cdot\cdot}{\underset{\cdot\cdot}{O}}: \longleftrightarrow \overset{-2}{:\underset{\cdot\cdot}{N}}-\overset{+1}{N}\equiv\overset{+1}{O}:$$

The second structures make the largest contribution to the resonance hybrid because the -1 formal charge is on the electronegative O.

Group 6A Elements (Section 22.8)

22.116 (a) O   (b) Te   (c) Po   (d) O

22.118 N_2 (14 e⁻), O_2 (16 e⁻), and Ar (18 e⁻) are all nonpolar with only dispersion forces. N_2, with fewest electrons would have the smallest dispersion forces and the lowest boiling point. The16 electrons in O_2 are spread over two atoms and are more polarizable than the 18 electrons in Ar, so O_2 has the largest dispersion forces and the highest boiling point. The order for increasing boiling point is: $N_2 < Ar < O_2$

22.120 (a) $4\,Li(s) + O_2(g) \rightarrow 2\,Li_2O(s)$ (b) $P_4(s) + 5\,O_2(g) \rightarrow P_4O_{10}(s)$
(c) $4\,Al(s) + 3\,O_2(g) \rightarrow 2\,Al_2O_3(s)$ (d) $Si(s) + O_2(g) \rightarrow SiO_2(s)$

22.122 An element that forms an acidic oxide is more likely to form a covalent hydride. C and N are examples.

22.124 $Li_2O < BeO < B_2O_3 < CO_2 < N_2O_5$ (see Figure 22.16)

22.126 $N_2O_5 < Al_2O_3 < K_2O < Cs_2O$ (see Figure 22.16)

22.128 (a) CrO_3 (higher Cr oxidation state)   (b) N_2O_5 (higher N oxidation state)
(c) SO_3 (higher S oxidation state)

22.130 (a) $Cl_2O_7(l) + H_2O(l) \rightarrow 2\,H^+(aq) + 2\,ClO_4^-(aq)$
(b) $K_2O(s) + H_2O(l) \rightarrow 2\,K^+(aq) + 2\,OH^-(aq)$
(c) $SO_3(l) + H_2O(l) \rightarrow H^+(aq) + HSO_4^-(aq)$

22.132 (a) $ZnO(s) + 2\,H^+(aq) \rightarrow Zn^{2+}(aq) + H_2O(l)$
(b) $ZnO(s) + 2\,OH^-(aq) + H_2O(l) \rightarrow Zn(OH)_4^{2-}(aq)$

22.134 (a) rhombic sulfur – yellow crystalline solid (mp 113 °C) that contains crown-shaped S_8 rings.
(b) monoclinic sulfur – an allotrope of sulfur in which the S_8 rings pack differently in the crystal.
(c) plastic sulfur – when sulfur is cooled rapidly, the sulfur forms disordered, tangled chains, yielding an amorphous, rubbery material called plastic sulfur.
(d) Liquid sulfur between 160 and 195 °C becomes dark reddish-brown and very viscous forming long polymer chains (S_n, n > 200,000).

22.136 (a) $Zn(s) + 2 H_3O^+(aq) \rightarrow Zn^{2+}(aq) + H_2(g) + 2 H_2O(l)$
(b) $BaSO_3(s) + 2 H_3O^+(aq) \rightarrow H_2SO_3(aq) + Ba^{2+}(aq) + 2 H_2O(l)$
(c) $Cu(s) + 2 H_2SO_4(l) \rightarrow Cu^{2+}(aq) + SO_4^{2-}(aq) + SO_2(g) + 2 H_2O(l)$
(d) $H_2S(aq) + I_2(aq) \rightarrow S(s) + 2 H^+(aq) + 2 I^-(aq)$

22.138 (a) Acid strength increases as the number of O atoms increases.
(b) In comparison with S, O is much too electronegative to form compounds of O in the +4 oxidation state. Also, an S atom is large enough to accommodate four bond pairs and a lone pair in its valence shell, but an O atom is too small to do so.
(c) Each S is sp^3 hybridized with two lone pairs of electrons. The bond angles are therefore 109.5°. A planar ring would require bond angles of 135°.

22.140 (a) H—S̈—H , bent.
(b) :Ö—S̈=Ö: ⟷ :Ö=S̈—Ö: , bent, S is sp^2 hybridized.
(c)

     :Ö: :Ö: :Ö
      | | ‖
:Ö—S=Ö: ⟷ :Ö=S—Ö: ⟷ :Ö—S—Ö:
trigonal planar, S is sp^2 hybridized.

22.142 (1) is CO_2. The molecule is linear because C has two charge clouds. There are two C=O double bonds.
(2) is SO_2. The molecule is bent because S has three charge clouds. There is one S–O single bond and one S=O double bond. SO_2 has two resonance structures so each S–O bond appears to be a bond and a half.
CO_2 has the stronger bonds.

Group 7A and 8A: The Halogens and Noble Gases (Sections 22.9–22.10)

22.144 (a) At is in Group 7A. The trend going down the group is gas → liquid → solid. At, being at the bottom of the group, should be a solid.
(b) At is likely to react with Na just like the other halogens, yielding NaAt.

22.146 $MnO_2(s) + 2 Br^-(aq) + 4 H^+(aq) \rightarrow Mn^{2+}(aq) + 2 H_2O(l) + Br_2(aq)$

22.148 (a) Assume a 100.0 g sample. From the percent composition data, a 100.0 g sample contains 25.25 g Ti, and 74.75 g Cl.

$$25.25 \text{ g Ti} \times \frac{1 \text{ mol Ti}}{47.87 \text{ g Ti}} = 0.5275 \text{ mol Ti}$$

$$74.75 \text{ g Cl} \times \frac{1 \text{ mol Cl}}{35.45 \text{ g Cl}} = 2.109 \text{ mol Cl}$$

$Ti_{0.5275}Cl_{2.109}$; divide each subscript by the smaller, 0.5275.
$Ti_{0.5275 / 0.5275}Cl_{2.109 / 0.5275}$
The empirical and molecular formula is $TiCl_4$, titanium tetrachloride.
(b) $Ti(s) + 2 Cl_2(g) \rightarrow TiCl_4(g)$
(c) $TiCl_4(l) + 2 Mg(s) \rightarrow Ti(s) + 2 MgCl_2(s)$

22.150 (a) $HBrO_3$, +5   (b) HIO, +1

22.152 (a) HIO_3 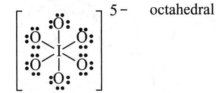 trigonal pyramidal

(b) ClO_2^-

$$\left[:\ddot{O}-\ddot{C}l-\ddot{O}: \right]^-$$ bent

(c) HOCl $H-\ddot{O}-\ddot{C}l:$ bent

(d) IO_6^{5-}

octahedral

22.154 Oxygen atoms are highly electronegative. Increasing the number of oxygen atoms increases the polarity of the O–H bond and increases the acid strength.

22.156 $I_2O_5(aq) + H_2O(l) \rightarrow 2\,HIO_3(aq)$; HIO_3 is iodic acid.

22.158 KrF_2, Kr^{2+} and F^-

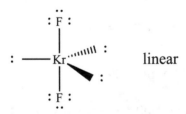 linear

Multiconcept Problems

22.160 (a) C as diamond
(b) $Cl_2(g) + H_2O(l) \rightarrow HOCl(aq) + H^+(aq) + Cl^-(aq)$
(c) NO   (d) NO_2 (e) BF_3 (f) Al_2O_3 (g) Si   (h) HNO_3
(i) C as diamond, graphite, and fullerene.

22.162 (a) $H_3PO_4(aq) + H_2O(l) \rightleftharpoons H_3O^+(aq) + H_2PO_4^-(aq)$
H_3PO_4 is a Brønsted-Lowry acid.
(b) $B(OH)_3(aq) + 2\,H_2O(l) \rightleftharpoons B(OH)_4^-(aq) + H_3O^+(aq)$
$B(OH)_3$ is a Lewis acid.

22.164 (a)

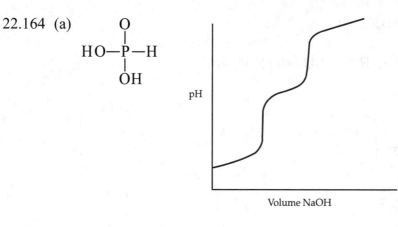

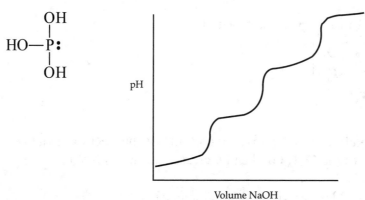

(b) $K_{a1} = 1.0 \times 10^{-2}$; $pK_{a1} = -\log(1.0 \times 10^{-2}) = 2.00$
     $K_{a2} = 2.6 \times 10^{-7}$; $pK_{a2} = -\log(2.6 \times 10^{-7}) = 6.59$

At the first equivalence point, pH $= \dfrac{pK_{a1} + pK_{a2}}{2} = \dfrac{2.00 + 6.59}{2} = 4.29$

mmol HPO_3^{2-} = (30.00 mL)(0.1240 mmol/mL) = 3.72 mmol HPO_3^{2-}
volume NaOH to reach second equivalence point

$$= 3.72 \text{ mmol } HPO_3^{2-} \times \frac{2 \text{ mmol NaOH}}{1 \text{ mmol } HPO_3^{2-}} \times \frac{1.00 \text{ mL}}{0.1000 \text{ mmol NaOH}} = 74.40 \text{ mL}$$

At the second equivalence point only Na_2HPO_3, a basic salt, is in solution.

$$[HPO_3^{2-}] = \frac{3.72 \text{ mmol}}{30.00 \text{ mL} + 74.40 \text{ mL}} = 0.0356 \text{ mmol/mL} = 0.0356 \text{ M}$$

$$K_b = \frac{K_w}{K_{a2}} = \frac{1.0 \times 10^{-14}}{2.6 \times 10^{-7}} = 3.8 \times 10^{-8}$$

	$HPO_3^{2-}(aq)$	+	$H_2O(l)$	⇌	$H_2PO_3^{2-}(aq)$	+	$OH^-(aq)$
initial (M)	0.0356				0		~0
change (M)	–x				+x		+x
equil (M)	0.0356 – x				x		x

$$K_b = \frac{[H_2PO_3^-][OH^-]}{[HPO_3^{2-}]} = 3.8 \times 10^{-8} = \frac{(x)(x)}{0.0356 - x} \approx \frac{x^2}{0.0356}$$

Solve for x.

$$x = [OH^-] = \sqrt{(3.8 \times 10^{-8})(0.0356)} = 3.68 \times 10^{-5} \text{ M}$$

$$[H_3O^+] = \frac{K_w}{[OH^-]} = \frac{1.0 \times 10^{-14}}{3.68 \times 10^{-5}} = 2.72 \times 10^{-10} \text{ M}$$

$$pH = -\log[H_3O^+] = -\log(2.72 \times 10^{-10}) = 9.57$$

22.166 (a) $P_4(s) + 5 O_2(g) \rightarrow P_4O_{10}(s)$
  $P_4O_{10}(s) + 6 H_2O(l) \rightarrow 4 H_3PO_4(aq)$
  (b) P_4, 123.90

$$\text{mol } H_3PO_4 = 5.00 \text{ g } P_4 \times \frac{1 \text{ mol } P_4}{123.90 \text{ g } P_4} \times \frac{1 \text{ mol } P_4O_{10}}{1 \text{ mol } P_4} \times \frac{4 \text{ mol } H_3PO_4}{1 \text{ mol } P_4O_{10}} = 0.1614 \text{ mol}$$

$$[H_3PO_4] = \frac{0.1614 \text{ mol}}{0.2500 \text{ L}} = 0.646 \text{ M}$$

For the dissociation of the first proton, the following equilibrium must be considered:

$$H_3PO_4(aq) + H_2O(l) \rightleftharpoons H_3O^+(aq) + H_2PO_4^-(aq)$$

	H_3PO_4	H_3O^+	$H_2PO_4^-$
initial (M)	0.646	~0	0
change (M)	–x	+x	+x
equil (M)	0.646 – x	x	x

$$K_{a1} = \frac{[H_3O^+][H_2PO_4^-]}{[H_3PO_4]} = 7.5 \times 10^{-3} = \frac{x^2}{0.646 - x}$$

$$x^2 + (7.5 \times 10^{-3})x - (4.84 \times 10^{-3}) = 0$$

Solve for x using the quadratic formula.

$$x = \frac{-(7.5 \times 10^{-3}) \pm \sqrt{(7.5 \times 10^{-3})^2 - (4)(1)(-4.84 \times 10^{-3})}}{2(1)} = \frac{(-7.5 \times 10^{-3}) \pm 0.139}{2}$$

x = 0.0658 and –0.0733;   Of the two solutions for x, only the positive value of x has physical meaning, because x is the $[H_3O^+]$.
x = 0.0658 M = $[H_2PO_4^-]$ = $[H_3O^+]$
Only the dissociation of the first proton contributes a significant amount of H_3O^+.
$pH = -\log[H_3O^+] = -\log(0.0658) = 1.18$

(c) $3 Ca^{2+}(aq) + 2 H_3PO_4(aq) \rightarrow Ca_3(PO_4)_2(s) + 6 H^+(aq)$
$Ca_3(PO_4)_2$, 310.18

$$\text{mass } Ca_3(PO_4)_2 = 0.1614 \text{ mol } H_3PO_4 \times \frac{1 \text{ mol } Ca_3(PO_4)_2}{2 \text{ mol } H_3PO_4} \times \frac{310.18 \text{ g } Ca_3(PO_4)_2}{1 \text{ mol } Ca_3(PO_4)_2} = 25.0 \text{ g}$$

(d) $Zn(s) + 2 H^+(aq) \rightarrow H_2(g) + Zn^{2+}(aq)$; the gas is H_2.

$$\text{mol } H_2 = 0.1614 \text{ mol } H_3PO_4 \times \frac{6 \text{ mol } H^+}{2 \text{ mol } H_3PO_4} \times \frac{1 \text{ mol } H_2}{2 \text{ mol } H^+} = 0.242 \text{ mol } H_2$$

$PV = nRT$; $20\ ^\circ C = 293\ K$

$$V = \frac{nRT}{P} = \frac{(0.242\ \text{mol})\left(0.082\ 06\ \dfrac{\text{L}\cdot\text{atm}}{\text{K}\cdot\text{mol}}\right)(293\ K)}{\left(742\ \text{mm Hg}\ \times\ \dfrac{1.00\ \text{atm}}{760\ \text{mm Hg}}\right)} = 5.96\ L$$

23.1 (a) 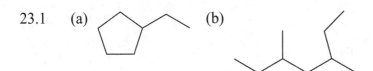 (b)

23.2 (a) C_8H_{16} (b) C_7H_{16}

23.3 Structures (a) and (c) are identical. They both contain a chain of six carbons with two –CH_3 branches at the fourth carbon and one –CH_3 branch at the second carbon. Structure (b) is different, having a chain of seven carbons.

23.4 $CH_3CH_2CH_2CH_2CH_2CH_3$

$$CH_3\underset{\underset{CH_3}{|}}{C}HCH_2CH_2CH_3$$

$$CH_3CH_2\underset{\underset{CH_3}{|}}{C}HCH_2CH_3$$

$$CH_3\underset{\underset{\underset{CH_3}{|}}{C}}{\overset{\overset{CH_3}{|}}{C}}H_2CH_3$$

$$CH_3\underset{\underset{CH_3}{|}}{C}H\overset{\overset{CH_3}{|}}{C}HCH_3$$

23.5   Ascorbic acid is chiral.

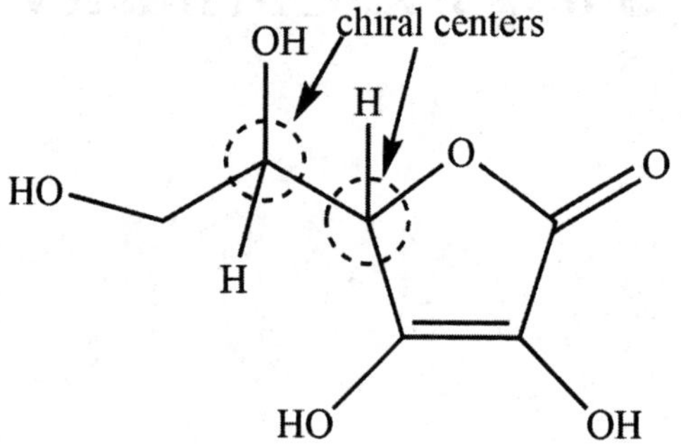

23.6   The four chiral centers are circled.

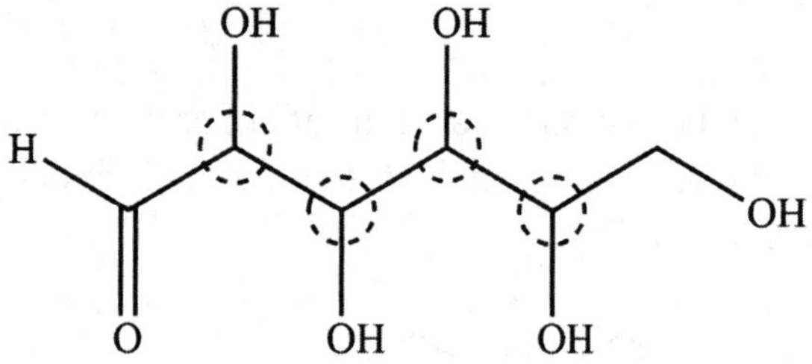

With four different chiral centers there are $2^4 = 16$ possible isomers.

23.7

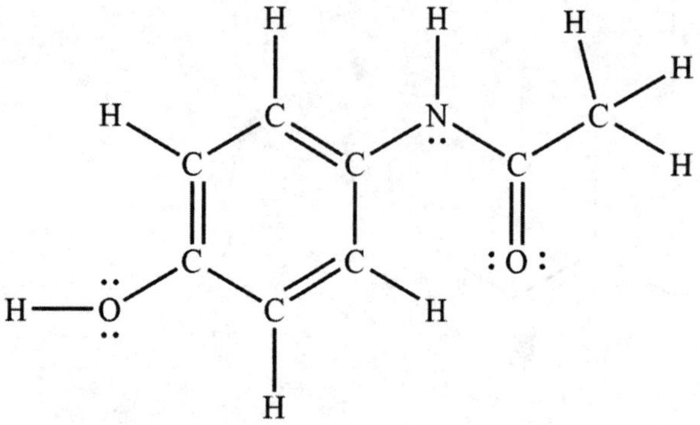

23.8

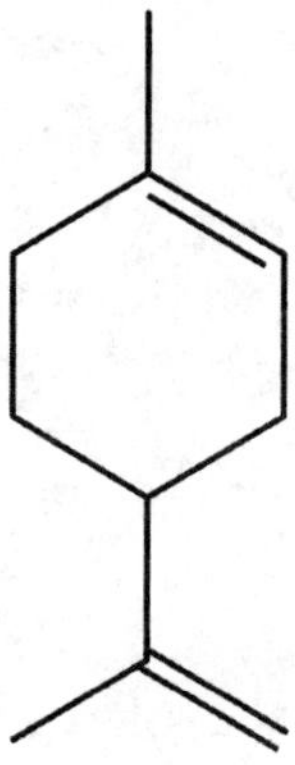

23.9

C–H bond formed
from sp³-1s overlap.

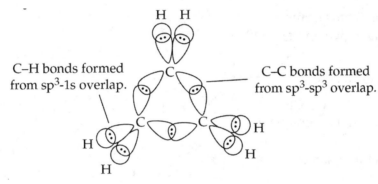

C–N bond formed
from sp³-sp³ overlap.

N–H bond formed
from sp³-1s overlap.

23.10   The molecule is unstable because the C–C bonds are relatively weak due to the less effective orbital overlap in a molecule with 60° bond angles.

C–H bonds formed
from sp³-1s overlap.

C–C bonds formed
from sp³-sp³ overlap.

23.11   (a)

Carbon–Oxygen double bond consists of one σ bond formed by head on overlap of sp² orbitals …

… and sideways overlap of p orbitals

(b) No   (c) Yes

23.12   (a)

$$\text{H} \diagdown \atop \text{H} \diagup \text{C}=\text{C}=\ddot{\text{O}}:$$

(b) Carbon–Carbon double bond consists of one σ bond formed by head on overlap of sp^2 and sp orbitals...

Carbon–Oxygen double bond consists of one σ bond formed by head on overlap of sp and sp^2 orbitals...

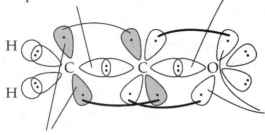

...and one π bond formed by sideways overlap of p orbitals

...and one π bond formed by sideways overlap of p orbitals

23.13 (a) Does not exhibit cis-trans isomerism.
(b) The cis isomer was shown in the problem.

The trans isomer is shown here:

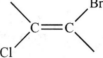

(c) Does not exhibit cis-trans isomerism.

23.14 (a) Fumaric acid is a trans isomer.
(b) Maleic acid

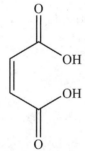

(c) No, because succinic acid does not have a carbon-carbon double bond.

23.15 (a)   (b)   (c)

23.16 (a)   (b) 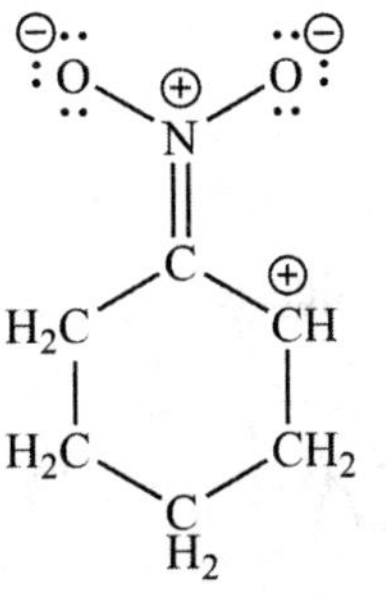   (c)

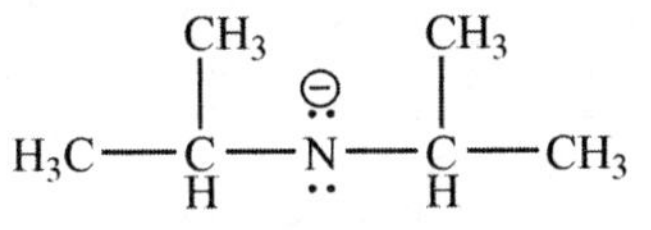

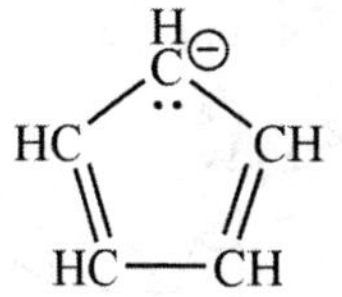

23.17 (a)

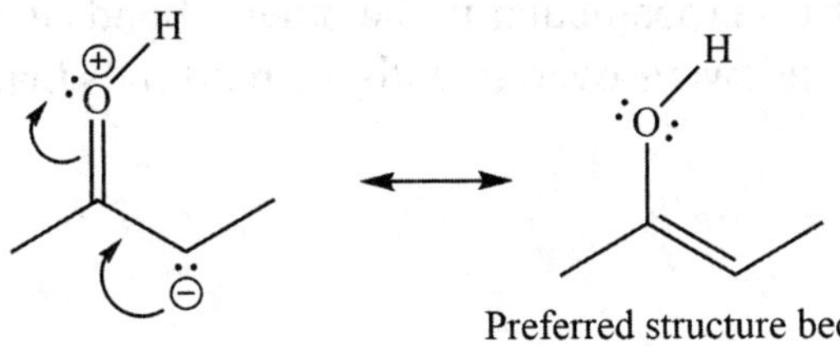

Preferred structure because
formal charges are zero

(b)   Moving electrons as indicated would result in
an incorrect electron-dot structure because oxygen would
have an expanded octet.

(c)

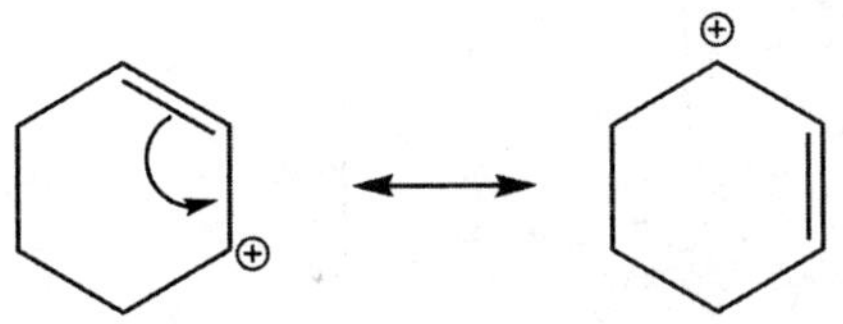

Both resonance structures are equivalent

(d)

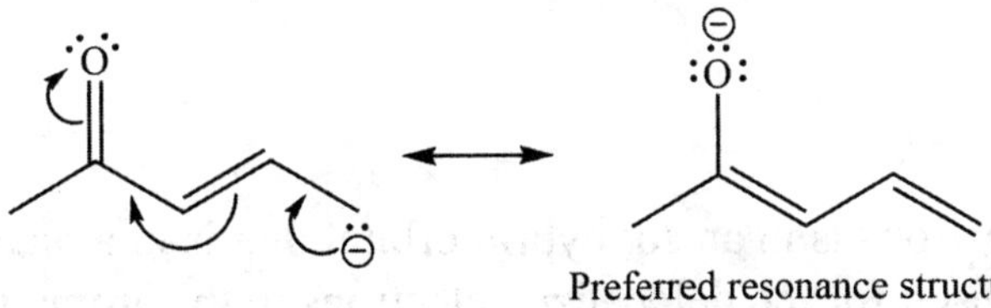

Preferred resonance structure
because negative formal charge is
on more electronegative oxygen

23.18 (a)   (b)

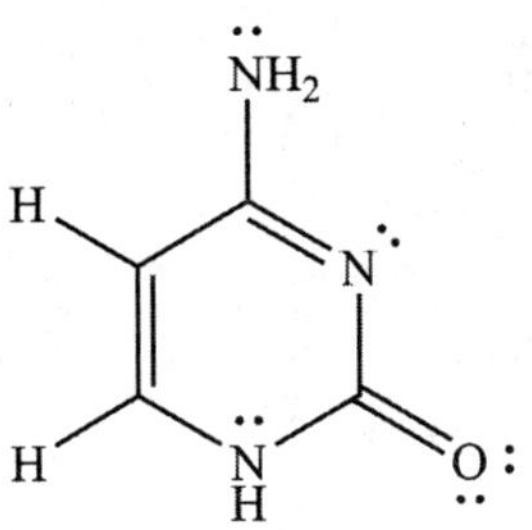

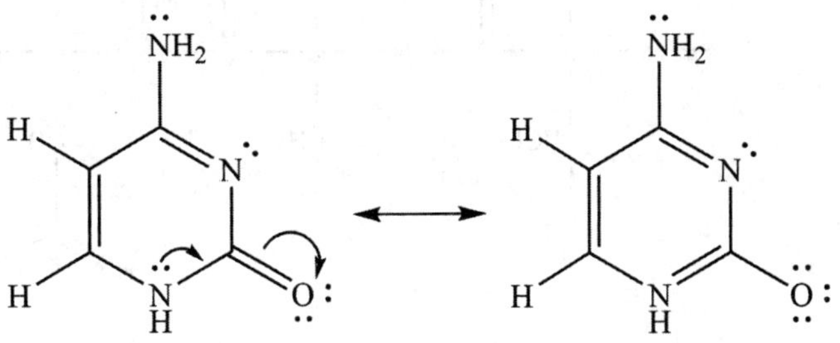

23.19 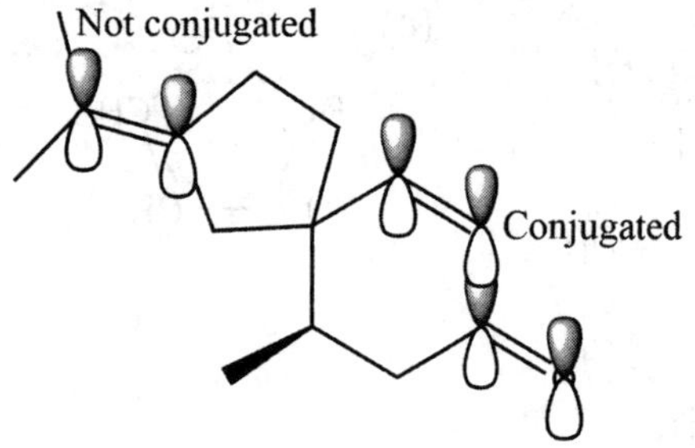
Not conjugated
Conjugated

23.20 In the triple bond one of the π bonds is perpendicular to the other π bonds in the molecule. Therefore, it cannot have sideways overlap with the p orbitals that make up the other π bonds.

23.21 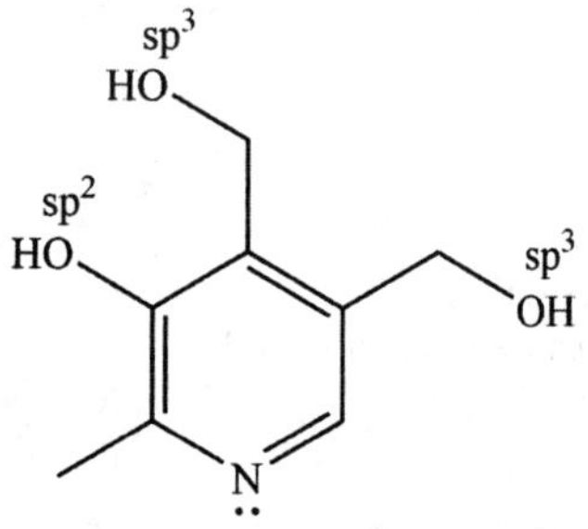
sp3
HO
sp2
HO
sp3
OH
N

23.22 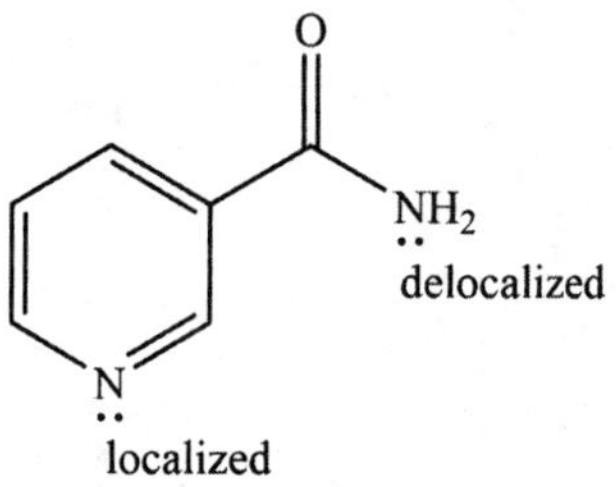
O
NH2
delocalized
N
localized

The localized lone pair of electrons is in an sp^2 hybrid orbital that is in a plane perpendicular to the orbitals used for the delocalized electrons in the aromatic ring.

23.23 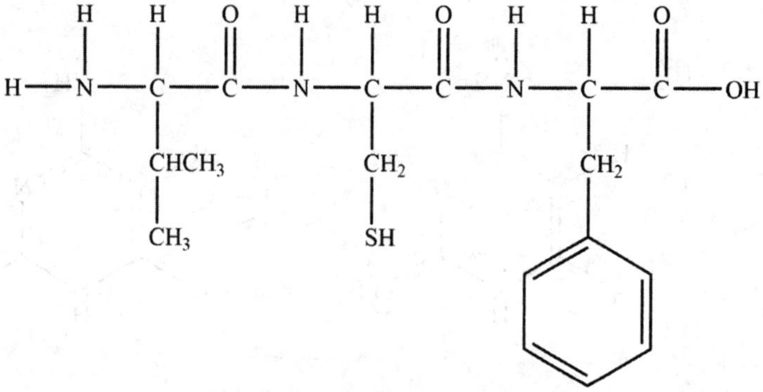
H H O H H O H H O
H N C C N C C N C C OH
CHCH3
CH3
CH2
SH
CH2

23.24 (a) 4 (b) Phe-nonpolar, Asn-polar, Trp-nonpolar, Ala-nonpolar

23.25   Proline is not aromatic because it does not have a ring of sp^2 hybridized atoms. Tyrosine is aromatic because it has a six-member ring of sp^2 hybridized carbon atoms that are part of a conjugated system. There are 6π electrons and satisfying the $4n + 2$ rule for aromaticity.

23.26   (a)

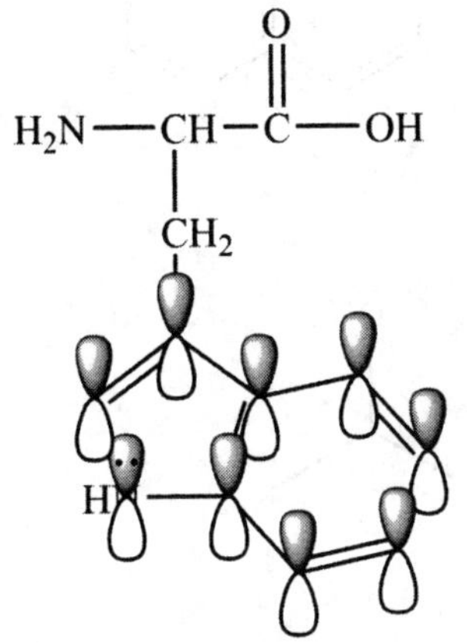

(b) All atoms in the two-ring system of the side chain of tryptophan are sp^2 hybridized and conjugated. Four double bonds contribute 8π electrons and one lone pair contributes 2π electrons for a total of 10π electrons, satisfying the $4n + 2$ rule.

23.27   (a) Only C_b is a chiral center.   (b) 16σ bonds and 2π bonds.
        (c) C_a and C_d are sp2 hybridized; C_b and C_c are sp3 hybridized.
        (d)

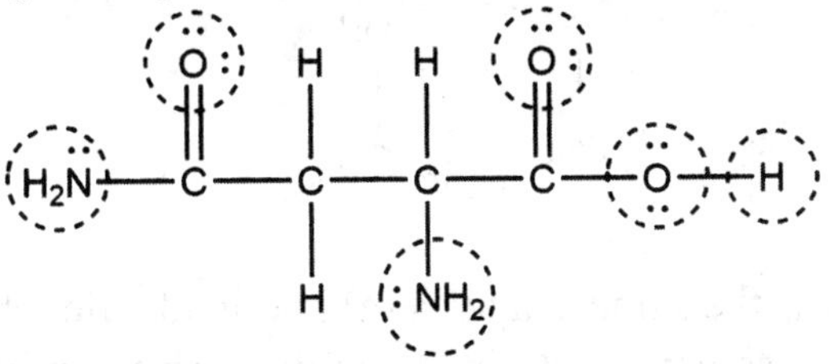

23.28   (a) $C_{10}H_{14}O$
        (b)

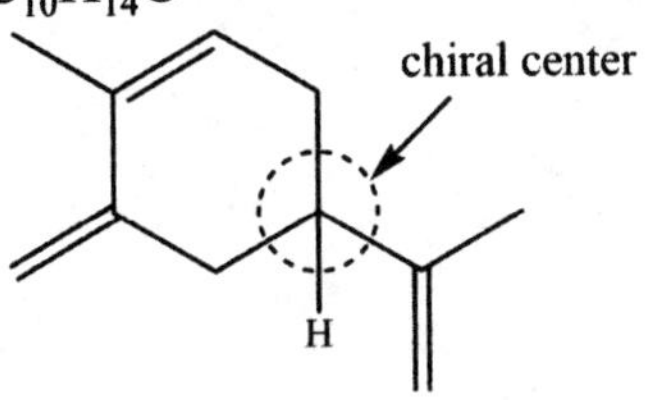

chiral center

(c) Carvone is not aromatic, the ring is not planar.

23.29   (a) and (d)

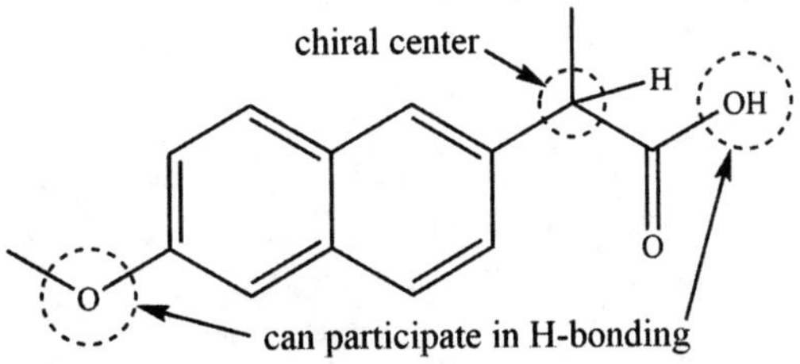

chiral center

OH

can participate in H-bonding

(b) Naproxen is aromatic. It is a cyclic, planar compound, and has adjacent p orbitals around a ring.

(c)

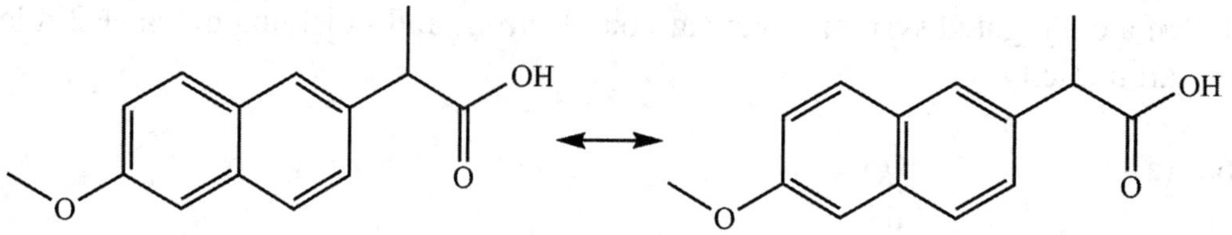
OH
O
O
OH
O
O

Conceptual Problems

23.30

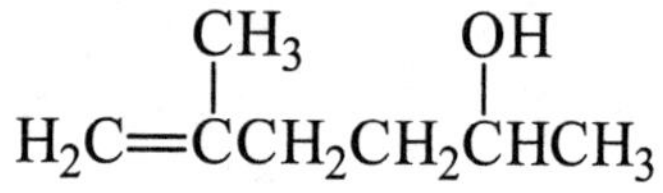
CH3
OH
H2C=CCH2CH2CHCH3

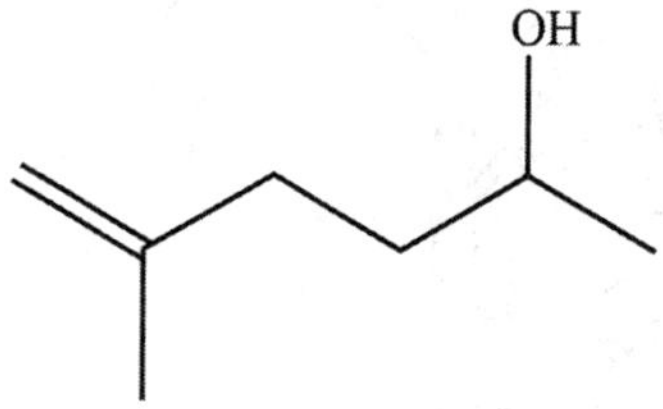
OH

Here are 2 isomers and their line drawings:

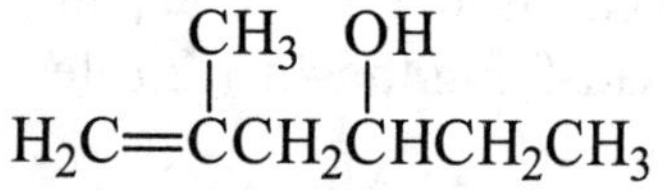
CH3 OH
H2C=CCH2CHCH2CH3

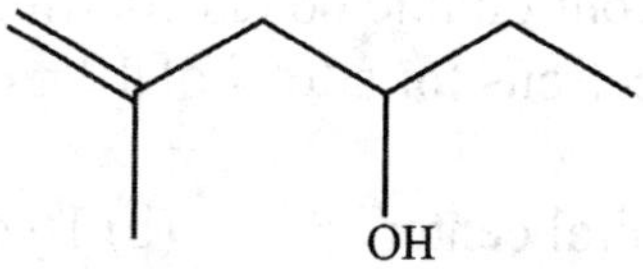
OH

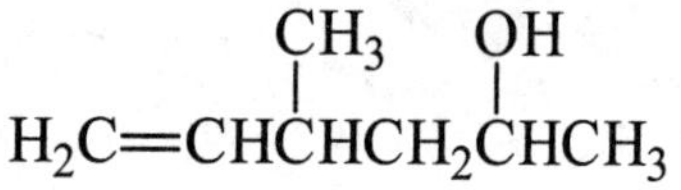
CH3 OH
H2C=CHCHCH2CHCH3

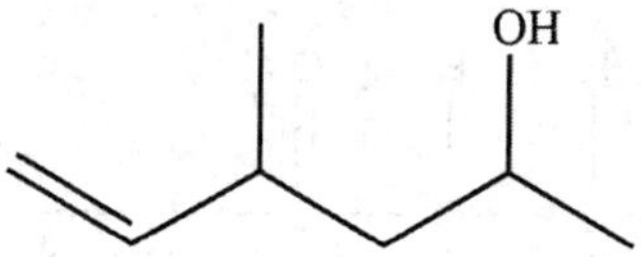
OH

23.32 The mirror image of molecule (a) has the same shape as (a) and is identical to it in all respects, so there is no handedness associated with it. The mirror image of molecule (b) is different than (b) so there is a handedness to this molecule.

23.34 ester, aromatic ring, and amine

23.36

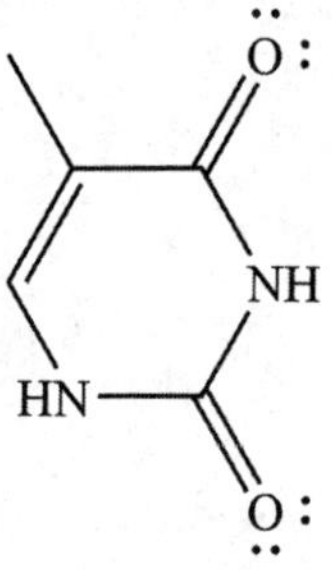
O
NH
HN
O

$C_5H_6N_2O_2$

23.38 (a) serine (b) methionine

Section Problems
Organic Molecules and Constitutional Isomers (Section 23.1)

23.40 In a straight-chain alkane, all the carbons are connected in a row. In a branched-chain alkane, there are branching connections of carbons along the carbon chain.

23.42 $CH_3CH_2CH_2CH_2CH_2CH_3$

23.44 (a) No, because they contain different numbers of carbons and hydrogens.
(b) They are isomers of each other.
(c) No, they are identical.

23.46 They have the same molecular formulas but different chemical structures. They are isomers.

Stereoisomers: Chiral Molecules (Section 23.2)

23.48 (b), (c), and (e) are chiral.

23.50 Only (d) is chiral. The chiral center is circled.

23.52 The chiral center is circled.

Functional Groups (Section 23.3)

23.54 A functional group is a part of a larger molecule and is composed of an atom or group of atoms that has a characteristic chemical behavior. They are important because their chemistry controls the chemistry in molecules that contain them.

23.56 (a)

$$CH_3CH_2\overset{\overset{\textstyle O}{\|}}{C}CH_2CH_3$$

(b)

$$CH_3CH_2CH_2\overset{\overset{\textstyle O}{\|}}{C}OCH_2CH_3$$

(c)

$$NH_2CH_2\overset{\overset{\textstyle O}{\|}}{C}OH$$

23.58 (a) C_3H_4 (b) C_4H_8 (c) C_5H_6

23.60 $CH_2{=}C{=}CHCH_2CH_3$ $CH_2{=}CHCH{=}CHCH_3$ $CH_2{=}CHCH_2CH{=}CH_2$

$CH_3CH{=}C{=}CHCH_3$ $\underset{\displaystyle CH_3}{CH_2{=}CCH{=}CH_2}$ $\underset{\displaystyle CH_3}{CH_2{=}C{=}CCH_3}$

23.62 C_3H_9 contains one more H than needed for an alkane.

23.64   (a) C_2H_6O (b) C_2H_7N

(c) $C_5H_{11}N$ (d) C_3H_8O

(e) C_3H_7F (f) C_3H_6O

23.66   (a) $C_3H_4O_3$ (b) C_6H_7NO

(c) $C_4H_8O_2$

23.68   Amine, ester, and arene

Carbohydrates (Section 23.4)

23.70   An aldose contains the aldehyde functional group while a ketose contains the ketone functional group.

23.72

$$\overset{\displaystyle O}{\underset{\displaystyle OH}{HOCH_2CH\overset{\displaystyle \|}{C}CH_2OH}}$$

23.74   (a) constitutional isomers   (b) anomers

Valence Bond Theory, Cis-Trans Isomers (Section 23.5)

23.76   (a) All C–H bond angles are ~109.5°.

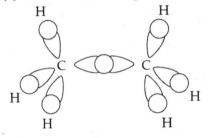

(b) Two C–H bond angles are ~120° and one is 90°.

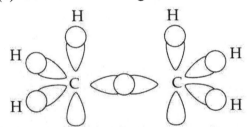

(c) In structure (b), the 90° bond angle introduces a larger repulsion and lower stability. Structure (a) is more favorable.

23.78

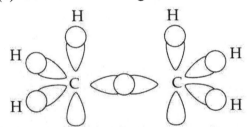

23.80

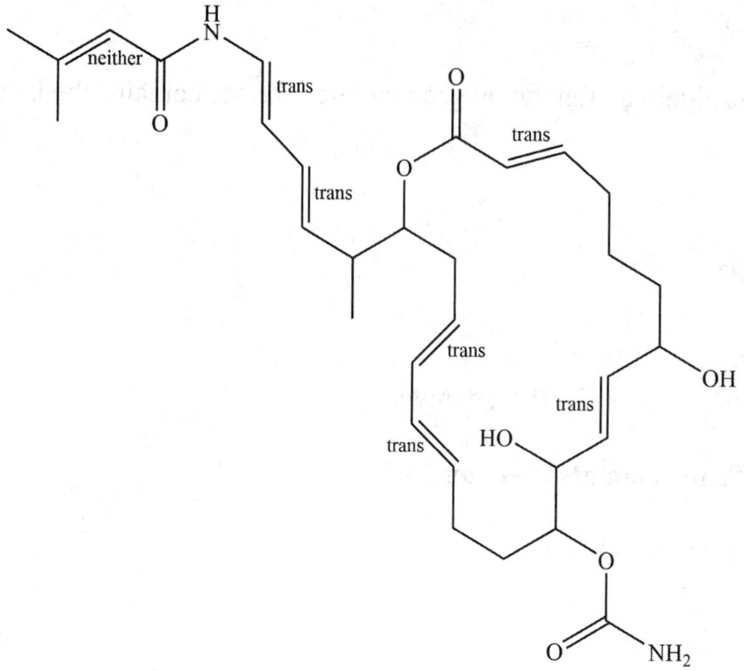

23.82 Compounds (b) and (c) exhibit cis–trans isomerism.

(b)

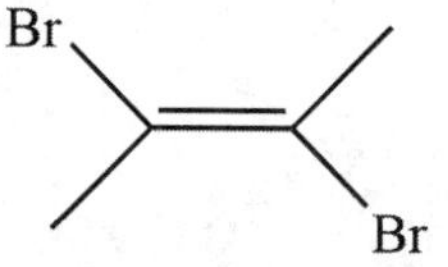

trans

(c)

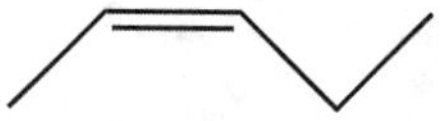

cis

23.84 (a) $CH_2{=}CHCH_2CH_2CH_2CH_3$ — This compound cannot form cis-trans isomers.

(b) $CH_3CH{=}CHCH_2CH_2CH_3$ — This compound can form cis-trans isomers because of the different groups on each double bond C.

(c) $CH_3CH_2CH{=}CHCH_2CH_3$ — This compound can form cis-trans isomers because of the different groups on each double bond C.

Lipids (Section 23.6)

23.86 Long-chain carboxylic acids are called fatty acids. Fatty acids are usually unbranched and have an even number of carbon atoms in the range of 12-22.

23.88

$$
\begin{array}{l}
CH_2O\overset{O}{\overset{\|}{C}}(CH_2)_{12}CH_3 \\
| \\
CHO\overset{O}{\overset{\|}{C}}(CH_2)_{12}CH_3 \\
| \\
CH_2O\overset{O}{\overset{\|}{C}}(CH_2)_{12}CH_3
\end{array}
$$

23.90

$$CH_3(CH_2)_{14}\overset{\overset{O}{\|}}{C}O(CH_2)_{15}CH_3$$

23.92 Only (c) is an unsaturated fatty acid. It has double bonds in a cis configuration. This geometry substantially disrupts intermolecular forces, leading to oils (liquids) at room temperature.

23.94

$$CH_3(CH_2)_{16}\overset{\overset{O}{\|}}{C}O(CH_2)_{21}CH_3$$

Formal Charge and Resonance (Section 23.7)

23.96 (a)

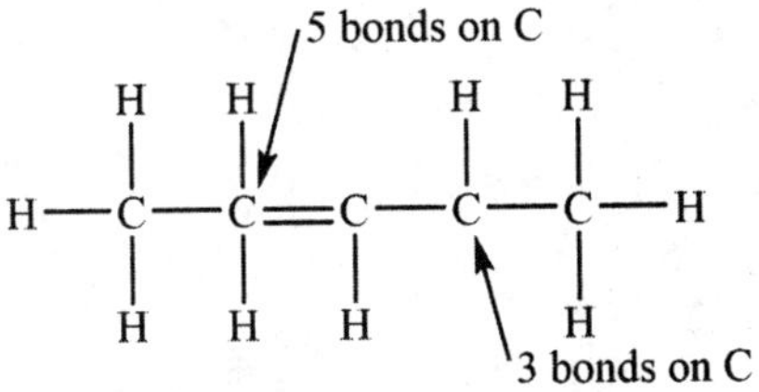

This structure is not valid.

(b)

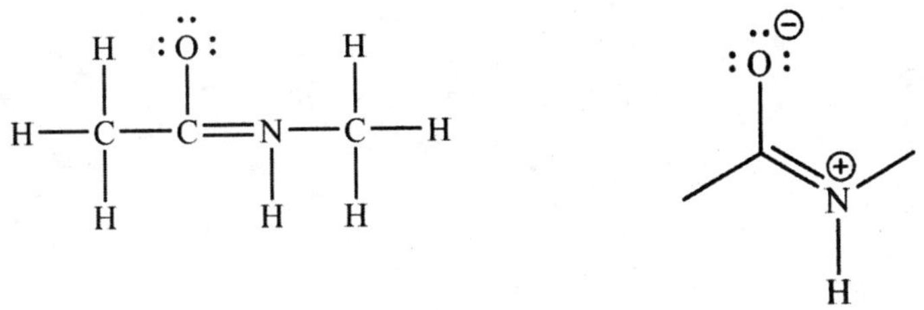

(c)

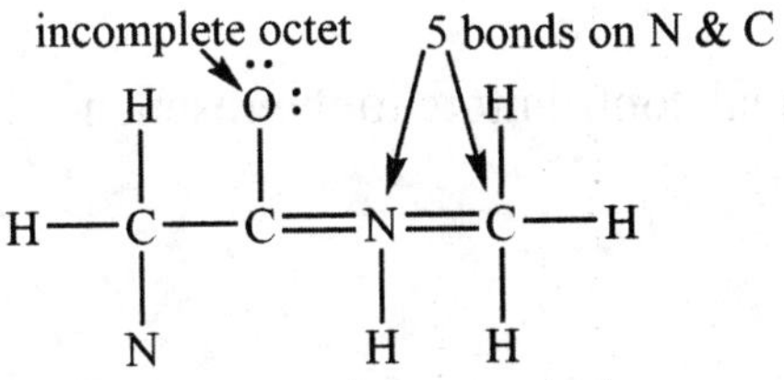

This structure is not valid.

23.98 (a)

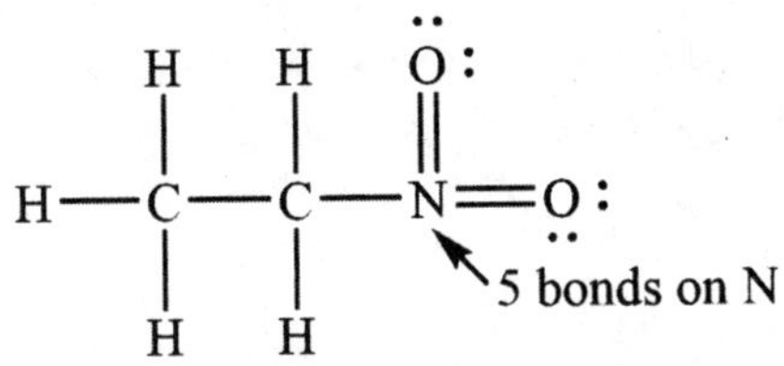

This structure is not valid.

(b)

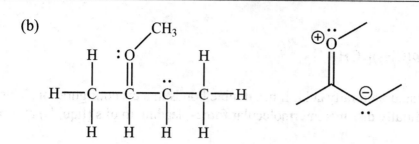

(c)

The original structure contributes more to the resonance hybrid because all formal charges are zero.

23.100 (a)

The original structure contributes more to the resonance hybrid because all formal charges are zero.

(b)

The two structures are identical and are equal contributors to the resonance hybrid.

23.102 (a)

The original structure contributes more to the resonance hybrid because all formal charges are zero.

(b)

The original structure contributes more to the resonance hybrid because all formal charges are zero.

23.104 They are resonance structures.

23.106 (a)

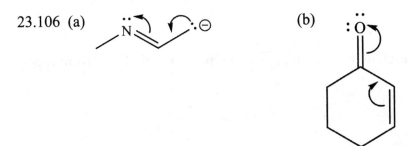

(b)

Conjugation Systems (Section 23.8)

23.108 (a)   (b)   (c)

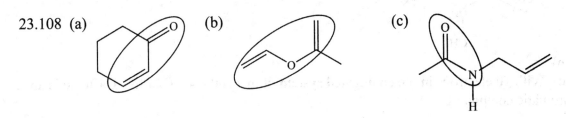

23.110

23.112   delocalized   localized

Proteins (Section 23.9)

23.114 (a) serine   (b) threonine   (c) proline   (d) phenylalanine   (e) cysteine

23.116 Pro-Glu-Asp

23.118 Met-Ile-Lys, Met-Lys-Ile, Ile-Met-Lys, Ile-Lys-Met, Lys-Met-Ile, Lys-Ile-Met

Aromaticity and Molecular Orbital Theory (Section 23.10)

23.120 Only (b) is aromatic.

23.122 There are 6 π-electrons, 2 each in the 2 π bonds and 2 in the p-orbital of the NH nitrogen.

23.124 (a)

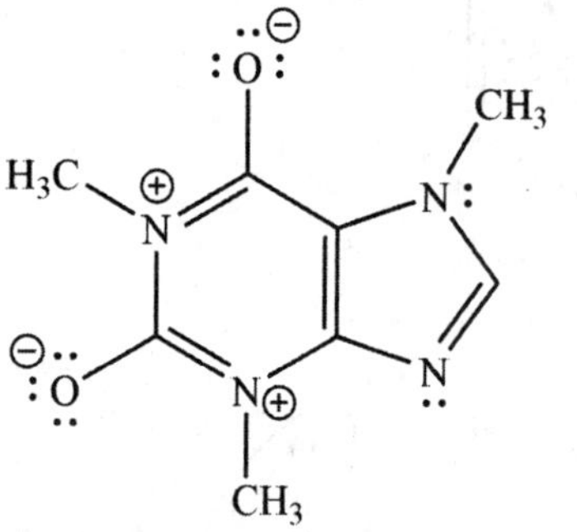

(b) sp^2
(c) With 10 electrons in the conjugated system, it meets the 4n + 2 rule. Caffeine is an aromatic compound.

Nucleic Acids (Section 23.11)

23.126 Just as proteins are polymers made of amino acid units, nucleic acids are polymers made up of nucleotide units linked together to form a long chain. Each nucleotide contains a phosphate group, an aldopentose sugar, and an amine base.

23.128

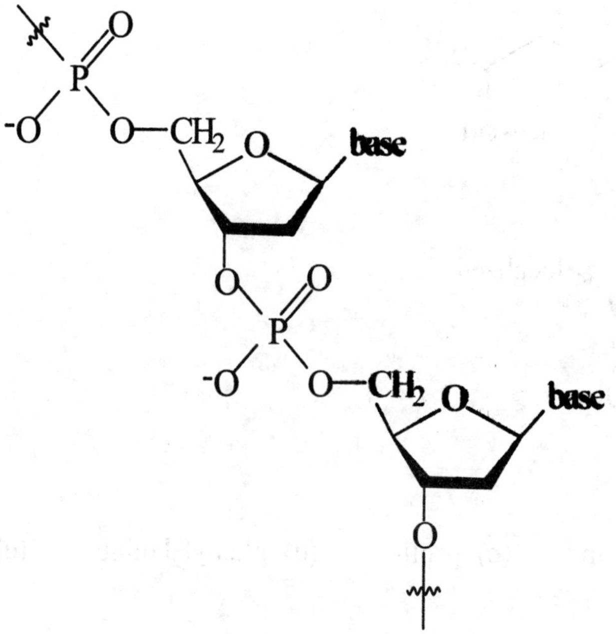

23.130 Original: T–A–C–C–G–A
          Complement: A–T–G–G–C–T

23.132

Multiconcept Problems

23.134 (a)

(b)

(c)

23.136 $\dfrac{55.847 \text{ u Fe}}{\text{mol mass cyt c}} \times 100\% = 0.43\%; \quad$ mol mass cyt c $= \dfrac{55.847 \text{ u Fe}}{0.0043} = 13{,}000$

23.138 (a) grams $I_2 = (0.0250 \text{ L})(0.200 \text{ mol/L})\left(\dfrac{253.81 \text{ g } I_2}{1 \text{ mol } I_2}\right) = 1.27 \text{ g } I_2$

(b) mol $Na_2S_2O_3 = (0.08199 \text{ L})(0.100 \text{ mol/L}) = 8.20 \times 10^{-3} \text{ mol}$

grams excess $I_2 = 8.20 \times 10^{-3} \text{ mol } Na_2S_2O_3 \times \dfrac{1 \text{ mol } I_2}{2 \text{ mol } Na_2S_2O_3} \times \dfrac{253.81 \text{ g } I_2}{1 \text{ mol } I_2} = 1.04 \text{ g } I_2$

grams I_2 reacted $= 1.27 \text{ g} - 1.04 \text{ g} = 0.23 \text{ g } I_2$

(c) iodine number $= \dfrac{0.23 \text{ g } I_2}{0.500 \text{ g milkfat}} \times 100 = 46$

(d) mol milkfat $= 0.500 \text{ g milkfat} \times \dfrac{1 \text{ mol milkfat}}{800 \text{ g milkfat}} = 6.25 \times 10^{-4} \text{ mol}$

mol I_2 reacted $= 0.23 \text{ g } I_2 \times \dfrac{1 \text{ mol } I_2}{253.81 \text{ g } I_2} = 9.06 \times 10^{-4} \text{ mol}$

number of double bonds per molecule $= \dfrac{9.06 \times 10^{-4} \text{ mol } I_2}{6.25 \times 10^{-4} \text{ mol milkfat}} = 1.4 \text{ double bonds}$